T0329121

# Dissolved Gas Concentration in Water

# Dissolved Gas Concentration in Water

## Computation as Functions of Temperature, Salinity and Pressure

***Second Edition***

***John Colt***
*Northwest Fisheries Science Center*
*National Marine Fisheries Service*
*Seattle, WA, USA*

AMSTERDAM • BOSTON • HEIDELBERG • LONDON • NEW YORK • OXFORD
PARIS • SAN DIEGO • SAN FRANCISCO • SINGAPORE • SYDNEY • TOKYO

Elsevier
32 Jamestown Road, London NW1 7BY
225 Wyman Street, Waltham, MA 02451, USA

First edition by American Fisheries Society, Bethesda, Maryland 1984
Second edition 2012



**Notices**
Knowledge and best practice in this field are constantly changing. As new research and experience broaden our understanding, changes in research methods, professional practices, or medical treatment may become necessary.

Practitioners and researchers must always rely on their own experience and knowledge in evaluating and using any information, methods, compounds, or experiments described herein. In using such information or methods they should be mindful of their own safety and the safety of others, including parties for whom they have a professional responsibility.

To the fullest extent of the law, neither the Publisher nor the authors, contributors, or editors, assume any liability for any injury and/or damage to persons or property as a matter of products liability, negligence or otherwise, or from any use or operation of any methods, products, instructions, or ideas contained in the material herein.

**British Library Cataloguing-in-Publication Data**
A catalogue record for this book is available from the British Library

**Library of Congress Cataloging-in-Publication Data**
A catalog record for this book is available from the Library of Congress

ISBN: 978-0-12-415916-7



This book has been manufactured using Print On Demand technology. Each copy is produced to order and is limited to black ink. The online version of this book will show color figures where appropriate.



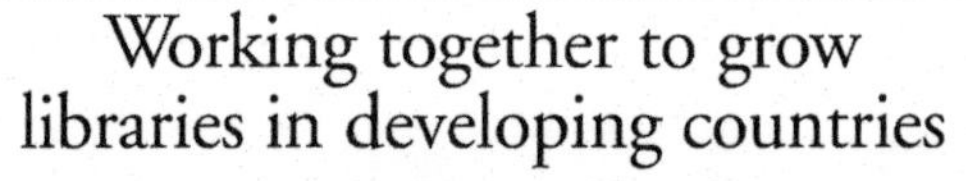

# Contents

# Guide to Gas Solubility Tables and Physical Properties

**Solubility in μmol/kg or nmol/kg**

| Gas | Freshwater | Seawater (0–40 g/kg) | Seawater (33–37 g/kg) | Brine (0–225 g/kg) |
|---|---|---|---|---|
| Oxygen ($O_2$) | 1.2{25} | 2.1{78} | 2.2{79} | 6.1{216}, 6.2{217} |
| Nitrogen ($N_2$) | 1.3{26} | 2.3{80} | 2.4{81} | |
| Argon (Ar) | 1.4{27} | 2.5{82} | 2.6{83} | |
| Carbon dioxide ($CO_2$) in 2010 | 1.5{28} | 2.7{84} | 2.8{85} | 6.3{218}, 6.4{219} |
| Carbon dioxide ($CO_2$) in 2030 | 1.6{29} | 2.9{87} | 2.10{88} | 6.5{220}, 6.6{211} |
| Helium (He) | 4.1{180} | 4.2{182} | 4.1{183} | |
| Neon (Ne) | 4.5{185} | 4.6{186} | 4.6{187} | |
| Krypton (Kr) | 4.9{189} | 4.10{190} | 4.10{191} | |
| Xenon (Xe) | 4.13{193} | 4.14{194} | 4.14{195} | |
| Hydrogen ($H_2$) | 5.1{200} | 5.2{201} | 5.2{202} | |
| Methane ($CH_4$) | 5.5{204} | 5.6{205} | 5.6{206} | |
| Nitrous oxide ($N_2O$) | 5.9{208} | 5.10{209} | 5.10{210} | |

**Solubility in mg/L**

| Gas | Freshwater | Seawater (0–40 g/kg) | Seawater (33–37 g/kg) | Brine (0–225 g/kg) |
|---|---|---|---|---|
| Oxygen ($O_2$) | 1.9{32} | 2.11{88} | 2.12{89} | 6.7{222}, 6.8{223} |
| Nitrogen ($N_2$) | 1.10{33} | 2.13{90} | 2.14{91} | |
| Argon (Ar) | 1.11{34} | 2.15{92} | 2.16{93} | |
| Carbon dioxide ($CO_2$) in 2010 | 1.12{35} | 2.17{94} | 2.18{95} | 6.9{224}, 6.10{225} |
| Carbon dioxide ($CO_2$) in 2030 | 1.13{36} | 2.19{96} | 2.20{97} | 6.11{226}, 6.12{227} |

## Solubility in mL/L

| Gas | Freshwater | Seawater (0–40 g/kg) | Seawater (33–37 g/kg) | Brine (0–225 g/kg) |
|---|---|---|---|---|
| Oxygen ($O_2$) | 1.14{37} | | | |
| Nitrogen ($N_2$) | 1.15{38} | | | |
| Argon (Ar) | 1.16{39} | | | |
| Carbon dioxide ($CO_2$) in 2010 | 1.17{40} | | | |
| Carbon dioxide ($CO_2$) in 2030 | 1.18{41} | | | |

## Gas Tension (mmHg/(mg/L))

| Gas | Freshwater | Seawater (0–40 g/kg) | Seawater (33–37 g/kg) | Brine (0–225 g/kg) |
|---|---|---|---|---|
| Oxygen ($O_2$) | 1.44{67} | 2.45{122} | 2.46{123} | |
| Nitrogen ($N_2$) | 1.45{68} | 2.47{124} | 2.48{125} | |
| Argon (Ar) | 1.46{69} | 2.49{126} | 2.50{127} | |
| Nitrogen + Argon ($N_2$ + Ar) | 1.47{70} | 2.51{128} | 2.52{129} | |
| Carbon dioxide ($CO_2$) | 1.48{71} | 2.53{130} | 2.54{131} | |

## Bunsen Coefficient (L/(L atm))

| Gas | Freshwater | Seawater (0–40 g/kg) | Seawater (33–37 g/kg) | Brine (0–225 g/kg) |
|---|---|---|---|---|
| Oxygen ($O_2$) | 1.32{55} | 2.25{102} | 2.26{103} | 6.13{228}, 6.14{229} |
| Nitrogen ($N_2$) | 1.33{56} | 2.27{104} | 2.28{105} | |
| Nitrogen + Argon ($N_2$ + Ar) | 3.3{153} | 3.4{154} | 3.5{155} | |
| Argon (Ar) | 1.34{57} | 2.29{106} | 2.30{107} | |
| Carbon dioxide ($CO_2$) | 1.35{58} | 2.31{108} | 2.32{109} | 6.15{230}, 6.16{231} |
| Helium (He) | 4.3{183} | 4.4{184} | {185} | |
| Neon (Ne) | 4.7{187} | 4.8{188} | {189} | |
| Krypton (Kr) | 4.11{191} | 4.12{192} | {193} | |
| Xenon (Xe) | 4.15{195} | 4.16{196} | {197} | |
| Hydrogen ($H_2$) | 5.3{202} | 5.4{203} | {204} | |
| Methane ($CH_4$) | 5.7{206} | 5.8{207} | {208} | |
| Nitrous oxide ($N_2O$) | 5.11{210} | 5.12{211} | {212} | |

## Bunsen Coefficient (L/(L mmHg))

| Gas | Freshwater | Seawater (0–40 g/kg) | Seawater (33–37 g/kg) | Brine (0–225 g/kg) |
|---|---|---|---|---|
| Oxygen ($O_2$) | 1.36{59} | 2.33{110} | 2.34{111} | |
| Nitrogen ($N_2$) | 1.37{60} | 2.35{112} | 2.36{113} | |
| Argon (Ar) | 1.38{61} | 2.37{114} | 2.38{115} | |
| Carbon dioxide ($CO_2$) | 1.39{62} | 2.39{116} | 2.40{117} | |

## Bunsen Coefficient (L/(L kPa))

| Gas | Freshwater | Seawater (0–40 g/kg) | Seawater (33–37 g/kg) | Brine (0–225 g/kg) |
|---|---|---|---|---|
| Oxygen ($O_2$) | 1.40{63} | | | |
| Nitrogen ($N_2$) | 1.41{64} | | | |
| Argon (Ar) | 1.42{65} | | | |
| Carbon dioxide ($CO_2$) | 1.43{66} | | | |

## Physical Properties of Water

| Parameter | Freshwater | Seawater (0–40 g/kg) | Seawater (33–37 g/kg) | Brine (0–225 g/kg) |
|---|---|---|---|---|
| Density | 7.1{238} | 7.2{239} | 7.3{240} | 6.19{234}, 6.20{235} |
| Specific weight | | | | |
| kN/m$^3$ | | 7.4{241} | | |
| mmHg/m | 1.29{52} | 2.43{118} | 2.43{120} | |
| kPa/m | 1.30{53} | 2.44{119} | 2.44{121} | |
| Vapor pressure | | | | |
| mmHg | 1.21{44} | 2.41{115}, 7.5{242} | 2.41{118}, 7.5{242} | 6.17{231}, 6.18{233} |
| kPa | 1.22{45} | 2.42{116} | 2.42{119} | |
| Atmosphere | 1.23{46} | | | |
| psi | 1.24{47} | | | |
| Heat capacity | | 7.6{243} | | |
| Viscosity | | 7.7{244} | | |
| Kinematic viscosity | | 7.8{245} | | |
| Surface tension | | 7.9{246} | | |
| Heat of vaporization | 7.10{247} | | | |

# Table and Page Index

# Example Problems Index

# Introduction

Gas solubility data are required for a variety of oceanographic, limnological, fisheries engineering, aquacultural, and engineering applications. The maintenance of an adequate concentration of dissolved oxygen is a major problem in the culture of aquatic animals. Low levels of dissolved oxygen can reduce the growth of cultured animals, decrease feed utilization, increase disease problems, and result in massive mortality. In very high intensity culture systems, the buildup of carbon dioxide must be controlled. Low dissolved oxygen concentrations are also a problem in lakes, streams, and marine conditions due to both natural and man-made causes. Under certain conditions, supersaturation of dissolved gases can be lethal to aquatic animals. The effects of gas supersaturation depend on the degree of supersaturation, the gas composition, and the position of the animal in the water column. In the marine environment, information on the supersaturation of inert gases can be used as powerful tracers of physical and biological processes.

Both the measurement and control of dissolved gas concentrations depend on an accurate knowledge of equilibrium concentrations. It is necessary to be able to compute the equilibrium concentration as a function of temperature, salinity, pressure, and gas composition. This book is divided into seven major sections:

1. Solubility of major atmospheric gases in freshwater
2. Solubility of major atmospheric gases in marine waters
3. Computation and reporting of gas supersaturation levels
4. Solubility of noble gases
5. Solubility of trace gases
6. Solubility of gases in brines
7. Physical properties of water.

The most accurate gas solubility relationships are computationally complex. This type of information is needed for some applications. For some engineering applications, however, the accuracy requirements are typically not as restrictive, so simpler solubility relationships can be used.

Solubility data are presented in the text in both equation and tabular form. With this information, the equilibrium concentrations of pure gases, air, or mixtures of gases can be computed. In most cases, interpolation should not be needed. Sample problems are included in each section.

Two additional stand-alone programs are provided to estimate gas solubility as a function of temperature, pressure, salinity, and gas composition. AIRSAT is an executable program for Windows computers that computes the standard air solubility concentration or air solubility of the 11 gases presented in this book. The second program, ARBSAT, computes the solubility of a gas as a function of an arbitrary mole fraction. Both programs can be downloaded from http://www.elsevierdirect.com/companion.jsp?ISBN=9780124159167.

An errata and update file can also be found at this site listed above: if improved solubility information is published, revised tables will be listed. If you have any suggestions, questions, or corrections, I can be reached at one of the following e-mail addresses:

john.colt@noaa.gov
johncolt@halcyon.com

# 1 Solubility of Atmospheric Gases in Freshwater

## General Gas Solubility Relationships

The solubility of gases in this book will be developed in terms of the following parameters:

| | |
|---|---|
| $C_o^\dagger$ or $C_o^*$ | Standard air saturation concentration. This is the saturation concentration for moist air at 1 atm total pressure. The superscripts "†" and "*" are used to denote concentrations expressed on a mass (kg) or a volume (L or $dm^3$) basis, respectively (Benson and Krause, 1984). |
| $C_p^\dagger$ or $C_p^*$ | Air saturation concentration. This is the saturation concentration for moist air at $p$ atm total pressure. |
| $C_{p,x}^\dagger$ or $C_{p,x}^*$ | Saturation concentration. This is the saturation concentration for moist gas at $p$ atm total pressure and a mole fraction of the gas of interest equal to $\chi$. |
| $\beta$ | Bunsen coefficient. The volume of gas at STP (standard temperature and pressure: pressure = 1 atm and temperature = 0°C) absorbed per unit volume of liquid when partial pressure (or fugacity) of the gas is 1 atm. Gas volumes can be expressed in terms of an ideal gas or a real gas. |

Standard air saturation concentration and air saturation concentration are equivalent to "unit standard atmospheric concentration" (USAC) and "standard atmospheric concentration" (SAC), respectively, as defined by Benson and Krause (1984). These definitions are a little wordy for routine use and do not convey that this is a saturation concentration.

$C_o^\dagger$, $C_o^*$, $C_p^\dagger$, and $C_p^*$ are based on the assumptions that the gas involved is air (20.946% oxygen, 78.084% nitrogen, 0.934% argon, and 0.0390% carbon dioxide (in year = 2010)) and that the air is saturated with water vapor (relative humidity = 100). Close to the air–water interface, where gas transfer occurs, the relative humidity is assumed to be 100%. The atmospheric concentration of carbon dioxide is changing, and computation of its atmospheric solubility will require a special approach. Changes in atmospheric carbon dioxide will have negligible impact on the composition of the other major atmospheric gases.

It is common to refer to a gas in terms of its degree of solubility (i.e., soluble, insoluble, and very insoluble), but there are no uniform criteria for assignment of this classification. Generally, the classification depends on which gases are under consideration. The solubilities of the gases considered in this book are summarized

Dissolved Gas Concentration in Water. DOI: 10.1016/B978-0-12-415916-7.00001-2

below for temperature = 20°C and salinity = 35 g/kg in terms of the Bunsen coefficient $\beta$ (mL/(L atm)) and $C_o^\dagger$ (nmol/kg):

| **Gas** | $\beta$ **(mL/(L atm))** | $C_o^\dagger$ **(nmol/kg)** |
|---|---|---|
| Oxygen ($O_2$) | 25.28 | 225,540 |
| Nitrogen ($N_2$) | 12.63 | 419,773 |
| Argon (Ar) | 27.84 | 11,075 |
| Carbon dioxide ($CO_2$) | 739.50 | 12,311 |
| Helium (He) | 7.46 | 2 |
| Neon (Ne) | 8.83 | 7 |
| Krypton (Kr) | 50.23 | 2 |
| Xenon (Xe) | 86.51 | 0.3 |
| Hydrogen ($H_2$) | 15.38 | 0.4 |
| Methane ($CH_4$) | 27.87 | 2 |
| Nitrous oxide ($N_2O$) | 532.80 | 7 |

$\beta$ is a measure of the solubility of a pure gas and is typically expressed in L/(L atm). In this table, $\beta$ has been multiplied by 1000 to give mL/(L atm), a unit that allows better visualization of the full solubility range of the gases. Carbon dioxide, nitrous oxide, and xenon are the most soluble of the gases considered in this book. Helium and neon gases are the most insoluble pure gases. $C_o^\dagger$ is the solubility of an individual gas when it is in equilibrium with moist air and therefore depends both on the solubility of the pure gas and on the mole fraction of gas in the atmosphere. $C_o^\dagger$ is proportional to the number of molecules of gas per kg of water. Note that the air solubility of nitrogen is twice the air solubility of oxygen, even though pure oxygen gas is twice as soluble as pure nitrogen gas. While pure xenon gas has the third highest solubility of the gases considered, its air solubility is the lowest because of its extremely low mole fraction (0.00000009 atm or 0.09 μatm) in the atmosphere.

The solubility of gases decreases with increasing temperature (Figure 1.1) and salinity (Figure 1.2). The impact of temperature is nonlinear and quite gas specific, especially for the noble gases (neon, krypton, helium, and xenon).

The "primary units" for air solubility of gases used in this book are μmol/kg or nmol/kg. This is in keeping with oceanographic work that requires the highest degree of accuracy. Less confusion will arise if concentrations are expressed per unit mass of solution because significant uncertainty is introduced when solution volume is reported without the relevant temperature and salinity data (Mortimer, 1981). It is also necessary to clearly distinguish between the original solubility equations and equations fitted to the original data. The original solubility equations are typically very complex and computationally intense, so many authors fit simpler equations to the original data as a function of temperature and salinity that can be used for routine work.

In the body of this book, the original solubility equations in terms of μmol/kg or nmol/kg are used to compute $C_o^\dagger$. All other solubility data presented in this book

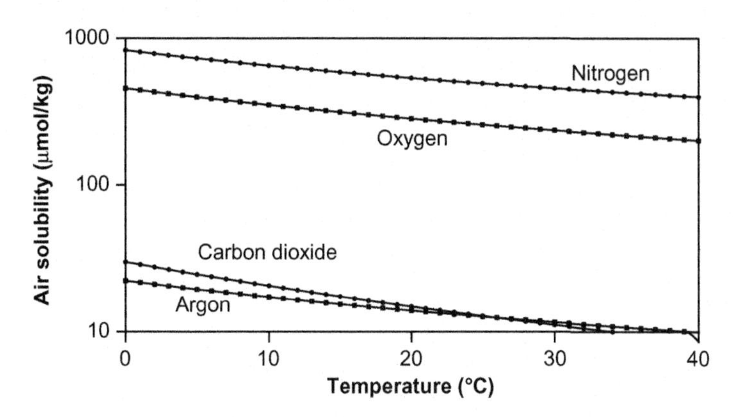


**Figure 1.1** Impact of temperature on air solubility of major atmospheric gases (freshwater).

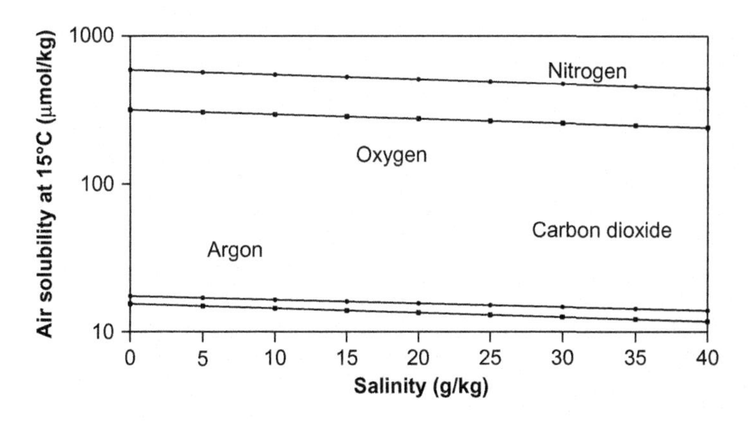


**Figure 1.2** Impact of salinity on air solubility of major atmospheric gases (15°C).

are based on this data and the appropriate conversions. When solubility information is present on a volume basis ($C_o^*$), it is based on $C_o^\dagger$ and the density of water, rather than on a separate expression for $C_o^*$. This may introduce small differences in the tabular values compared to the use of separate expressions or equations fitted to the original data. The fitted solubility equations can be used in computer programs, but the basis of the relationships used should be clearly specified. Several different equations for the vapor pressure of water are needed to match what was used in the original solubility equations.

While the "primary units" for gas solubility are μmol/kg or nmol/kg, these are not commonly used in many fields. More commonly used units are mg/L and mL real gas/L; and less commonly used units are mL ideal gas/L or μg-atom/L. The concentration in terms of volume is related to mass by the following relationship:

$$C^* = \rho C^\dagger \tag{1.1}$$

where

$C^*$ = concentration (μmol/L or nmol/L)
$\rho$ = density of water (kg/L)
$C^\dagger$ = concentration (μmol/kg or nmol/kg).

Conversions between moles/mass and moles/volume and other alternative units are presented in Table 1.1{04}. When solubility is expressed in mL or L, it is important to realize that this is a hypothetical volume of either real or ideal gas at STP and does not represent the actual volume of the gas in solution. In general oceanographic work,

**Table 1.1** Conversion of Gas Concentrations to Alternate Units

**Concentrations on a Mass Basis**[a]

| Gas | Base Unit | To Convert to Alternate Units, Multiply $C_o^\dagger$ or $C^\dagger$ by: | | | |
|---|---|---|---|---|---|
| | $C_o^\dagger$ or $C^\dagger$ | mg/kg | mL real gas/kg | mL ideal gas/kg | μg-atom/kg |
| $O_2$ | μmol/kg | $31.998 \times 10^{-3}$ | $22.392 \times 10^{-3}$ | $22.414 \times 10^{-3}$ | 2.000 |
| $N_2$ | μmol/kg | $28.014 \times 10^{-3}$ | $22.404 \times 10^{-3}$ | $22.414 \times 10^{-3}$ | 2.000 |
| Ar | μmol/kg | $39.948 \times 10^{-3}$ | $22.393 \times 10^{-3}$ | $22.414 \times 10^{-3}$ | — |
| $CO_2$ | μmol/kg | $44.009 \times 10^{-3}$ | $22.263 \times 10^{-3}$ | $22.414 \times 10^{-3}$ | — |
| He | nmol/kg | $4.0026 \times 10^{-6}$ | $22.426 \times 10^{-6}$ | $22.414 \times 10^{-6}$ | — |
| Ne | nmol/kg | $20.180 \times 10^{-6}$ | $22.424 \times 10^{-6}$ | $22.414 \times 10^{-6}$ | — |
| Kr | nmol/kg | $83.800 \times 10^{-6}$ | $22.351 \times 10^{-6}$ | $22.414 \times 10^{-6}$ | — |
| Xe | nmol/kg | $131.29 \times 10^{-6}$ | $22.260 \times 10^{-6}$ | $22.414 \times 10^{-6}$ | — |
| $H_2$ | nmol/kg | $2.0158 \times 10^{-6}$ | $22.428 \times 10^{-6}$ | $22.414 \times 10^{-6}$ | — |
| $CH_4$ | nmol/kg | $16.043 \times 10^{-6}$ | $22.360 \times 10^{-6}$ | $22.414 \times 10^{-6}$ | — |
| $N_2O$ | nmol/kg | $44.013 \times 10^{-6}$ | $22.243 \times 10^{-6}$ | $22.414 \times 10^{-6}$ | — |

**Concentrations on a Volume Basis**[a]

| Gas | Base Unit | To Convert to Alternate Units, Multiply $C_o^\dagger$ or $C^\dagger$ by: | | | |
|---|---|---|---|---|---|
| | $C_o^\dagger$ or $C^\dagger$ | mg/L | mL real gas/L | mL ideal gas/L | μg-atom/L |
| $O_2$ | μmol/kg | $\rho\, 31.999 \times 10^{-3}$ | $\rho\, 22.392 \times 10^{-3}$ | $\rho\, 22.414 \times 10^{-3}$ | 2.000 |
| $N_2$ | μmol/kg | $\rho\, 28.013 \times 10^{-3}$ | $\rho\, 22.404 \times 10^{-3}$ | $\rho\, 22.414 \times 10^{-3}$ | 2.000 |
| Ar | μmol/kg | $\rho\, 39.948 \times 10^{-3}$ | $\rho\, 22.393 \times 10^{-3}$ | $\rho\, 22.414 \times 10^{-3}$ | — |
| $CO_2$ | μmol/kg | $\rho\, 44.010 \times 10^{-3}$ | $\rho\, 22.263 \times 10^{-3}$ | $\rho\, 22.414 \times 10^{-3}$ | — |
| He | nmol/kg | $\rho\, 4.0026 \times 10^{-6}$ | $\rho\, 22.426 \times 10^{-6}$ | $\rho\, 22.414 \times 10^{-6}$ | — |
| Ne | nmol/kg | $\rho\, 20.180 \times 10^{-6}$ | $\rho\, 22.424 \times 10^{-6}$ | $\rho\, 22.414 \times 10^{-6}$ | — |
| Kr | nmol/kg | $\rho\, 83.800 \times 10^{-6}$ | $\rho\, 22.351 \times 10^{-6}$ | $\rho\, 22.414 \times 10^{-6}$ | — |
| Xe | nmol/kg | $\rho\, 131.29 \times 10^{-6}$ | $\rho\, 22.260 \times 10^{-6}$ | $\rho\, 22.414 \times 10^{-6}$ | — |
| $H_2$ | nmol/kg | $\rho\, 2.0158 \times 10^{-6}$ | $\rho\, 22.428 \times 10^{-6}$ | $\rho\, 22.414 \times 10^{-6}$ | — |
| $CH_4$ | nmol/kg | $\rho\, 16.043 \times 10^{-6}$ | $\rho\, 22.360 \times 10^{-6}$ | $\rho\, 22.414 \times 10^{-6}$ | — |
| $N_2O$ | nmol/kg | $\rho\, 44.013 \times 10^{-6}$ | $\rho\, 22.243 \times 10^{-6}$ | $\rho\, 22.414 \times 10^{-6}$ | |

The shaded areas indicate that tabular data is presented for these gases in this book. Note that 1 mL = 1 cm$^3$, and 1 L = 1 dm$^3$.

[a]$C_o^* = \rho C_o^\dagger$ or $C_o^\dagger = C_o^*/\rho$ where $\rho$ is the density of water in kg/L or (kg/m$^3$)/1000. The density of water depends on both temperature and salinity (see Tables 7.1–7.3{238, 239, and 240}).

the unit of pressure is commonly the standard atmosphere (760 mmHg, 101.325 kPa). While the units of millimeter of mercury are not SI units, they are especially useful in total gas supersaturation and physiological work. Conversions between the different pressure units are presented in Appendix D. The hydrostatic pressure in the ocean increases approximately 1 bar (100,000 Pa) per 10 m water depth. Therefore, the pressure in decibars (10,000 Pa) is approximately equal to the water depth in meters. In oceanographic work, pressure in decibars refers to hydrostatic pressure, and the pressure at the water surface is 0 decibars (e.g., see Lewis and Wallace, 1998).

The vapor pressure of water will be presented in a number of units to allow a transition to SI units. The conventional units of liter and milliliter will be used instead of the SI units of $dm^3$ or $cm^3$. Water density may be expressed either as $kg/m^3$ or kg/L. The range of temperature and salinities presented in most tables are 0−40°C and 0−40 g/kg. If the original data did not cover this full range of temperature and salinity, this data will be shaded to indicate that it has been extrapolated. Use this extrapolated data with care since the accuracy is not well defined.

The tabular values in this book were generated by an Absoft FORTRAN compiler on a 32-bit Microsoft Windows computer. Comparison of these results with those determined by a calculator may show a difference in the last digit due to use of more digits in a computer. These differences in the 5th or 6th decimal place are not significant.

The details of the equations used in the writing of this book are presented in Appendix A for gas solubility and in Appendix B for physical properties of water. General information on the computer programs developed for this book is presented in Appendix C. These programs can be downloaded from the following website: http://www.elsevierdirect.com/companion.jsp?ISBN=9780124159167. Properties of gases, key symbols, and units and conversions in this book are presented in Appendix D.

The maximum uncertainty in the solubility of each major atmospheric gas near 1 atm total pressure is:

| Gas | Uncertainty (%) | Source |
|---|---|---|
| $O_2$ | ±0.10 | Benson and Krause (1984) |
| $N_2$ | ±0.14 | Hamme and Emerson (2004) |
| Ar | ±0.13 | Hamme and Emerson (2004) |
| $CO_2$ | ±0.20 | Weiss (1974) |

Of the four major atmospheric gases, carbon dioxide is many times more soluble than the other three gases. Its departure from ideal gas approximations is large compared to the accuracy with which its solubility can be measured (Weiss, 1974). The equations presented for carbon dioxide for freshwater and seawater are valid for total pressures up to 10 atm, but higher pressures require the use of more sophisticated equations of state (Weiss, 1974).

For engineering and culture applications, where an accuracy of 1−2% is adequate, the atmospheric gases can be treated as ideal gases with a significant reduction in computational complexity. Under these applications, the difference between μmol/kg and μmol/L or between mg/kg and mg/L may be small and ignored, especially for freshwater conditions.

## Standard Air Solubility Concentration in μmol/kg ($C_o^\dagger$)—Freshwater

The standard air solubility concentration ($C_o^\dagger$) of the major atmospheric gases are presented in terms of μmol/kg for freshwater in the following tables:

| Table | Gas | Page |
|---|---|---|
| 1.2 | Oxygen | 25 |
| 1.3 | Nitrogen | 26 |
| 1.4 | Argon | 27 |
| 1.5 | Carbon dioxide for year 2010 | 28 |
| 1.6 | Carbon dioxide for year 2030 | 29 |

The carbon dioxide concentration is based on an assumed mole fraction of 390 μatm for 2010 and 440 μatm for 2030. A separate approach will be presented in the following section to allow computation of the standard air solubility of carbon dioxide gas as a function of mole fraction.

### Example 1-1

Convert the standard air solubility concentration of oxygen in μmol/kg for 30.3°C to mg/L, mL real gas/L, mL ideal gas/L, and μg-atom/L. Use density of water (Table 7.1) and the conversion factors listed in Table 1.1.

| Parameter | Value | Source |
|---|---|---|
| $C_o^\dagger$ (30.3°C) | 236.05 μmol/kg | Table 1.2 |
| $\rho$ (30.3°C) | 0.995560 kg/L | Table 7.1 |
| μmol/L → mg/L | $31.998 \times 10^{-3}$ | Table 1.1 |
| μmol/L → mL real gas/L | $22.392 \times 10^{-3}$ | Table 1.1 |
| μmol/L → mL ideal gas/L | $22.414 \times 10^{-3}$ | Table 1.1 |
| μmol/L → μg-atom/L | 2.000 | Table 1.1 |

It is first necessary to convert $C_o^\dagger$ to $C_o^*$:

$$C_o^* = \rho_w C_o^\dagger$$

$$C_o^* = (0.995560\ \text{kg/L})(236.05\ \mu\text{mol/kg})$$

$C_o^* = 235.002\ \mu\text{mol/L}$
$C_o^*(\text{mg/L}) = (235.002)(31.998 \times 10^{-3}) = 7.52\ \text{mg/L}$
$C_o^*(\text{mL real gas/L}) = (235.002)(22.392 \times 10^{-3}) = 5.26\ \text{mL/L}$
$C_o^*(\text{mL ideal gas/L}) = (235.002)(22.414 \times 10^{-3}) = 5.27\ \text{mL/L}$
$C_o^*(\mu\text{g-atom/L}) = (235.002)(2.000) = 470.004\ \mu\text{g-atom}$

**Example 1-2**

Compute the standard air solubility concentration of the four major atmospheric gases in μmol/kg, mg/L, and mL real gas/L for 14.1°C and 38.9°C and freshwater conditions. Use Tables 1.2–1.5, 1.9–1.12, and 1.14–1.17 and atmospheric carbon dioxide for year equals 2010.

**μmol/kg**

| Gas | 14.1°C | 38.9°C | Source |
|---|---|---|---|
| $O_2$ | 321.61 | 205.44 | Table 1.2 |
| $N_2$ | 600.08 | 405.64 | Table 1.3 |
| Ar | 15.744 | 10.109 | Table 1.4 |
| $CO_2$ | 17.927 | 8.855 | Table 1.5 |
| Total | 955.361 | 630.044 | |

**mg/L**

| Gas | 14.1°C | 38.9°C | Source |
|---|---|---|---|
| $O_2$ | 10.283 | 6.525 | Table 1.9 |
| $N_2$ | 16.798 | 11.280 | Table 1.10 |
| Ar | 0.6284 | 0.4009 | Table 1.11 |
| $CO_2$ | 0.7883 | 0.3868 | Table 1.12 |
| Total | 28.4977 | 18.593 | |

**mL real gas/L**

| Gas | 14.1°C | 38.9°C | Source |
|---|---|---|---|
| $O_2$ | 7.196 | 4.566 | Table 1.14 |
| $N_2$ | 13.434 | 9.021 | Table 1.15 |
| Ar | 0.3523 | 0.2247 | Table 1.16 |
| $CO_2$ | 0.3988 | 0.1957 | Table 1.17 |
| Total | 21.381 | 14.007 | |

## Standard Air Solubility Concentration for Carbon Dioxide as a Function of Mole Fraction—Freshwater

Weiss and Price (1980) developed an approach for computing the standard air solubility concentration for carbon dioxide as a function of the mole fraction of carbon dioxide in the atmosphere ($\chi_{CO_2}$):

$$C_o^\dagger = F^\dagger \chi_{CO_2} \tag{1.2}$$

$$C_o^* = F^* \chi_{CO_2} \tag{1.3}$$

For freshwater conditions, the values of $F^\dagger$ and $F^*$ depend only on temperature and are presented in Tables 1.7{30} and 1.8{31} for $C^\dagger$ and $C^*$, respectively. Equations (1.2) and (1.3) and Tables 1.7{30} and 1.8{31} can be used to compute the standard air solubility concentration of carbon dioxide as a function of an arbitrary mole fraction. This approach is limited to conditions where $\chi_{CO_2} \ll 1$. The computation of air solubility of carbon dioxide for large values of $\chi_{CO_2}$ will be presented in a later section as a function of the Bunsen coefficient ($\beta$).

Due to burning of fossil fuels and land use changes (IPCC, 2007) the mole fraction of carbon dioxide gas in the atmosphere is increasing (Figure 1.3). Imposed on this long-term rise is a seasonal variation due to the uptake of carbon dioxide by plants. The best long-term information on atmospheric carbon dioxide is from Mauna Loa in the Hawaiian Islands. Information on historical and current mole fraction data for carbon dioxide at this site can be found at:

| | |
|---|---|
| Yearly $CO_2$ Data | ftp://ftp.cmdl.noaa.gov/ccg/co2/trends/co2_annmean_mlo.txt |
| Monthly $CO_2$ Data | ftp://ftp.cmdl.noaa.gov/ccg/co2/trends/co2_mm_mlo.txt |
| Measurement Data | http://www.esrl.noaa.gov/gmd/ccgg/about/co2_measurements.html |

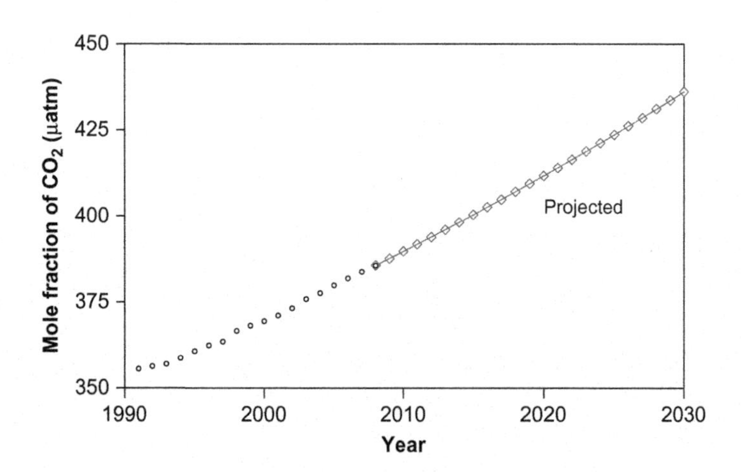


**Figure 1.3** Variation of mole fraction of carbon dioxide at Mauna Loa in the Hawaiian Islands (http://www.ipcc-data.org/ ancilliary/ipcc_ddc_co2_mauna_loa.txt). Projected values based on second-order polynomial.

Figure 1.3 is based on the yearly data from this site. A simple second-order polynomial equation was fitted to this data and used to project carbon dioxide levels from 2008 to 2030:

| Year | μatm | Year | μatm |
|---|---|---|---|
| 1991 | 355.54 | 2011 | ***391.8*** |
| 1992 | 356.29 | 2012 | ***393.9*** |
| 1993 | 356.97 | 2013 | ***396.0*** |
| 1994 | 358.69 | 2014 | ***398.2*** |
| 1995 | 360.71 | 2015 | ***400.4*** |
| 1996 | 362.41 | 2016 | ***402.6*** |
| 1997 | 363.53 | 2017 | ***404.8*** |
| 1998 | 366.64 | 2018 | ***407.1*** |
| 1999 | 368.16 | 2019 | ***409.4*** |
| 2000 | 369.45 | 2020 | ***411.7*** |
| 2001 | 371.12 | 2021 | ***414.0*** |
| 2002 | 373.24 | 2022 | ***416.4*** |
| 2003 | 375.88 | 2023 | ***418.8*** |
| 2004 | 377.60 | 2024 | ***421.2*** |
| 2005 | 379.87 | 2025 | ***423.6*** |
| 2006 | 381.89 | 2026 | ***426.1*** |
| 2007 | 383.79 | 2027 | ***428.5*** |
| 2008 | ***385.6*** | 2028 | ***431.1*** |
| 2009 | ***387.6*** | 2029 | ***433.6*** |
| 2010 | ***389.7*** | 2030 | ***436.1*** |

The above projections can be used to roughly estimate $\chi_{CO_2}$ in the future and solubility concentrations (Eqs (1.2) and (1.3)). Because of uncertainty in future carbon dioxide emission rates, more accurate estimates can be obtained from the previously listed web sites or from determination of local $\chi_{CO_2}$ value.

## Example 1-3

Compute the standard air solubility concentration of carbon dioxide in μmol/kg for 11.5°C and 21.9°C, freshwater conditions and years 2010, 2020, and 2030. Use mole fraction data presented on page 9 and in Eq. (1.1); compare with Table 1.6 for year = 2030.

$\chi_{CO_2}$

| Year | μatm | Source |
|---|---|---|
| 2010 | 389.7 | Page 9 |
| 2020 | 411.7 | Page 9 |
| 2030 | 436.1 | Page 9 |

$\boldsymbol{F_o^\dagger}$

| Gas | 11.5°C | 21.9°C | Source |
|---|---|---|---|
| $F_o^\dagger$ | $5.0154 \times 10^{-2}$ | $3.6009 \times 10^{-2}$ | Table 1.7 |

From Eq. (1.1):

$$C_o^\dagger = F_o^\dagger \chi_{CO_2}$$

For year 2010 and 11.5°C, $\chi_{CO_2} = 389.7 \times 10^{-6}$ and $F_o^\dagger = 5.0153 \times 10^{-2}$ mol/(kg atm) (above)

$$C_o^\dagger = (5.0154 \times 10^{-2})(389.7 \times 10^{-6})(10^6\ \mu\text{mol/mol})$$

$$C_o^\dagger = 19.5450\ \mu\text{mol/kg}$$

**μmol/kg**

| Year | 11.5°C | 21.9°C | Source |
|---|---|---|---|
| 2010 | **19.545** | 14.033 | Eq. (1.1) |
| 2020 | 20.648 | 14.825 | Eq. (1.1) |
| 2030 | 21.872 | 15.704 | Eq. (1.1) |

Compare these results with the values presented in Table 1.6:

| Source | 11.5°C | 21.9°C |
|---|---|---|
| Table 1.6 | 22.068 | 15.844 |
| Above | 21.872 | 15.704 |

Table 1.6 values are slightly larger because they are based on $\chi_{CO_2} = 440$ rather than the $\chi_{CO_2} = 436.1$ used above.

## Standard Air Solubility Concentration in Conventional Units ($C_o^*$)—Freshwater

The standard air solubility concentration ($C_o^*$) of the major atmospheric gases are presented in terms of conventional units for freshwater in the following tables:

| Gas | mg/L | | mL real gas at STP/L | |
|---|---|---|---|---|
| | Table | Page | Table | Page |
| Oxygen | 1.9 | 32 | 1.14 | 37 |
| Nitrogen | 1.10 | 33 | 1.15 | 38 |
| Argon | 1.11 | 34 | 1.16 | 39 |
| Carbon dioxide, year 2010 | 1.12 | 35 | 1.17 | 40 |
| Carbon dioxide, year 2030 | 1.13 | 36 | 1.18 | 41 |

The carbon dioxide concentrations are based on an assumed mole fraction of 390 μatm for 2010 and 440 μatm for 2030. Conversion factors for other units are presented in Table 1.1{04}; Mortimer (1981) also presented useful relationships to convert between the various units.

## Computation of Air Saturation Concentrations ($C_p^\dagger$ or $C_p^*$)—Freshwater

While it is possible to compute $C_p^\dagger$ or $C_p^*$ from the original equations, it is more convenient to develop a procedure based on the standard air solubility concentration and pressure adjustment factor based on the local barometric pressure. Benson and Krause (1984) developed the following equation for oxygen:

$$C_p = C_o \text{BP}\left[\frac{(1 - P_{wv}/\text{BP})(1 - \theta_o \text{BP})}{(1 - P_{wv})(1 - \theta_o)}\right] \quad (1.4)$$

where

$C_p$ = air solubility concentration (either on a mass or volume basis)
$C_o$ = standard air solubility concentration (either on a mass or volume basis)
BP = local barometric pressure (atm)
$P_{wv}$ = vapor pressure of water (atm)
$\theta_o$ = a constant that depends on the second virial coefficient of oxygen (atm).

This equation can be used on units based on either mass or volume and for standard units or alternate units. The units of $C_p$ will correspond to the units of $C_o$, as the unit of BP[ ] in Eq. (1.4) is dimensionless. Equation (1.4) has been adopted by Standard Methods (2005) and Mortimer (1981).

For an ideal gas, $\theta_o = 0$, and Eq. (1.4) reduces to

$$C_p = C_o\left[\frac{(\mathrm{BP} - P_{wv})}{(1 - P_{wv})}\right] \tag{1.5}$$

This equation is the more common relationship used to compute air saturation concentration as a function of the standard air saturation concentration and local barometric pressure (Hutchinson, 1957). The value of 1 in the denominator in Eq. (1.5) is only equal to 1 for pressures measured in atmospheres. In reality, it is the standard atmospheric pressure $P_{\mathrm{standard}}$; so, for example, for pressures measured in kPa, this equation would be rewritten as:

$$C_p = C_o\left[\frac{(\mathrm{BP} - P_{wv})}{(101.325\ \mathrm{kPa} - P_{wv})}\right] \tag{1.6}$$

where both BP and $P_{wv}$ must be expressed in kPa.

---

### Example 1-4

If the barometric pressure drops from 760 to 456 mmHg (temperature = 13.0°C, freshwater, year = 2010), compute the decrease in the saturation concentration of the major atmospheric gases. Use Table 1.20. Compute the air saturation concentration for oxygen (barometric pressure = 456 mmHg, temperature = 13.0°C, freshwater, year = 2010) using Eq. (1.5) and compare to previous result.

**Input Data**

| Parameter | Value | Source |
|---|---|---|
| $C_o^*$ for $O_2$ | 10.536 mg/L | Table 1.9 |
| $C_o^*$ for $N_2$ | 17.175 mg/L | Table 1.10 |
| $C_o^*$ for Ar | 0.6438 mg/L | Table 1.11 |
| $C_o^*$ for $CO_2$ | 0.8178 mg/L | Table 1.12 |
| PAF | 0.5940 | Table 1.20 |
| $P_{wv}$ | 1.4974 kPa | Table 1.22 |

From Eq. (1.5):

$$C_p = C_o\left[\frac{(\mathrm{BP} - P_{wv})}{(1 - P_{wv})}\right]$$

The pressure adjustment factor in Table 1.20 is equal to the terms within the bracket, so this equation reduces to:

$C_p = C_o$ [pressure adjustment factor in Table 1.20]

and

$$\text{Decrease (\%)} = \left[\frac{\text{PAF} \times C_o^* - C_o^*}{C_o^*}\right]100$$

$$\text{Decrease (mg/L)} = \text{PAF} \times C_o^* - C_o^*$$

For oxygen, the decreases are equal to:

$$\text{Decrease (\%)} = \left[\frac{0.5940 \times 10.536 - 10.536}{10.536}\right]100 = -40.59\%$$

$$\text{Decrease (mg/L)} = 0.5940 \times 10.536 - 10.536 = -4.28 \text{ mg/L}$$

| Parameter | Decrease (%) | Decrease (mg/L) |
|---|---|---|
| $O_2$ | **−40.59** | **−4.28** |
| $N_2$ | −40.59 | −6.97 |
| Ar | −40.59 | −0.26 |
| $CO_2$ | −40.59 | −0.33 |

**Compute air saturation concentration from Eq.** (1.5)
Convert water vapor in mmHg to kPa (760 mmHg = 101.325 kPa)

$$\left[\frac{456 \text{ mmHg}}{760 \text{ mmHg}}\right]101.325 \text{ kPa} = 60.795 \text{ kPa}$$

Equation (1.5):

$$C_p = C_o\left[\frac{(\text{BP} - P_{wv})}{(101.325 \text{ kPa} - P_{wv})}\right]$$

$$C_p = 10.536\left[\frac{(60.795 - 1.4974)}{(101.325 - 1.4974)}\right] = 6.258 \text{ mg/L}$$

and

$C_p = C_o$ [pressure adjustment factor in Table 1.19]

$C_p = 10.536\ [0.5940] = 6.258$ mg/L

Pressure adjustment factors based on Eqs (1.4) and (1.5) are presented in Tables 1.19{42} and 1.20{43} as functions of temperature and barometric pressure.

Barometric pressure can be estimated in at least four different ways:

1. Value based on precision laboratory barometer.
2. Field value based on portable electronic barometer calibrated against laboratory unit.
3. Field value estimated from reading at base station and an estimate of site elevation (Stringer, 1972).
4. Field value estimated from standard atmospheric pressure models (Mortimer, 1981).

The barometric pressures reported by weather stations may be corrected to sea level and cannot be directly used. The actual pressure used in Eqs (1.4) and (1.5) and Tables 1.19{42} and 1.20{43} can depend on the time frame of the process under consideration. For evaluation of gas transfer systems, a point estimate of the local barometric pressure is appropriate. For seasonal studies of oxygen in a lake, a long term or seasonally averaged local barometric pressure should be used (Mortimer, 1981). The use of standard atmosphere models introduce up to a 1–3% error, as this approach ignores latitudinal and seasonal pressure variations (Mortimer, 1981).

For highly accurate work, the value of the pressure adjustment factor for oxygen (Eq. (1.4)) can be computed directly from $\theta_o$ (see Appendix A) and $P_{wv}$. Equation (1.5) only requires information on the vapor pressure of water ($P_{wv}$) at the local temperature and the local barometric pressure (BP). The choice between Eqs (1.4) and (1.5) for oxygen may also depend on the accuracy of the corresponding temperature and pressure measurements. The changes resulting from $\pm 0.001°C$ and $\pm 0.01$ mmHg are compared to the differences between Eqs (1.4) and (1.5) in the following table for oxygen:

| Temperature (°C) | Absolute Value of the Percent Change in Oxygen Solubility | | |
|---|---|---|---|
| | **Temperature (±0.001°C)** | **Pressure (±0.01 mmHg) (BP = 760 mmHg)** | **Equations (1.4) versus (1.3) (BP = 740 mmHg)** |
| 0 | 0.002 844 | 0.001 323 | 0.001 282 |
| 10 | 0.002 352 | 0.001 329 | 0.001 105 |
| 20 | 0.001 967 | 0.001 347 | 0.000 943 |
| 30 | 0.001 694 | 0.001 373 | 0.000 793 |
| 40 | 0.001 545 | 0.001 418 | 0.000 665 |

For routine work, the use of the pressure adjustment factor based on ideal gas (Eq. (1.5)) is acceptable.

## Vapor Pressure of Water ($P_{wv}$)—Freshwater

Values of the vapor pressure of water are presented in the following tables for various units:

| Table | Units | Page |
|---|---|---|
| 1.21 | mmHg | 44 |
| 1.22 | kPa | 45 |
| 1.23 | atm | 46 |
| 1.24 | psi | 47 |

As long as the 1 atm standard pressure, barometric pressure, and vapor pressure of water are all in the same units, any pressure units can be used in Eq. (1.5). This is also true for Eq. (1.4) as long as the value of $\theta_o$ is converted to the same pressure units.

## Computation of Air Saturation Concentrations as Function of Elevation—Freshwater

The solubility of gases at any elevation can be computed from standard air saturation concentrations and either Eqs (1.4) or (1.5) if the barometric pressure is known. If barometric pressure cannot be measured directly, it can be calculated by reference to a second locality at which elevation and barometric pressure are known (Stringer, 1972):

$$\log_{10} \mathrm{BP} = \log_{10} \mathrm{BP}_o - \frac{h - h_o}{kT_a} \tag{1.7}$$

where

$\mathrm{BP}$ = barometric pressure of station (mmHg)
$\mathrm{BP}_o$ = barometric pressure of reference station (mmHg)
$h$ = elevation in meters above sea level for station
$h_o$ = elevation in meters above sea level for reference station
$k = 67.4$
$T_a$ = average of the air temperature between the two stations in K (273.15 + °C).

The reference station most commonly used is sea level, for which,

$h_o = 0$ (sea level)
$\mathrm{BP}_o = 760$ mmHg
$T_a = 288.15$ K (15°C).

For these conditions,

$$\log_{10} \text{BP} = 2.880814 - \frac{h}{19,421.3} \tag{1.8}$$

In the United States, most topographic maps show elevations in feet. These could be converted to meters (Appendix D), for which Eq. (1.8) would be valid, but it may be more convenient to use a modification of this equation. In this case, $h_o$, $\text{BP}_o$, and $T_a$ do not change, but $k$ is different; and the resulting formula is

$$\log_{10} \text{BP} = 2.880814 - \frac{h'}{63,718.2} \tag{1.9}$$

where $h'$ = elevation of station in feet.

Once the atmospheric pressure has been computed from Eqs (1.8) or (1.9), air solubility can be computed from the previously developed equations. The solubility of oxygen is presented as functions of temperature and elevation in the following tables:

| Table | Values | Page |
|---|---|---|
| 1.25 | 0–1800 m | 48 |
| 1.26 | 2000–3800 m | 49 |
| 1.27 | 0–4500 ft | 50 |
| 1.28 | 5000–9500 ft | 51 |

Both Eqs (1.8) and (1.9) and Tables 1.25–1.28{48, 49, 50, and 51} are based on the assumption that the barometric pressure at sea level is equal to 760 mmHg and the average air temperature between the two stations is 15°C. This approach is adequate for routine work, but direct measure of barometric pressure is necessary for highest accuracy work. If needed, similar tables could be developed for the other gases from the standard air saturation concentrations and Eqs (1.8) or (1.9).

**Example 1-5**

Compute the difference in the air solubility concentration of oxygen between 200 and 1800 m (water temperature = 19°C, BP = 760 mmHg). Use Tables 1.21 and 1.25 and compare with Eqs (1.4) and (1.7).

From Tables 1.21 and 1.25:
**Input Data**

| Parameter | Value | Source |
|---|---|---|
| $C_o^*$ for $O_2$ at 0 m | 9.276 mg/L | Table 1.25 |
| $C_o^*$ for $O_2$ at 200 m | 9.054 mg/L | Table 1.25 |
| $C_o^*$ for $O_2$ at 1800 m | 7.455 mg/L | Table 1.25 |
| $P_{wv}$ | 16.482 mmHg | Table 1.21 |

**From** Table 1.25

$$\Delta DO = DO_{200} - DO_{1800}$$

$$\Delta DO = 9.054 - 7.455 = 1.599 \text{ mg/L}$$

**From Eqs (1.7) and (1.4)**

$$\log_{10} \text{BP} = 2.880814 - \frac{h}{19,421.3}$$

$$\log_{10} \text{BP}_{200} = 2.880814 - \frac{200}{19,421.3} = 742.192 \text{ mmHg}$$

$$\log_{10} \text{BP}_{1800} = 2.880814 - \frac{1800}{19,421.3} = 613.949 \text{ mmHg}$$

$$C_p = C_o \left[\frac{(\text{BP} - P_{wv})}{(1 - P_{wv})}\right]$$

$$DO_{200} = 9.276 \left[\frac{(742.192 - 16.482)}{(760 - 16.482)}\right] = 9.054 \text{ mg/L}$$

$$DO_{200} = 9.276 \left[\frac{(613.949 - 16.482)}{(760 - 16.482)}\right] = 7.454 \text{ mg/L}$$

$$\Delta DO = 9.054 - 7.454 = 1.600 \text{ mg}/L$$

## Computation of Air Saturation Concentrations as Function of Water Depth—Freshwater

To compute the air saturation concentration for a bubble at depth $z$, the actual pressure inside the bubble is equal to the sum of the barometric and hydrostatic pressures. The values computed in this section are based on the assumption that the bubble and dissolved gases in the surrounding waters are at equilibrium. This very rarely occurs because the bubble rises toward the surface much faster than it takes to achieve equilibrium. In addition, transfer of gases into and out of the bubble will change the composition of the gases inside the bubble. Therefore, neither the standard air saturation concentration nor the air saturation concentration can be used to compute the saturation concentration. The computation of saturation for these conditions must be based on the Bunsen coefficient (presented in the following section). Nevertheless, these values are useful because they show how the efficiency of aeration devices increases at greater depths or how gas supersaturation can be produced by bubble entrainment.

The total pressure, $P_t$, at depth $Z$ is:

$$P_t = BP + \rho g Z \tag{1.10}$$

where

$P_t$ = total pressure (kPa)
BP = barometric pressure (atm)
$\rho$ = density of water (kg/m$^3$)
$g$ = acceleration due to gravity (9.80655 m/s$^2$)
$Z$ = depth in meters below the surface (m).

As written, Eq. (1.10) would have units of kPa. Since total gas supersaturation is reported in mmHg, it is convenient to express $\rho g$ in terms of mmHg/m (Table 1.29 {52}) or kPa/m (Table 1.30{53}).

Once the total pressure has been computed (Eq. (1.10)), the air saturation concentration can be computed from Eqs (1.4) or (1.5). The air solubility of oxygen, nitrogen, argon, and carbon dioxide for depths ranging from 0 to 40 m are presented in Table 1.31{54}.

## Computation of Bunsen Coefficients and Gas Solubility of an Arbitrary Mole Fraction—Freshwater

The Bunsen coefficient ($\beta$) historically represents the solubility of a real gas in liters at STP when the partial pressure in the gas phase is equal to 1 standard atmosphere (1 atm). The partial pressure of the $i$th gas in the atmospheres is equal to:

$$\text{Partial pressure (atm)} = \chi_i(BP - P_{wv}) \tag{1.11}$$

where

$\chi_i$ = mole fraction of $i$th gas (atm)
BP = barometric pressure (atm)
$P_{wv}$ = vapor pressure of water (atm).

The saturation concentration ($C_{p,\chi}$) in units of L gas at STP/L of water is therefore equal to:

$$C_{p,\chi} = \beta_i \chi_i (BP - P_{wv}) \tag{1.12}$$

for a pure gas at $BP = 1 + P_{wv}$, $C_{p,\chi} = \beta_i$. Equation (1.12) can be used to compute the solubility concentration of a gas at an arbitrary pressure, temperature, and gas composition. While the Bunsen coefficient has historically been expressed in terms of L/(L atm), this is not the most convenient unit for atmospheric gases. Equation (1.12) can be converted to mg/L by multiplying the right side by $1000\ \text{mL/L} \times K_i$ (Appendix D) and to pressures in mmHg:

$$C_{p,\chi} = 1000\ K_i \beta_i \chi_i \left[\frac{\text{BP} - P_{wv}}{760}\right] \tag{1.13}$$

In terms of pressures in kPa, Eq. (1.13) can be written as:

$$C_{p,\chi} = 1000\ K_i \beta_i \chi_i \left[\frac{\text{BP} - P_{wv}}{101.325}\right] \tag{1.14}$$

Equations (1.13) and (1.14) can be rearranged to:

$$C_{p,\chi} = \left[\frac{1000\ K_i \beta_i}{760}\right]_i \chi_i(\text{BP} - P_{wv}) \tag{1.15}$$

$$C_{p,\chi} = \left[\frac{1000\ K_i \beta_i}{101.325}\right]_i \chi_i(\text{BP} - P_{wv}) \tag{1.16}$$

Letting the terms within the brackets in Eqs (1.15) and (1.16) be equal to $\beta_i'$ (mg/(L mmHg)) and $\beta_i''$ (mg/(L kPa)), Eqs (1.15) and (1.16) can be rewritten as:

$$C_{p,\chi} = \beta_i' \chi_i(\text{BP} - P_{wv}) \quad (\text{BP and } P_{wv} \text{ in mmHg}) \tag{1.17}$$

$$C_{p,\chi} = \beta_i'' \chi_i(\text{BP} - P_{wv}) \quad (\text{BP and } P_{wv} \text{ in kPa}) \tag{1.18}$$

Equations (1.17) and (1.18) are more convenient to use than Eqs (1.13) and (1.14).

### Example 1-6

Compute the solubility of the major atmospheric gases when $\chi = 1.00$, barometric pressure = 450 mmHg, temperature = 35°C, moist air, and freshwater. Use Eq. (1.12).

**Input Data**

| Parameter | Value | Source |
|---|---|---|
| $\beta_{O_2}$ | 0.02458 L/(L atm) | Table 1.32 |
| $\beta_{N_2}$ | 0.01289 L/(L atm) | Table 1.33 |
| $\beta_{Ar}$ | 0.02710 L/(L atm) | Table 1.34 |
| $\beta_{CO_2}$ | 0.1306 L/(L atm) | Table 1.35 |
| $P_{wv}$ | 42.201 mmHg | Table 1.21 |
| $K_{O_2}$ | 1.42899 mg/mL | D-1 |
| $K_{N_2}$ | 1.25040 mg/mL | D-1 |
| $K_{Ar}$ | 1.78395 mg/mL | D-1 |
| $K_{O_2}$ | 1.97678 mg/mL | D-1 |

From Eq. (1.12):

$$C_{p,\chi} = 1000\ K_i \beta_i \chi_i \left[\frac{\text{BP} - P_{wv}}{760}\right]$$

$$C_{O_2} = 1000 \times 1.42899 \times 0.02458 \times 1.000 \left[\frac{450 - 42.201}{760}\right] = 18.847\ \text{mg/L}$$

$$C_{N_2} = 1000 \times 1.25040 \times 0.01289 \times 1.000 \left[\frac{450 - 42.201}{760}\right] = 8.648\ \text{mg/L}$$

$$C_{Ar} = 1000 \times 1.78395 \times 0.02710 \times 1.000 \left[\frac{450 - 42.201}{760}\right] = 25.941\ \text{mg/L}$$

$$C_{CO_2} = 1000 \times 1.97678 \times 0.1306 \times 1.000 \left[\frac{450 - 42.201}{760}\right] = 138.527\ \text{mg/L}$$

The Bunsen coefficients for oxygen, nitrogen, argon, and carbon dioxide are presented at 0.1°C temperature intervals for the range of 0–40°C in the following tables:

| Gas | Bunsen Coefficient ($\beta$) (L real gas/(L atm)) | | Bunsen Coefficient ($\beta'$) (mg/(L mmHg)) | | Bunsen Coefficient ($\beta''$) (mg/(L kPa)) | |
|---|---|---|---|---|---|---|
| | Table | Page | Table | Page | Table | Page |
| $O_2$ | 1.32 | 55 | 1.36 | 59 | 1.40 | 63 |
| $N_2$ | 1.33 | 56 | 1.37 | 60 | 1.41 | 64 |
| Ar | 1.34 | 57 | 1.38 | 61 | 1.42 | 65 |
| $CO_2$ | 1.35 | 58 | 1.39 | 62 | 1.43 | 66 |

The use of Bunsen coefficients (Eqs (1.12), (1.17), or (1.18)) to directly compute the solubility of a gas of arbitrary composition is useful for many engineering applications. These equations should not be used when the highest accurate solubility information is needed, especially for mole fractions near atmospheric values.

### Example 1-7

Recompute the solubilities in Example 1-6 if the mole fractions are reduced to 0.500 and the gauge pressure is equal to 15.0 m of water. Assume that the barometric pressure is equal to 760 mmHg, temperature = 35°C, moist air, and freshwater. Use Eqs (1.9) and (1.12).

**Input Data**

| Parameter | Value | Source |
|---|---|---|
| $\rho g$ | 73.117 mmHg/m | Table 1.29 |
| See Example 1-6 for remaining input data | | |

From Eq. (1.9):

$$P_t = \text{BP} + \rho g Z$$

$$P_t = 760 + 73.117 \times 15.0 = 1856.755 \text{ mmHg}$$

From Eq. (1.12):

$$C_{p,\chi} = 1000\ K_i \beta_i \chi_i \left[\frac{\text{BP} - P_{wv}}{760}\right]$$

$$C_{O_2} = 1000 \times 1.42899 \times 0.02458 \times 0.5000 \left[\frac{1856.755 - 42.201}{760}\right] = 41.932 \text{ mg/L}$$

$$C_{N_2} = 1000 \times 1.25040 \times 0.01289 \times 0.500 \left[\frac{1856.755 - 42.201}{760}\right] = 19.241 \text{ mg/L}$$

$$C_{Ar} = 1000 \times 1.78395 \times 0.02710 \times 0.500 \left[\frac{1856.755 - 42.201}{760}\right] = 55.714 \text{ mg/L}$$

$$C_{CO_2} = 1000 \times 1.97678 \times 0.1306 \times 0.500 \left[\frac{1856.755 - 42.201}{760}\right] = 308.200 \text{ mg/L}$$

## Example 1-8

Compute the solubility of the major atmospheric gases in air at 20°C, 760 mmHg, moist air, freshwater, and year = 2010. Use Eq. (1.16) and compare results with values in Tables 1.9–1.12 in terms of mg/L and percent.

**Input Data**

| Parameter | Value | Source |
|---|---|---|
| $\beta'_{O_2}$ | 0.05846 L/(L atm) | Table 1.36 |
| $\beta'_{N_2}$ | 0.02592 L/(L atm) | Table 1.37 |
| $\beta'_{Ar}$ | 0.08021 L/(L atm) | Table 1.38 |
| $\beta'_{O_2}$ | 2.2641 L/(L atm) | Table 1.39 |
| $P_{wv}$ | 17.539 mmHg | Table 1.21 |
| $\chi_{O_2}$ | 0.20946 | D-1 |
| $\chi_{N_2}$ | 0.78084 | D-1 |
| $\chi_{Ar}$ | 0.00934 | D-1 |
| $\chi_{O_2}$ | 0.000390 | Page 9, year = 2010 |

From Eq. (1.16):

$$C_{p,\chi} = \beta'_i \chi_i (BP - P_{wv})$$

$$C_{O_2} = 0.05846 \times 0.20946\ [760 - 17.539] = 9.091 \text{ mg/L}$$

$$C_{N_2} = 0.02592 \times 0.78084\ [760 - 17.539] = 15.027 \text{ mg/L}$$

$$C_{Ar} = 0.08021 \times 0.00934\ [760 - 17.539] = 0.5562 \text{ mg/L}$$

$$C_{CO_2} = 2.2641 \times 0.000390\ [760 - 17.539] = 0.6556 \text{ mg/L}$$

**Comparison of Solubility Results (mg/L)**

| Parameter | Based on Eq. (1.16) | Tables 1.9–1.13 | Difference (mg/L) | Difference (%) |
|---|---|---|---|---|
| $O_2$ | 9.091 | 9.092 | −0.001 | −0.01 |
| $N_2$ | 15.027 | 15.028 | −0.001 | −0.01 |
| Ar | 0.5562 | 0.5562 | +0.000 | 0.00 |
| $CO_2$ | 0.6556 | 0.6533 | +0.002 | +0.35 |

The relationships presented in this section are based on the assumption of an ideal gas. Therefore, the fugacity of the gas is equal to its partial pressure (Eq. (1.11)). While the Bunsen coefficient has historically been defined at partial pressure = 1 atm, it has also been defined at a fugacity = 1 atm (Benson and Krause, 1980; Weiss, 1974).

## Computation of Gas Tension, mmHg—Freshwater

The partial pressure of a dissolved gas is commonly called gas tension and is expressed as mmHg. At equilibrium, the partial pressure of a gas in the gas phase is equal to the partial pressure of the same gas in the liquid phase. The gas tension is equal to the partial pressure in the gas phase that would be in equilibrium with the measured gas concentration. Equation (1.13) can be rearranged to:

$$\text{Gas tension (mmHg)} = \frac{C}{\beta_i}\left[\frac{760}{1000\,K_i}\right] \tag{1.19}$$

$C$ = concentration of the $i$th gas (mg/L)
$\beta_i$ = Bunsen coefficient of the $i$th gas (L/(L atm))
$K_i$ = ratio of molecular weight to volume (mg/mL at STP).

Letting the value of the terms within the bracket in Eq. (1.19) equal to $A_i$, this equation can be written as:

$$\text{Gas tension (mmHg)} = C\left[\frac{A_i}{\beta_i}\right] \tag{1.20}$$

Values of $A_i$ are presented in Table D-1{287} in Appendix D. The term with the bracket in Eq. (1.20) is the conversion factor between mg/L and mmHg. Values of this factor for freshwater are presented for the following five gases as a function of temperature in the following tables:

| Table | Table | Page |
|---|---|---|
| $O_2$ | 1.44 | 67 |
| $N_2$ | 1.45 | 68 |
| Ar | 1.46 | 69 |
| $N_2$ + Ar | 1.47 | 70 |
| $CO_2$ | 1.48 | 71 |

The gas represented by $N_2$ + Ar is the sum of nitrogen + argon gas. This parameter is measured by some types of total gas monitoring equipment and will be discussed in the following section.

### Example 1-9

Compute the partial pressure of the major atmospheric gases in the liquid phase (gas tension) if each gas has a concentration of 10.5 mg/L, temperature = 32°C, and freshwater. Use Eq. (1.19) and the tabular data in Tables 1.44–1.48.

**Input Data**

| Parameter | Value | Source |
|---|---|---|
| $O_2$ | 20.771 mmHg/mg/L | Table 1.44 |
| $N_2$ | 45.601 mmHg/mg/L | Table 1.45 |
| Ar | 15.105 mmHg/mg/L | Table 1.46 |
| $CO_2$ | 0.60807 mmHg/mg/L | Table 1.48 |

From Eq. (1.19):

$$\text{Gas tension (mmHg)} = C\left[\frac{A_i}{\beta_i}\right]$$

$$P_{O_2} = 10.5 \times 20.771 = 218.1 \text{ mmHg}$$

$$P_{N_2} = 10.5 \times 45.601 = 478.8 \text{ mmHg}$$

$$P_{Ar} = 10.5 \times 15.105 = 158.6 \text{ mmHg}$$

$$P_{CO_2} = 10.5 \times 0.60807 = 6.4 \text{ mmHg}$$

**Table 1.2** Standard Air Saturation Concentration of Oxygen as a Function of Temperature in μmol/kg (freshwater, 1 atm moist air)

| **Temperature (°C)** | **Δt (°C)** | | | | | | | | | |
|---|---|---|---|---|---|---|---|---|---|---|
| | **0.0** | **0.1** | **0.2** | **0.3** | **0.4** | **0.5** | **0.6** | **0.7** | **0.8** | **0.9** |
| **0** | 457.00 | 455.71 | 454.42 | 453.13 | 451.85 | 450.58 | 449.32 | 448.06 | 446.80 | 445.56 |
| **1** | 444.31 | 443.08 | 441.85 | 440.62 | 439.40 | 438.19 | 436.98 | 435.78 | 434.59 | 433.40 |
| **2** | 432.21 | 431.03 | 429.86 | 428.69 | 427.53 | 426.37 | 425.22 | 424.07 | 422.93 | 421.79 |
| **3** | 420.66 | 419.54 | 418.42 | 417.30 | 416.19 | 415.08 | 413.98 | 412.89 | 411.80 | 410.71 |
| **4** | 409.63 | 408.56 | 407.49 | 406.42 | 405.36 | 404.30 | 403.25 | 402.21 | 401.17 | 400.13 |
| **5** | 399.10 | 398.07 | 397.04 | 396.03 | 395.01 | 394.00 | 393.00 | 392.00 | 391.00 | 390.01 |
| **6** | 389.02 | 388.04 | 387.06 | 386.08 | 385.11 | 384.15 | 383.19 | 382.23 | 381.28 | 380.33 |
| **7** | 379.38 | 378.44 | 377.51 | 376.57 | 375.65 | 374.72 | 373.80 | 372.88 | 371.97 | 371.06 |
| **8** | 370.16 | 369.26 | 368.36 | 367.47 | 366.58 | 365.70 | 364.81 | 363.94 | 363.06 | 362.19 |
| **9** | 361.33 | 360.46 | 359.60 | 358.75 | 357.90 | 357.05 | 356.20 | 355.36 | 354.53 | 353.69 |
| **10** | 352.86 | 352.03 | 351.21 | 350.39 | 349.57 | 348.76 | 347.95 | 347.15 | 346.34 | 345.54 |
| **11** | 344.75 | 343.95 | 343.16 | 342.38 | 341.59 | 340.81 | 340.04 | 339.26 | 338.49 | 337.73 |
| **12** | 336.96 | 336.20 | 335.44 | 334.69 | 333.94 | 333.19 | 332.44 | 331.70 | 330.96 | 330.22 |
| **13** | 329.49 | 328.76 | 328.03 | 327.31 | 326.58 | 325.86 | 325.15 | 324.43 | 323.72 | 323.02 |
| **14** | 322.31 | 321.61 | 320.91 | 320.21 | 319.52 | 318.83 | 318.14 | 317.46 | 316.77 | 316.09 |
| **15** | 315.42 | 314.74 | 314.07 | 313.40 | 312.73 | 312.07 | 311.41 | 310.75 | 310.09 | 309.44 |
| **16** | 308.79 | 308.14 | 307.49 | 306.85 | 306.20 | 305.56 | 304.93 | 304.29 | 303.66 | 303.03 |
| **17** | 302.41 | 301.78 | 301.16 | 300.54 | 299.92 | 299.31 | 298.69 | 298.08 | 297.47 | 296.87 |
| **18** | 296.27 | 295.66 | 295.06 | 294.47 | 293.87 | 293.28 | 292.69 | 292.10 | 291.52 | 290.93 |
| **19** | 290.35 | 289.77 | 289.19 | 288.62 | 288.05 | 287.47 | 286.91 | 286.34 | 285.77 | 285.21 |
| **20** | 284.65 | 284.09 | 283.53 | 282.98 | 282.43 | 281.88 | 281.33 | 280.78 | 280.24 | 279.69 |
| **21** | 279.15 | 278.61 | 278.08 | 277.54 | 277.01 | 276.48 | 275.95 | 275.42 | 274.89 | 274.37 |
| **22** | 273.85 | 273.33 | 272.81 | 272.29 | 271.78 | 271.27 | 270.75 | 270.24 | 269.74 | 269.23 |
| **23** | 268.73 | 268.22 | 267.72 | 267.22 | 266.73 | 266.23 | 265.74 | 265.25 | 264.75 | 264.27 |
| **24** | 263.78 | 263.29 | 262.81 | 262.33 | 261.85 | 261.37 | 260.89 | 260.41 | 259.94 | 259.47 |
| **25** | 258.99 | 258.52 | 258.06 | 257.59 | 257.12 | 256.66 | 256.20 | 255.74 | 255.28 | 254.82 |
| **26** | 254.37 | 253.91 | 253.46 | 253.01 | 252.56 | 252.11 | 251.66 | 251.21 | 250.77 | 250.33 |
| **27** | 249.88 | 249.44 | 249.00 | 248.57 | 248.13 | 247.70 | 247.26 | 246.83 | 246.40 | 245.97 |
| **28** | 245.54 | 245.12 | 244.69 | 244.27 | 243.84 | 243.42 | 243.00 | 242.58 | 242.17 | 241.75 |
| **29** | 241.33 | 240.92 | 240.51 | 240.10 | 239.69 | 239.28 | 238.87 | 238.46 | 238.06 | 237.65 |
| **30** | 237.25 | 236.85 | 236.45 | 236.05 | 235.65 | 235.25 | 234.86 | 234.46 | 234.07 | 233.68 |
| **31** | 233.29 | 232.90 | 232.51 | 232.12 | 231.73 | 231.35 | 230.96 | 230.58 | 230.19 | 229.81 |
| **32** | 229.43 | 229.05 | 228.68 | 228.30 | 227.92 | 227.55 | 227.17 | 226.80 | 226.43 | 226.06 |
| **33** | 225.69 | 225.32 | 224.95 | 224.58 | 224.22 | 223.85 | 223.49 | 223.12 | 222.76 | 222.40 |
| **34** | 222.04 | 221.68 | 221.32 | 220.96 | 220.61 | 220.25 | 219.90 | 219.54 | 219.19 | 218.84 |
| **35** | 218.49 | 218.14 | 217.79 | 217.44 | 217.09 | 216.75 | 216.40 | 216.05 | 215.71 | 215.37 |
| **36** | 215.02 | 214.68 | 214.34 | 214.00 | 213.66 | 213.32 | 212.99 | 212.65 | 212.31 | 211.98 |
| **37** | 211.65 | 211.31 | 210.98 | 210.65 | 210.32 | 209.99 | 209.66 | 209.33 | 209.00 | 208.67 |
| **38** | 208.34 | 208.02 | 207.69 | 207.37 | 207.05 | 206.72 | 206.40 | 206.08 | 205.76 | 205.44 |
| **39** | 205.12 | 204.80 | 204.48 | 204.16 | 203.85 | 203.53 | 203.22 | 202.90 | 202.59 | 202.27 |
| **40** | 201.96 | 201.65 | 201.34 | 201.03 | 200.72 | 200.41 | 200.10 | 199.79 | 199.48 | 199.17 |

*Source*: Based on Eq. 22, Benson and Krause (1984).

**Table 1.3** Standard Air Saturation Concentration of Nitrogen as a Function of Temperature in μmol/kg (freshwater, 1 atm moist air)

| Temperature (°C) | Δt (°C) | | | | | | | | | |
|---|---|---|---|---|---|---|---|---|---|---|
| | **0.0** | **0.1** | **0.2** | **0.3** | **0.4** | **0.5** | **0.6** | **0.7** | **0.8** | **0.9** |
| **0** | 830.45 | 828.25 | 826.05 | 823.87 | 821.70 | 819.54 | 817.38 | 815.24 | 813.11 | 810.99 |
| **1** | 808.87 | 806.77 | 804.68 | 802.59 | 800.52 | 798.46 | 796.40 | 794.36 | 792.32 | 790.30 |
| **2** | 788.28 | 786.28 | 784.28 | 782.29 | 780.31 | 778.34 | 776.38 | 774.43 | 772.49 | 770.55 |
| **3** | 768.63 | 766.71 | 764.80 | 762.90 | 761.01 | 759.13 | 757.26 | 755.40 | 753.54 | 751.69 |
| **4** | 749.85 | 748.02 | 746.20 | 744.39 | 742.58 | 740.78 | 738.99 | 737.21 | 735.44 | 733.67 |
| **5** | 731.91 | 730.16 | 728.42 | 726.69 | 724.96 | 723.24 | 721.53 | 719.83 | 718.13 | 716.44 |
| **6** | 714.76 | 713.09 | 711.42 | 709.76 | 708.11 | 706.47 | 704.83 | 703.20 | 701.58 | 699.96 |
| **7** | 698.35 | 696.75 | 695.16 | 693.57 | 691.99 | 690.42 | 688.85 | 687.29 | 685.74 | 684.19 |
| **8** | 682.65 | 681.12 | 679.59 | 678.07 | 676.56 | 675.05 | 673.55 | 672.06 | 670.57 | 669.09 |
| **9** | 667.61 | 666.14 | 664.68 | 663.23 | 661.78 | 660.33 | 658.90 | 657.47 | 656.04 | 654.62 |
| **10** | 653.21 | 651.80 | 650.40 | 649.01 | 647.62 | 646.23 | 644.85 | 643.48 | 642.12 | 640.76 |
| **11** | 639.40 | 638.05 | 636.71 | 635.37 | 634.04 | 632.71 | 631.39 | 630.08 | 628.77 | 627.46 |
| **12** | 626.16 | 624.87 | 623.58 | 622.30 | 621.02 | 619.75 | 618.48 | 617.22 | 615.96 | 614.71 |
| **13** | 613.46 | 612.22 | 610.98 | 609.75 | 608.52 | 607.30 | 606.09 | 604.87 | 603.67 | 602.46 |
| **14** | 601.27 | 600.08 | 598.89 | 597.71 | 596.53 | 595.35 | 594.19 | 593.02 | 591.86 | 590.71 |
| **15** | 589.56 | 588.41 | 587.27 | 586.14 | 585.00 | 583.88 | 582.75 | 581.64 | 580.52 | 579.41 |
| **16** | 578.31 | 577.21 | 576.11 | 575.02 | 573.93 | 572.85 | 571.77 | 570.69 | 569.62 | 568.55 |
| **17** | 567.49 | 566.43 | 565.38 | 564.33 | 563.28 | 562.24 | 561.20 | 560.17 | 559.14 | 558.11 |
| **18** | 557.09 | 556.07 | 555.06 | 554.05 | 553.04 | 552.04 | 551.04 | 550.04 | 549.05 | 548.06 |
| **19** | 547.08 | 546.10 | 545.12 | 544.15 | 543.18 | 542.22 | 541.25 | 540.30 | 539.34 | 538.39 |
| **20** | 537.44 | 536.50 | 535.56 | 534.62 | 533.69 | 532.76 | 531.83 | 530.91 | 529.99 | 529.07 |
| **21** | 528.16 | 527.25 | 526.34 | 525.44 | 524.54 | 523.65 | 522.75 | 521.86 | 520.98 | 520.09 |
| **22** | 519.21 | 518.34 | 517.46 | 516.59 | 515.72 | 514.86 | 514.00 | 513.14 | 512.29 | 511.43 |
| **23** | 510.59 | 509.74 | 508.90 | 508.06 | 507.22 | 506.39 | 505.56 | 504.73 | 503.90 | 503.08 |
| **24** | 502.26 | 501.45 | 500.63 | 499.82 | 499.01 | 498.21 | 497.41 | 496.61 | 495.81 | 495.02 |
| **25** | 494.23 | 493.44 | 492.65 | 491.87 | 491.09 | 490.31 | 489.54 | 488.77 | 488.00 | 487.23 |
| **26** | 486.47 | 485.70 | 484.95 | 484.19 | 483.44 | 482.68 | 481.94 | 481.19 | 480.45 | 479.70 |
| **27** | 478.97 | 478.23 | 477.50 | 476.76 | 476.04 | 475.31 | 474.58 | 473.86 | 473.14 | 472.43 |
| **28** | 471.71 | 471.00 | 470.29 | 469.58 | 468.88 | 468.18 | 467.47 | 466.78 | 466.08 | 465.39 |
| **29** | 464.70 | 464.01 | 463.32 | 462.63 | 461.95 | 461.27 | 460.59 | 459.92 | 459.24 | 458.57 |
| **30** | 457.90 | 457.23 | 456.57 | 455.90 | 455.24 | 454.58 | 453.93 | 453.27 | 452.62 | 451.97 |
| **31** | 451.32 | 450.67 | 450.03 | 449.38 | 448.74 | 448.11 | 447.47 | 446.83 | 446.20 | 445.57 |
| **32** | 444.94 | 444.31 | 443.69 | 443.06 | 442.44 | 441.82 | 441.20 | 440.59 | 439.97 | 439.36 |
| **33** | 438.75 | 438.14 | 437.54 | 436.93 | 436.33 | 435.73 | 435.13 | 434.53 | 433.93 | 433.34 |
| **34** | 432.74 | 432.15 | 431.56 | 430.98 | 430.39 | 429.81 | 429.22 | 428.64 | 428.06 | 427.49 |
| **35** | 426.91 | 426.34 | 425.76 | 425.19 | 424.62 | 424.05 | 423.49 | 422.92 | 422.36 | 421.80 |
| **36** | 421.24 | 420.68 | 420.12 | 419.57 | 419.01 | 418.46 | 417.91 | 417.36 | 416.81 | 416.27 |
| **37** | 415.72 | 415.18 | 414.64 | 414.10 | 413.56 | 413.02 | 412.48 | 411.95 | 411.42 | 410.88 |
| **38** | 410.35 | 409.82 | 409.30 | 408.77 | 408.25 | 407.72 | 407.20 | 406.68 | 406.16 | 405.64 |
| **39** | 405.12 | 404.61 | 404.09 | 403.58 | 403.07 | 402.56 | 402.05 | 401.54 | 401.03 | 400.53 |
| **40** | 400.02 | 399.52 | 399.02 | 398.52 | 398.02 | 397.52 | 397.02 | 396.53 | 396.03 | 395.54 |

*Source*: Based on Eq. 1, Hamme and Emerson (2004).

**Table 1.4** Standard Air Saturation Concentration of Argon as a Function of Temperature in μmol/kg (freshwater, 1 atm moist air)

| Temperature (°C) | Δt (°C) | | | | | | | | | |
|---|---|---|---|---|---|---|---|---|---|---|
| | **0.0** | **0.1** | **0.2** | **0.3** | **0.4** | **0.5** | **0.6** | **0.7** | **0.8** | **0.9** |
| **0** | 22.301 | 22.238 | 22.176 | 22.114 | 22.052 | 21.991 | 21.930 | 21.869 | 21.808 | 21.748 |
| **1** | 21.688 | 21.628 | 21.569 | 21.509 | 21.450 | 21.392 | 21.333 | 21.275 | 21.218 | 21.160 |
| **2** | 21.103 | 21.046 | 20.989 | 20.932 | 20.876 | 20.820 | 20.764 | 20.709 | 20.654 | 20.599 |
| **3** | 20.544 | 20.490 | 20.435 | 20.381 | 20.328 | 20.274 | 20.221 | 20.168 | 20.115 | 20.063 |
| **4** | 20.010 | 19.958 | 19.907 | 19.855 | 19.804 | 19.752 | 19.702 | 19.651 | 19.600 | 19.550 |
| **5** | 19.500 | 19.451 | 19.401 | 19.352 | 19.303 | 19.254 | 19.205 | 19.156 | 19.108 | 19.060 |
| **6** | 19.012 | 18.965 | 18.917 | 18.870 | 18.823 | 18.776 | 18.730 | 18.683 | 18.637 | 18.591 |
| **7** | 18.546 | 18.500 | 18.455 | 18.409 | 18.364 | 18.320 | 18.275 | 18.231 | 18.186 | 18.142 |
| **8** | 18.099 | 18.055 | 18.011 | 17.968 | 17.925 | 17.882 | 17.839 | 17.797 | 17.755 | 17.712 |
| **9** | 17.670 | 17.629 | 17.587 | 17.545 | 17.504 | 17.463 | 17.422 | 17.381 | 17.341 | 17.300 |
| **10** | 17.260 | 17.220 | 17.180 | 17.140 | 17.101 | 17.061 | 17.022 | 16.983 | 16.944 | 16.905 |
| **11** | 16.866 | 16.828 | 16.790 | 16.751 | 16.713 | 16.676 | 16.638 | 16.600 | 16.563 | 16.526 |
| **12** | 16.489 | 16.452 | 16.415 | 16.378 | 16.342 | 16.305 | 16.269 | 16.233 | 16.197 | 16.162 |
| **13** | 16.126 | 16.091 | 16.055 | 16.020 | 15.985 | 15.950 | 15.915 | 15.881 | 15.846 | 15.812 |
| **14** | 15.778 | 15.744 | 15.710 | 15.676 | 15.642 | 15.609 | 15.575 | 15.542 | 15.509 | 15.476 |
| **15** | 15.443 | 15.410 | 15.377 | 15.345 | 15.313 | 15.280 | 15.248 | 15.216 | 15.184 | 15.152 |
| **16** | 15.121 | 15.089 | 15.058 | 15.027 | 14.995 | 14.964 | 14.934 | 14.903 | 14.872 | 14.841 |
| **17** | 14.811 | 14.781 | 14.750 | 14.720 | 14.690 | 14.660 | 14.631 | 14.601 | 14.571 | 14.542 |
| **18** | 14.513 | 14.484 | 14.454 | 14.425 | 14.397 | 14.368 | 14.339 | 14.310 | 14.282 | 14.254 |
| **19** | 14.225 | 14.197 | 14.169 | 14.141 | 14.113 | 14.086 | 14.058 | 14.030 | 14.003 | 13.976 |
| **20** | 13.948 | 13.921 | 13.894 | 13.867 | 13.840 | 13.814 | 13.787 | 13.760 | 13.734 | 13.708 |
| **21** | 13.681 | 13.655 | 13.629 | 13.603 | 13.577 | 13.551 | 13.526 | 13.500 | 13.474 | 13.449 |
| **22** | 13.424 | 13.398 | 13.373 | 13.348 | 13.323 | 13.298 | 13.273 | 13.248 | 13.224 | 13.199 |
| **23** | 13.175 | 13.150 | 13.126 | 13.102 | 13.078 | 13.053 | 13.029 | 13.006 | 12.982 | 12.958 |
| **24** | 12.934 | 12.911 | 12.887 | 12.864 | 12.840 | 12.817 | 12.794 | 12.771 | 12.748 | 12.725 |
| **25** | 12.702 | 12.679 | 12.656 | 12.634 | 12.611 | 12.588 | 12.566 | 12.544 | 12.521 | 12.499 |
| **26** | 12.477 | 12.455 | 12.433 | 12.411 | 12.389 | 12.367 | 12.346 | 12.324 | 12.302 | 12.281 |
| **27** | 12.259 | 12.238 | 12.217 | 12.195 | 12.174 | 12.153 | 12.132 | 12.111 | 12.090 | 12.069 |
| **28** | 12.049 | 12.028 | 12.007 | 11.987 | 11.966 | 11.946 | 11.925 | 11.905 | 11.885 | 11.864 |
| **29** | 11.844 | 11.824 | 11.804 | 11.784 | 11.764 | 11.744 | 11.725 | 11.705 | 11.685 | 11.666 |
| **30** | 11.646 | 11.627 | 11.607 | 11.588 | 11.569 | 11.549 | 11.530 | 11.511 | 11.492 | 11.473 |
| **31** | 11.454 | 11.435 | 11.416 | 11.397 | 11.379 | 11.360 | 11.341 | 11.323 | 11.304 | 11.286 |
| **32** | 11.267 | 11.249 | 11.231 | 11.212 | 11.194 | 11.176 | 11.158 | 11.140 | 11.122 | 11.104 |
| **33** | 11.086 | 11.068 | 11.050 | 11.032 | 11.015 | 10.997 | 10.979 | 10.962 | 10.944 | 10.927 |
| **34** | 10.909 | 10.892 | 10.875 | 10.857 | 10.840 | 10.823 | 10.806 | 10.789 | 10.772 | 10.755 |
| **35** | 10.738 | 10.721 | 10.704 | 10.687 | 10.670 | 10.654 | 10.637 | 10.620 | 10.604 | 10.587 |
| **36** | 10.571 | 10.554 | 10.538 | 10.521 | 10.505 | 10.489 | 10.473 | 10.456 | 10.440 | 10.424 |
| **37** | 10.408 | 10.392 | 10.376 | 10.360 | 10.344 | 10.328 | 10.312 | 10.296 | 10.281 | 10.265 |
| **38** | 10.249 | 10.233 | 10.218 | 10.202 | 10.187 | 10.171 | 10.156 | 10.140 | 10.125 | 10.109 |
| **39** | 10.094 | 10.079 | 10.064 | 10.048 | 10.033 | 10.018 | 10.003 | 9.988 | 9.973 | 9.958 |
| **40** | 9.943 | 9.928 | 9.913 | 9.898 | 9.883 | 9.868 | 9.854 | 9.839 | 9.824 | 9.809 |

*Source*: Based on Eq. 1, Hamme and Emerson (2004).

**Table 1.5** Standard Air Saturation Concentration of Carbon Dioxide as a Function of Temperature in μmol/kg (freshwater, 1 atm moist air; mole fraction = 390 μatm)

| Temperature (°C) | Δ*t* (°C) | | | | | | | | | |
|---|---|---|---|---|---|---|---|---|---|---|
| | **0.0** | **0.1** | **0.2** | **0.3** | **0.4** | **0.5** | **0.6** | **0.7** | **0.8** | **0.9** |
| **0** | 29.940 | 29.820 | 29.701 | 29.583 | 29.465 | 29.348 | 29.231 | 29.115 | 29.000 | 28.885 |
| **1** | 28.771 | 28.658 | 28.545 | 28.433 | 28.321 | 28.210 | 28.099 | 27.989 | 27.880 | 27.771 |
| **2** | 27.663 | 27.556 | 27.449 | 27.342 | 27.236 | 27.131 | 27.026 | 26.922 | 26.819 | 26.715 |
| **3** | 26.613 | 26.511 | 26.409 | 26.308 | 26.208 | 26.108 | 26.009 | 25.910 | 25.811 | 25.714 |
| **4** | 25.616 | 25.520 | 25.423 | 25.327 | 25.232 | 25.137 | 25.043 | 24.949 | 24.856 | 24.763 |
| **5** | 24.670 | 24.578 | 24.487 | 24.396 | 24.305 | 24.215 | 24.126 | 24.037 | 23.948 | 23.860 |
| **6** | 23.772 | 23.685 | 23.598 | 23.511 | 23.425 | 23.340 | 23.254 | 23.170 | 23.085 | 23.001 |
| **7** | 22.918 | 22.835 | 22.752 | 22.670 | 22.588 | 22.507 | 22.426 | 22.345 | 22.265 | 22.185 |
| **8** | 22.106 | 22.027 | 21.948 | 21.870 | 21.792 | 21.715 | 21.638 | 21.561 | 21.485 | 21.409 |
| **9** | 21.334 | 21.258 | 21.184 | 21.109 | 21.035 | 20.961 | 20.888 | 20.815 | 20.742 | 20.670 |
| **10** | 20.598 | 20.527 | 20.455 | 20.384 | 20.314 | 20.244 | 20.174 | 20.104 | 20.035 | 19.966 |
| **11** | 19.898 | 19.829 | 19.762 | 19.694 | 19.627 | 19.560 | 19.493 | 19.427 | 19.361 | 19.295 |
| **12** | 19.230 | 19.165 | 19.100 | 19.036 | 18.972 | 18.908 | 18.845 | 18.781 | 18.718 | 18.656 |
| **13** | 18.594 | 18.532 | 18.470 | 18.408 | 18.347 | 18.286 | 18.226 | 18.166 | 18.105 | 18.046 |
| **14** | 17.986 | 17.927 | 17.868 | 17.810 | 17.751 | 17.693 | 17.635 | 17.578 | 17.520 | 17.463 |
| **15** | 17.407 | 17.350 | 17.294 | 17.238 | 17.182 | 17.127 | 17.071 | 17.016 | 16.962 | 16.907 |
| **16** | 16.853 | 16.799 | 16.745 | 16.692 | 16.638 | 16.585 | 16.533 | 16.480 | 16.428 | 16.376 |
| **17** | 16.324 | 16.272 | 16.221 | 16.170 | 16.119 | 16.068 | 16.017 | 15.967 | 15.917 | 15.867 |
| **18** | 15.818 | 15.768 | 15.719 | 15.670 | 15.622 | 15.573 | 15.525 | 15.477 | 15.429 | 15.381 |
| **19** | 15.334 | 15.287 | 15.240 | 15.193 | 15.146 | 15.100 | 15.053 | 15.007 | 14.962 | 14.916 |
| **20** | 14.871 | 14.825 | 14.780 | 14.735 | 14.691 | 14.646 | 14.602 | 14.558 | 14.514 | 14.470 |
| **21** | 14.427 | 14.383 | 14.340 | 14.297 | 14.255 | 14.212 | 14.170 | 14.127 | 14.085 | 14.043 |
| **22** | 14.002 | 13.960 | 13.919 | 13.878 | 13.837 | 13.796 | 13.755 | 13.715 | 13.674 | 13.634 |
| **23** | 13.594 | 13.554 | 13.515 | 13.475 | 13.436 | 13.397 | 13.358 | 13.319 | 13.280 | 13.241 |
| **24** | 13.203 | 13.165 | 13.127 | 13.089 | 13.051 | 13.013 | 12.976 | 12.939 | 12.902 | 12.865 |
| **25** | 12.828 | 12.791 | 12.754 | 12.718 | 12.682 | 12.646 | 12.610 | 12.574 | 12.538 | 12.503 |
| **26** | 12.467 | 12.432 | 12.397 | 12.362 | 12.327 | 12.292 | 12.258 | 12.223 | 12.189 | 12.155 |
| **27** | 12.121 | 12.087 | 12.053 | 12.020 | 11.986 | 11.953 | 11.920 | 11.887 | 11.854 | 11.821 |
| **28** | 11.788 | 11.755 | 11.723 | 11.691 | 11.658 | 11.626 | 11.594 | 11.563 | 11.531 | 11.499 |
| **29** | 11.468 | 11.436 | 11.405 | 11.374 | 11.343 | 11.312 | 11.281 | 11.251 | 11.220 | 11.190 |
| **30** | 11.159 | 11.129 | 11.099 | 11.069 | 11.039 | 11.010 | 10.980 | 10.950 | 10.921 | 10.892 |
| **31** | 10.862 | 10.833 | 10.804 | 10.776 | 10.747 | 10.718 | 10.690 | 10.661 | 10.633 | 10.605 |
| **32** | 10.576 | 10.548 | 10.520 | 10.493 | 10.465 | 10.437 | 10.410 | 10.382 | 10.355 | 10.328 |
| **33** | 10.300 | 10.273 | 10.246 | 10.220 | 10.193 | 10.166 | 10.140 | 10.113 | 10.087 | 10.060 |
| **34** | 10.034 | 10.008 | 9.982 | 9.956 | 9.930 | 9.905 | 9.879 | 9.853 | 9.828 | 9.803 |
| **35** | 9.777 | 9.752 | 9.727 | 9.702 | 9.677 | 9.652 | 9.627 | 9.603 | 9.578 | 9.554 |
| **36** | 9.529 | 9.505 | 9.480 | 9.456 | 9.432 | 9.408 | 9.384 | 9.360 | 9.336 | 9.313 |
| **37** | 9.289 | 9.266 | 9.242 | 9.219 | 9.195 | 9.172 | 9.149 | 9.126 | 9.103 | 9.080 |
| **38** | 9.057 | 9.034 | 9.012 | 8.989 | 8.966 | 8.944 | 8.921 | 8.899 | 8.877 | 8.855 |
| **39** | 8.832 | 8.810 | 8.788 | 8.766 | 8.745 | 8.723 | 8.701 | 8.679 | 8.658 | 8.636 |
| **40** | 8.615 | 8.593 | 8.572 | 8.551 | 8.530 | 8.509 | 8.488 | 8.467 | 8.446 | 8.425 |

*Source*: Based on Eqs. 5 and 9, Weiss (1974); mole fraction estimated from Mauna Loa data for 2010.

**Table 1.6** Standard Air Saturation Concentration of Carbon Dioxide as a Function of Temperature in μmol/kg (freshwater, 1 atm moist air; mole fraction = 440 μatm)

| Temperature (°C) | Δt (°C) | | | | | | | | | |
|---|---|---|---|---|---|---|---|---|---|---|
| | **0.0** | **0.1** | **0.2** | **0.3** | **0.4** | **0.5** | **0.6** | **0.7** | **0.8** | **0.9** |
| **0** | 33.779 | 33.644 | 33.509 | 33.376 | 33.243 | 33.110 | 32.979 | 32.848 | 32.718 | 32.588 |
| **1** | 32.460 | 32.332 | 32.204 | 32.078 | 31.952 | 31.826 | 31.702 | 31.578 | 31.455 | 31.332 |
| **2** | 31.210 | 31.088 | 30.968 | 30.848 | 30.728 | 30.609 | 30.491 | 30.374 | 30.257 | 30.140 |
| **3** | 30.025 | 29.910 | 29.795 | 29.681 | 29.568 | 29.455 | 29.343 | 29.232 | 29.121 | 29.010 |
| **4** | 28.900 | 28.791 | 28.683 | 28.574 | 28.467 | 28.360 | 28.254 | 28.148 | 28.042 | 27.937 |
| **5** | 27.833 | 27.730 | 27.626 | 27.524 | 27.422 | 27.320 | 27.219 | 27.118 | 27.018 | 26.919 |
| **6** | 26.820 | 26.721 | 26.623 | 26.525 | 26.428 | 26.332 | 26.236 | 26.140 | 26.045 | 25.950 |
| **7** | 25.856 | 25.763 | 25.669 | 25.577 | 25.484 | 25.392 | 25.301 | 25.210 | 25.120 | 25.030 |
| **8** | 24.940 | 24.851 | 24.762 | 24.674 | 24.586 | 24.499 | 24.412 | 24.326 | 24.239 | 24.154 |
| **9** | 24.069 | 23.984 | 23.899 | 23.815 | 23.732 | 23.649 | 23.566 | 23.484 | 23.402 | 23.320 |
| **10** | 23.239 | 23.158 | 23.078 | 22.998 | 22.918 | 22.839 | 22.760 | 22.682 | 22.604 | 22.526 |
| **11** | 22.449 | 22.372 | 22.295 | 22.219 | 22.143 | 22.068 | 21.992 | 21.918 | 21.843 | 21.769 |
| **12** | 21.696 | 21.622 | 21.549 | 21.477 | 21.404 | 21.332 | 21.261 | 21.189 | 21.118 | 21.048 |
| **13** | 20.977 | 20.907 | 20.838 | 20.768 | 20.699 | 20.631 | 20.562 | 20.494 | 20.427 | 20.359 |
| **14** | 20.292 | 20.225 | 20.159 | 20.093 | 20.027 | 19.961 | 19.896 | 19.831 | 19.767 | 19.702 |
| **15** | 19.638 | 19.574 | 19.511 | 19.448 | 19.385 | 19.322 | 19.260 | 19.198 | 19.136 | 19.075 |
| **16** | 19.013 | 18.953 | 18.892 | 18.832 | 18.771 | 18.712 | 18.652 | 18.593 | 18.534 | 18.475 |
| **17** | 18.417 | 18.358 | 18.300 | 18.243 | 18.185 | 18.128 | 18.071 | 18.014 | 17.958 | 17.902 |
| **18** | 17.846 | 17.790 | 17.735 | 17.679 | 17.624 | 17.570 | 17.515 | 17.461 | 17.407 | 17.353 |
| **19** | 17.300 | 17.246 | 17.193 | 17.140 | 17.088 | 17.035 | 16.983 | 16.931 | 16.880 | 16.828 |
| **20** | 16.777 | 16.726 | 16.675 | 16.625 | 16.574 | 16.524 | 16.474 | 16.424 | 16.375 | 16.326 |
| **21** | 16.276 | 16.228 | 16.179 | 16.130 | 16.082 | 16.034 | 15.986 | 15.939 | 15.891 | 15.844 |
| **22** | 15.797 | 15.750 | 15.703 | 15.657 | 15.610 | 15.564 | 15.519 | 15.473 | 15.427 | 15.382 |
| **23** | 15.337 | 15.292 | 15.247 | 15.203 | 15.158 | 15.114 | 15.070 | 15.026 | 14.983 | 14.939 |
| **24** | 14.896 | 14.853 | 14.810 | 14.767 | 14.724 | 14.682 | 14.640 | 14.598 | 14.556 | 14.514 |
| **25** | 14.472 | 14.431 | 14.390 | 14.349 | 14.308 | 14.267 | 14.226 | 14.186 | 14.146 | 14.106 |
| **26** | 14.066 | 14.026 | 13.986 | 13.947 | 13.908 | 13.868 | 13.829 | 13.791 | 13.752 | 13.713 |
| **27** | 13.675 | 13.637 | 13.599 | 13.561 | 13.523 | 13.485 | 13.448 | 13.410 | 13.373 | 13.336 |
| **28** | 13.299 | 13.263 | 13.226 | 13.189 | 13.153 | 13.117 | 13.081 | 13.045 | 13.009 | 12.973 |
| **29** | 12.938 | 12.903 | 12.867 | 12.832 | 12.797 | 12.762 | 12.728 | 12.693 | 12.659 | 12.624 |
| **30** | 12.590 | 12.556 | 12.522 | 12.488 | 12.455 | 12.421 | 12.388 | 12.354 | 12.321 | 12.288 |
| **31** | 12.255 | 12.222 | 12.190 | 12.157 | 12.125 | 12.092 | 12.060 | 12.028 | 11.996 | 11.964 |
| **32** | 11.932 | 11.901 | 11.869 | 11.838 | 11.806 | 11.775 | 11.744 | 11.713 | 11.682 | 11.652 |
| **33** | 11.621 | 11.591 | 11.560 | 11.530 | 11.500 | 11.470 | 11.440 | 11.410 | 11.380 | 11.350 |
| **34** | 11.321 | 11.291 | 11.262 | 11.233 | 11.204 | 11.175 | 11.146 | 11.117 | 11.088 | 11.059 |
| **35** | 11.031 | 11.002 | 10.974 | 10.946 | 10.918 | 10.890 | 10.862 | 10.834 | 10.806 | 10.778 |
| **36** | 10.751 | 10.723 | 10.696 | 10.669 | 10.641 | 10.614 | 10.587 | 10.560 | 10.533 | 10.507 |
| **37** | 10.480 | 10.453 | 10.427 | 10.401 | 10.374 | 10.348 | 10.322 | 10.296 | 10.270 | 10.244 |
| **38** | 10.218 | 10.192 | 10.167 | 10.141 | 10.116 | 10.090 | 10.065 | 10.040 | 10.015 | 9.990 |
| **39** | 9.965 | 9.940 | 9.915 | 9.890 | 9.866 | 9.841 | 9.817 | 9.792 | 9.768 | 9.744 |
| **40** | 9.719 | 9.695 | 9.671 | 9.647 | 9.623 | 9.599 | 9.576 | 9.552 | 9.528 | 9.505 |

*Source:* Based on Eqs. 5 and 9, Weiss (1974); mole fraction estimated from Mauna Loa data for 2030.

**Table 1.7** $F^{\dagger} \times 10^{+2}$ for Carbon Dioxide as a Function of Temperature in mol/(kg atm) (at 10°C, the table value is 5.2816; the $F^{\dagger}$ value is 5.2816/100 or 0.052816 mol/(kg atm))

| Temperature (°C) | Δt (°C) | | | | | | | | | |
|---|---|---|---|---|---|---|---|---|---|---|
| | 0.0 | 0.1 | 0.2 | 0.3 | 0.4 | 0.5 | 0.6 | 0.7 | 0.8 | 0.9 |
| 0 | 7.6770 | 7.6463 | 7.6157 | 7.5853 | 7.5551 | 7.5251 | 7.4952 | 7.4655 | 7.4359 | 7.4065 |
| 1 | 7.3772 | 7.3481 | 7.3192 | 7.2904 | 7.2618 | 7.2333 | 7.2050 | 7.1768 | 7.1488 | 7.1209 |
| 2 | 7.0931 | 7.0656 | 7.0381 | 7.0108 | 6.9837 | 6.9567 | 6.9298 | 6.9031 | 6.8765 | 6.8501 |
| 3 | 6.8238 | 6.7977 | 6.7716 | 6.7457 | 6.7200 | 6.6944 | 6.6689 | 6.6435 | 6.6183 | 6.5932 |
| 4 | 6.5683 | 6.5435 | 6.5188 | 6.4942 | 6.4698 | 6.4454 | 6.4213 | 6.3972 | 6.3732 | 6.3494 |
| 5 | 6.3257 | 6.3022 | 6.2787 | 6.2554 | 6.2322 | 6.2091 | 6.1861 | 6.1632 | 6.1405 | 6.1179 |
| 6 | 6.0954 | 6.0730 | 6.0507 | 6.0285 | 6.0064 | 5.9845 | 5.9627 | 5.9409 | 5.9193 | 5.8978 |
| 7 | 5.8764 | 5.8551 | 5.8339 | 5.8129 | 5.7919 | 5.7710 | 5.7502 | 5.7296 | 5.7090 | 5.6886 |
| 8 | 5.6682 | 5.6480 | 5.6278 | 5.6078 | 5.5878 | 5.5680 | 5.5482 | 5.5285 | 5.5090 | 5.4895 |
| 9 | 5.4701 | 5.4509 | 5.4317 | 5.4126 | 5.3936 | 5.3747 | 5.3559 | 5.3372 | 5.3186 | 5.3000 |
| 10 | 5.2816 | 5.2632 | 5.2450 | 5.2268 | 5.2087 | 5.1907 | 5.1728 | 5.1549 | 5.1372 | 5.1195 |
| 11 | 5.1020 | 5.0845 | 5.0671 | 5.0498 | 5.0325 | 5.0154 | 4.9983 | 4.9813 | 4.9644 | 4.9476 |
| 12 | 4.9308 | 4.9141 | 4.8975 | 4.8810 | 4.8646 | 4.8482 | 4.8319 | 4.8157 | 4.7996 | 4.7836 |
| 13 | 4.7676 | 4.7517 | 4.7359 | 4.7201 | 4.7044 | 4.6888 | 4.6733 | 4.6578 | 4.6424 | 4.6271 |
| 14 | 4.6119 | 4.5967 | 4.5816 | 4.5665 | 4.5516 | 4.5367 | 4.5219 | 4.5071 | 4.4924 | 4.4778 |
| 15 | 4.4632 | 4.4487 | 4.4343 | 4.4199 | 4.4057 | 4.3914 | 4.3773 | 4.3632 | 4.3491 | 4.3352 |
| 16 | 4.3212 | 4.3074 | 4.2936 | 4.2799 | 4.2662 | 4.2526 | 4.2391 | 4.2256 | 4.2122 | 4.1989 |
| 17 | 4.1856 | 4.1723 | 4.1592 | 4.1460 | 4.1330 | 4.1200 | 4.1070 | 4.0942 | 4.0813 | 4.0686 |
| 18 | 4.0558 | 4.0432 | 4.0306 | 4.0180 | 4.0055 | 3.9931 | 3.9807 | 3.9684 | 3.9561 | 3.9439 |
| 19 | 3.9317 | 3.9196 | 3.9076 | 3.8956 | 3.8836 | 3.8717 | 3.8598 | 3.8480 | 3.8363 | 3.8246 |
| 20 | 3.8130 | 3.8014 | 3.7898 | 3.7783 | 3.7669 | 3.7555 | 3.7441 | 3.7328 | 3.7216 | 3.7103 |
| 21 | 3.6992 | 3.6881 | 3.6770 | 3.6660 | 3.6550 | 3.6441 | 3.6332 | 3.6224 | 3.6116 | 3.6009 |
| 22 | 3.5902 | 3.5795 | 3.5689 | 3.5584 | 3.5478 | 3.5374 | 3.5269 | 3.5166 | 3.5062 | 3.4959 |
| 23 | 3.4857 | 3.4754 | 3.4653 | 3.4551 | 3.4451 | 3.4350 | 3.4250 | 3.4150 | 3.4051 | 3.3952 |
| 24 | 3.3854 | 3.3756 | 3.3658 | 3.3561 | 3.3464 | 3.3368 | 3.3272 | 3.3176 | 3.3081 | 3.2986 |
| 25 | 3.2892 | 3.2798 | 3.2704 | 3.2610 | 3.2517 | 3.2425 | 3.2333 | 3.2241 | 3.2149 | 3.2058 |
| 26 | 3.1967 | 3.1877 | 3.1787 | 3.1697 | 3.1608 | 3.1519 | 3.1430 | 3.1342 | 3.1254 | 3.1167 |
| 27 | 3.1079 | 3.0993 | 3.0906 | 3.0820 | 3.0734 | 3.0648 | 3.0563 | 3.0478 | 3.0394 | 3.0310 |
| 28 | 3.0226 | 3.0142 | 3.0059 | 2.9976 | 2.9893 | 2.9811 | 2.9729 | 2.9647 | 2.9566 | 2.9485 |
| 29 | 2.9404 | 2.9324 | 2.9244 | 2.9164 | 2.9085 | 2.9005 | 2.8926 | 2.8848 | 2.8770 | 2.8692 |
| 30 | 2.8614 | 2.8536 | 2.8459 | 2.8382 | 2.8306 | 2.8230 | 2.8154 | 2.8078 | 2.8003 | 2.7927 |
| 31 | 2.7853 | 2.7778 | 2.7704 | 2.7630 | 2.7556 | 2.7482 | 2.7409 | 2.7336 | 2.7263 | 2.7191 |
| 32 | 2.7119 | 2.7047 | 2.6975 | 2.6904 | 2.6833 | 2.6762 | 2.6691 | 2.6621 | 2.6551 | 2.6481 |
| 33 | 2.6411 | 2.6342 | 2.6273 | 2.6204 | 2.6136 | 2.6067 | 2.5999 | 2.5931 | 2.5864 | 2.5796 |
| 34 | 2.5729 | 2.5662 | 2.5595 | 2.5529 | 2.5463 | 2.5397 | 2.5331 | 2.5265 | 2.5200 | 2.5135 |
| 35 | 2.5070 | 2.5005 | 2.4941 | 2.4877 | 2.4813 | 2.4749 | 2.4686 | 2.4622 | 2.4559 | 2.4496 |
| 36 | 2.4434 | 2.4371 | 2.4309 | 2.4247 | 2.4185 | 2.4123 | 2.4062 | 2.4001 | 2.3940 | 2.3879 |
| 37 | 2.3818 | 2.3758 | 2.3698 | 2.3638 | 2.3578 | 2.3518 | 2.3459 | 2.3400 | 2.3341 | 2.3282 |
| 38 | 2.3223 | 2.3165 | 2.3107 | 2.3048 | 2.2991 | 2.2933 | 2.2875 | 2.2818 | 2.2761 | 2.2704 |
| 39 | 2.2647 | 2.2591 | 2.2534 | 2.2478 | 2.2422 | 2.2366 | 2.2310 | 2.2255 | 2.2200 | 2.2144 |
| 40 | 2.2089 | 2.2035 | 2.1980 | 2.1925 | 2.1871 | 2.1817 | 2.1763 | 2.1709 | 2.1656 | 2.1602 |

*Source*: Based on Weiss (1974), Weiss and Price (1980).

**Table 1.8** $F^* \times 10^{+2}$ for Carbon Dioxide as a Function of Temperature in mol/(L atm) (at 10°C, the table value is 5.2800; the $F^*$ value is 5.2800/100 or 0. 052800 mol/(L atm))

| Temperature (°C) | Δt (°C) | | | | | | | | | |
|---|---|---|---|---|---|---|---|---|---|---|
| | 0.0 | 0.1 | 0.2 | 0.3 | 0.4 | 0.5 | 0.6 | 0.7 | 0.8 | 0.9 |
| 0 | 7.6758 | 7.6451 | 7.6146 | 7.5843 | 7.5541 | 7.5241 | 7.4943 | 7.4646 | 7.4351 | 7.4057 |
| 1 | 7.3765 | 7.3474 | 7.3185 | 7.2898 | 7.2612 | 7.2327 | 7.2044 | 7.1763 | 7.1483 | 7.1204 |
| 2 | 7.0927 | 7.0652 | 7.0378 | 7.0105 | 6.9834 | 6.9564 | 6.9296 | 6.9029 | 6.8763 | 6.8499 |
| 3 | 6.8236 | 6.7974 | 6.7714 | 6.7456 | 6.7198 | 6.6942 | 6.6687 | 6.6434 | 6.6182 | 6.5931 |
| 4 | 6.5681 | 6.5433 | 6.5186 | 6.4940 | 6.4696 | 6.4453 | 6.4211 | 6.3970 | 6.3731 | 6.3492 |
| 5 | 6.3255 | 6.3019 | 6.2785 | 6.2551 | 6.2319 | 6.2088 | 6.1858 | 6.1629 | 6.1402 | 6.1175 |
| 6 | 6.0950 | 6.0726 | 6.0503 | 6.0281 | 6.0060 | 5.9840 | 5.9622 | 5.9404 | 5.9188 | 5.8973 |
| 7 | 5.8758 | 5.8545 | 5.8333 | 5.8122 | 5.7912 | 5.7703 | 5.7495 | 5.7288 | 5.7082 | 5.6878 |
| 8 | 5.6674 | 5.6471 | 5.6269 | 5.6068 | 5.5868 | 5.5669 | 5.5472 | 5.5275 | 5.5079 | 5.4884 |
| 9 | 5.4690 | 5.4496 | 5.4304 | 5.4113 | 5.3923 | 5.3733 | 5.3545 | 5.3357 | 5.3171 | 5.2985 |
| 10 | 5.2800 | 5.2616 | 5.2433 | 5.2251 | 5.2069 | 5.1889 | 5.1709 | 5.1531 | 5.1353 | 5.1176 |
| 11 | 5.1000 | 5.0824 | 5.0650 | 5.0476 | 5.0303 | 5.0131 | 4.9960 | 4.9790 | 4.9620 | 4.9451 |
| 12 | 4.9283 | 4.9116 | 4.8950 | 4.8784 | 4.8619 | 4.8455 | 4.8292 | 4.8129 | 4.7968 | 4.7807 |
| 13 | 4.7646 | 4.7487 | 4.7328 | 4.7170 | 4.7013 | 4.6856 | 4.6700 | 4.6545 | 4.6391 | 4.6237 |
| 14 | 4.6084 | 4.5932 | 4.5780 | 4.5629 | 4.5479 | 4.5329 | 4.5181 | 4.5032 | 4.4885 | 4.4738 |
| 15 | 4.4592 | 4.4447 | 4.4302 | 4.4158 | 4.4014 | 4.3871 | 4.3729 | 4.3588 | 4.3447 | 4.3307 |
| 16 | 4.3167 | 4.3028 | 4.2889 | 4.2752 | 4.2615 | 4.2478 | 4.2342 | 4.2207 | 4.2072 | 4.1938 |
| 17 | 4.1805 | 4.1672 | 4.1539 | 4.1408 | 4.1276 | 4.1146 | 4.1016 | 4.0886 | 4.0758 | 4.0629 |
| 18 | 4.0502 | 4.0374 | 4.0248 | 4.0122 | 3.9996 | 3.9871 | 3.9747 | 3.9623 | 3.9500 | 3.9377 |
| 19 | 3.9255 | 3.9133 | 3.9012 | 3.8891 | 3.8771 | 3.8652 | 3.8532 | 3.8414 | 3.8296 | 3.8178 |
| 20 | 3.8061 | 3.7945 | 3.7829 | 3.7713 | 3.7598 | 3.7483 | 3.7369 | 3.7256 | 3.7142 | 3.7030 |
| 21 | 3.6918 | 3.6806 | 3.6695 | 3.6584 | 3.6474 | 3.6364 | 3.6255 | 3.6146 | 3.6037 | 3.5929 |
| 22 | 3.5822 | 3.5715 | 3.5608 | 3.5502 | 3.5396 | 3.5291 | 3.5186 | 3.5082 | 3.4978 | 3.4874 |
| 23 | 3.4771 | 3.4668 | 3.4566 | 3.4464 | 3.4363 | 3.4261 | 3.4161 | 3.4061 | 3.3961 | 3.3861 |
| 24 | 3.3763 | 3.3664 | 3.3566 | 3.3468 | 3.3371 | 3.3274 | 3.3177 | 3.3081 | 3.2985 | 3.2890 |
| 25 | 3.2795 | 3.2700 | 3.2606 | 3.2512 | 3.2418 | 3.2325 | 3.2232 | 3.2140 | 3.2048 | 3.1956 |
| 26 | 3.1865 | 3.1774 | 3.1683 | 3.1593 | 3.1503 | 3.1414 | 3.1324 | 3.1236 | 3.1147 | 3.1059 |
| 27 | 3.0971 | 3.0884 | 3.0797 | 3.0710 | 3.0623 | 3.0537 | 3.0452 | 3.0366 | 3.0281 | 3.0196 |
| 28 | 3.0112 | 3.0028 | 2.9944 | 2.9861 | 2.9777 | 2.9695 | 2.9612 | 2.9530 | 2.9448 | 2.9367 |
| 29 | 2.9285 | 2.9204 | 2.9124 | 2.9043 | 2.8963 | 2.8884 | 2.8804 | 2.8725 | 2.8646 | 2.8568 |
| 30 | 2.8489 | 2.8411 | 2.8334 | 2.8256 | 2.8179 | 2.8103 | 2.8026 | 2.7950 | 2.7874 | 2.7798 |
| 31 | 2.7723 | 2.7648 | 2.7573 | 2.7498 | 2.7424 | 2.7350 | 2.7276 | 2.7203 | 2.7130 | 2.7057 |
| 32 | 2.6984 | 2.6912 | 2.6840 | 2.6768 | 2.6696 | 2.6625 | 2.6554 | 2.6483 | 2.6412 | 2.6342 |
| 33 | 2.6272 | 2.6202 | 2.6132 | 2.6063 | 2.5994 | 2.5925 | 2.5856 | 2.5788 | 2.5720 | 2.5652 |
| 34 | 2.5584 | 2.5517 | 2.5450 | 2.5383 | 2.5316 | 2.5250 | 2.5183 | 2.5117 | 2.5051 | 2.4986 |
| 35 | 2.4921 | 2.4855 | 2.4791 | 2.4726 | 2.4661 | 2.4597 | 2.4533 | 2.4469 | 2.4406 | 2.4342 |
| 36 | 2.4279 | 2.4216 | 2.4154 | 2.4091 | 2.4029 | 2.3967 | 2.3905 | 2.3843 | 2.3782 | 2.3721 |
| 37 | 2.3660 | 2.3599 | 2.3538 | 2.3478 | 2.3417 | 2.3357 | 2.3297 | 2.3238 | 2.3178 | 2.3119 |
| 38 | 2.3060 | 2.3001 | 2.2942 | 2.2884 | 2.2826 | 2.2767 | 2.2709 | 2.2652 | 2.2594 | 2.2537 |
| 39 | 2.2480 | 2.2423 | 2.2366 | 2.2309 | 2.2253 | 2.2196 | 2.2140 | 2.2084 | 2.2029 | 2.1973 |
| 40 | 2.1918 | 2.1862 | 2.1807 | 2.1752 | 2.1698 | 2.1643 | 2.1589 | 2.1534 | 2.1480 | 2.1426 |

*Source*: Based on Weiss (1974), Weiss and Price (1980).

**Table 1.9** Air Saturation Concentration of Oxygen as a Function of Temperature in mg/L (freshwater, 1 atm moist air)

| Temperature (°C) | Δt (°C) | | | | | | | | | |
|---|---|---|---|---|---|---|---|---|---|---|
| | **0.0** | **0.1** | **0.2** | **0.3** | **0.4** | **0.5** | **0.6** | **0.7** | **0.8** | **0.9** |
| **0** | 14.621 | 14.579 | 14.538 | 14.497 | 14.457 | 14.416 | 14.376 | 14.335 | 14.295 | 14.255 |
| **1** | 14.216 | 14.176 | 14.137 | 14.098 | 14.059 | 14.020 | 13.982 | 13.943 | 13.905 | 13.867 |
| **2** | 13.829 | 13.791 | 13.754 | 13.717 | 13.679 | 13.642 | 13.606 | 13.569 | 13.532 | 13.496 |
| **3** | 13.460 | 13.424 | 13.388 | 13.352 | 13.317 | 13.282 | 13.246 | 13.211 | 13.176 | 13.142 |
| **4** | 13.107 | 13.073 | 13.038 | 13.004 | 12.970 | 12.937 | 12.903 | 12.869 | 12.836 | 12.803 |
| **5** | 12.770 | 12.737 | 12.704 | 12.672 | 12.639 | 12.607 | 12.575 | 12.542 | 12.511 | 12.479 |
| **6** | 12.447 | 12.416 | 12.384 | 12.353 | 12.322 | 12.291 | 12.260 | 12.230 | 12.199 | 12.169 |
| **7** | 12.138 | 12.108 | 12.078 | 12.048 | 12.019 | 11.989 | 11.959 | 11.930 | 11.901 | 11.872 |
| **8** | 11.843 | 11.814 | 11.785 | 11.756 | 11.728 | 11.699 | 11.671 | 11.643 | 11.615 | 11.587 |
| **9** | 11.559 | 11.532 | 11.504 | 11.476 | 11.449 | 11.422 | 11.395 | 11.368 | 11.341 | 11.314 |
| **10** | 11.288 | 11.261 | 11.235 | 11.208 | 11.182 | 11.156 | 11.130 | 11.104 | 11.078 | 11.052 |
| **11** | 11.027 | 11.001 | 10.976 | 10.951 | 10.926 | 10.900 | 10.876 | 10.851 | 10.826 | 10.801 |
| **12** | 10.777 | 10.752 | 10.728 | 10.704 | 10.679 | 10.655 | 10.631 | 10.607 | 10.584 | 10.560 |
| **13** | 10.536 | 10.513 | 10.490 | 10.466 | 10.443 | 10.420 | 10.397 | 10.374 | 10.351 | 10.328 |
| **14** | 10.306 | 10.283 | 10.260 | 10.238 | 10.216 | 10.194 | 10.171 | 10.149 | 10.127 | 10.105 |
| **15** | 10.084 | 10.062 | 10.040 | 10.019 | 9.997 | 9.976 | 9.955 | 9.933 | 9.912 | 9.891 |
| **16** | 9.870 | 9.849 | 9.828 | 9.808 | 9.787 | 9.766 | 9.746 | 9.725 | 9.705 | 9.685 |
| **17** | 9.665 | 9.644 | 9.624 | 9.604 | 9.584 | 9.565 | 9.545 | 9.525 | 9.506 | 9.486 |
| **18** | 9.467 | 9.447 | 9.428 | 9.409 | 9.389 | 9.370 | 9.351 | 9.332 | 9.313 | 9.295 |
| **19** | 9.276 | 9.257 | 9.239 | 9.220 | 9.201 | 9.183 | 9.165 | 9.146 | 9.128 | 9.110 |
| **20** | 9.092 | 9.074 | 9.056 | 9.038 | 9.020 | 9.002 | 8.985 | 8.967 | 8.949 | 8.932 |
| **21** | 8.914 | 8.897 | 8.880 | 8.862 | 8.845 | 8.828 | 8.811 | 8.794 | 8.777 | 8.760 |
| **22** | 8.743 | 8.726 | 8.710 | 8.693 | 8.676 | 8.660 | 8.643 | 8.627 | 8.610 | 8.594 |
| **23** | 8.578 | 8.561 | 8.545 | 8.529 | 8.513 | 8.497 | 8.481 | 8.465 | 8.449 | 8.433 |
| **24** | 8.418 | 8.402 | 8.386 | 8.371 | 8.355 | 8.340 | 8.324 | 8.309 | 8.293 | 8.278 |
| **25** | 8.263 | 8.248 | 8.232 | 8.217 | 8.202 | 8.187 | 8.172 | 8.157 | 8.143 | 8.128 |
| **26** | 8.113 | 8.098 | 8.084 | 8.069 | 8.054 | 8.040 | 8.025 | 8.011 | 7.997 | 7.982 |
| **27** | 7.968 | 7.954 | 7.939 | 7.925 | 7.911 | 7.897 | 7.883 | 7.869 | 7.855 | 7.841 |
| **28** | 7.827 | 7.813 | 7.800 | 7.786 | 7.772 | 7.759 | 7.745 | 7.731 | 7.718 | 7.704 |
| **29** | 7.691 | 7.677 | 7.664 | 7.651 | 7.637 | 7.624 | 7.611 | 7.598 | 7.585 | 7.572 |
| **30** | 7.559 | 7.545 | 7.533 | 7.520 | 7.507 | 7.494 | 7.481 | 7.468 | 7.455 | 7.443 |
| **31** | 7.430 | 7.417 | 7.405 | 7.392 | 7.379 | 7.367 | 7.354 | 7.342 | 7.330 | 7.317 |
| **32** | 7.305 | 7.293 | 7.280 | 7.268 | 7.256 | 7.244 | 7.232 | 7.219 | 7.207 | 7.195 |
| **33** | 7.183 | 7.171 | 7.159 | 7.147 | 7.136 | 7.124 | 7.112 | 7.100 | 7.088 | 7.077 |
| **34** | 7.065 | 7.053 | 7.042 | 7.030 | 7.018 | 7.007 | 6.995 | 6.984 | 6.972 | 6.961 |
| **35** | 6.949 | 6.938 | 6.927 | 6.915 | 6.904 | 6.893 | 6.882 | 6.870 | 6.859 | 6.848 |
| **36** | 6.837 | 6.826 | 6.815 | 6.804 | 6.793 | 6.782 | 6.771 | 6.760 | 6.749 | 6.738 |
| **37** | 6.727 | 6.716 | 6.705 | 6.695 | 6.684 | 6.673 | 6.662 | 6.652 | 6.641 | 6.630 |
| **38** | 6.620 | 6.609 | 6.599 | 6.588 | 6.577 | 6.567 | 6.556 | 6.546 | 6.536 | 6.525 |
| **39** | 6.515 | 6.504 | 6.494 | 6.484 | 6.473 | 6.463 | 6.453 | 6.443 | 6.432 | 6.422 |
| **40** | 6.412 | 6.402 | 6.392 | 6.382 | 6.372 | 6.361 | 6.351 | 6.341 | 6.331 | 6.321 |

*Source*: Based on Eq. 22, Benson and Krause (1984), Millero and Poisson (1981).

**Table 1.10** Air Saturation Concentration of Nitrogen as a Function of Temperature in mg/L (freshwater, 1 atm moist air)

| Temperature (°C) | Δt (°C) | | | | | | | | | |
|---|---|---|---|---|---|---|---|---|---|---|
| | **0.0** | **0.1** | **0.2** | **0.3** | **0.4** | **0.5** | **0.6** | **0.7** | **0.8** | **0.9** |
| **0** | 23.261 | 23.199 | 23.138 | 23.077 | 23.016 | 22.956 | 22.895 | 22.836 | 22.776 | 22.717 |
| **1** | 22.658 | 22.599 | 22.540 | 22.482 | 22.424 | 22.366 | 22.309 | 22.252 | 22.195 | 22.138 |
| **2** | 22.082 | 22.026 | 21.970 | 21.914 | 21.859 | 21.804 | 21.749 | 21.694 | 21.640 | 21.586 |
| **3** | 21.532 | 21.478 | 21.425 | 21.371 | 21.318 | 21.266 | 21.213 | 21.161 | 21.109 | 21.057 |
| **4** | 21.006 | 20.955 | 20.904 | 20.853 | 20.802 | 20.752 | 20.702 | 20.652 | 20.602 | 20.552 |
| **5** | 20.503 | 20.454 | 20.405 | 20.357 | 20.308 | 20.260 | 20.212 | 20.164 | 20.117 | 20.069 |
| **6** | 20.022 | 19.975 | 19.928 | 19.882 | 19.836 | 19.789 | 19.744 | 19.698 | 19.652 | 19.607 |
| **7** | 19.562 | 19.517 | 19.472 | 19.428 | 19.383 | 19.339 | 19.295 | 19.251 | 19.208 | 19.164 |
| **8** | 19.121 | 19.078 | 19.035 | 18.992 | 18.950 | 18.907 | 18.865 | 18.823 | 18.782 | 18.740 |
| **9** | 18.698 | 18.657 | 18.616 | 18.575 | 18.534 | 18.494 | 18.453 | 18.413 | 18.373 | 18.333 |
| **10** | 18.294 | 18.254 | 18.215 | 18.175 | 18.136 | 18.097 | 18.059 | 18.020 | 17.982 | 17.943 |
| **11** | 17.905 | 17.867 | 17.829 | 17.792 | 17.754 | 17.717 | 17.680 | 17.643 | 17.606 | 17.569 |
| **12** | 17.533 | 17.496 | 17.460 | 17.424 | 17.388 | 17.352 | 17.316 | 17.281 | 17.245 | 17.210 |
| **13** | 17.175 | 17.140 | 17.105 | 17.070 | 17.036 | 17.001 | 16.967 | 16.933 | 16.899 | 16.865 |
| **14** | 16.831 | 16.798 | 16.764 | 16.731 | 16.698 | 16.665 | 16.632 | 16.599 | 16.566 | 16.533 |
| **15** | 16.501 | 16.469 | 16.437 | 16.405 | 16.373 | 16.341 | 16.309 | 16.278 | 16.246 | 16.215 |
| **16** | 16.184 | 16.153 | 16.122 | 16.091 | 16.060 | 16.029 | 15.999 | 15.969 | 15.938 | 15.908 |
| **17** | 15.878 | 15.848 | 15.819 | 15.789 | 15.759 | 15.730 | 15.701 | 15.671 | 15.642 | 15.613 |
| **18** | 15.584 | 15.556 | 15.527 | 15.498 | 15.470 | 15.442 | 15.413 | 15.385 | 15.357 | 15.329 |
| **19** | 15.302 | 15.274 | 15.246 | 15.219 | 15.191 | 15.164 | 15.137 | 15.110 | 15.083 | 15.056 |
| **20** | 15.029 | 15.002 | 14.976 | 14.949 | 14.923 | 14.896 | 14.870 | 14.844 | 14.818 | 14.792 |
| **21** | 14.766 | 14.740 | 14.715 | 14.689 | 14.664 | 14.638 | 14.613 | 14.588 | 14.563 | 14.538 |
| **22** | 14.513 | 14.488 | 14.463 | 14.439 | 14.414 | 14.390 | 14.365 | 14.341 | 14.317 | 14.292 |
| **23** | 14.268 | 14.244 | 14.221 | 14.197 | 14.173 | 14.149 | 14.126 | 14.102 | 14.079 | 14.056 |
| **24** | 14.032 | 14.009 | 13.986 | 13.963 | 13.940 | 13.917 | 13.895 | 13.872 | 13.849 | 13.827 |
| **25** | 13.804 | 13.782 | 13.760 | 13.738 | 13.715 | 13.693 | 13.671 | 13.649 | 13.628 | 13.606 |
| **26** | 13.584 | 13.562 | 13.541 | 13.519 | 13.498 | 13.477 | 13.455 | 13.434 | 13.413 | 13.392 |
| **27** | 13.371 | 13.350 | 13.329 | 13.308 | 13.288 | 13.267 | 13.246 | 13.226 | 13.206 | 13.185 |
| **28** | 13.165 | 13.145 | 13.124 | 13.104 | 13.084 | 13.064 | 13.044 | 13.024 | 13.005 | 12.985 |
| **29** | 12.965 | 12.946 | 12.926 | 12.907 | 12.887 | 12.868 | 12.848 | 12.829 | 12.810 | 12.791 |
| **30** | 12.772 | 12.753 | 12.734 | 12.715 | 12.696 | 12.677 | 12.659 | 12.640 | 12.621 | 12.603 |
| **31** | 12.584 | 12.566 | 12.548 | 12.529 | 12.511 | 12.493 | 12.475 | 12.457 | 12.439 | 12.421 |
| **32** | 12.403 | 12.385 | 12.367 | 12.349 | 12.331 | 12.314 | 12.296 | 12.279 | 12.261 | 12.244 |
| **33** | 12.226 | 12.209 | 12.191 | 12.174 | 12.157 | 12.140 | 12.123 | 12.106 | 12.089 | 12.072 |
| **34** | 12.055 | 12.038 | 12.021 | 12.004 | 11.988 | 11.971 | 11.954 | 11.938 | 11.921 | 11.905 |
| **35** | 11.888 | 11.872 | 11.855 | 11.839 | 11.823 | 11.807 | 11.790 | 11.774 | 11.758 | 11.742 |
| **36** | 11.726 | 11.710 | 11.694 | 11.678 | 11.663 | 11.647 | 11.631 | 11.615 | 11.600 | 11.584 |
| **37** | 11.568 | 11.553 | 11.537 | 11.522 | 11.506 | 11.491 | 11.476 | 11.460 | 11.445 | 11.430 |
| **38** | 11.415 | 11.400 | 11.385 | 11.370 | 11.354 | 11.340 | 11.325 | 11.310 | 11.295 | 11.280 |
| **39** | 11.265 | 11.250 | 11.236 | 11.221 | 11.206 | 11.192 | 11.177 | 11.163 | 11.148 | 11.134 |
| **40** | 11.119 | 11.105 | 11.090 | 11.076 | 11.062 | 11.047 | 11.033 | 11.019 | 11.005 | 10.991 |

*Source*: Based on Eq. 1, Hamme and Emerson (2004), Millero and Poisson (1981).

**Table 1.11** Air Saturation Concentration of Argon as a Function of Temperature in mg/L (freshwater, 1 atm moist air)

| Temperature (°C) | Δt (°C) | | | | | | | | | |
|---|---|---|---|---|---|---|---|---|---|---|
| | **0.0** | **0.1** | **0.2** | **0.3** | **0.4** | **0.5** | **0.6** | **0.7** | **0.8** | **0.9** |
| **0** | 0.8907 | 0.8882 | 0.8858 | 0.8833 | 0.8808 | 0.8784 | 0.8759 | 0.8735 | 0.8711 | 0.8687 |
| **1** | 0.8663 | 0.8639 | 0.8615 | 0.8592 | 0.8568 | 0.8545 | 0.8522 | 0.8498 | 0.8475 | 0.8452 |
| **2** | 0.8430 | 0.8407 | 0.8384 | 0.8362 | 0.8339 | 0.8317 | 0.8295 | 0.8272 | 0.8250 | 0.8229 |
| **3** | 0.8207 | 0.8185 | 0.8163 | 0.8142 | 0.8120 | 0.8099 | 0.8078 | 0.8056 | 0.8035 | 0.8014 |
| **4** | 0.7994 | 0.7973 | 0.7952 | 0.7931 | 0.7911 | 0.7890 | 0.7870 | 0.7850 | 0.7830 | 0.7810 |
| **5** | 0.7790 | 0.7770 | 0.7750 | 0.7730 | 0.7711 | 0.7691 | 0.7672 | 0.7652 | 0.7633 | 0.7614 |
| **6** | 0.7595 | 0.7576 | 0.7557 | 0.7538 | 0.7519 | 0.7500 | 0.7482 | 0.7463 | 0.7445 | 0.7426 |
| **7** | 0.7408 | 0.7390 | 0.7371 | 0.7353 | 0.7335 | 0.7317 | 0.7300 | 0.7282 | 0.7264 | 0.7246 |
| **8** | 0.7229 | 0.7211 | 0.7194 | 0.7177 | 0.7159 | 0.7142 | 0.7125 | 0.7108 | 0.7091 | 0.7074 |
| **9** | 0.7057 | 0.7041 | 0.7024 | 0.7007 | 0.6991 | 0.6974 | 0.6958 | 0.6942 | 0.6925 | 0.6909 |
| **10** | 0.6893 | 0.6877 | 0.6861 | 0.6845 | 0.6829 | 0.6813 | 0.6797 | 0.6782 | 0.6766 | 0.6751 |
| **11** | 0.6735 | 0.6720 | 0.6704 | 0.6689 | 0.6674 | 0.6659 | 0.6643 | 0.6628 | 0.6613 | 0.6598 |
| **12** | 0.6584 | 0.6569 | 0.6554 | 0.6539 | 0.6525 | 0.6510 | 0.6496 | 0.6481 | 0.6467 | 0.6452 |
| **13** | 0.6438 | 0.6424 | 0.6410 | 0.6395 | 0.6381 | 0.6367 | 0.6353 | 0.6339 | 0.6326 | 0.6312 |
| **14** | 0.6298 | 0.6284 | 0.6271 | 0.6257 | 0.6244 | 0.6230 | 0.6217 | 0.6203 | 0.6190 | 0.6177 |
| **15** | 0.6164 | 0.6150 | 0.6137 | 0.6124 | 0.6111 | 0.6098 | 0.6085 | 0.6072 | 0.6060 | 0.6047 |
| **16** | 0.6034 | 0.6021 | 0.6009 | 0.5996 | 0.5984 | 0.5971 | 0.5959 | 0.5946 | 0.5934 | 0.5922 |
| **17** | 0.5909 | 0.5897 | 0.5885 | 0.5873 | 0.5861 | 0.5849 | 0.5837 | 0.5825 | 0.5813 | 0.5801 |
| **18** | 0.5789 | 0.5778 | 0.5766 | 0.5754 | 0.5743 | 0.5731 | 0.5719 | 0.5708 | 0.5697 | 0.5685 |
| **19** | 0.5674 | 0.5662 | 0.5651 | 0.5640 | 0.5629 | 0.5617 | 0.5606 | 0.5595 | 0.5584 | 0.5573 |
| **20** | 0.5562 | 0.5551 | 0.5540 | 0.5529 | 0.5519 | 0.5508 | 0.5497 | 0.5486 | 0.5476 | 0.5465 |
| **21** | 0.5454 | 0.5444 | 0.5433 | 0.5423 | 0.5412 | 0.5402 | 0.5392 | 0.5381 | 0.5371 | 0.5361 |
| **22** | 0.5350 | 0.5340 | 0.5330 | 0.5320 | 0.5310 | 0.5300 | 0.5290 | 0.5280 | 0.5270 | 0.5260 |
| **23** | 0.5250 | 0.5240 | 0.5230 | 0.5221 | 0.5211 | 0.5201 | 0.5191 | 0.5182 | 0.5172 | 0.5163 |
| **24** | 0.5153 | 0.5143 | 0.5134 | 0.5125 | 0.5115 | 0.5106 | 0.5096 | 0.5087 | 0.5078 | 0.5068 |
| **25** | 0.5059 | 0.5050 | 0.5041 | 0.5032 | 0.5022 | 0.5013 | 0.5004 | 0.4995 | 0.4986 | 0.4977 |
| **26** | 0.4968 | 0.4959 | 0.4950 | 0.4942 | 0.4933 | 0.4924 | 0.4915 | 0.4906 | 0.4898 | 0.4889 |
| **27** | 0.4880 | 0.4872 | 0.4863 | 0.4854 | 0.4846 | 0.4837 | 0.4829 | 0.4820 | 0.4812 | 0.4803 |
| **28** | 0.4795 | 0.4787 | 0.4778 | 0.4770 | 0.4762 | 0.4753 | 0.4745 | 0.4737 | 0.4729 | 0.4721 |
| **29** | 0.4712 | 0.4704 | 0.4696 | 0.4688 | 0.4680 | 0.4672 | 0.4664 | 0.4656 | 0.4648 | 0.4640 |
| **30** | 0.4632 | 0.4624 | 0.4616 | 0.4609 | 0.4601 | 0.4593 | 0.4585 | 0.4577 | 0.4570 | 0.4562 |
| **31** | 0.4554 | 0.4547 | 0.4539 | 0.4531 | 0.4524 | 0.4516 | 0.4509 | 0.4501 | 0.4494 | 0.4486 |
| **32** | 0.4479 | 0.4471 | 0.4464 | 0.4456 | 0.4449 | 0.4442 | 0.4434 | 0.4427 | 0.4420 | 0.4412 |
| **33** | 0.4405 | 0.4398 | 0.4391 | 0.4383 | 0.4376 | 0.4369 | 0.4362 | 0.4355 | 0.4348 | 0.4341 |
| **34** | 0.4334 | 0.4327 | 0.4320 | 0.4313 | 0.4306 | 0.4299 | 0.4292 | 0.4285 | 0.4278 | 0.4271 |
| **35** | 0.4264 | 0.4257 | 0.4250 | 0.4243 | 0.4237 | 0.4230 | 0.4223 | 0.4216 | 0.4210 | 0.4203 |
| **36** | 0.4196 | 0.4189 | 0.4183 | 0.4176 | 0.4169 | 0.4163 | 0.4156 | 0.4150 | 0.4143 | 0.4137 |
| **37** | 0.4130 | 0.4123 | 0.4117 | 0.4110 | 0.4104 | 0.4098 | 0.4091 | 0.4085 | 0.4078 | 0.4072 |
| **38** | 0.4066 | 0.4059 | 0.4053 | 0.4046 | 0.4040 | 0.4034 | 0.4028 | 0.4021 | 0.4015 | 0.4009 |
| **39** | 0.4003 | 0.3996 | 0.3990 | 0.3984 | 0.3978 | 0.3972 | 0.3965 | 0.3959 | 0.3953 | 0.3947 |
| **40** | 0.3941 | 0.3935 | 0.3929 | 0.3923 | 0.3917 | 0.3911 | 0.3905 | 0.3899 | 0.3893 | 0.3887 |

*Source*: Based on Eq. 1, Hamme and Emerson (2004), Millero and Poisson (1981).

**Table 1.12** Air Saturation Concentration of Carbon Dioxide as a Function of Temperature in mg/L (freshwater, 1 atm moist air, mole fraction = 390 μatm)

| Temperature (°C) | Δt (°C) | | | | | | | | | |
|---|---|---|---|---|---|---|---|---|---|---|
| | **0.0** | **0.1** | **0.2** | **0.3** | **0.4** | **0.5** | **0.6** | **0.7** | **0.8** | **0.9** |
| **0** | 1.3174 | 1.3122 | 1.3069 | 1.3017 | 1.2966 | 1.2914 | 1.2863 | 1.2812 | 1.2761 | 1.2711 |
| **1** | 1.2661 | 1.2611 | 1.2561 | 1.2512 | 1.2463 | 1.2414 | 1.2365 | 1.2317 | 1.2269 | 1.2221 |
| **2** | 1.2174 | 1.2126 | 1.2079 | 1.2032 | 1.1986 | 1.1940 | 1.1894 | 1.1848 | 1.1802 | 1.1757 |
| **3** | 1.1712 | 1.1667 | 1.1622 | 1.1578 | 1.1534 | 1.1490 | 1.1446 | 1.1402 | 1.1359 | 1.1316 |
| **4** | 1.1273 | 1.1231 | 1.1188 | 1.1146 | 1.1104 | 1.1062 | 1.1021 | 1.0979 | 1.0938 | 1.0898 |
| **5** | 1.0857 | 1.0816 | 1.0776 | 1.0736 | 1.0696 | 1.0656 | 1.0617 | 1.0578 | 1.0539 | 1.0500 |
| **6** | 1.0461 | 1.0423 | 1.0384 | 1.0346 | 1.0308 | 1.0271 | 1.0233 | 1.0196 | 1.0159 | 1.0122 |
| **7** | 1.0085 | 1.0048 | 1.0012 | 0.9976 | 0.9940 | 0.9904 | 0.9868 | 0.9833 | 0.9797 | 0.9762 |
| **8** | 0.9727 | 0.9692 | 0.9658 | 0.9623 | 0.9589 | 0.9555 | 0.9521 | 0.9487 | 0.9453 | 0.9420 |
| **9** | 0.9387 | 0.9353 | 0.9321 | 0.9288 | 0.9255 | 0.9223 | 0.9190 | 0.9158 | 0.9126 | 0.9094 |
| **10** | 0.9062 | 0.9031 | 0.8999 | 0.8968 | 0.8937 | 0.8906 | 0.8875 | 0.8844 | 0.8814 | 0.8784 |
| **11** | 0.8753 | 0.8723 | 0.8693 | 0.8663 | 0.8634 | 0.8604 | 0.8575 | 0.8546 | 0.8517 | 0.8488 |
| **12** | 0.8459 | 0.8430 | 0.8401 | 0.8373 | 0.8345 | 0.8317 | 0.8289 | 0.8261 | 0.8233 | 0.8205 |
| **13** | 0.8178 | 0.8150 | 0.8123 | 0.8096 | 0.8069 | 0.8042 | 0.8015 | 0.7989 | 0.7962 | 0.7936 |
| **14** | 0.7910 | 0.7883 | 0.7857 | 0.7832 | 0.7806 | 0.7780 | 0.7755 | 0.7729 | 0.7704 | 0.7679 |
| **15** | 0.7654 | 0.7629 | 0.7604 | 0.7579 | 0.7554 | 0.7530 | 0.7505 | 0.7481 | 0.7457 | 0.7433 |
| **16** | 0.7409 | 0.7385 | 0.7361 | 0.7338 | 0.7314 | 0.7291 | 0.7267 | 0.7244 | 0.7221 | 0.7198 |
| **17** | 0.7175 | 0.7152 | 0.7130 | 0.7107 | 0.7084 | 0.7062 | 0.7040 | 0.7018 | 0.6995 | 0.6973 |
| **18** | 0.6951 | 0.6930 | 0.6908 | 0.6886 | 0.6865 | 0.6843 | 0.6822 | 0.6801 | 0.6780 | 0.6758 |
| **19** | 0.6738 | 0.6717 | 0.6696 | 0.6675 | 0.6654 | 0.6634 | 0.6614 | 0.6593 | 0.6573 | 0.6553 |
| **20** | 0.6533 | 0.6513 | 0.6493 | 0.6473 | 0.6453 | 0.6433 | 0.6414 | 0.6394 | 0.6375 | 0.6356 |
| **21** | 0.6336 | 0.6317 | 0.6298 | 0.6279 | 0.6260 | 0.6241 | 0.6223 | 0.6204 | 0.6185 | 0.6167 |
| **22** | 0.6148 | 0.6130 | 0.6112 | 0.6093 | 0.6075 | 0.6057 | 0.6039 | 0.6021 | 0.6003 | 0.5986 |
| **23** | 0.5968 | 0.5950 | 0.5933 | 0.5915 | 0.5898 | 0.5880 | 0.5863 | 0.5846 | 0.5829 | 0.5812 |
| **24** | 0.5795 | 0.5778 | 0.5761 | 0.5744 | 0.5728 | 0.5711 | 0.5694 | 0.5678 | 0.5661 | 0.5645 |
| **25** | 0.5629 | 0.5612 | 0.5596 | 0.5580 | 0.5564 | 0.5548 | 0.5532 | 0.5516 | 0.5501 | 0.5485 |
| **26** | 0.5469 | 0.5453 | 0.5438 | 0.5422 | 0.5407 | 0.5392 | 0.5376 | 0.5361 | 0.5346 | 0.5331 |
| **27** | 0.5316 | 0.5301 | 0.5286 | 0.5271 | 0.5256 | 0.5241 | 0.5227 | 0.5212 | 0.5197 | 0.5183 |
| **28** | 0.5168 | 0.5154 | 0.5139 | 0.5125 | 0.5111 | 0.5097 | 0.5082 | 0.5068 | 0.5054 | 0.5040 |
| **29** | 0.5026 | 0.5012 | 0.4999 | 0.4985 | 0.4971 | 0.4957 | 0.4944 | 0.4930 | 0.4917 | 0.4903 |
| **30** | 0.4890 | 0.4876 | 0.4863 | 0.4850 | 0.4837 | 0.4823 | 0.4810 | 0.4797 | 0.4784 | 0.4771 |
| **31** | 0.4758 | 0.4745 | 0.4732 | 0.4720 | 0.4707 | 0.4694 | 0.4682 | 0.4669 | 0.4656 | 0.4644 |
| **32** | 0.4631 | 0.4619 | 0.4607 | 0.4594 | 0.4582 | 0.4570 | 0.4558 | 0.4545 | 0.4533 | 0.4521 |
| **33** | 0.4509 | 0.4497 | 0.4485 | 0.4473 | 0.4461 | 0.4450 | 0.4438 | 0.4426 | 0.4414 | 0.4403 |
| **34** | 0.4391 | 0.4380 | 0.4368 | 0.4357 | 0.4345 | 0.4334 | 0.4322 | 0.4311 | 0.4300 | 0.4288 |
| **35** | 0.4277 | 0.4266 | 0.4255 | 0.4244 | 0.4233 | 0.4222 | 0.4211 | 0.4200 | 0.4189 | 0.4178 |
| **36** | 0.4167 | 0.4156 | 0.4146 | 0.4135 | 0.4124 | 0.4114 | 0.4103 | 0.4092 | 0.4082 | 0.4071 |
| **37** | 0.4061 | 0.4050 | 0.4040 | 0.4030 | 0.4019 | 0.4009 | 0.3999 | 0.3988 | 0.3978 | 0.3968 |
| **38** | 0.3958 | 0.3948 | 0.3938 | 0.3928 | 0.3918 | 0.3908 | 0.3898 | 0.3888 | 0.3878 | 0.3868 |
| **39** | 0.3858 | 0.3849 | 0.3839 | 0.3829 | 0.3819 | 0.3810 | 0.3800 | 0.3790 | 0.3781 | 0.3771 |
| **40** | 0.3762 | 0.3752 | 0.3743 | 0.3733 | 0.3724 | 0.3715 | 0.3705 | 0.3696 | 0.3687 | 0.3678 |

*Source*: Based on Eqs. 5 and 9, Weiss (1974), Millero and Poisson (1981); mole fraction estimated from Mauna Loa data for 2010.

**Table 1.13** Air Saturation Concentration of Carbon Dioxide as a Function of Temperature in mg/L (freshwater, 1 atm moist air, mole fraction = 440 μatm)

| Temperature (°C) | Δt (°C) | | | | | | | | | |
|---|---|---|---|---|---|---|---|---|---|---|
| | 0.0 | 0.1 | 0.2 | 0.3 | 0.4 | 0.5 | 0.6 | 0.7 | 0.8 | 0.9 |
| 0 | 1.4863 | 1.4804 | 1.4745 | 1.4686 | 1.4628 | 1.4570 | 1.4512 | 1.4454 | 1.4397 | 1.4340 |
| 1 | 1.4284 | 1.4228 | 1.4172 | 1.4116 | 1.4061 | 1.4005 | 1.3951 | 1.3896 | 1.3842 | 1.3788 |
| 2 | 1.3734 | 1.3681 | 1.3628 | 1.3575 | 1.3523 | 1.3470 | 1.3418 | 1.3367 | 1.3315 | 1.3264 |
| 3 | 1.3213 | 1.3163 | 1.3112 | 1.3062 | 1.3012 | 1.2963 | 1.2913 | 1.2864 | 1.2815 | 1.2767 |
| 4 | 1.2718 | 1.2670 | 1.2623 | 1.2575 | 1.2528 | 1.2481 | 1.2434 | 1.2387 | 1.2341 | 1.2295 |
| 5 | 1.2249 | 1.2203 | 1.2158 | 1.2112 | 1.2067 | 1.2023 | 1.1978 | 1.1934 | 1.1890 | 1.1846 |
| 6 | 1.1802 | 1.1759 | 1.1716 | 1.1673 | 1.1630 | 1.1587 | 1.1545 | 1.1503 | 1.1461 | 1.1419 |
| 7 | 1.1378 | 1.1337 | 1.1296 | 1.1255 | 1.1214 | 1.1174 | 1.1133 | 1.1093 | 1.1053 | 1.1014 |
| 8 | 1.0974 | 1.0935 | 1.0896 | 1.0857 | 1.0818 | 1.0780 | 1.0741 | 1.0703 | 1.0665 | 1.0628 |
| 9 | 1.0590 | 1.0553 | 1.0515 | 1.0478 | 1.0442 | 1.0405 | 1.0368 | 1.0332 | 1.0296 | 1.0260 |
| 10 | 1.0224 | 1.0189 | 1.0153 | 1.0118 | 1.0083 | 1.0048 | 1.0013 | 0.9978 | 0.9944 | 0.9910 |
| 11 | 0.9876 | 0.9842 | 0.9808 | 0.9774 | 0.9741 | 0.9707 | 0.9674 | 0.9641 | 0.9608 | 0.9576 |
| 12 | 0.9543 | 0.9511 | 0.9479 | 0.9447 | 0.9415 | 0.9383 | 0.9351 | 0.9320 | 0.9288 | 0.9257 |
| 13 | 0.9226 | 0.9195 | 0.9165 | 0.9134 | 0.9104 | 0.9073 | 0.9043 | 0.9013 | 0.8983 | 0.8953 |
| 14 | 0.8924 | 0.8894 | 0.8865 | 0.8836 | 0.8807 | 0.8778 | 0.8749 | 0.8720 | 0.8692 | 0.8663 |
| 15 | 0.8635 | 0.8607 | 0.8579 | 0.8551 | 0.8523 | 0.8495 | 0.8468 | 0.8440 | 0.8413 | 0.8386 |
| 16 | 0.8359 | 0.8332 | 0.8305 | 0.8278 | 0.8252 | 0.8225 | 0.8199 | 0.8173 | 0.8147 | 0.8121 |
| 17 | 0.8095 | 0.8069 | 0.8044 | 0.8018 | 0.7993 | 0.7967 | 0.7942 | 0.7917 | 0.7892 | 0.7867 |
| 18 | 0.7843 | 0.7818 | 0.7794 | 0.7769 | 0.7745 | 0.7721 | 0.7697 | 0.7673 | 0.7649 | 0.7625 |
| 19 | 0.7601 | 0.7578 | 0.7554 | 0.7531 | 0.7508 | 0.7484 | 0.7461 | 0.7438 | 0.7416 | 0.7393 |
| 20 | 0.7370 | 0.7348 | 0.7325 | 0.7303 | 0.7280 | 0.7258 | 0.7236 | 0.7214 | 0.7192 | 0.7170 |
| 21 | 0.7149 | 0.7127 | 0.7106 | 0.7084 | 0.7063 | 0.7042 | 0.7020 | 0.6999 | 0.6978 | 0.6957 |
| 22 | 0.6937 | 0.6916 | 0.6895 | 0.6875 | 0.6854 | 0.6834 | 0.6813 | 0.6793 | 0.6773 | 0.6753 |
| 23 | 0.6733 | 0.6713 | 0.6693 | 0.6674 | 0.6654 | 0.6634 | 0.6615 | 0.6595 | 0.6576 | 0.6557 |
| 24 | 0.6538 | 0.6519 | 0.6500 | 0.6481 | 0.6462 | 0.6443 | 0.6424 | 0.6406 | 0.6387 | 0.6369 |
| 25 | 0.6350 | 0.6332 | 0.6314 | 0.6296 | 0.6277 | 0.6259 | 0.6241 | 0.6224 | 0.6206 | 0.6188 |
| 26 | 0.6170 | 0.6153 | 0.6135 | 0.6118 | 0.6100 | 0.6083 | 0.6066 | 0.6048 | 0.6031 | 0.6014 |
| 27 | 0.5997 | 0.5980 | 0.5963 | 0.5947 | 0.5930 | 0.5913 | 0.5897 | 0.5880 | 0.5864 | 0.5847 |
| 28 | 0.5831 | 0.5815 | 0.5798 | 0.5782 | 0.5766 | 0.5750 | 0.5734 | 0.5718 | 0.5702 | 0.5687 |
| 29 | 0.5671 | 0.5655 | 0.5639 | 0.5624 | 0.5608 | 0.5593 | 0.5578 | 0.5562 | 0.5547 | 0.5532 |
| 30 | 0.5517 | 0.5502 | 0.5487 | 0.5472 | 0.5457 | 0.5442 | 0.5427 | 0.5412 | 0.5397 | 0.5383 |
| 31 | 0.5368 | 0.5354 | 0.5339 | 0.5325 | 0.5310 | 0.5296 | 0.5282 | 0.5268 | 0.5253 | 0.5239 |
| 32 | 0.5225 | 0.5211 | 0.5197 | 0.5183 | 0.5169 | 0.5156 | 0.5142 | 0.5128 | 0.5114 | 0.5101 |
| 33 | 0.5087 | 0.5074 | 0.5060 | 0.5047 | 0.5033 | 0.5020 | 0.5007 | 0.4994 | 0.4980 | 0.4967 |
| 34 | 0.4954 | 0.4941 | 0.4928 | 0.4915 | 0.4902 | 0.4889 | 0.4876 | 0.4864 | 0.4851 | 0.4838 |
| 35 | 0.4826 | 0.4813 | 0.4800 | 0.4788 | 0.4775 | 0.4763 | 0.4751 | 0.4738 | 0.4726 | 0.4714 |
| 36 | 0.4701 | 0.4689 | 0.4677 | 0.4665 | 0.4653 | 0.4641 | 0.4629 | 0.4617 | 0.4605 | 0.4593 |
| 37 | 0.4581 | 0.4570 | 0.4558 | 0.4546 | 0.4535 | 0.4523 | 0.4511 | 0.4500 | 0.4488 | 0.4477 |
| 38 | 0.4465 | 0.4454 | 0.4443 | 0.4431 | 0.4420 | 0.4409 | 0.4397 | 0.4386 | 0.4375 | 0.4364 |
| 39 | 0.4353 | 0.4342 | 0.4331 | 0.4320 | 0.4309 | 0.4298 | 0.4287 | 0.4276 | 0.4266 | 0.4255 |
| 40 | 0.4244 | 0.4233 | 0.4223 | 0.4212 | 0.4202 | 0.4191 | 0.4180 | 0.4170 | 0.4159 | 0.4149 |

*Source*: Based on Eqs. 5 and 9, Weiss (1974), Millero and Poisson (1981); mole fraction estimated from Mauna Loa data for 2030.

**Table 1.14** Air Saturation Concentration of Oxygen as a Function of Temperature in mL real gas at STP/L (freshwater, 1 atm moist air.)

| Temperature (°C) | Δt (°C) | | | | | | | | | |
|---|---|---|---|---|---|---|---|---|---|---|
| | 0.0 | 0.1 | 0.2 | 0.3 | 0.4 | 0.5 | 0.6 | 0.7 | 0.8 | 0.9 |
| 0 | 10.232 | 10.203 | 10.174 | 10.145 | 10.117 | 10.088 | 10.060 | 10.032 | 10.004 | 9.976 |
| 1 | 9.948 | 9.920 | 9.893 | 9.866 | 9.838 | 9.811 | 9.784 | 9.757 | 9.731 | 9.704 |
| 2 | 9.678 | 9.651 | 9.625 | 9.599 | 9.573 | 9.547 | 9.521 | 9.495 | 9.470 | 9.444 |
| 3 | 9.419 | 9.394 | 9.369 | 9.344 | 9.319 | 9.294 | 9.270 | 9.245 | 9.221 | 9.196 |
| 4 | 9.172 | 9.148 | 9.124 | 9.100 | 9.077 | 9.053 | 9.029 | 9.006 | 8.983 | 8.959 |
| 5 | 8.936 | 8.913 | 8.890 | 8.867 | 8.845 | 8.822 | 8.800 | 8.777 | 8.755 | 8.733 |
| 6 | 8.710 | 8.688 | 8.666 | 8.645 | 8.623 | 8.601 | 8.580 | 8.558 | 8.537 | 8.516 |
| 7 | 8.494 | 8.473 | 8.452 | 8.431 | 8.410 | 8.390 | 8.369 | 8.349 | 8.328 | 8.308 |
| 8 | 8.287 | 8.267 | 8.247 | 8.227 | 8.207 | 8.187 | 8.167 | 8.148 | 8.128 | 8.109 |
| 9 | 8.089 | 8.070 | 8.050 | 8.031 | 8.012 | 7.993 | 7.974 | 7.955 | 7.936 | 7.918 |
| 10 | 7.899 | 7.880 | 7.862 | 7.843 | 7.825 | 7.807 | 7.789 | 7.770 | 7.752 | 7.734 |
| 11 | 7.717 | 7.699 | 7.681 | 7.663 | 7.646 | 7.628 | 7.611 | 7.593 | 7.576 | 7.559 |
| 12 | 7.541 | 7.524 | 7.507 | 7.490 | 7.473 | 7.457 | 7.440 | 7.423 | 7.406 | 7.390 |
| 13 | 7.373 | 7.357 | 7.341 | 7.324 | 7.308 | 7.292 | 7.276 | 7.260 | 7.244 | 7.228 |
| 14 | 7.212 | 7.196 | 7.180 | 7.165 | 7.149 | 7.133 | 7.118 | 7.102 | 7.087 | 7.072 |
| 15 | 7.056 | 7.041 | 7.026 | 7.011 | 6.996 | 6.981 | 6.966 | 6.951 | 6.936 | 6.922 |
| 16 | 6.907 | 6.892 | 6.878 | 6.863 | 6.849 | 6.834 | 6.820 | 6.806 | 6.792 | 6.777 |
| 17 | 6.763 | 6.749 | 6.735 | 6.721 | 6.707 | 6.693 | 6.679 | 6.666 | 6.652 | 6.638 |
| 18 | 6.625 | 6.611 | 6.598 | 6.584 | 6.571 | 6.557 | 6.544 | 6.531 | 6.517 | 6.504 |
| 19 | 6.491 | 6.478 | 6.465 | 6.452 | 6.439 | 6.426 | 6.413 | 6.401 | 6.388 | 6.375 |
| 20 | 6.362 | 6.350 | 6.337 | 6.325 | 6.312 | 6.300 | 6.287 | 6.275 | 6.263 | 6.250 |
| 21 | 6.238 | 6.226 | 6.214 | 6.202 | 6.190 | 6.178 | 6.166 | 6.154 | 6.142 | 6.130 |
| 22 | 6.118 | 6.107 | 6.095 | 6.083 | 6.072 | 6.060 | 6.048 | 6.037 | 6.025 | 6.014 |
| 23 | 6.003 | 5.991 | 5.980 | 5.969 | 5.957 | 5.946 | 5.935 | 5.924 | 5.913 | 5.902 |
| 24 | 5.891 | 5.880 | 5.869 | 5.858 | 5.847 | 5.836 | 5.825 | 5.814 | 5.804 | 5.793 |
| 25 | 5.782 | 5.772 | 5.761 | 5.750 | 5.740 | 5.729 | 5.719 | 5.709 | 5.698 | 5.688 |
| 26 | 5.677 | 5.667 | 5.657 | 5.647 | 5.636 | 5.626 | 5.616 | 5.606 | 5.596 | 5.586 |
| 27 | 5.576 | 5.566 | 5.556 | 5.546 | 5.536 | 5.526 | 5.517 | 5.507 | 5.497 | 5.487 |
| 28 | 5.478 | 5.468 | 5.458 | 5.449 | 5.439 | 5.429 | 5.420 | 5.410 | 5.401 | 5.391 |
| 29 | 5.382 | 5.373 | 5.363 | 5.354 | 5.345 | 5.335 | 5.326 | 5.317 | 5.308 | 5.299 |
| 30 | 5.289 | 5.280 | 5.271 | 5.262 | 5.253 | 5.244 | 5.235 | 5.226 | 5.217 | 5.208 |
| 31 | 5.199 | 5.191 | 5.182 | 5.173 | 5.164 | 5.155 | 5.147 | 5.138 | 5.129 | 5.121 |
| 32 | 5.112 | 5.103 | 5.095 | 5.086 | 5.078 | 5.069 | 5.061 | 5.052 | 5.044 | 5.035 |
| 33 | 5.027 | 5.018 | 5.010 | 5.002 | 4.993 | 4.985 | 4.977 | 4.969 | 4.960 | 4.952 |
| 34 | 4.944 | 4.936 | 4.928 | 4.920 | 4.911 | 4.903 | 4.895 | 4.887 | 4.879 | 4.871 |
| 35 | 4.863 | 4.855 | 4.847 | 4.839 | 4.831 | 4.824 | 4.816 | 4.808 | 4.800 | 4.792 |
| 36 | 4.784 | 4.777 | 4.769 | 4.761 | 4.753 | 4.746 | 4.738 | 4.730 | 4.723 | 4.715 |
| 37 | 4.708 | 4.700 | 4.692 | 4.685 | 4.677 | 4.670 | 4.662 | 4.655 | 4.647 | 4.640 |
| 38 | 4.632 | 4.625 | 4.618 | 4.610 | 4.603 | 4.596 | 4.588 | 4.581 | 4.574 | 4.566 |
| 39 | 4.559 | 4.552 | 4.545 | 4.537 | 4.530 | 4.523 | 4.516 | 4.509 | 4.501 | 4.494 |
| 40 | 4.487 | 4.480 | 4.473 | 4.466 | 4.459 | 4.452 | 4.445 | 4.438 | 4.431 | 4.424 |

*Source*: Based on Eq. 22, Benson and Krause (1984), Millero and Poisson (1981).

**Table 1.15** Air Saturation Concentration of Nitrogen as a Function of Temperature in mL real gas at STP/L (freshwater, 1 atm moist air)

| Temperature (°C) | Δ*t* (°C) | | | | | | | | | |
|---|---|---|---|---|---|---|---|---|---|---|
| | **0.0** | **0.1** | **0.2** | **0.3** | **0.4** | **0.5** | **0.6** | **0.7** | **0.8** | **0.9** |
| **0** | 18.603 | 18.553 | 18.504 | 18.455 | 18.407 | 18.359 | 18.310 | 18.263 | 18.215 | 18.167 |
| **1** | 18.120 | 18.073 | 18.026 | 17.980 | 17.933 | 17.887 | 17.841 | 17.796 | 17.750 | 17.705 |
| **2** | 17.660 | 17.615 | 17.570 | 17.526 | 17.481 | 17.437 | 17.393 | 17.350 | 17.306 | 17.263 |
| **3** | 17.220 | 17.177 | 17.134 | 17.092 | 17.049 | 17.007 | 16.965 | 16.923 | 16.882 | 16.840 |
| **4** | 16.799 | 16.758 | 16.717 | 16.677 | 16.636 | 16.596 | 16.556 | 16.516 | 16.476 | 16.437 |
| **5** | 16.397 | 16.358 | 16.319 | 16.280 | 16.241 | 16.203 | 16.164 | 16.126 | 16.088 | 16.050 |
| **6** | 16.013 | 15.975 | 15.938 | 15.900 | 15.863 | 15.826 | 15.790 | 15.753 | 15.717 | 15.680 |
| **7** | 15.644 | 15.608 | 15.573 | 15.537 | 15.502 | 15.466 | 15.431 | 15.396 | 15.361 | 15.326 |
| **8** | 15.292 | 15.257 | 15.223 | 15.189 | 15.155 | 15.121 | 15.087 | 15.054 | 15.020 | 14.987 |
| **9** | 14.954 | 14.921 | 14.888 | 14.855 | 14.823 | 14.790 | 14.758 | 14.726 | 14.694 | 14.662 |
| **10** | 14.630 | 14.598 | 14.567 | 14.536 | 14.504 | 14.473 | 14.442 | 14.411 | 14.381 | 14.350 |
| **11** | 14.320 | 14.289 | 14.259 | 14.229 | 14.199 | 14.169 | 14.139 | 14.110 | 14.080 | 14.051 |
| **12** | 14.022 | 13.992 | 13.963 | 13.934 | 13.906 | 13.877 | 13.848 | 13.820 | 13.792 | 13.764 |
| **13** | 13.735 | 13.707 | 13.680 | 13.652 | 13.624 | 13.597 | 13.569 | 13.542 | 13.515 | 13.488 |
| **14** | 13.461 | 13.434 | 13.407 | 13.380 | 13.354 | 13.327 | 13.301 | 13.275 | 13.249 | 13.223 |
| **15** | 13.197 | 13.171 | 13.145 | 13.119 | 13.094 | 13.068 | 13.043 | 13.018 | 12.993 | 12.968 |
| **16** | 12.943 | 12.918 | 12.893 | 12.868 | 12.844 | 12.819 | 12.795 | 12.771 | 12.747 | 12.723 |
| **17** | 12.699 | 12.675 | 12.651 | 12.627 | 12.603 | 12.580 | 12.556 | 12.533 | 12.510 | 12.487 |
| **18** | 12.464 | 12.441 | 12.418 | 12.395 | 12.372 | 12.349 | 12.327 | 12.304 | 12.282 | 12.260 |
| **19** | 12.237 | 12.215 | 12.193 | 12.171 | 12.149 | 12.127 | 12.106 | 12.084 | 12.062 | 12.041 |
| **20** | 12.019 | 11.998 | 11.977 | 11.955 | 11.934 | 11.913 | 11.892 | 11.871 | 11.851 | 11.830 |
| **21** | 11.809 | 11.789 | 11.768 | 11.748 | 11.727 | 11.707 | 11.687 | 11.667 | 11.646 | 11.626 |
| **22** | 11.607 | 11.587 | 11.567 | 11.547 | 11.528 | 11.508 | 11.488 | 11.469 | 11.450 | 11.430 |
| **23** | 11.411 | 11.392 | 11.373 | 11.354 | 11.335 | 11.316 | 11.297 | 11.278 | 11.260 | 11.241 |
| **24** | 11.222 | 11.204 | 11.185 | 11.167 | 11.149 | 11.130 | 11.112 | 11.094 | 11.076 | 11.058 |
| **25** | 11.040 | 11.022 | 11.004 | 10.986 | 10.969 | 10.951 | 10.934 | 10.916 | 10.899 | 10.881 |
| **26** | 10.864 | 10.846 | 10.829 | 10.812 | 10.795 | 10.778 | 10.761 | 10.744 | 10.727 | 10.710 |
| **27** | 10.693 | 10.677 | 10.660 | 10.643 | 10.627 | 10.610 | 10.594 | 10.577 | 10.561 | 10.545 |
| **28** | 10.528 | 10.512 | 10.496 | 10.480 | 10.464 | 10.448 | 10.432 | 10.416 | 10.400 | 10.385 |
| **29** | 10.369 | 10.353 | 10.338 | 10.322 | 10.306 | 10.291 | 10.275 | 10.260 | 10.245 | 10.229 |
| **30** | 10.214 | 10.199 | 10.184 | 10.169 | 10.154 | 10.139 | 10.124 | 10.109 | 10.094 | 10.079 |
| **31** | 10.064 | 10.050 | 10.035 | 10.020 | 10.006 | 9.991 | 9.977 | 9.962 | 9.948 | 9.933 |
| **32** | 9.919 | 9.905 | 9.890 | 9.876 | 9.862 | 9.848 | 9.834 | 9.820 | 9.806 | 9.792 |
| **33** | 9.778 | 9.764 | 9.750 | 9.736 | 9.722 | 9.709 | 9.695 | 9.681 | 9.668 | 9.654 |
| **34** | 9.641 | 9.627 | 9.614 | 9.600 | 9.587 | 9.574 | 9.560 | 9.547 | 9.534 | 9.521 |
| **35** | 9.507 | 9.494 | 9.481 | 9.468 | 9.455 | 9.442 | 9.429 | 9.416 | 9.404 | 9.391 |
| **36** | 9.378 | 9.365 | 9.352 | 9.340 | 9.327 | 9.314 | 9.302 | 9.289 | 9.277 | 9.264 |
| **37** | 9.252 | 9.239 | 9.227 | 9.215 | 9.202 | 9.190 | 9.178 | 9.165 | 9.153 | 9.141 |
| **38** | 9.129 | 9.117 | 9.105 | 9.093 | 9.081 | 9.069 | 9.057 | 9.045 | 9.033 | 9.021 |
| **39** | 9.009 | 8.997 | 8.986 | 8.974 | 8.962 | 8.950 | 8.939 | 8.927 | 8.916 | 8.904 |
| **40** | 8.892 | 8.881 | 8.869 | 8.858 | 8.846 | 8.835 | 8.824 | 8.812 | 8.801 | 8.790 |

*Source*: Based on Eq. 1, Hamme and Emerson (2004), Millero and Poisson (1981).

**Table 1.16** Air Saturation Concentration of Argon as a Function of Temperature in mL real gas at STP/L (freshwater, 1 atm moist air)

| Temperature (°C) | Δt (°C) | | | | | | | | | |
|---|---|---|---|---|---|---|---|---|---|---|
| | **0.0** | **0.1** | **0.2** | **0.3** | **0.4** | **0.5** | **0.6** | **0.7** | **0.8** | **0.9** |
| **0** | 0.4993 | 0.4979 | 0.4965 | 0.4951 | 0.4937 | 0.4924 | 0.4910 | 0.4896 | 0.4883 | 0.4869 |
| **1** | 0.4856 | 0.4843 | 0.4829 | 0.4816 | 0.4803 | 0.4790 | 0.4777 | 0.4764 | 0.4751 | 0.4738 |
| **2** | 0.4725 | 0.4712 | 0.4700 | 0.4687 | 0.4675 | 0.4662 | 0.4650 | 0.4637 | 0.4625 | 0.4613 |
| **3** | 0.4600 | 0.4588 | 0.4576 | 0.4564 | 0.4552 | 0.4540 | 0.4528 | 0.4516 | 0.4504 | 0.4493 |
| **4** | 0.4481 | 0.4469 | 0.4458 | 0.4446 | 0.4434 | 0.4423 | 0.4412 | 0.4400 | 0.4389 | 0.4378 |
| **5** | 0.4367 | 0.4355 | 0.4344 | 0.4333 | 0.4322 | 0.4311 | 0.4300 | 0.4290 | 0.4279 | 0.4268 |
| **6** | 0.4257 | 0.4247 | 0.4236 | 0.4225 | 0.4215 | 0.4204 | 0.4194 | 0.4183 | 0.4173 | 0.4163 |
| **7** | 0.4153 | 0.4142 | 0.4132 | 0.4122 | 0.4112 | 0.4102 | 0.4092 | 0.4082 | 0.4072 | 0.4062 |
| **8** | 0.4052 | 0.4042 | 0.4033 | 0.4023 | 0.4013 | 0.4004 | 0.3994 | 0.3984 | 0.3975 | 0.3966 |
| **9** | 0.3956 | 0.3947 | 0.3937 | 0.3928 | 0.3919 | 0.3909 | 0.3900 | 0.3891 | 0.3882 | 0.3873 |
| **10** | 0.3864 | 0.3855 | 0.3846 | 0.3837 | 0.3828 | 0.3819 | 0.3810 | 0.3802 | 0.3793 | 0.3784 |
| **11** | 0.3775 | 0.3767 | 0.3758 | 0.3750 | 0.3741 | 0.3732 | 0.3724 | 0.3716 | 0.3707 | 0.3699 |
| **12** | 0.3690 | 0.3682 | 0.3674 | 0.3666 | 0.3657 | 0.3649 | 0.3641 | 0.3633 | 0.3625 | 0.3617 |
| **13** | 0.3609 | 0.3601 | 0.3593 | 0.3585 | 0.3577 | 0.3569 | 0.3561 | 0.3554 | 0.3546 | 0.3538 |
| **14** | 0.3530 | 0.3523 | 0.3515 | 0.3507 | 0.3500 | 0.3492 | 0.3485 | 0.3477 | 0.3470 | 0.3462 |
| **15** | 0.3455 | 0.3448 | 0.3440 | 0.3433 | 0.3426 | 0.3418 | 0.3411 | 0.3404 | 0.3397 | 0.3390 |
| **16** | 0.3382 | 0.3375 | 0.3368 | 0.3361 | 0.3354 | 0.3347 | 0.3340 | 0.3333 | 0.3326 | 0.3319 |
| **17** | 0.3313 | 0.3306 | 0.3299 | 0.3292 | 0.3285 | 0.3279 | 0.3272 | 0.3265 | 0.3259 | 0.3252 |
| **18** | 0.3245 | 0.3239 | 0.3232 | 0.3226 | 0.3219 | 0.3213 | 0.3206 | 0.3200 | 0.3193 | 0.3187 |
| **19** | 0.3180 | 0.3174 | 0.3168 | 0.3161 | 0.3155 | 0.3149 | 0.3143 | 0.3136 | 0.3130 | 0.3124 |
| **20** | 0.3118 | 0.3112 | 0.3106 | 0.3100 | 0.3093 | 0.3087 | 0.3081 | 0.3075 | 0.3069 | 0.3063 |
| **21** | 0.3058 | 0.3052 | 0.3046 | 0.3040 | 0.3034 | 0.3028 | 0.3022 | 0.3017 | 0.3011 | 0.3005 |
| **22** | 0.2999 | 0.2994 | 0.2988 | 0.2982 | 0.2976 | 0.2971 | 0.2965 | 0.2960 | 0.2954 | 0.2948 |
| **23** | 0.2943 | 0.2937 | 0.2932 | 0.2926 | 0.2921 | 0.2916 | 0.2910 | 0.2905 | 0.2899 | 0.2894 |
| **24** | 0.2889 | 0.2883 | 0.2878 | 0.2873 | 0.2867 | 0.2862 | 0.2857 | 0.2852 | 0.2846 | 0.2841 |
| **25** | 0.2836 | 0.2831 | 0.2826 | 0.2820 | 0.2815 | 0.2810 | 0.2805 | 0.2800 | 0.2795 | 0.2790 |
| **26** | 0.2785 | 0.2780 | 0.2775 | 0.2770 | 0.2765 | 0.2760 | 0.2755 | 0.2750 | 0.2745 | 0.2741 |
| **27** | 0.2736 | 0.2731 | 0.2726 | 0.2721 | 0.2716 | 0.2712 | 0.2707 | 0.2702 | 0.2697 | 0.2693 |
| **28** | 0.2688 | 0.2683 | 0.2678 | 0.2674 | 0.2669 | 0.2665 | 0.2660 | 0.2655 | 0.2651 | 0.2646 |
| **29** | 0.2642 | 0.2637 | 0.2632 | 0.2628 | 0.2623 | 0.2619 | 0.2614 | 0.2610 | 0.2605 | 0.2601 |
| **30** | 0.2597 | 0.2592 | 0.2588 | 0.2583 | 0.2579 | 0.2575 | 0.2570 | 0.2566 | 0.2562 | 0.2557 |
| **31** | 0.2553 | 0.2549 | 0.2544 | 0.2540 | 0.2536 | 0.2532 | 0.2527 | 0.2523 | 0.2519 | 0.2515 |
| **32** | 0.2511 | 0.2506 | 0.2502 | 0.2498 | 0.2494 | 0.2490 | 0.2486 | 0.2482 | 0.2477 | 0.2473 |
| **33** | 0.2469 | 0.2465 | 0.2461 | 0.2457 | 0.2453 | 0.2449 | 0.2445 | 0.2441 | 0.2437 | 0.2433 |
| **34** | 0.2429 | 0.2425 | 0.2421 | 0.2417 | 0.2413 | 0.2410 | 0.2406 | 0.2402 | 0.2398 | 0.2394 |
| **35** | 0.2390 | 0.2386 | 0.2382 | 0.2379 | 0.2375 | 0.2371 | 0.2367 | 0.2363 | 0.2360 | 0.2356 |
| **36** | 0.2352 | 0.2348 | 0.2345 | 0.2341 | 0.2337 | 0.2334 | 0.2330 | 0.2326 | 0.2322 | 0.2319 |
| **37** | 0.2315 | 0.2311 | 0.2308 | 0.2304 | 0.2301 | 0.2297 | 0.2293 | 0.2290 | 0.2286 | 0.2283 |
| **38** | 0.2279 | 0.2275 | 0.2272 | 0.2268 | 0.2265 | 0.2261 | 0.2258 | 0.2254 | 0.2251 | 0.2247 |
| **39** | 0.2244 | 0.2240 | 0.2237 | 0.2233 | 0.2230 | 0.2226 | 0.2223 | 0.2219 | 0.2216 | 0.2213 |
| **40** | 0.2209 | 0.2206 | 0.2202 | 0.2199 | 0.2196 | 0.2192 | 0.2189 | 0.2185 | 0.2182 | 0.2179 |

*Source*: Based on Eq. 1, Hamme and Emerson (2004), Millero and Poisson (1981).

**Table 1.17** Air Saturation Concentration of Carbon Dioxide as a Function of Temperature in mL real gas at STP/L (freshwater, 1 atm moist air, mole fraction = 390 μatm)

| **Temperature (°C)** | $\Delta t$ (°C) | | | | | | | | | |
|---|---|---|---|---|---|---|---|---|---|---|
| | **0.0** | **0.1** | **0.2** | **0.3** | **0.4** | **0.5** | **0.6** | **0.7** | **0.8** | **0.9** |
| **0** | 0.6665 | 0.6638 | 0.6611 | 0.6585 | 0.6559 | 0.6533 | 0.6507 | 0.6481 | 0.6456 | 0.6430 |
| **1** | 0.6405 | 0.6379 | 0.6354 | 0.6329 | 0.6305 | 0.6280 | 0.6255 | 0.6231 | 0.6207 | 0.6182 |
| **2** | 0.6158 | 0.6134 | 0.6111 | 0.6087 | 0.6063 | 0.6040 | 0.6017 | 0.5993 | 0.5970 | 0.5947 |
| **3** | 0.5925 | 0.5902 | 0.5879 | 0.5857 | 0.5835 | 0.5812 | 0.5790 | 0.5768 | 0.5746 | 0.5724 |
| **4** | 0.5703 | 0.5681 | 0.5660 | 0.5638 | 0.5617 | 0.5596 | 0.5575 | 0.5554 | 0.5533 | 0.5513 |
| **5** | 0.5492 | 0.5472 | 0.5451 | 0.5431 | 0.5411 | 0.5391 | 0.5371 | 0.5351 | 0.5331 | 0.5312 |
| **6** | 0.5292 | 0.5273 | 0.5253 | 0.5234 | 0.5215 | 0.5196 | 0.5177 | 0.5158 | 0.5139 | 0.5120 |
| **7** | 0.5102 | 0.5083 | 0.5065 | 0.5046 | 0.5028 | 0.5010 | 0.4992 | 0.4974 | 0.4956 | 0.4938 |
| **8** | 0.4921 | 0.4903 | 0.4886 | 0.4868 | 0.4851 | 0.4834 | 0.4816 | 0.4799 | 0.4782 | 0.4765 |
| **9** | 0.4748 | 0.4732 | 0.4715 | 0.4698 | 0.4682 | 0.4665 | 0.4649 | 0.4633 | 0.4617 | 0.4600 |
| **10** | 0.4584 | 0.4568 | 0.4553 | 0.4537 | 0.4521 | 0.4505 | 0.4490 | 0.4474 | 0.4459 | 0.4443 |
| **11** | 0.4428 | 0.4413 | 0.4398 | 0.4383 | 0.4368 | 0.4353 | 0.4338 | 0.4323 | 0.4308 | 0.4294 |
| **12** | 0.4279 | 0.4265 | 0.4250 | 0.4236 | 0.4221 | 0.4207 | 0.4193 | 0.4179 | 0.4165 | 0.4151 |
| **13** | 0.4137 | 0.4123 | 0.4109 | 0.4096 | 0.4082 | 0.4068 | 0.4055 | 0.4041 | 0.4028 | 0.4015 |
| **14** | 0.4001 | 0.3988 | 0.3975 | 0.3962 | 0.3949 | 0.3936 | 0.3923 | 0.3910 | 0.3897 | 0.3884 |
| **15** | 0.3872 | 0.3859 | 0.3847 | 0.3834 | 0.3822 | 0.3809 | 0.3797 | 0.3785 | 0.3772 | 0.3760 |
| **16** | 0.3748 | 0.3736 | 0.3724 | 0.3712 | 0.3700 | 0.3688 | 0.3676 | 0.3665 | 0.3653 | 0.3641 |
| **17** | 0.3630 | 0.3618 | 0.3607 | 0.3595 | 0.3584 | 0.3573 | 0.3561 | 0.3550 | 0.3539 | 0.3528 |
| **18** | 0.3517 | 0.3506 | 0.3495 | 0.3484 | 0.3473 | 0.3462 | 0.3451 | 0.3440 | 0.3430 | 0.3419 |
| **19** | 0.3408 | 0.3398 | 0.3387 | 0.3377 | 0.3366 | 0.3356 | 0.3346 | 0.3335 | 0.3325 | 0.3315 |
| **20** | 0.3305 | 0.3295 | 0.3284 | 0.3274 | 0.3264 | 0.3255 | 0.3245 | 0.3235 | 0.3225 | 0.3215 |
| **21** | 0.3205 | 0.3196 | 0.3186 | 0.3176 | 0.3167 | 0.3157 | 0.3148 | 0.3138 | 0.3129 | 0.3120 |
| **22** | 0.3110 | 0.3101 | 0.3092 | 0.3082 | 0.3073 | 0.3064 | 0.3055 | 0.3046 | 0.3037 | 0.3028 |
| **23** | 0.3019 | 0.3010 | 0.3001 | 0.2992 | 0.2984 | 0.2975 | 0.2966 | 0.2957 | 0.2949 | 0.2940 |
| **24** | 0.2931 | 0.2923 | 0.2914 | 0.2906 | 0.2897 | 0.2889 | 0.2881 | 0.2872 | 0.2864 | 0.2856 |
| **25** | 0.2847 | 0.2839 | 0.2831 | 0.2823 | 0.2815 | 0.2807 | 0.2799 | 0.2791 | 0.2783 | 0.2775 |
| **26** | 0.2767 | 0.2759 | 0.2751 | 0.2743 | 0.2735 | 0.2728 | 0.2720 | 0.2712 | 0.2704 | 0.2697 |
| **27** | 0.2689 | 0.2682 | 0.2674 | 0.2666 | 0.2659 | 0.2651 | 0.2644 | 0.2637 | 0.2629 | 0.2622 |
| **28** | 0.2614 | 0.2607 | 0.2600 | 0.2593 | 0.2585 | 0.2578 | 0.2571 | 0.2564 | 0.2557 | 0.2550 |
| **29** | 0.2543 | 0.2536 | 0.2529 | 0.2522 | 0.2515 | 0.2508 | 0.2501 | 0.2494 | 0.2487 | 0.2480 |
| **30** | 0.2474 | 0.2467 | 0.2460 | 0.2453 | 0.2447 | 0.2440 | 0.2433 | 0.2427 | 0.2420 | 0.2414 |
| **31** | 0.2407 | 0.2401 | 0.2394 | 0.2388 | 0.2381 | 0.2375 | 0.2368 | 0.2362 | 0.2356 | 0.2349 |
| **32** | 0.2343 | 0.2337 | 0.2330 | 0.2324 | 0.2318 | 0.2312 | 0.2306 | 0.2299 | 0.2293 | 0.2287 |
| **33** | 0.2281 | 0.2275 | 0.2269 | 0.2263 | 0.2257 | 0.2251 | 0.2245 | 0.2239 | 0.2233 | 0.2227 |
| **34** | 0.2221 | 0.2216 | 0.2210 | 0.2204 | 0.2198 | 0.2192 | 0.2187 | 0.2181 | 0.2175 | 0.2169 |
| **35** | 0.2164 | 0.2158 | 0.2152 | 0.2147 | 0.2141 | 0.2136 | 0.2130 | 0.2125 | 0.2119 | 0.2114 |
| **36** | 0.2108 | 0.2103 | 0.2097 | 0.2092 | 0.2086 | 0.2081 | 0.2076 | 0.2070 | 0.2065 | 0.2060 |
| **37** | 0.2054 | 0.2049 | 0.2044 | 0.2038 | 0.2033 | 0.2028 | 0.2023 | 0.2018 | 0.2012 | 0.2007 |
| **38** | 0.2002 | 0.1997 | 0.1992 | 0.1987 | 0.1982 | 0.1977 | 0.1972 | 0.1967 | 0.1962 | 0.1957 |
| **39** | 0.1952 | 0.1947 | 0.1942 | 0.1937 | 0.1932 | 0.1927 | 0.1922 | 0.1917 | 0.1913 | 0.1908 |
| **40** | 0.1903 | 0.1898 | 0.1893 | 0.1889 | 0.1884 | 0.1879 | 0.1874 | 0.1870 | 0.1865 | 0.1860 |

*Source*: Based on Eqs. 5 and 9, Weiss (1974), Millero and Poisson (1981); mole fraction estimated from Mauna Loa data for 2010.

**Table 1.18** Air Saturation Concentration of Carbon Dioxide as a Function of Temperature in mL real gas at STP/L (freshwater, 1 atm moist air, mole fraction = 440 μatm)

| Temperature (°C) | Δt (°C) | | | | | | | | | |
|---|---|---|---|---|---|---|---|---|---|---|
| | 0.0 | 0.1 | 0.2 | 0.3 | 0.4 | 0.5 | 0.6 | 0.7 | 0.8 | 0.9 |
| 0 | 0.7519 | 0.7489 | 0.7459 | 0.7429 | 0.7400 | 0.7370 | 0.7341 | 0.7312 | 0.7283 | 0.7254 |
| 1 | 0.7226 | 0.7197 | 0.7169 | 0.7141 | 0.7113 | 0.7085 | 0.7057 | 0.7030 | 0.7002 | 0.6975 |
| 2 | 0.6948 | 0.6921 | 0.6894 | 0.6867 | 0.6841 | 0.6814 | 0.6788 | 0.6762 | 0.6736 | 0.6710 |
| 3 | 0.6684 | 0.6659 | 0.6633 | 0.6608 | 0.6583 | 0.6557 | 0.6532 | 0.6508 | 0.6483 | 0.6458 |
| 4 | 0.6434 | 0.6410 | 0.6385 | 0.6361 | 0.6337 | 0.6314 | 0.6290 | 0.6266 | 0.6243 | 0.6220 |
| 5 | 0.6196 | 0.6173 | 0.6150 | 0.6127 | 0.6105 | 0.6082 | 0.6059 | 0.6037 | 0.6015 | 0.5993 |
| 6 | 0.5970 | 0.5949 | 0.5927 | 0.5905 | 0.5883 | 0.5862 | 0.5840 | 0.5819 | 0.5798 | 0.5777 |
| 7 | 0.5756 | 0.5735 | 0.5714 | 0.5693 | 0.5673 | 0.5652 | 0.5632 | 0.5612 | 0.5592 | 0.5572 |
| 8 | 0.5552 | 0.5532 | 0.5512 | 0.5492 | 0.5473 | 0.5453 | 0.5434 | 0.5415 | 0.5395 | 0.5376 |
| 9 | 0.5357 | 0.5338 | 0.5319 | 0.5301 | 0.5282 | 0.5264 | 0.5245 | 0.5227 | 0.5208 | 0.5190 |
| 10 | 0.5172 | 0.5154 | 0.5136 | 0.5118 | 0.5101 | 0.5083 | 0.5065 | 0.5048 | 0.5030 | 0.5013 |
| 11 | 0.4996 | 0.4979 | 0.4962 | 0.4945 | 0.4928 | 0.4911 | 0.4894 | 0.4877 | 0.4861 | 0.4844 |
| 12 | 0.4828 | 0.4811 | 0.4795 | 0.4779 | 0.4763 | 0.4747 | 0.4731 | 0.4715 | 0.4699 | 0.4683 |
| 13 | 0.4667 | 0.4652 | 0.4636 | 0.4621 | 0.4605 | 0.4590 | 0.4575 | 0.4559 | 0.4544 | 0.4529 |
| 14 | 0.4514 | 0.4499 | 0.4484 | 0.4470 | 0.4455 | 0.4440 | 0.4426 | 0.4411 | 0.4397 | 0.4382 |
| 15 | 0.4368 | 0.4354 | 0.4340 | 0.4326 | 0.4312 | 0.4298 | 0.4284 | 0.4270 | 0.4256 | 0.4242 |
| 16 | 0.4229 | 0.4215 | 0.4201 | 0.4188 | 0.4174 | 0.4161 | 0.4148 | 0.4134 | 0.4121 | 0.4108 |
| 17 | 0.4095 | 0.4082 | 0.4069 | 0.4056 | 0.4043 | 0.4031 | 0.4018 | 0.4005 | 0.3992 | 0.3980 |
| 18 | 0.3967 | 0.3955 | 0.3943 | 0.3930 | 0.3918 | 0.3906 | 0.3894 | 0.3881 | 0.3869 | 0.3857 |
| 19 | 0.3845 | 0.3833 | 0.3821 | 0.3810 | 0.3798 | 0.3786 | 0.3775 | 0.3763 | 0.3751 | 0.3740 |
| 20 | 0.3728 | 0.3717 | 0.3706 | 0.3694 | 0.3683 | 0.3672 | 0.3661 | 0.3649 | 0.3638 | 0.3627 |
| 21 | 0.3616 | 0.3605 | 0.3595 | 0.3584 | 0.3573 | 0.3562 | 0.3551 | 0.3541 | 0.3530 | 0.3520 |
| 22 | 0.3509 | 0.3499 | 0.3488 | 0.3478 | 0.3467 | 0.3457 | 0.3447 | 0.3436 | 0.3426 | 0.3416 |
| 23 | 0.3406 | 0.3396 | 0.3386 | 0.3376 | 0.3366 | 0.3356 | 0.3346 | 0.3336 | 0.3327 | 0.3317 |
| 24 | 0.3307 | 0.3298 | 0.3288 | 0.3278 | 0.3269 | 0.3259 | 0.3250 | 0.3241 | 0.3231 | 0.3222 |
| 25 | 0.3212 | 0.3203 | 0.3194 | 0.3185 | 0.3176 | 0.3166 | 0.3157 | 0.3148 | 0.3139 | 0.3130 |
| 26 | 0.3121 | 0.3112 | 0.3104 | 0.3095 | 0.3086 | 0.3077 | 0.3068 | 0.3060 | 0.3051 | 0.3042 |
| 27 | 0.3034 | 0.3025 | 0.3017 | 0.3008 | 0.3000 | 0.2991 | 0.2983 | 0.2975 | 0.2966 | 0.2958 |
| 28 | 0.2950 | 0.2941 | 0.2933 | 0.2925 | 0.2917 | 0.2909 | 0.2901 | 0.2893 | 0.2885 | 0.2877 |
| 29 | 0.2869 | 0.2861 | 0.2853 | 0.2845 | 0.2837 | 0.2829 | 0.2822 | 0.2814 | 0.2806 | 0.2798 |
| 30 | 0.2791 | 0.2783 | 0.2776 | 0.2768 | 0.2760 | 0.2753 | 0.2745 | 0.2738 | 0.2730 | 0.2723 |
| 31 | 0.2716 | 0.2708 | 0.2701 | 0.2694 | 0.2686 | 0.2679 | 0.2672 | 0.2665 | 0.2658 | 0.2650 |
| 32 | 0.2643 | 0.2636 | 0.2629 | 0.2622 | 0.2615 | 0.2608 | 0.2601 | 0.2594 | 0.2587 | 0.2580 |
| 33 | 0.2574 | 0.2567 | 0.2560 | 0.2553 | 0.2546 | 0.2540 | 0.2533 | 0.2526 | 0.2519 | 0.2513 |
| 34 | 0.2506 | 0.2500 | 0.2493 | 0.2486 | 0.2480 | 0.2473 | 0.2467 | 0.2460 | 0.2454 | 0.2448 |
| 35 | 0.2441 | 0.2435 | 0.2428 | 0.2422 | 0.2416 | 0.2409 | 0.2403 | 0.2397 | 0.2391 | 0.2385 |
| 36 | 0.2378 | 0.2372 | 0.2366 | 0.2360 | 0.2354 | 0.2348 | 0.2342 | 0.2336 | 0.2330 | 0.2324 |
| 37 | 0.2318 | 0.2312 | 0.2306 | 0.2300 | 0.2294 | 0.2288 | 0.2282 | 0.2276 | 0.2270 | 0.2265 |
| 38 | 0.2259 | 0.2253 | 0.2247 | 0.2242 | 0.2236 | 0.2230 | 0.2225 | 0.2219 | 0.2213 | 0.2208 |
| 39 | 0.2202 | 0.2196 | 0.2191 | 0.2185 | 0.2180 | 0.2174 | 0.2169 | 0.2163 | 0.2158 | 0.2152 |
| 40 | 0.2147 | 0.2142 | 0.2136 | 0.2131 | 0.2125 | 0.2120 | 0.2115 | 0.2109 | 0.2104 | 0.2099 |

*Source*: Based on Eqs. 5 and 9, Weiss (1974), Millero and Poisson (1981); mole fraction estimated from Mauna Loa data for 2030.

**Table 1.19** Pressure Adjustment Factor for Oxygen as a Function of Temperature and Barometric Pressure

| | **Barometric Pressure** | | | | | | | | | |
|---|---|---|---|---|---|---|---|---|---|---|
| **atm→** | **1.10** | **1.00** | **0.95** | **0.90** | **0.85** | **0.80** | **0.75** | **0.70** | **0.65** | **0.60** |
| **kPa→** | **111.46** | **101.33** | **96.26** | **91.19** | **86.13** | **81.06** | **75.99** | **70.93** | **65.86** | **60.80** |
| **mmHg→** | **836.0** | **760.0** | **722.0** | **684.0** | **646.0** | **608.0** | **570.0** | **532.0** | **494.0** | **456.0** |
| **millibar→** | **1115** | **1013** | **963** | **912** | **861** | **811** | **760** | **709** | **659** | **608** |
| **Temperature (°C)** | | | | | | | | | | |
| **0** | 1.1005 | 1.0000 | 0.9497 | 0.8995 | 0.8492 | 0.7989 | 0.7487 | 0.6984 | 0.6481 | 0.5978 |
| **1** | 1.1005 | 1.0000 | 0.9497 | 0.8994 | 0.8491 | 0.7989 | 0.7486 | 0.6982 | 0.6479 | 0.5976 |
| **2** | 1.1006 | 1.0000 | 0.9497 | 0.8994 | 0.8491 | 0.7988 | 0.7484 | 0.6981 | 0.6478 | 0.5974 |
| **3** | 1.1006 | 1.0000 | 0.9497 | 0.8993 | 0.8490 | 0.7986 | 0.7483 | 0.6979 | 0.6476 | 0.5972 |
| **4** | 1.1007 | 1.0000 | 0.9496 | 0.8993 | 0.8489 | 0.7985 | 0.7482 | 0.6978 | 0.6474 | 0.5970 |
| **5** | 1.1008 | 1.0000 | 0.9496 | 0.8992 | 0.8488 | 0.7984 | 0.7480 | 0.6976 | 0.6472 | 0.5967 |
| **6** | 1.1008 | 1.0000 | 0.9496 | 0.8991 | 0.8487 | 0.7983 | 0.7478 | 0.6974 | 0.6469 | 0.5965 |
| **7** | 1.1009 | 1.0000 | 0.9495 | 0.8991 | 0.8486 | 0.7981 | 0.7477 | 0.6972 | 0.6467 | 0.5962 |
| **8** | 1.1010 | 1.0000 | 0.9495 | 0.8990 | 0.8485 | 0.7980 | 0.7475 | 0.6970 | 0.6465 | 0.5959 |
| **9** | 1.1011 | 1.0000 | 0.9495 | 0.8989 | 0.8484 | 0.7978 | 0.7473 | 0.6967 | 0.6462 | 0.5956 |
| **10** | 1.1011 | 1.0000 | 0.9494 | 0.8989 | 0.8483 | 0.7977 | 0.7471 | 0.6965 | 0.6459 | 0.5953 |
| **11** | 1.1012 | 1.0000 | 0.9494 | 0.8988 | 0.8481 | 0.7975 | 0.7469 | 0.6962 | 0.6456 | 0.5950 |
| **12** | 1.1013 | 1.0000 | 0.9493 | 0.8987 | 0.8480 | 0.7973 | 0.7466 | 0.6960 | 0.6453 | 0.5946 |
| **13** | 1.1014 | 1.0000 | 0.9493 | 0.8986 | 0.8479 | 0.7971 | 0.7464 | 0.6957 | 0.6449 | 0.5942 |
| **14** | 1.1015 | 1.0000 | 0.9492 | 0.8985 | 0.8477 | 0.7969 | 0.7461 | 0.6954 | 0.6446 | 0.5938 |
| **15** | 1.1016 | 1.0000 | 0.9492 | 0.8984 | 0.8475 | 0.7967 | 0.7459 | 0.6950 | 0.6442 | 0.5933 |
| **16** | 1.1017 | 1.0000 | 0.9491 | 0.8982 | 0.8474 | 0.7965 | 0.7456 | 0.6947 | 0.6438 | 0.5929 |
| **17** | 1.1019 | 1.0000 | 0.9491 | 0.8981 | 0.8472 | 0.7962 | 0.7453 | 0.6943 | 0.6434 | 0.5924 |
| **18** | 1.1020 | 1.0000 | 0.9490 | 0.8980 | 0.8470 | 0.7960 | 0.7449 | 0.6939 | 0.6429 | 0.5919 |
| **19** | 1.1021 | 1.0000 | 0.9489 | 0.8979 | 0.8468 | 0.7957 | 0.7446 | 0.6935 | 0.6424 | 0.5913 |
| **20** | 1.1023 | 1.0000 | 0.9489 | 0.8977 | 0.8466 | 0.7954 | 0.7442 | 0.6931 | 0.6419 | 0.5907 |
| **21** | 1.1024 | 1.0000 | 0.9488 | 0.8975 | 0.8463 | 0.7951 | 0.7438 | 0.6926 | 0.6414 | 0.5901 |
| **22** | 1.1026 | 1.0000 | 0.9487 | 0.8974 | 0.8461 | 0.7948 | 0.7434 | 0.6921 | 0.6408 | 0.5895 |
| **23** | 1.1028 | 1.0000 | 0.9486 | 0.8972 | 0.8458 | 0.7944 | 0.7430 | 0.6916 | 0.6402 | 0.5888 |
| **24** | 1.1030 | 1.0000 | 0.9485 | 0.8970 | 0.8455 | 0.7940 | 0.7425 | 0.6910 | 0.6395 | 0.5880 |
| **25** | 1.1032 | 1.0000 | 0.9484 | 0.8968 | 0.8452 | 0.7937 | 0.7421 | 0.6905 | 0.6389 | 0.5873 |
| **26** | 1.1034 | 1.0000 | 0.9483 | 0.8966 | 0.8449 | 0.7932 | 0.7415 | 0.6898 | 0.6381 | 0.5864 |
| **27** | 1.1036 | 1.0000 | 0.9482 | 0.8964 | 0.8446 | 0.7928 | 0.7410 | 0.6892 | 0.6374 | 0.5856 |
| **28** | 1.1038 | 1.0000 | 0.9481 | 0.8962 | 0.8443 | 0.7924 | 0.7404 | 0.6885 | 0.6366 | 0.5847 |
| **29** | 1.1040 | 1.0000 | 0.9480 | 0.8959 | 0.8439 | 0.7919 | 0.7398 | 0.6878 | 0.6357 | 0.5837 |
| **30** | 1.1043 | 1.0000 | 0.9478 | 0.8957 | 0.8435 | 0.7914 | 0.7392 | 0.6870 | 0.6348 | 0.5827 |
| **31** | 1.1046 | 1.0000 | 0.9477 | 0.8954 | 0.8431 | 0.7908 | 0.7385 | 0.6862 | 0.6339 | 0.5816 |
| **32** | 1.1049 | 1.0000 | 0.9476 | 0.8951 | 0.8427 | 0.7903 | 0.7378 | 0.6854 | 0.6329 | 0.5805 |
| **33** | 1.1052 | 1.0000 | 0.9474 | 0.8948 | 0.8422 | 0.7896 | 0.7371 | 0.6845 | 0.6319 | 0.5792 |
| **34** | 1.1055 | 1.0000 | 0.9473 | 0.8945 | 0.8418 | 0.7890 | 0.7363 | 0.6835 | 0.6307 | 0.5780 |
| **35** | 1.1058 | 1.0000 | 0.9471 | 0.8942 | 0.8413 | 0.7883 | 0.7354 | 0.6825 | 0.6296 | 0.5766 |
| **36** | 1.1062 | 1.0000 | 0.9469 | 0.8938 | 0.8407 | 0.7876 | 0.7345 | 0.6814 | 0.6283 | 0.5752 |
| **37** | 1.1065 | 1.0000 | 0.9467 | 0.8934 | 0.8402 | 0.7869 | 0.7336 | 0.6803 | 0.6270 | 0.5737 |
| **38** | 1.1069 | 1.0000 | 0.9465 | 0.8931 | 0.8396 | 0.7861 | 0.7326 | 0.6791 | 0.6256 | 0.5721 |
| **39** | 1.1074 | 1.0000 | 0.9463 | 0.8926 | 0.8390 | 0.7853 | 0.7316 | 0.6779 | 0.6242 | 0.5705 |
| **40** | 1.1078 | 1.0000 | 0.9461 | 0.8922 | 0.8383 | 0.7844 | 0.7305 | 0.6766 | 0.6226 | 0.5687 |

The PAF includes both the BP and terms within the bracket in Eq. (1.4).
*Source*: Based on Eq. 24, Benson and Krause (1984).

**Table 1.20** Pressure Adjustment Factor for Ideal Gas as a Function of Temperature and Barometric Pressure

| | Barometric Pressure | | | | | | | | | |
|---|---|---|---|---|---|---|---|---|---|---|
| **atm →** | **1.10** | **1.00** | **0.95** | **0.90** | **0.85** | **0.80** | **0.75** | **0.70** | **0.65** | **0.60** |
| **kPa →** | **111.46** | **101.33** | **96.26** | **91.19** | **86.13** | **81.06** | **75.99** | **70.93** | **65.86** | **60.80** |
| **mmHg →** | **836.0** | **760.0** | **722.0** | **684.0** | **646.0** | **608.0** | **570.0** | **532.0** | **494.0** | **456.0** |
| **millibar →** | **1115** | **1013** | **963** | **912** | **861** | **811** | **760** | **709** | **659** | **608** |
| **Temperature (°C)** | | | | | | | | | | |
| **0** | 1.1006 | 1.0000 | 0.9497 | 0.8994 | 0.8491 | 0.7988 | 0.7485 | 0.6982 | 0.6479 | 0.5976 |
| **1** | 1.1007 | 1.0000 | 0.9497 | 0.8993 | 0.8490 | 0.7987 | 0.7484 | 0.6980 | 0.6477 | 0.5974 |
| **2** | 1.1007 | 1.0000 | 0.9496 | 0.8993 | 0.8489 | 0.7986 | 0.7482 | 0.6979 | 0.6475 | 0.5972 |
| **3** | 1.1008 | 1.0000 | 0.9496 | 0.8992 | 0.8489 | 0.7985 | 0.7481 | 0.6977 | 0.6474 | 0.5970 |
| **4** | 1.1008 | 1.0000 | 0.9496 | 0.8992 | 0.8488 | 0.7984 | 0.7480 | 0.6976 | 0.6472 | 0.5968 |
| **5** | 1.1009 | 1.0000 | 0.9496 | 0.8991 | 0.8487 | 0.7983 | 0.7478 | 0.6974 | 0.6470 | 0.5965 |
| **6** | 1.1009 | 1.0000 | 0.9495 | 0.8991 | 0.8486 | 0.7981 | 0.7477 | 0.6972 | 0.6467 | 0.5963 |
| **7** | 1.1010 | 1.0000 | 0.9495 | 0.8990 | 0.8485 | 0.7980 | 0.7475 | 0.6970 | 0.6465 | 0.5960 |
| **8** | 1.1011 | 1.0000 | 0.9495 | 0.8989 | 0.8484 | 0.7979 | 0.7473 | 0.6968 | 0.6463 | 0.5957 |
| **9** | 1.1011 | 1.0000 | 0.9494 | 0.8989 | 0.8483 | 0.7977 | 0.7471 | 0.6966 | 0.6460 | 0.5954 |
| **10** | 1.1012 | 1.0000 | 0.9494 | 0.8988 | 0.8482 | 0.7975 | 0.7469 | 0.6963 | 0.6457 | 0.5951 |
| **11** | 1.1013 | 1.0000 | 0.9493 | 0.8987 | 0.8480 | 0.7974 | 0.7467 | 0.6961 | 0.6454 | 0.5948 |
| **12** | 1.1014 | 1.0000 | 0.9493 | 0.8986 | 0.8479 | 0.7972 | 0.7465 | 0.6958 | 0.6451 | 0.5944 |
| **13** | 1.1015 | 1.0000 | 0.9493 | 0.8985 | 0.8478 | 0.7970 | 0.7463 | 0.6955 | 0.6448 | 0.5940 |
| **14** | 1.1016 | 1.0000 | 0.9492 | 0.8984 | 0.8476 | 0.7968 | 0.7460 | 0.6952 | 0.6444 | 0.5936 |
| **15** | 1.1017 | 1.0000 | 0.9491 | 0.8983 | 0.8474 | 0.7966 | 0.7457 | 0.6949 | 0.6440 | 0.5932 |
| **16** | 1.1018 | 1.0000 | 0.9491 | 0.8982 | 0.8473 | 0.7963 | 0.7454 | 0.6945 | 0.6436 | 0.5927 |
| **17** | 1.1019 | 1.0000 | 0.9490 | 0.8981 | 0.8471 | 0.7961 | 0.7451 | 0.6942 | 0.6432 | 0.5922 |
| **18** | 1.1021 | 1.0000 | 0.9490 | 0.8979 | 0.8469 | 0.7958 | 0.7448 | 0.6938 | 0.6427 | 0.5917 |
| **19** | 1.1022 | 1.0000 | 0.9489 | 0.8978 | 0.8467 | 0.7956 | 0.7445 | 0.6934 | 0.6423 | 0.5911 |
| **20** | 1.1024 | 1.0000 | 0.9488 | 0.8976 | 0.8465 | 0.7953 | 0.7441 | 0.6929 | 0.6417 | 0.5906 |
| **21** | 1.1025 | 1.0000 | 0.9487 | 0.8975 | 0.8462 | 0.7950 | 0.7437 | 0.6925 | 0.6412 | 0.5899 |
| **22** | 1.1027 | 1.0000 | 0.9487 | 0.8973 | 0.8460 | 0.7946 | 0.7433 | 0.6920 | 0.6406 | 0.5893 |
| **23** | 1.1028 | 1.0000 | 0.9486 | 0.8972 | 0.8457 | 0.7943 | 0.7429 | 0.6915 | 0.6400 | 0.5886 |
| **24** | 1.1030 | 1.0000 | 0.9485 | 0.8970 | 0.8455 | 0.7939 | 0.7424 | 0.6909 | 0.6394 | 0.5879 |
| **25** | 1.1032 | 1.0000 | 0.9484 | 0.8968 | 0.8452 | 0.7936 | 0.7419 | 0.6903 | 0.6387 | 0.5871 |
| **26** | 1.1034 | 1.0000 | 0.9483 | 0.8966 | 0.8449 | 0.7931 | 0.7414 | 0.6897 | 0.6380 | 0.5863 |
| **27** | 1.1036 | 1.0000 | 0.9482 | 0.8964 | 0.8445 | 0.7927 | 0.7409 | 0.6891 | 0.6372 | 0.5854 |
| **28** | 1.1039 | 1.0000 | 0.9481 | 0.8961 | 0.8442 | 0.7923 | 0.7403 | 0.6884 | 0.6364 | 0.5845 |
| **29** | 1.1041 | 1.0000 | 0.9479 | 0.8959 | 0.8438 | 0.7918 | 0.7397 | 0.6877 | 0.6356 | 0.5835 |
| **30** | 1.1044 | 1.0000 | 0.9478 | 0.8956 | 0.8434 | 0.7913 | 0.7391 | 0.6869 | 0.6347 | 0.5825 |
| **31** | 1.1046 | 1.0000 | 0.9477 | 0.8954 | 0.8430 | 0.7907 | 0.7384 | 0.6861 | 0.6338 | 0.5815 |
| **32** | 1.1049 | 1.0000 | 0.9475 | 0.8951 | 0.8426 | 0.7902 | 0.7377 | 0.6852 | 0.6328 | 0.5803 |
| **33** | 1.1052 | 1.0000 | 0.9474 | 0.8948 | 0.8422 | 0.7896 | 0.7369 | 0.6843 | 0.6317 | 0.5791 |
| **34** | 1.1055 | 1.0000 | 0.9472 | 0.8945 | 0.8417 | 0.7889 | 0.7362 | 0.6834 | 0.6306 | 0.5778 |
| **35** | 1.1059 | 1.0000 | 0.9471 | 0.8941 | 0.8412 | 0.7883 | 0.7353 | 0.6824 | 0.6294 | 0.5765 |
| **36** | 1.1062 | 1.0000 | 0.9469 | 0.8938 | 0.8407 | 0.7875 | 0.7344 | 0.6813 | 0.6282 | 0.5751 |
| **37** | 1.1066 | 1.0000 | 0.9467 | 0.8934 | 0.8401 | 0.7868 | 0.7335 | 0.6802 | 0.6269 | 0.5736 |
| **38** | 1.1070 | 1.0000 | 0.9465 | 0.8930 | 0.8395 | 0.7860 | 0.7325 | 0.6790 | 0.6255 | 0.5720 |
| **39** | 1.1074 | 1.0000 | 0.9463 | 0.8926 | 0.8389 | 0.7852 | 0.7315 | 0.6778 | 0.6241 | 0.5704 |
| **40** | 1.1078 | 1.0000 | 0.9461 | 0.8922 | 0.8382 | 0.7843 | 0.7304 | 0.6765 | 0.6225 | 0.5686 |

The PAF includes both the BP and terms within the bracket in Eq. (1.4).
*Source:* Based on Hutchinson (1957).

**Table 1.21** Vapor Pressure of Freshwater as a Function of Temperature in mmHg

| Temperature (°C) | Δt (°C) | | | | | | | | | |
|---|---|---|---|---|---|---|---|---|---|---|
| | 0.0 | 0.1 | 0.2 | 0.3 | 0.4 | 0.5 | 0.6 | 0.7 | 0.8 | 0.9 |
| 0 | 4.581 | 4.614 | 4.648 | 4.681 | 4.716 | 4.750 | 4.784 | 4.819 | 4.854 | 4.889 |
| 1 | 4.925 | 4.960 | 4.996 | 5.032 | 5.069 | 5.105 | 5.142 | 5.179 | 5.216 | 5.254 |
| 2 | 5.291 | 5.329 | 5.368 | 5.406 | 5.445 | 5.484 | 5.523 | 5.562 | 5.602 | 5.642 |
| 3 | 5.682 | 5.723 | 5.763 | 5.804 | 5.845 | 5.887 | 5.929 | 5.971 | 6.013 | 6.055 |
| 4 | 6.098 | 6.141 | 6.184 | 6.228 | 6.272 | 6.316 | 6.360 | 6.405 | 6.450 | 6.495 |
| 5 | 6.541 | 6.587 | 6.633 | 6.679 | 6.726 | 6.773 | 6.820 | 6.867 | 6.915 | 6.963 |
| 6 | 7.012 | 7.060 | 7.109 | 7.159 | 7.208 | 7.258 | 7.308 | 7.359 | 7.410 | 7.461 |
| 7 | 7.512 | 7.564 | 7.616 | 7.668 | 7.721 | 7.774 | 7.827 | 7.881 | 7.935 | 7.989 |
| 8 | 8.044 | 8.099 | 8.154 | 8.210 | 8.266 | 8.322 | 8.379 | 8.436 | 8.493 | 8.550 |
| 9 | 8.608 | 8.667 | 8.726 | 8.785 | 8.844 | 8.904 | 8.964 | 9.024 | 9.085 | 9.146 |
| 10 | 9.208 | 9.270 | 9.332 | 9.395 | 9.458 | 9.521 | 9.585 | 9.649 | 9.713 | 9.778 |
| 11 | 9.843 | 9.909 | 9.975 | 10.041 | 10.108 | 10.175 | 10.243 | 10.311 | 10.379 | 10.448 |
| 12 | 10.517 | 10.587 | 10.657 | 10.727 | 10.798 | 10.869 | 10.941 | 11.013 | 11.085 | 11.158 |
| 13 | 11.232 | 11.305 | 11.379 | 11.454 | 11.529 | 11.604 | 11.680 | 11.757 | 11.833 | 11.910 |
| 14 | 11.988 | 12.066 | 12.145 | 12.224 | 12.303 | 12.383 | 12.463 | 12.544 | 12.625 | 12.707 |
| 15 | 12.789 | 12.872 | 12.955 | 13.038 | 13.122 | 13.207 | 13.292 | 13.377 | 13.463 | 13.550 |
| 16 | 13.636 | 13.724 | 13.812 | 13.900 | 13.989 | 14.078 | 14.168 | 14.259 | 14.350 | 14.441 |
| 17 | 14.533 | 14.625 | 14.718 | 14.812 | 14.906 | 15.000 | 15.095 | 15.191 | 15.287 | 15.383 |
| 18 | 15.480 | 15.578 | 15.676 | 15.775 | 15.874 | 15.974 | 16.074 | 16.175 | 16.277 | 16.379 |
| 19 | 16.482 | 16.585 | 16.689 | 16.793 | 16.898 | 17.003 | 17.109 | 17.216 | 17.323 | 17.431 |
| 20 | 17.539 | 17.648 | 17.758 | 17.868 | 17.978 | 18.090 | 18.202 | 18.314 | 18.427 | 18.541 |
| 21 | 18.656 | 18.770 | 18.886 | 19.002 | 19.119 | 19.237 | 19.355 | 19.473 | 19.593 | 19.713 |
| 22 | 19.834 | 19.955 | 20.077 | 20.199 | 20.323 | 20.447 | 20.571 | 20.696 | 20.822 | 20.949 |
| 23 | 21.076 | 21.204 | 21.333 | 21.462 | 21.592 | 21.722 | 21.854 | 21.986 | 22.119 | 22.252 |
| 24 | 22.386 | 22.521 | 22.656 | 22.793 | 22.930 | 23.067 | 23.206 | 23.345 | 23.485 | 23.625 |
| 25 | 23.767 | 23.909 | 24.051 | 24.195 | 24.339 | 24.484 | 24.630 | 24.776 | 24.924 | 25.072 |
| 26 | 25.221 | 25.370 | 25.521 | 25.672 | 25.824 | 25.976 | 26.130 | 26.284 | 26.439 | 26.595 |
| 27 | 26.752 | 26.909 | 27.067 | 27.227 | 27.386 | 27.547 | 27.709 | 27.871 | 28.034 | 28.198 |
| 28 | 28.363 | 28.529 | 28.695 | 28.863 | 29.031 | 29.200 | 29.370 | 29.541 | 29.712 | 29.885 |
| 29 | 30.058 | 30.233 | 30.408 | 30.584 | 30.761 | 30.939 | 31.117 | 31.297 | 31.477 | 31.659 |
| 30 | 31.841 | 32.024 | 32.208 | 32.393 | 32.579 | 32.766 | 32.954 | 33.143 | 33.333 | 33.523 |
| 31 | 33.715 | 33.907 | 34.101 | 34.295 | 34.491 | 34.687 | 34.885 | 35.083 | 35.282 | 35.483 |
| 32 | 35.684 | 35.886 | 36.089 | 36.294 | 36.499 | 36.705 | 36.912 | 37.121 | 37.330 | 37.540 |
| 33 | 37.752 | 37.964 | 38.178 | 38.392 | 38.608 | 38.824 | 39.042 | 39.260 | 39.480 | 39.701 |
| 34 | 39.923 | 40.146 | 40.370 | 40.595 | 40.821 | 41.048 | 41.277 | 41.506 | 41.737 | 41.969 |
| 35 | 42.201 | 42.435 | 42.670 | 42.907 | 43.144 | 43.382 | 43.622 | 43.863 | 44.105 | 44.348 |
| 36 | 44.592 | 44.837 | 45.084 | 45.331 | 45.580 | 45.830 | 46.082 | 46.334 | 46.588 | 46.842 |
| 37 | 47.098 | 47.356 | 47.614 | 47.874 | 48.135 | 48.397 | 48.660 | 48.925 | 49.191 | 49.458 |
| 38 | 49.726 | 49.996 | 50.267 | 50.539 | 50.812 | 51.087 | 51.363 | 51.640 | 51.919 | 52.198 |
| 39 | 52.480 | 52.762 | 53.046 | 53.331 | 53.617 | 53.905 | 54.194 | 54.485 | 54.776 | 55.069 |
| 40 | 55.364 | 55.660 | 55.957 | 56.255 | 56.555 | 56.857 | 57.159 | 57.463 | 57.769 | 58.076 |

*Source*: Freshwater, based on Ambrose and Lawrenson (1972).

**Table 1.22** Vapor Pressure of Freshwater as a Function of Temperature in kPa

| Temperature (°C) | Δt (°C) | | | | | | | | | |
|---|---|---|---|---|---|---|---|---|---|---|
| | **0.0** | **0.1** | **0.2** | **0.3** | **0.4** | **0.5** | **0.6** | **0.7** | **0.8** | **0.9** |
| **0** | 0.6107 | 0.6151 | 0.6196 | 0.6241 | 0.6287 | 0.6333 | 0.6379 | 0.6425 | 0.6472 | 0.6518 |
| **1** | 0.6566 | 0.6613 | 0.6661 | 0.6709 | 0.6758 | 0.6806 | 0.6855 | 0.6905 | 0.6954 | 0.7004 |
| **2** | 0.7055 | 0.7105 | 0.7156 | 0.7207 | 0.7259 | 0.7311 | 0.7363 | 0.7416 | 0.7469 | 0.7522 |
| **3** | 0.7576 | 0.7629 | 0.7684 | 0.7738 | 0.7793 | 0.7849 | 0.7904 | 0.7960 | 0.8016 | 0.8073 |
| **4** | 0.8130 | 0.8188 | 0.8245 | 0.8303 | 0.8362 | 0.8421 | 0.8480 | 0.8539 | 0.8599 | 0.8660 |
| **5** | 0.8720 | 0.8781 | 0.8843 | 0.8905 | 0.8967 | 0.9029 | 0.9092 | 0.9156 | 0.9219 | 0.9284 |
| **6** | 0.9348 | 0.9413 | 0.9478 | 0.9544 | 0.9610 | 0.9677 | 0.9744 | 0.9811 | 0.9879 | 0.9947 |
| **7** | 1.0015 | 1.0084 | 1.0154 | 1.0224 | 1.0294 | 1.0364 | 1.0436 | 1.0507 | 1.0579 | 1.0651 |
| **8** | 1.0724 | 1.0798 | 1.0871 | 1.0945 | 1.1020 | 1.1095 | 1.1171 | 1.1246 | 1.1323 | 1.1400 |
| **9** | 1.1477 | 1.1555 | 1.1633 | 1.1712 | 1.1791 | 1.1871 | 1.1951 | 1.2031 | 1.2112 | 1.2194 |
| **10** | 1.2276 | 1.2358 | 1.2442 | 1.2525 | 1.2609 | 1.2693 | 1.2778 | 1.2864 | 1.2950 | 1.3036 |
| **11** | 1.3123 | 1.3211 | 1.3299 | 1.3388 | 1.3477 | 1.3566 | 1.3656 | 1.3747 | 1.3838 | 1.3930 |
| **12** | 1.4022 | 1.4115 | 1.4208 | 1.4302 | 1.4396 | 1.4491 | 1.4587 | 1.4683 | 1.4779 | 1.4877 |
| **13** | 1.4974 | 1.5073 | 1.5171 | 1.5271 | 1.5371 | 1.5471 | 1.5572 | 1.5674 | 1.5776 | 1.5879 |
| **14** | 1.5983 | 1.6087 | 1.6192 | 1.6297 | 1.6403 | 1.6509 | 1.6616 | 1.6724 | 1.6832 | 1.6941 |
| **15** | 1.7051 | 1.7161 | 1.7271 | 1.7383 | 1.7495 | 1.7608 | 1.7721 | 1.7835 | 1.7949 | 1.8065 |
| **16** | 1.8180 | 1.8297 | 1.8414 | 1.8532 | 1.8650 | 1.8770 | 1.8889 | 1.9010 | 1.9131 | 1.9253 |
| **17** | 1.9375 | 1.9499 | 1.9623 | 1.9747 | 1.9872 | 1.9998 | 2.0125 | 2.0252 | 2.0380 | 2.0509 |
| **18** | 2.0639 | 2.0769 | 2.0900 | 2.1032 | 2.1164 | 2.1297 | 2.1431 | 2.1565 | 2.1701 | 2.1837 |
| **19** | 2.1974 | 2.2111 | 2.2250 | 2.2389 | 2.2528 | 2.2669 | 2.2810 | 2.2952 | 2.3095 | 2.3239 |
| **20** | 2.3384 | 2.3529 | 2.3675 | 2.3822 | 2.3969 | 2.4118 | 2.4267 | 2.4417 | 2.4568 | 2.4719 |
| **21** | 2.4872 | 2.5025 | 2.5179 | 2.5334 | 2.5490 | 2.5647 | 2.5804 | 2.5963 | 2.6122 | 2.6282 |
| **22** | 2.6443 | 2.6604 | 2.6767 | 2.6930 | 2.7095 | 2.7260 | 2.7426 | 2.7593 | 2.7761 | 2.7930 |
| **23** | 2.8099 | 2.8270 | 2.8441 | 2.8613 | 2.8787 | 2.8961 | 2.9136 | 2.9312 | 2.9489 | 2.9667 |
| **24** | 2.9846 | 3.0025 | 3.0206 | 3.0388 | 3.0570 | 3.0754 | 3.0938 | 3.1124 | 3.1310 | 3.1498 |
| **25** | 3.1686 | 3.1875 | 3.2066 | 3.2257 | 3.2450 | 3.2643 | 3.2837 | 3.3033 | 3.3229 | 3.3426 |
| **26** | 3.3625 | 3.3824 | 3.4025 | 3.4226 | 3.4429 | 3.4632 | 3.4837 | 3.5043 | 3.5249 | 3.5457 |
| **27** | 3.5666 | 3.5876 | 3.6087 | 3.6299 | 3.6512 | 3.6727 | 3.6942 | 3.7158 | 3.7376 | 3.7595 |
| **28** | 3.7814 | 3.8035 | 3.8257 | 3.8480 | 3.8705 | 3.8930 | 3.9157 | 3.9384 | 3.9613 | 3.9843 |
| **29** | 4.0074 | 4.0307 | 4.0540 | 4.0775 | 4.1011 | 4.1248 | 4.1486 | 4.1726 | 4.1966 | 4.2208 |
| **30** | 4.2451 | 4.2695 | 4.2941 | 4.3188 | 4.3436 | 4.3685 | 4.3935 | 4.4187 | 4.4440 | 4.4694 |
| **31** | 4.4949 | 4.5206 | 4.5464 | 4.5723 | 4.5984 | 4.6246 | 4.6509 | 4.6773 | 4.7039 | 4.7306 |
| **32** | 4.7574 | 4.7844 | 4.8115 | 4.8387 | 4.8661 | 4.8936 | 4.9213 | 4.9490 | 4.9769 | 5.0050 |
| **33** | 5.0332 | 5.0615 | 5.0899 | 5.1185 | 5.1473 | 5.1761 | 5.2051 | 5.2343 | 5.2636 | 5.2930 |
| **34** | 5.3226 | 5.3523 | 5.3822 | 5.4122 | 5.4424 | 5.4727 | 5.5031 | 5.5337 | 5.5645 | 5.5954 |
| **35** | 5.6264 | 5.6576 | 5.6889 | 5.7204 | 5.7521 | 5.7838 | 5.8158 | 5.8479 | 5.8801 | 5.9125 |
| **36** | 5.9451 | 5.9778 | 6.0107 | 6.0437 | 6.0769 | 6.1102 | 6.1437 | 6.1774 | 6.2112 | 6.2451 |
| **37** | 6.2793 | 6.3136 | 6.3480 | 6.3827 | 6.4174 | 6.4524 | 6.4875 | 6.5228 | 6.5582 | 6.5938 |
| **38** | 6.6296 | 6.6656 | 6.7017 | 6.7379 | 6.7744 | 6.8110 | 6.8478 | 6.8848 | 6.9219 | 6.9592 |
| **39** | 6.9967 | 7.0344 | 7.0722 | 7.1102 | 7.1484 | 7.1867 | 7.2253 | 7.2640 | 7.3029 | 7.3420 |
| **40** | 7.3812 | 7.4207 | 7.4603 | 7.5001 | 7.5401 | 7.5802 | 7.6206 | 7.6611 | 7.7019 | 7.7428 |

*Source*: Freshwater, based on Ambrose and Lawrenson (1972).

**Table 1.23** Vapor Pressure of Freshwater as a Function of Temperature in Atmosphere

| **Temperature (°C)** | $\Delta t$ (°C) | | | | | | | | | |
|---|---|---|---|---|---|---|---|---|---|---|
| | **0.0** | **0.1** | **0.2** | **0.3** | **0.4** | **0.5** | **0.6** | **0.7** | **0.8** | **0.9** |
| **0** | .00603 | .00607 | .00612 | .00616 | .00620 | .00625 | .00630 | .00634 | .00639 | .00643 |
| **1** | .00648 | .00653 | .00657 | .00662 | .00667 | .00672 | .00677 | .00681 | .00686 | .00691 |
| **2** | .00696 | .00701 | .00706 | .00711 | .00716 | .00722 | .00727 | .00732 | .00737 | .00742 |
| **3** | .00748 | .00753 | .00758 | .00764 | .00769 | .00775 | .00780 | .00786 | .00791 | .00797 |
| **4** | .00802 | .00808 | .00814 | .00819 | .00825 | .00831 | .00837 | .00843 | .00849 | .00855 |
| **5** | .00861 | .00867 | .00873 | .00879 | .00885 | .00891 | .00897 | .00904 | .00910 | .00916 |
| **6** | .00923 | .00929 | .00935 | .00942 | .00948 | .00955 | .00962 | .00968 | .00975 | .00982 |
| **7** | .00988 | .00995 | .01002 | .01009 | .01016 | .01023 | .01030 | .01037 | .01044 | .01051 |
| **8** | .01058 | .01066 | .01073 | .01080 | .01088 | .01095 | .01102 | .01110 | .01117 | .01125 |
| **9** | .01133 | .01140 | .01148 | .01156 | .01164 | .01172 | .01179 | .01187 | .01195 | .01203 |
| **10** | .01212 | .01220 | .01228 | .01236 | .01244 | .01253 | .01261 | .01270 | .01278 | .01287 |
| **11** | .01295 | .01304 | .01313 | .01321 | .01330 | .01339 | .01348 | .01357 | .01366 | .01375 |
| **12** | .01384 | .01393 | .01402 | .01411 | .01421 | .01430 | .01440 | .01449 | .01459 | .01468 |
| **13** | .01478 | .01488 | .01497 | .01507 | .01517 | .01527 | .01537 | .01547 | .01557 | .01567 |
| **14** | .01577 | .01588 | .01598 | .01608 | .01619 | .01629 | .01640 | .01651 | .01661 | .01672 |
| **15** | .01683 | .01694 | .01705 | .01716 | .01727 | .01738 | .01749 | .01760 | .01771 | .01783 |
| **16** | .01794 | .01806 | .01817 | .01829 | .01841 | .01852 | .01864 | .01876 | .01888 | .01900 |
| **17** | .01912 | .01924 | .01937 | .01949 | .01961 | .01974 | .01986 | .01999 | .02011 | .02024 |
| **18** | .02037 | .02050 | .02063 | .02076 | .02089 | .02102 | .02115 | .02128 | .02142 | .02155 |
| **19** | .02169 | .02182 | .02196 | .02210 | .02223 | .02237 | .02251 | .02265 | .02279 | .02294 |
| **20** | .02308 | .02322 | .02337 | .02351 | .02366 | .02380 | .02395 | .02410 | .02425 | .02440 |
| **21** | .02455 | .02470 | .02485 | .02500 | .02516 | .02531 | .02547 | .02562 | .02578 | .02594 |
| **22** | .02610 | .02626 | .02642 | .02658 | .02674 | .02690 | .02707 | .02723 | .02740 | .02756 |
| **23** | .02773 | .02790 | .02807 | .02824 | .02841 | .02858 | .02875 | .02893 | .02910 | .02928 |
| **24** | .02946 | .02963 | .02981 | .02999 | .03017 | .03035 | .03053 | .03072 | .03090 | .03109 |
| **25** | .03127 | .03146 | .03165 | .03184 | .03203 | .03222 | .03241 | .03260 | .03279 | .03299 |
| **26** | .03319 | .03338 | .03358 | .03378 | .03398 | .03418 | .03438 | .03458 | .03479 | .03499 |
| **27** | .03520 | .03541 | .03562 | .03582 | .03603 | .03625 | .03646 | .03667 | .03689 | .03710 |
| **28** | .03732 | .03754 | .03776 | .03798 | .03820 | .03842 | .03864 | .03887 | .03910 | .03932 |
| **29** | .03955 | .03978 | .04001 | .04024 | .04047 | .04071 | .04094 | .04118 | .04142 | .04166 |
| **30** | .04190 | .04214 | .04238 | .04262 | .04287 | .04311 | .04336 | .04361 | .04386 | .04411 |
| **31** | .04436 | .04462 | .04487 | .04513 | .04538 | .04564 | .04590 | .04616 | .04642 | .04669 |
| **32** | .04695 | .04722 | .04749 | .04775 | .04802 | .04830 | .04857 | .04884 | .04912 | .04940 |
| **33** | .04967 | .04995 | .05023 | .05052 | .05080 | .05108 | .05137 | .05166 | .05195 | .05224 |
| **34** | .05253 | .05282 | .05312 | .05341 | .05371 | .05401 | .05431 | .05461 | .05492 | .05522 |
| **35** | .05553 | .05584 | .05615 | .05646 | .05677 | .05708 | .05740 | .05771 | .05803 | .05835 |
| **36** | .05867 | .05900 | .05932 | .05965 | .05997 | .06030 | .06063 | .06097 | .06130 | .06163 |
| **37** | .06197 | .06231 | .06265 | .06299 | .06334 | .06368 | .06403 | .06437 | .06472 | .06508 |
| **38** | .06543 | .06578 | .06614 | .06650 | .06686 | .06722 | .06758 | .06795 | .06831 | .06868 |
| **39** | .06905 | .06942 | .06980 | .07017 | .07055 | .07093 | .07131 | .07169 | .07207 | .07246 |
| **40** | .07285 | .07324 | .07363 | .07402 | .07441 | .07481 | .07521 | .07561 | .07601 | .07642 |

*Source*: Freshwater, based on Ambrose and Lawrenson (1972).

**Table 1.24** Vapor Pressure of Freshwater as a Function of Temperature in psi

| Temperature (°C) | Δt (°C) | | | | | | | | | |
|---|---|---|---|---|---|---|---|---|---|---|
| | **0.0** | **0.1** | **0.2** | **0.3** | **0.4** | **0.5** | **0.6** | **0.7** | **0.8** | **0.9** |
| **0** | 0.0886 | 0.0892 | 0.0899 | 0.0905 | 0.0912 | 0.0918 | 0.0925 | 0.0932 | 0.0939 | 0.0945 |
| **1** | 0.0952 | 0.0959 | 0.0966 | 0.0973 | 0.0980 | 0.0987 | 0.0994 | 0.1001 | 0.1009 | 0.1016 |
| **2** | 0.1023 | 0.1031 | 0.1038 | 0.1045 | 0.1053 | 0.1060 | 0.1068 | 0.1076 | 0.1083 | 0.1091 |
| **3** | 0.1099 | 0.1107 | 0.1114 | 0.1122 | 0.1130 | 0.1138 | 0.1146 | 0.1155 | 0.1163 | 0.1171 |
| **4** | 0.1179 | 0.1188 | 0.1196 | 0.1204 | 0.1213 | 0.1221 | 0.1230 | 0.1239 | 0.1247 | 0.1256 |
| **5** | 0.1265 | 0.1274 | 0.1283 | 0.1292 | 0.1301 | 0.1310 | 0.1319 | 0.1328 | 0.1337 | 0.1346 |
| **6** | 0.1356 | 0.1365 | 0.1375 | 0.1384 | 0.1394 | 0.1403 | 0.1413 | 0.1423 | 0.1433 | 0.1443 |
| **7** | 0.1453 | 0.1463 | 0.1473 | 0.1483 | 0.1493 | 0.1503 | 0.1514 | 0.1524 | 0.1534 | 0.1545 |
| **8** | 0.1555 | 0.1566 | 0.1577 | 0.1587 | 0.1598 | 0.1609 | 0.1620 | 0.1631 | 0.1642 | 0.1653 |
| **9** | 0.1665 | 0.1676 | 0.1687 | 0.1699 | 0.1710 | 0.1722 | 0.1733 | 0.1745 | 0.1757 | 0.1769 |
| **10** | 0.1780 | 0.1792 | 0.1804 | 0.1817 | 0.1829 | 0.1841 | 0.1853 | 0.1866 | 0.1878 | 0.1891 |
| **11** | 0.1903 | 0.1916 | 0.1929 | 0.1942 | 0.1955 | 0.1968 | 0.1981 | 0.1994 | 0.2007 | 0.2020 |
| **12** | 0.2034 | 0.2047 | 0.2061 | 0.2074 | 0.2088 | 0.2102 | 0.2116 | 0.2130 | 0.2144 | 0.2158 |
| **13** | 0.2172 | 0.2186 | 0.2200 | 0.2215 | 0.2229 | 0.2244 | 0.2259 | 0.2273 | 0.2288 | 0.2303 |
| **14** | 0.2318 | 0.2333 | 0.2348 | 0.2364 | 0.2379 | 0.2394 | 0.2410 | 0.2426 | 0.2441 | 0.2457 |
| **15** | 0.2473 | 0.2489 | 0.2505 | 0.2521 | 0.2537 | 0.2554 | 0.2570 | 0.2587 | 0.2603 | 0.2620 |
| **16** | 0.2637 | 0.2654 | 0.2671 | 0.2688 | 0.2705 | 0.2722 | 0.2740 | 0.2757 | 0.2775 | 0.2792 |
| **17** | 0.2810 | 0.2828 | 0.2846 | 0.2864 | 0.2882 | 0.2901 | 0.2919 | 0.2937 | 0.2956 | 0.2975 |
| **18** | 0.2993 | 0.3012 | 0.3031 | 0.3050 | 0.3070 | 0.3089 | 0.3108 | 0.3128 | 0.3147 | 0.3167 |
| **19** | 0.3187 | 0.3207 | 0.3227 | 0.3247 | 0.3267 | 0.3288 | 0.3308 | 0.3329 | 0.3350 | 0.3371 |
| **20** | 0.3391 | 0.3413 | 0.3434 | 0.3455 | 0.3476 | 0.3498 | 0.3520 | 0.3541 | 0.3563 | 0.3585 |
| **21** | 0.3607 | 0.3630 | 0.3652 | 0.3674 | 0.3697 | 0.3720 | 0.3743 | 0.3766 | 0.3789 | 0.3812 |
| **22** | 0.3835 | 0.3859 | 0.3882 | 0.3906 | 0.3930 | 0.3954 | 0.3978 | 0.4002 | 0.4026 | 0.4051 |
| **23** | 0.4075 | 0.4100 | 0.4125 | 0.4150 | 0.4175 | 0.4200 | 0.4226 | 0.4251 | 0.4277 | 0.4303 |
| **24** | 0.4329 | 0.4355 | 0.4381 | 0.4407 | 0.4434 | 0.4460 | 0.4487 | 0.4514 | 0.4541 | 0.4568 |
| **25** | 0.4596 | 0.4623 | 0.4651 | 0.4678 | 0.4706 | 0.4734 | 0.4763 | 0.4791 | 0.4819 | 0.4848 |
| **26** | 0.4877 | 0.4906 | 0.4935 | 0.4964 | 0.4993 | 0.5023 | 0.5053 | 0.5082 | 0.5112 | 0.5143 |
| **27** | 0.5173 | 0.5203 | 0.5234 | 0.5265 | 0.5296 | 0.5327 | 0.5358 | 0.5389 | 0.5421 | 0.5453 |
| **28** | 0.5484 | 0.5517 | 0.5549 | 0.5581 | 0.5614 | 0.5646 | 0.5679 | 0.5712 | 0.5745 | 0.5779 |
| **29** | 0.5812 | 0.5846 | 0.5880 | 0.5914 | 0.5948 | 0.5982 | 0.6017 | 0.6052 | 0.6087 | 0.6122 |
| **30** | 0.6157 | 0.6192 | 0.6228 | 0.6264 | 0.6300 | 0.6336 | 0.6372 | 0.6409 | 0.6445 | 0.6482 |
| **31** | 0.6519 | 0.6557 | 0.6594 | 0.6632 | 0.6669 | 0.6707 | 0.6746 | 0.6784 | 0.6822 | 0.6861 |
| **32** | 0.6900 | 0.6939 | 0.6978 | 0.7018 | 0.7058 | 0.7098 | 0.7138 | 0.7178 | 0.7218 | 0.7259 |
| **33** | 0.7300 | 0.7341 | 0.7382 | 0.7424 | 0.7465 | 0.7507 | 0.7549 | 0.7592 | 0.7634 | 0.7677 |
| **34** | 0.7720 | 0.7763 | 0.7806 | 0.7850 | 0.7893 | 0.7937 | 0.7982 | 0.8026 | 0.8071 | 0.8115 |
| **35** | 0.8160 | 0.8206 | 0.8251 | 0.8297 | 0.8343 | 0.8389 | 0.8435 | 0.8482 | 0.8528 | 0.8575 |
| **36** | 0.8623 | 0.8670 | 0.8718 | 0.8766 | 0.8814 | 0.8862 | 0.8911 | 0.8959 | 0.9009 | 0.9058 |
| **37** | 0.9107 | 0.9157 | 0.9207 | 0.9257 | 0.9308 | 0.9358 | 0.9409 | 0.9460 | 0.9512 | 0.9564 |
| **38** | 0.9615 | 0.9668 | 0.9720 | 0.9773 | 0.9825 | 0.9879 | 0.9932 | 0.9985 | 1.0039 | 1.0093 |
| **39** | 1.0148 | 1.0202 | 1.0257 | 1.0312 | 1.0368 | 1.0423 | 1.0479 | 1.0536 | 1.0592 | 1.0649 |
| **40** | 1.0706 | 1.0763 | 1.0820 | 1.0878 | 1.0936 | 1.0994 | 1.1053 | 1.1112 | 1.1171 | 1.1230 |

*Source*: Freshwater, based on Ambrose and Lawrenson (1972).

**Table 1.25** The Air Solubility of Oxygen in mg/L as Functions of Temperature and Elevation in Meters for 0–1800 m

| Temperature (°C) | Elevation Above Sea Level (m) | | | | | | | | | |
|---|---|---|---|---|---|---|---|---|---|---|
| | **0** | **200** | **400** | **600** | **800** | **1000** | **1200** | **1400** | **1600** | **1800** |
| **0** | 14.621 | 14.276 | 13.940 | 13.612 | 13.291 | 12.978 | 12.672 | 12.373 | 12.081 | 11.796 |
| **1** | 14.216 | 13.881 | 13.554 | 13.234 | 12.922 | 12.617 | 12.320 | 12.029 | 11.745 | 11.468 |
| **2** | 13.829 | 13.503 | 13.185 | 12.874 | 12.570 | 12.273 | 11.984 | 11.701 | 11.425 | 11.155 |
| **3** | 13.460 | 13.142 | 12.832 | 12.530 | 12.234 | 11.945 | 11.663 | 11.387 | 11.118 | 10.856 |
| **4** | 13.107 | 12.798 | 12.496 | 12.201 | 11.912 | 11.631 | 11.356 | 11.088 | 10.826 | 10.570 |
| **5** | 12.770 | 12.468 | 12.174 | 11.886 | 11.605 | 11.331 | 11.063 | 10.801 | 10.546 | 10.296 |
| **6** | 12.447 | 12.153 | 11.866 | 11.585 | 11.311 | 11.044 | 10.782 | 10.527 | 10.278 | 10.035 |
| **7** | 12.138 | 11.851 | 11.571 | 11.297 | 11.030 | 10.769 | 10.514 | 10.265 | 10.022 | 9.784 |
| **8** | 11.843 | 11.562 | 11.289 | 11.021 | 10.760 | 10.505 | 10.256 | 10.013 | 9.776 | 9.544 |
| **9** | 11.559 | 11.285 | 11.018 | 10.757 | 10.502 | 10.253 | 10.010 | 9.772 | 9.540 | 9.314 |
| **10** | 11.288 | 11.020 | 10.759 | 10.504 | 10.254 | 10.011 | 9.773 | 9.541 | 9.315 | 9.093 |
| **11** | 11.027 | 10.765 | 10.510 | 10.260 | 10.017 | 9.779 | 9.546 | 9.319 | 9.098 | 8.881 |
| **12** | 10.777 | 10.521 | 10.271 | 10.027 | 9.789 | 9.556 | 9.329 | 9.107 | 8.890 | 8.678 |
| **13** | 10.536 | 10.286 | 10.041 | 9.803 | 9.569 | 9.342 | 9.119 | 8.902 | 8.690 | 8.483 |
| **14** | 10.306 | 10.060 | 9.821 | 9.587 | 9.359 | 9.136 | 8.918 | 8.705 | 8.498 | 8.295 |
| **15** | 10.084 | 9.843 | 9.609 | 9.380 | 9.156 | 8.938 | 8.724 | 8.516 | 8.313 | 8.114 |
| **16** | 9.870 | 9.635 | 9.405 | 9.180 | 8.961 | 8.747 | 8.538 | 8.334 | 8.135 | 7.940 |
| **17** | 9.665 | 9.434 | 9.209 | 8.988 | 8.774 | 8.564 | 8.359 | 8.159 | 7.963 | 7.772 |
| **18** | 9.467 | 9.240 | 9.019 | 8.804 | 8.593 | 8.387 | 8.186 | 7.990 | 7.798 | 7.611 |
| **19** | 9.276 | 9.054 | 8.837 | 8.625 | 8.418 | 8.216 | 8.019 | 7.827 | 7.639 | 7.455 |
| **20** | 9.092 | 8.874 | 8.661 | 8.453 | 8.250 | 8.052 | 7.858 | 7.669 | 7.485 | 7.304 |
| **21** | 8.914 | 8.700 | 8.491 | 8.287 | 8.088 | 7.893 | 7.703 | 7.518 | 7.336 | 7.159 |
| **22** | 8.743 | 8.533 | 8.328 | 8.127 | 7.931 | 7.740 | 7.553 | 7.371 | 7.193 | 7.019 |
| **23** | 8.578 | 8.371 | 8.169 | 7.972 | 7.780 | 7.592 | 7.408 | 7.229 | 7.054 | 6.883 |
| **24** | 8.418 | 8.214 | 8.016 | 7.822 | 7.633 | 7.449 | 7.268 | 7.092 | 6.920 | 6.752 |
| **25** | 8.263 | 8.063 | 7.868 | 7.678 | 7.491 | 7.310 | 7.132 | 6.959 | 6.790 | 6.625 |
| **26** | 8.113 | 7.917 | 7.725 | 7.537 | 7.354 | 7.175 | 7.001 | 6.830 | 6.664 | 6.501 |
| **27** | 7.968 | 7.775 | 7.586 | 7.401 | 7.221 | 7.045 | 6.873 | 6.706 | 6.542 | 6.382 |
| **28** | 7.827 | 7.637 | 7.451 | 7.269 | 7.092 | 6.919 | 6.750 | 6.584 | 6.423 | 6.266 |
| **29** | 7.691 | 7.503 | 7.320 | 7.141 | 6.967 | 6.796 | 6.630 | 6.467 | 6.308 | 6.153 |
| **30** | 7.559 | 7.374 | 7.193 | 7.017 | 6.845 | 6.677 | 6.513 | 6.353 | 6.196 | 6.043 |
| **31** | 7.430 | 7.248 | 7.070 | 6.896 | 6.727 | 6.561 | 6.399 | 6.241 | 6.087 | 5.937 |
| **32** | 7.305 | 7.125 | 6.950 | 6.779 | 6.612 | 6.448 | 6.289 | 6.133 | 5.981 | 5.833 |
| **33** | 7.183 | 7.006 | 6.833 | 6.665 | 6.500 | 6.339 | 6.181 | 6.028 | 5.878 | 5.731 |
| **34** | 7.065 | 6.890 | 6.720 | 6.553 | 6.390 | 6.232 | 6.077 | 5.925 | 5.777 | 5.633 |
| **35** | 6.949 | 6.777 | 6.609 | 6.445 | 6.284 | 6.127 | 5.974 | 5.825 | 5.679 | 5.536 |
| **36** | 6.837 | 6.667 | 6.501 | 6.338 | 6.180 | 6.025 | 5.874 | 5.727 | 5.583 | 5.442 |
| **37** | 6.727 | 6.559 | 6.395 | 6.235 | 6.078 | 5.926 | 5.776 | 5.631 | 5.489 | 5.350 |
| **38** | 6.620 | 6.454 | 6.292 | 6.134 | 5.979 | 5.828 | 5.681 | 5.537 | 5.396 | 5.259 |
| **39** | 6.515 | 6.351 | 6.191 | 6.035 | 5.882 | 5.733 | 5.587 | 5.445 | 5.306 | 5.171 |
| **40** | 6.412 | 6.250 | 6.092 | 5.937 | 5.787 | 5.639 | 5.495 | 5.355 | 5.218 | 5.084 |

*Source*: Based on Eq. (1.7).

**Table 1.26** The Air Solubility of Oxygen in mg/L as Functions of Temperature and Elevation in Meters for 2000–3800 m

| Temperature (°C) | Elevation Above Sea Level (m) | | | | | | | | | |
|---|---|---|---|---|---|---|---|---|---|---|
| | **2000** | **2200** | **2400** | **2600** | **2800** | **3000** | **3200** | **3400** | **3600** | **3800** |
| **0** | 11.518 | 11.246 | 10.981 | 10.722 | 10.468 | 10.221 | 9.980 | 9.744 | 9.514 | 9.289 |
| **1** | 11.197 | 10.933 | 10.675 | 10.423 | 10.177 | 9.936 | 9.701 | 9.472 | 9.248 | 9.029 |
| **2** | 10.891 | 10.634 | 10.383 | 10.137 | 9.898 | 9.664 | 9.435 | 9.212 | 8.994 | 8.781 |
| **3** | 10.599 | 10.349 | 10.104 | 9.865 | 9.632 | 9.404 | 9.181 | 8.964 | 8.751 | 8.544 |
| **4** | 10.320 | 10.076 | 9.837 | 9.604 | 9.377 | 9.155 | 8.938 | 8.726 | 8.519 | 8.317 |
| **5** | 10.053 | 9.815 | 9.582 | 9.355 | 9.134 | 8.917 | 8.706 | 8.499 | 8.298 | 8.101 |
| **6** | 9.797 | 9.565 | 9.338 | 9.117 | 8.901 | 8.689 | 8.483 | 8.282 | 8.085 | 7.893 |
| **7** | 9.552 | 9.326 | 9.104 | 8.888 | 8.677 | 8.471 | 8.270 | 8.074 | 7.882 | 7.694 |
| **8** | 9.318 | 9.096 | 8.880 | 8.670 | 8.464 | 8.262 | 8.066 | 7.874 | 7.687 | 7.504 |
| **9** | 9.093 | 8.877 | 8.666 | 8.460 | 8.258 | 8.062 | 7.870 | 7.683 | 7.500 | 7.321 |
| **10** | 8.877 | 8.666 | 8.460 | 8.258 | 8.062 | 7.870 | 7.682 | 7.499 | 7.320 | 7.145 |
| **11** | 8.670 | 8.464 | 8.262 | 8.065 | 7.873 | 7.685 | 7.502 | 7.323 | 7.148 | 6.977 |
| **12** | 8.471 | 8.269 | 8.072 | 7.880 | 7.691 | 7.508 | 7.328 | 7.153 | 6.982 | 6.815 |
| **13** | 8.280 | 8.083 | 7.890 | 7.701 | 7.517 | 7.337 | 7.162 | 6.990 | 6.823 | 6.659 |
| **14** | 8.097 | 7.903 | 7.714 | 7.530 | 7.349 | 7.173 | 7.002 | 6.834 | 6.670 | 6.510 |
| **15** | 7.920 | 7.730 | 7.545 | 7.364 | 7.188 | 7.016 | 6.847 | 6.683 | 6.522 | 6.365 |
| **16** | 7.750 | 7.564 | 7.383 | 7.205 | 7.033 | 6.864 | 6.699 | 6.538 | 6.380 | 6.227 |
| **17** | 7.586 | 7.404 | 7.226 | 7.052 | 6.883 | 6.717 | 6.555 | 6.397 | 6.243 | 6.093 |
| **18** | 7.428 | 7.249 | 7.075 | 6.905 | 6.738 | 6.576 | 6.417 | 6.262 | 6.111 | 5.963 |
| **19** | 7.275 | 7.100 | 6.929 | 6.762 | 6.599 | 6.439 | 6.284 | 6.132 | 5.983 | 5.838 |
| **20** | 7.128 | 6.956 | 6.788 | 6.624 | 6.464 | 6.308 | 6.155 | 6.006 | 5.860 | 5.718 |
| **21** | 6.986 | 6.817 | 6.653 | 6.491 | 6.334 | 6.181 | 6.031 | 5.884 | 5.741 | 5.601 |
| **22** | 6.849 | 6.683 | 6.521 | 6.363 | 6.208 | 6.057 | 5.910 | 5.766 | 5.626 | 5.488 |
| **23** | 6.716 | 6.553 | 6.394 | 6.238 | 6.087 | 5.938 | 5.794 | 5.652 | 5.514 | 5.379 |
| **24** | 6.588 | 6.427 | 6.271 | 6.118 | 5.969 | 5.823 | 5.681 | 5.542 | 5.406 | 5.273 |
| **25** | 6.463 | 6.306 | 6.152 | 6.001 | 5.854 | 5.711 | 5.571 | 5.434 | 5.301 | 5.170 |
| **26** | 6.342 | 6.187 | 6.036 | 5.888 | 5.744 | 5.603 | 5.465 | 5.330 | 5.199 | 5.071 |
| **27** | 6.225 | 6.073 | 5.924 | 5.778 | 5.636 | 5.497 | 5.362 | 5.229 | 5.100 | 4.974 |
| **28** | 6.112 | 5.962 | 5.815 | 5.671 | 5.532 | 5.395 | 5.261 | 5.131 | 5.004 | 4.879 |
| **29** | 6.001 | 5.853 | 5.709 | 5.568 | 5.430 | 5.295 | 5.164 | 5.035 | 4.910 | 4.788 |
| **30** | 5.894 | 5.748 | 5.606 | 5.467 | 5.331 | 5.198 | 5.069 | 4.942 | 4.819 | 4.698 |
| **31** | 5.789 | 5.646 | 5.505 | 5.368 | 5.235 | 5.104 | 4.976 | 4.852 | 4.730 | 4.611 |
| **32** | 5.688 | 5.546 | 5.408 | 5.273 | 5.141 | 5.012 | 4.886 | 4.763 | 4.643 | 4.526 |
| **33** | 5.588 | 5.449 | 5.312 | 5.179 | 5.049 | 4.922 | 4.798 | 4.677 | 4.558 | 4.443 |
| **34** | 5.492 | 5.354 | 5.219 | 5.088 | 4.959 | 4.834 | 4.712 | 4.592 | 4.475 | 4.361 |
| **35** | 5.397 | 5.261 | 5.128 | 4.998 | 4.872 | 4.748 | 4.627 | 4.509 | 4.394 | 4.282 |
| **36** | 5.304 | 5.170 | 5.039 | 4.911 | 4.786 | 4.664 | 4.545 | 4.428 | 4.315 | 4.204 |
| **37** | 5.214 | 5.081 | 4.952 | 4.825 | 4.702 | 4.582 | 4.464 | 4.349 | 4.237 | 4.127 |
| **38** | 5.125 | 4.994 | 4.866 | 4.742 | 4.620 | 4.501 | 4.384 | 4.271 | 4.160 | 4.052 |
| **39** | 5.038 | 4.909 | 4.783 | 4.659 | 4.539 | 4.421 | 4.306 | 4.194 | 4.084 | 3.977 |
| **40** | 4.953 | 4.825 | 4.700 | 4.578 | 4.459 | 4.343 | 4.229 | 4.119 | 4.010 | 3.905 |

*Source*: Based on Eq. (1.7).

**Table 1.27** The Air Solubility of Oxygen in mg/L as Functions of Temperature and Elevation in Feet for 0–4500 ft

| Temperature (°C) | Elevation Above Sea Level (ft) | | | | | | | | | |
|---|---|---|---|---|---|---|---|---|---|---|
| | 0 | 500 | 1000 | 1500 | 2000 | 2500 | 3000 | 3500 | 4000 | 4500 |
| 0 | 14.621 | 14.358 | 14.099 | 13.845 | 13.596 | 13.351 | 13.111 | 12.875 | 12.643 | 12.415 |
| 1 | 14.216 | 13.960 | 13.708 | 13.461 | 13.219 | 12.981 | 12.747 | 12.517 | 12.292 | 12.070 |
| 2 | 13.829 | 13.580 | 13.335 | 13.095 | 12.859 | 12.627 | 12.400 | 12.176 | 11.956 | 11.741 |
| 3 | 13.460 | 13.217 | 12.979 | 12.745 | 12.515 | 12.289 | 12.068 | 11.850 | 11.636 | 11.426 |
| 4 | 13.107 | 12.871 | 12.639 | 12.411 | 12.187 | 11.967 | 11.751 | 11.539 | 11.330 | 11.126 |
| 5 | 12.770 | 12.539 | 12.313 | 12.091 | 11.872 | 11.658 | 11.448 | 11.241 | 11.038 | 10.838 |
| 6 | 12.447 | 12.222 | 12.002 | 11.785 | 11.572 | 11.363 | 11.157 | 10.956 | 10.758 | 10.563 |
| 7 | 12.138 | 11.919 | 11.704 | 11.492 | 11.284 | 11.080 | 10.880 | 10.683 | 10.490 | 10.300 |
| 8 | 11.843 | 11.628 | 11.418 | 11.212 | 11.009 | 10.809 | 10.614 | 10.422 | 10.233 | 10.048 |
| 9 | 11.559 | 11.350 | 11.145 | 10.943 | 10.745 | 10.550 | 10.359 | 10.171 | 9.987 | 9.806 |
| 10 | 11.288 | 11.083 | 10.882 | 10.685 | 10.491 | 10.301 | 10.114 | 9.931 | 9.751 | 9.574 |
| 11 | 11.027 | 10.827 | 10.631 | 10.438 | 10.249 | 10.063 | 9.880 | 9.701 | 9.524 | 9.351 |
| 12 | 10.777 | 10.581 | 10.389 | 10.201 | 10.015 | 9.833 | 9.655 | 9.479 | 9.307 | 9.138 |
| 13 | 10.536 | 10.345 | 10.157 | 9.973 | 9.791 | 9.613 | 9.438 | 9.267 | 9.098 | 8.932 |
| 14 | 10.306 | 10.118 | 9.934 | 9.754 | 9.576 | 9.402 | 9.231 | 9.062 | 8.897 | 8.735 |
| 15 | 10.084 | 9.900 | 9.720 | 9.543 | 9.369 | 9.198 | 9.031 | 8.866 | 8.704 | 8.545 |
| 16 | 9.870 | 9.690 | 9.514 | 9.340 | 9.170 | 9.003 | 8.838 | 8.677 | 8.518 | 8.363 |
| 17 | 9.665 | 9.488 | 9.315 | 9.145 | 8.978 | 8.814 | 8.653 | 8.495 | 8.339 | 8.187 |
| 18 | 9.467 | 9.294 | 9.124 | 8.957 | 8.793 | 8.632 | 8.474 | 8.319 | 8.167 | 8.017 |
| 19 | 9.276 | 9.106 | 8.940 | 8.776 | 8.615 | 8.457 | 8.302 | 8.150 | 8.001 | 7.854 |
| 20 | 9.092 | 8.925 | 8.762 | 8.601 | 8.443 | 8.289 | 8.136 | 7.987 | 7.840 | 7.696 |
| 21 | 8.914 | 8.751 | 8.590 | 8.433 | 8.278 | 8.126 | 7.976 | 7.829 | 7.685 | 7.544 |
| 22 | 8.743 | 8.582 | 8.425 | 8.270 | 8.118 | 7.968 | 7.821 | 7.677 | 7.536 | 7.397 |
| 23 | 8.578 | 8.420 | 8.265 | 8.112 | 7.963 | 7.816 | 7.672 | 7.530 | 7.391 | 7.254 |
| 24 | 8.418 | 8.262 | 8.110 | 7.960 | 7.813 | 7.669 | 7.527 | 7.388 | 7.251 | 7.117 |
| 25 | 8.263 | 8.110 | 7.960 | 7.813 | 7.668 | 7.526 | 7.387 | 7.250 | 7.116 | 6.983 |
| 26 | 8.113 | 7.963 | 7.815 | 7.671 | 7.528 | 7.389 | 7.251 | 7.117 | 6.984 | 6.854 |
| 27 | 7.968 | 7.820 | 7.675 | 7.532 | 7.392 | 7.255 | 7.120 | 6.987 | 6.857 | 6.729 |
| 28 | 7.827 | 7.682 | 7.539 | 7.399 | 7.261 | 7.125 | 6.992 | 6.862 | 6.734 | 6.608 |
| 29 | 7.691 | 7.548 | 7.407 | 7.269 | 7.133 | 7.000 | 6.869 | 6.740 | 6.614 | 6.490 |
| 30 | 7.559 | 7.417 | 7.279 | 7.143 | 7.009 | 6.877 | 6.748 | 6.622 | 6.497 | 6.375 |
| 31 | 7.430 | 7.291 | 7.154 | 7.020 | 6.888 | 6.759 | 6.632 | 6.507 | 6.384 | 6.264 |
| 32 | 7.305 | 7.168 | 7.033 | 6.901 | 6.771 | 6.643 | 6.518 | 6.395 | 6.274 | 6.155 |
| 33 | 7.183 | 7.048 | 6.915 | 6.785 | 6.657 | 6.531 | 6.407 | 6.286 | 6.167 | 6.049 |
| 34 | 7.065 | 6.931 | 6.800 | 6.672 | 6.545 | 6.421 | 6.299 | 6.179 | 6.062 | 5.946 |
| 35 | 6.949 | 6.818 | 6.688 | 6.561 | 6.437 | 6.314 | 6.194 | 6.076 | 5.960 | 5.846 |
| 36 | 6.837 | 6.707 | 6.579 | 6.454 | 6.331 | 6.210 | 6.091 | 5.974 | 5.860 | 5.747 |
| 37 | 6.727 | 6.599 | 6.473 | 6.349 | 6.227 | 6.108 | 5.991 | 5.875 | 5.762 | 5.651 |
| 38 | 6.620 | 6.493 | 6.368 | 6.246 | 6.126 | 6.008 | 5.892 | 5.779 | 5.667 | 5.557 |
| 39 | 6.515 | 6.390 | 6.267 | 6.146 | 6.027 | 5.911 | 5.796 | 5.684 | 5.573 | 5.465 |
| 40 | 6.412 | 6.288 | 6.167 | 6.047 | 5.930 | 5.815 | 5.702 | 5.591 | 5.482 | 5.375 |

*Source*: Based on Eq. (1.8).

**Table 1.28** The Air Solubility of Oxygen in mg/L as Functions of Temperature and Elevation in Feet for 4500−9500 ft

| Temperature (°C) | Elevation Above Sea Level (ft) | | | | | | | | | |
|---|---|---|---|---|---|---|---|---|---|---|
| | **5000** | **5500** | **6000** | **6500** | **7000** | **7500** | **8000** | **8500** | **9000** | **9500** |
| **0** | 12.191 | 11.972 | 11.756 | 11.544 | 11.336 | 11.131 | 10.931 | 10.733 | 10.540 | 10.350 |
| **1** | 11.852 | 11.639 | 11.429 | 11.223 | 11.020 | 10.821 | 10.626 | 10.434 | 10.246 | 10.061 |
| **2** | 11.529 | 11.321 | 11.117 | 10.916 | 10.719 | 10.525 | 10.335 | 10.149 | 9.965 | 9.785 |
| **3** | 11.220 | 11.017 | 10.818 | 10.623 | 10.431 | 10.243 | 10.058 | 9.876 | 9.697 | 9.522 |
| **4** | 10.925 | 10.727 | 10.533 | 10.343 | 10.156 | 9.972 | 9.792 | 9.615 | 9.441 | 9.270 |
| **5** | 10.642 | 10.450 | 10.261 | 10.075 | 9.893 | 9.714 | 9.538 | 9.366 | 9.196 | 9.029 |
| **6** | 10.372 | 10.184 | 10.000 | 9.819 | 9.641 | 9.467 | 9.295 | 9.127 | 8.961 | 8.799 |
| **7** | 10.113 | 9.930 | 9.750 | 9.574 | 9.400 | 9.230 | 9.063 | 8.898 | 8.737 | 8.578 |
| **8** | 9.865 | 9.687 | 9.511 | 9.339 | 9.169 | 9.003 | 8.840 | 8.679 | 8.522 | 8.367 |
| **9** | 9.628 | 9.453 | 9.282 | 9.113 | 8.948 | 8.785 | 8.626 | 8.469 | 8.315 | 8.164 |
| **10** | 9.400 | 9.229 | 9.062 | 8.897 | 8.735 | 8.577 | 8.421 | 8.268 | 8.117 | 7.969 |
| **11** | 9.181 | 9.015 | 8.851 | 8.690 | 8.532 | 8.376 | 8.224 | 8.074 | 7.927 | 7.783 |
| **12** | 8.972 | 8.808 | 8.648 | 8.490 | 8.336 | 8.184 | 8.035 | 7.888 | 7.744 | 7.603 |
| **13** | 8.770 | 8.610 | 8.453 | 8.299 | 8.148 | 7.999 | 7.853 | 7.710 | 7.569 | 7.431 |
| **14** | 8.576 | 8.419 | 8.266 | 8.115 | 7.967 | 7.821 | 7.678 | 7.538 | 7.400 | 7.265 |
| **15** | 8.389 | 8.236 | 8.086 | 7.938 | 7.793 | 7.650 | 7.510 | 7.373 | 7.238 | 7.105 |
| **16** | 8.210 | 8.060 | 7.912 | 7.767 | 7.625 | 7.485 | 7.348 | 7.214 | 7.081 | 6.951 |
| **17** | 8.037 | 7.890 | 7.745 | 7.603 | 7.464 | 7.327 | 7.192 | 7.060 | 6.930 | 6.803 |
| **18** | 7.870 | 7.726 | 7.584 | 7.445 | 7.308 | 7.174 | 7.042 | 6.912 | 6.785 | 6.660 |
| **19** | 7.710 | 7.568 | 7.429 | 7.292 | 7.158 | 7.026 | 6.897 | 6.770 | 6.645 | 6.522 |
| **20** | 7.554 | 7.415 | 7.279 | 7.145 | 7.013 | 6.884 | 6.757 | 6.632 | 6.509 | 6.389 |
| **21** | 7.405 | 7.268 | 7.134 | 7.002 | 6.873 | 6.746 | 6.621 | 6.499 | 6.378 | 6.260 |
| **22** | 7.260 | 7.126 | 6.994 | 6.865 | 6.738 | 6.613 | 6.490 | 6.370 | 6.252 | 6.136 |
| **23** | 7.120 | 6.988 | 6.859 | 6.732 | 6.607 | 6.484 | 6.364 | 6.246 | 6.129 | 6.015 |
| **24** | 6.985 | 6.855 | 6.728 | 6.603 | 6.480 | 6.360 | 6.241 | 6.125 | 6.011 | 5.899 |
| **25** | 6.854 | 6.726 | 6.601 | 6.478 | 6.357 | 6.239 | 6.122 | 6.008 | 5.896 | 5.786 |
| **26** | 6.727 | 6.601 | 6.478 | 6.357 | 6.238 | 6.122 | 6.007 | 5.895 | 5.784 | 5.676 |
| **27** | 6.604 | 6.480 | 6.359 | 6.240 | 6.123 | 6.008 | 5.896 | 5.785 | 5.676 | 5.569 |
| **28** | 6.484 | 6.363 | 6.243 | 6.126 | 6.011 | 5.898 | 5.787 | 5.678 | 5.571 | 5.466 |
| **29** | 6.368 | 6.248 | 6.131 | 6.015 | 5.902 | 5.791 | 5.681 | 5.574 | 5.469 | 5.365 |
| **30** | 6.255 | 6.137 | 6.022 | 5.908 | 5.796 | 5.687 | 5.579 | 5.473 | 5.369 | 5.267 |
| **31** | 6.145 | 6.029 | 5.915 | 5.803 | 5.693 | 5.585 | 5.479 | 5.375 | 5.272 | 5.172 |
| **32** | 6.039 | 5.924 | 5.812 | 5.701 | 5.593 | 5.486 | 5.381 | 5.279 | 5.178 | 5.079 |
| **33** | 5.934 | 5.822 | 5.711 | 5.602 | 5.495 | 5.390 | 5.286 | 5.185 | 5.086 | 4.988 |
| **34** | 5.833 | 5.721 | 5.612 | 5.505 | 5.399 | 5.295 | 5.194 | 5.094 | 4.996 | 4.899 |
| **35** | 5.734 | 5.624 | 5.516 | 5.410 | 5.306 | 5.203 | 5.103 | 5.004 | 4.907 | 4.812 |
| **36** | 5.637 | 5.528 | 5.422 | 5.317 | 5.214 | 5.113 | 5.014 | 4.917 | 4.821 | 4.727 |
| **37** | 5.542 | 5.435 | 5.330 | 5.226 | 5.125 | 5.025 | 4.927 | 4.831 | 4.737 | 4.644 |
| **38** | 5.449 | 5.344 | 5.240 | 5.138 | 5.037 | 4.939 | 4.842 | 4.747 | 4.654 | 4.562 |
| **39** | 5.359 | 5.254 | 5.151 | 5.050 | 4.951 | 4.854 | 4.759 | 4.665 | 4.573 | 4.482 |
| **40** | 5.269 | 5.166 | 5.065 | 4.965 | 4.867 | 4.771 | 4.676 | 4.584 | 4.493 | 4.403 |

*Source*: Based on Eq. (1.7).

**Table 1.29** The Hydrostatic Pressure of Water in mmHg/m as a Function of Temperature

| Temperature (°C) | Δt (°C) | | | | | | | | | |
|---|---|---|---|---|---|---|---|---|---|---|
| | **0.0** | **0.1** | **0.2** | **0.3** | **0.4** | **0.5** | **0.6** | **0.7** | **0.8** | **0.9** |
| **0** | 73.544 | 73.545 | 73.545 | 73.546 | 73.546 | 73.547 | 73.547 | 73.548 | 73.548 | 73.548 |
| **1** | 73.549 | 73.549 | 73.549 | 73.550 | 73.550 | 73.550 | 73.551 | 73.551 | 73.551 | 73.551 |
| **2** | 73.552 | 73.552 | 73.552 | 73.552 | 73.553 | 73.553 | 73.553 | 73.553 | 73.553 | 73.553 |
| **3** | 73.554 | 73.554 | 73.554 | 73.554 | 73.554 | 73.554 | 73.554 | 73.554 | 73.554 | 73.554 |
| **4** | 73.554 | 73.554 | 73.554 | 73.554 | 73.554 | 73.554 | 73.554 | 73.554 | 73.554 | 73.554 |
| **5** | 73.553 | 73.553 | 73.553 | 73.553 | 73.553 | 73.553 | 73.553 | 73.552 | 73.552 | 73.552 |
| **6** | 73.552 | 73.551 | 73.551 | 73.551 | 73.551 | 73.550 | 73.550 | 73.550 | 73.550 | 73.549 |
| **7** | 73.549 | 73.549 | 73.548 | 73.548 | 73.547 | 73.547 | 73.547 | 73.546 | 73.546 | 73.545 |
| **8** | 73.545 | 73.545 | 73.544 | 73.544 | 73.543 | 73.543 | 73.542 | 73.542 | 73.541 | 73.541 |
| **9** | 73.540 | 73.539 | 73.539 | 73.538 | 73.538 | 73.537 | 73.537 | 73.536 | 73.535 | 73.535 |
| **10** | 73.534 | 73.533 | 73.533 | 73.532 | 73.531 | 73.531 | 73.530 | 73.529 | 73.529 | 73.528 |
| **11** | 73.527 | 73.526 | 73.526 | 73.525 | 73.524 | 73.523 | 73.522 | 73.522 | 73.521 | 73.520 |
| **12** | 73.519 | 73.518 | 73.517 | 73.517 | 73.516 | 73.515 | 73.514 | 73.513 | 73.512 | 73.511 |
| **13** | 73.510 | 73.509 | 73.508 | 73.507 | 73.506 | 73.505 | 73.505 | 73.504 | 73.503 | 73.502 |
| **14** | 73.500 | 73.499 | 73.498 | 73.497 | 73.496 | 73.495 | 73.494 | 73.493 | 73.492 | 73.491 |
| **15** | 73.490 | 73.489 | 73.488 | 73.486 | 73.485 | 73.484 | 73.483 | 73.482 | 73.481 | 73.480 |
| **16** | 73.478 | 73.477 | 73.476 | 73.475 | 73.473 | 73.472 | 73.471 | 73.470 | 73.468 | 73.467 |
| **17** | 73.466 | 73.465 | 73.463 | 73.462 | 73.461 | 73.459 | 73.458 | 73.457 | 73.455 | 73.454 |
| **18** | 73.453 | 73.451 | 73.450 | 73.449 | 73.447 | 73.446 | 73.444 | 73.443 | 73.442 | 73.440 |
| **19** | 73.439 | 73.437 | 73.436 | 73.434 | 73.433 | 73.431 | 73.430 | 73.429 | 73.427 | 73.425 |
| **20** | 73.424 | 73.422 | 73.421 | 73.419 | 73.418 | 73.416 | 73.415 | 73.413 | 73.412 | 73.410 |
| **21** | 73.408 | 73.407 | 73.405 | 73.404 | 73.402 | 73.400 | 73.399 | 73.397 | 73.395 | 73.394 |
| **22** | 73.392 | 73.390 | 73.389 | 73.387 | 73.385 | 73.384 | 73.382 | 73.380 | 73.379 | 73.377 |
| **23** | 73.375 | 73.373 | 73.372 | 73.370 | 73.368 | 73.366 | 73.364 | 73.363 | 73.361 | 73.359 |
| **24** | 73.357 | 73.355 | 73.354 | 73.352 | 73.350 | 73.348 | 73.346 | 73.344 | 73.343 | 73.341 |
| **25** | 73.339 | 73.337 | 73.335 | 73.333 | 73.331 | 73.329 | 73.327 | 73.325 | 73.323 | 73.322 |
| **26** | 73.320 | 73.318 | 73.316 | 73.314 | 73.312 | 73.310 | 73.308 | 73.306 | 73.304 | 73.302 |
| **27** | 73.300 | 73.298 | 73.296 | 73.294 | 73.292 | 73.290 | 73.287 | 73.285 | 73.283 | 73.281 |
| **28** | 73.279 | 73.277 | 73.275 | 73.273 | 73.271 | 73.269 | 73.266 | 73.264 | 73.262 | 73.260 |
| **29** | 73.258 | 73.256 | 73.254 | 73.251 | 73.249 | 73.247 | 73.245 | 73.243 | 73.240 | 73.238 |
| **30** | 73.236 | 73.234 | 73.232 | 73.229 | 73.227 | 73.225 | 73.223 | 73.220 | 73.218 | 73.216 |
| **31** | 73.214 | 73.211 | 73.209 | 73.207 | 73.204 | 73.202 | 73.200 | 73.197 | 73.195 | 73.193 |
| **32** | 73.190 | 73.188 | 73.186 | 73.183 | 73.181 | 73.179 | 73.176 | 73.174 | 73.171 | 73.169 |
| **33** | 73.167 | 73.164 | 73.162 | 73.159 | 73.157 | 73.154 | 73.152 | 73.150 | 73.147 | 73.145 |
| **34** | 73.142 | 73.140 | 73.137 | 73.135 | 73.132 | 73.130 | 73.127 | 73.125 | 73.122 | 73.120 |
| **35** | 73.117 | 73.115 | 73.112 | 73.110 | 73.107 | 73.105 | 73.102 | 73.099 | 73.097 | 73.094 |
| **36** | 73.092 | 73.089 | 73.086 | 73.084 | 73.081 | 73.079 | 73.076 | 73.073 | 73.071 | 73.068 |
| **37** | 73.066 | 73.063 | 73.060 | 73.058 | 73.055 | 73.052 | 73.050 | 73.047 | 73.044 | 73.041 |
| **38** | 73.039 | 73.036 | 73.033 | 73.031 | 73.028 | 73.025 | 73.022 | 73.020 | 73.017 | 73.014 |
| **39** | 73.012 | 73.009 | 73.006 | 73.003 | 73.000 | 72.998 | 72.995 | 72.992 | 72.989 | 72.986 |
| **40** | 72.984 | 72.981 | 72.978 | 72.975 | 72.972 | 72.970 | 72.967 | 72.964 | 72.961 | 72.958 |

*Source*: Based on Millero and Poisson (1981).

**Table 1.30** The Hydrostatic Pressure of Water in kPa/m as a Function of Temperature

| **Temperature (°C)** | $\Delta t$ (°C) | | | | | | | | | |
|---|---|---|---|---|---|---|---|---|---|---|
| | **0.0** | **0.1** | **0.2** | **0.3** | **0.4** | **0.5** | **0.6** | **0.7** | **0.8** | **0.9** |
| **0** | 9.8051 | 9.8052 | 9.8052 | 9.8053 | 9.8054 | 9.8054 | 9.8055 | 9.8055 | 9.8056 | 9.8056 |
| **1** | 9.8057 | 9.8057 | 9.8058 | 9.8058 | 9.8059 | 9.8059 | 9.8059 | 9.8060 | 9.8060 | 9.8061 |
| **2** | 9.8061 | 9.8061 | 9.8062 | 9.8062 | 9.8062 | 9.8062 | 9.8063 | 9.8063 | 9.8063 | 9.8063 |
| **3** | 9.8063 | 9.8063 | 9.8064 | 9.8064 | 9.8064 | 9.8064 | 9.8064 | 9.8064 | 9.8064 | 9.8064 |
| **4** | 9.8064 | 9.8064 | 9.8064 | 9.8064 | 9.8064 | 9.8064 | 9.8064 | 9.8064 | 9.8064 | 9.8063 |
| **5** | 9.8063 | 9.8063 | 9.8063 | 9.8063 | 9.8062 | 9.8062 | 9.8062 | 9.8062 | 9.8061 | 9.8061 |
| **6** | 9.8061 | 9.8061 | 9.8060 | 9.8060 | 9.8060 | 9.8059 | 9.8059 | 9.8058 | 9.8058 | 9.8058 |
| **7** | 9.8057 | 9.8057 | 9.8056 | 9.8056 | 9.8055 | 9.8055 | 9.8054 | 9.8054 | 9.8053 | 9.8052 |
| **8** | 9.8052 | 9.8051 | 9.8051 | 9.8050 | 9.8049 | 9.8049 | 9.8048 | 9.8047 | 9.8047 | 9.8046 |
| **9** | 9.8045 | 9.8045 | 9.8044 | 9.8043 | 9.8042 | 9.8041 | 9.8041 | 9.8040 | 9.8039 | 9.8038 |
| **10** | 9.8037 | 9.8036 | 9.8036 | 9.8035 | 9.8034 | 9.8033 | 9.8032 | 9.8031 | 9.8030 | 9.8029 |
| **11** | 9.8028 | 9.8027 | 9.8026 | 9.8025 | 9.8024 | 9.8023 | 9.8022 | 9.8021 | 9.8020 | 9.8019 |
| **12** | 9.8017 | 9.8016 | 9.8015 | 9.8014 | 9.8013 | 9.8012 | 9.8010 | 9.8009 | 9.8008 | 9.8007 |
| **13** | 9.8006 | 9.8004 | 9.8003 | 9.8002 | 9.8001 | 9.7999 | 9.7998 | 9.7997 | 9.7995 | 9.7994 |
| **14** | 9.7993 | 9.7991 | 9.7990 | 9.7988 | 9.7987 | 9.7986 | 9.7984 | 9.7983 | 9.7981 | 9.7980 |
| **15** | 9.7978 | 9.7977 | 9.7975 | 9.7974 | 9.7972 | 9.7971 | 9.7969 | 9.7968 | 9.7966 | 9.7965 |
| **16** | 9.7963 | 9.7961 | 9.7960 | 9.7958 | 9.7957 | 9.7955 | 9.7953 | 9.7952 | 9.7950 | 9.7948 |
| **17** | 9.7947 | 9.7945 | 9.7943 | 9.7941 | 9.7940 | 9.7938 | 9.7936 | 9.7934 | 9.7933 | 9.7931 |
| **18** | 9.7929 | 9.7927 | 9.7925 | 9.7923 | 9.7922 | 9.7920 | 9.7918 | 9.7916 | 9.7914 | 9.7912 |
| **19** | 9.7910 | 9.7908 | 9.7906 | 9.7905 | 9.7903 | 9.7901 | 9.7899 | 9.7897 | 9.7895 | 9.7893 |
| **20** | 9.7891 | 9.7889 | 9.7887 | 9.7884 | 9.7882 | 9.7880 | 9.7878 | 9.7876 | 9.7874 | 9.7872 |
| **21** | 9.7870 | 9.7868 | 9.7866 | 9.7863 | 9.7861 | 9.7859 | 9.7857 | 9.7855 | 9.7853 | 9.7850 |
| **22** | 9.7848 | 9.7846 | 9.7844 | 9.7841 | 9.7839 | 9.7837 | 9.7835 | 9.7832 | 9.7830 | 9.7828 |
| **23** | 9.7825 | 9.7823 | 9.7821 | 9.7818 | 9.7816 | 9.7814 | 9.7811 | 9.7809 | 9.7806 | 9.7804 |
| **24** | 9.7802 | 9.7799 | 9.7797 | 9.7794 | 9.7792 | 9.7789 | 9.7787 | 9.7785 | 9.7782 | 9.7780 |
| **25** | 9.7777 | 9.7774 | 9.7772 | 9.7769 | 9.7767 | 9.7764 | 9.7762 | 9.7759 | 9.7757 | 9.7754 |
| **26** | 9.7751 | 9.7749 | 9.7746 | 9.7744 | 9.7741 | 9.7738 | 9.7736 | 9.7733 | 9.7730 | 9.7728 |
| **27** | 9.7725 | 9.7722 | 9.7719 | 9.7717 | 9.7714 | 9.7711 | 9.7709 | 9.7706 | 9.7703 | 9.7700 |
| **28** | 9.7697 | 9.7695 | 9.7692 | 9.7689 | 9.7686 | 9.7683 | 9.7681 | 9.7678 | 9.7675 | 9.7672 |
| **29** | 9.7669 | 9.7666 | 9.7663 | 9.7661 | 9.7658 | 9.7655 | 9.7652 | 9.7649 | 9.7646 | 9.7643 |
| **30** | 9.7640 | 9.7637 | 9.7634 | 9.7631 | 9.7628 | 9.7625 | 9.7622 | 9.7619 | 9.7616 | 9.7613 |
| **31** | 9.7610 | 9.7607 | 9.7604 | 9.7601 | 9.7598 | 9.7595 | 9.7592 | 9.7588 | 9.7585 | 9.7582 |
| **32** | 9.7579 | 9.7576 | 9.7573 | 9.7570 | 9.7567 | 9.7563 | 9.7560 | 9.7557 | 9.7554 | 9.7551 |
| **33** | 9.7547 | 9.7544 | 9.7541 | 9.7538 | 9.7535 | 9.7531 | 9.7528 | 9.7525 | 9.7521 | 9.7518 |
| **34** | 9.7515 | 9.7512 | 9.7508 | 9.7505 | 9.7502 | 9.7498 | 9.7495 | 9.7492 | 9.7488 | 9.7485 |
| **35** | 9.7482 | 9.7478 | 9.7475 | 9.7471 | 9.7468 | 9.7465 | 9.7461 | 9.7458 | 9.7454 | 9.7451 |
| **36** | 9.7448 | 9.7444 | 9.7441 | 9.7437 | 9.7434 | 9.7430 | 9.7427 | 9.7423 | 9.7420 | 9.7416 |
| **37** | 9.7413 | 9.7409 | 9.7406 | 9.7402 | 9.7399 | 9.7395 | 9.7391 | 9.7388 | 9.7384 | 9.7381 |
| **38** | 9.7377 | 9.7373 | 9.7370 | 9.7366 | 9.7363 | 9.7359 | 9.7355 | 9.7352 | 9.7348 | 9.7344 |
| **39** | 9.7341 | 9.7337 | 9.7333 | 9.7330 | 9.7326 | 9.7322 | 9.7319 | 9.7315 | 9.7311 | 9.7307 |
| **40** | 9.7304 | 9.7300 | 9.7296 | 9.7292 | 9.7289 | 9.7285 | 9.7281 | 9.7277 | 9.7273 | 9.7270 |

*Source*: Based on Millero and Poisson (1981).

**Table 1.31** Air Solubility in mg/L as a Function of Depth (Moist air, temperature = 20.0, salinity = 0.0 g/kg, barometric pressure = 760 mmHg)

| Depth (m) | Gases (mg/L) | | | | |
|---|---|---|---|---|---|
| | $O_2$ | $N_2$ | Ar | $CO_2$ | Total |
| **0** | 9.092 | 15.029 | 0.556 | 0.653 | 25.330 |
| **1** | 9.991 | 16.515 | 0.611 | 0.718 | 27.835 |
| **2** | 10.890 | 18.001 | 0.666 | 0.782 | 30.340 |
| **3** | 11.789 | 19.488 | 0.721 | 0.847 | 32.845 |
| **4** | 12.688 | 20.974 | 0.776 | 0.912 | 35.350 |
| **5** | 13.588 | 22.460 | 0.831 | 0.976 | 37.855 |
| **6** | 14.487 | 23.946 | 0.886 | 1.041 | 40.360 |
| **7** | 15.386 | 25.433 | 0.941 | 1.105 | 42.865 |
| **8** | 16.285 | 26.919 | 0.996 | 1.170 | 45.370 |
| **9** | 17.184 | 28.405 | 1.051 | 1.235 | 47.875 |
| **10** | 18.083 | 29.891 | 1.106 | 1.299 | 50.380 |
| **11** | 18.982 | 31.378 | 1.161 | 1.364 | 52.885 |
| **12** | 19.881 | 32.864 | 1.216 | 1.428 | 55.390 |
| **13** | 20.781 | 34.350 | 1.271 | 1.493 | 57.895 |
| **14** | 21.680 | 35.836 | 1.326 | 1.558 | 60.400 |
| **15** | 22.579 | 37.323 | 1.381 | 1.622 | 62.905 |
| **16** | 23.478 | 38.809 | 1.436 | 1.687 | 65.410 |
| **17** | 24.377 | 40.295 | 1.491 | 1.752 | 67.915 |
| **18** | 25.276 | 41.781 | 1.546 | 1.816 | 70.420 |
| **19** | 26.175 | 43.268 | 1.601 | 1.881 | 72.925 |
| **20** | 27.074 | 44.754 | 1.656 | 1.945 | 75.430 |
| **21** | 27.973 | 46.240 | 1.711 | 2.010 | 77.935 |
| **22** | 28.873 | 47.726 | 1.766 | 2.075 | 80.440 |
| **23** | 29.772 | 49.213 | 1.821 | 2.139 | 82.945 |
| **24** | 30.671 | 50.699 | 1.876 | 2.204 | 85.450 |
| **25** | 31.570 | 52.185 | 1.931 | 2.268 | 87.955 |
| **26** | 32.469 | 53.671 | 1.986 | 2.333 | 90.460 |
| **27** | 33.368 | 55.158 | 2.041 | 2.398 | 92.965 |
| **28** | 34.267 | 56.644 | 2.096 | 2.462 | 95.470 |
| **29** | 35.166 | 58.130 | 2.151 | 2.527 | 97.975 |
| **30** | 36.066 | 59.616 | 2.206 | 2.591 | 100.480 |
| **31** | 36.965 | 61.103 | 2.261 | 2.656 | 102.985 |
| **32** | 37.864 | 62.589 | 2.316 | 2.721 | 105.490 |
| **33** | 38.763 | 64.075 | 2.371 | 2.785 | 107.995 |
| **34** | 39.662 | 65.561 | 2.426 | 2.850 | 110.500 |
| **35** | 40.561 | 67.048 | 2.481 | 2.914 | 113.005 |
| **36** | 41.460 | 68.534 | 2.536 | 2.979 | 115.510 |
| **37** | 42.359 | 70.020 | 2.591 | 3.044 | 118.015 |
| **38** | 43.259 | 71.506 | 2.646 | 3.108 | 120.520 |
| **39** | 44.158 | 72.993 | 2.701 | 3.173 | 123.025 |
| **40** | 45.057 | 74.479 | 2.756 | 3.237 | 125.530 |

**Table 1.32** Bunsen Coefficient ($\beta$) for Oxygen as a Function of Temperature (L real gas/(L atm))

| Temperature (°C) | Δt (°C) | | | | | | | | | |
|---|---|---|---|---|---|---|---|---|---|---|
| | **0.0** | **0.1** | **0.2** | **0.3** | **0.4** | **0.5** | **0.6** | **0.7** | **0.8** | **0.9** |
| **0** | .04914 | .04901 | .04887 | .04873 | .04860 | .04847 | .04833 | .04820 | .04807 | .04793 |
| **1** | .04780 | .04767 | .04754 | .04741 | .04728 | .04716 | .04703 | .04690 | .04678 | .04665 |
| **2** | .04653 | .04640 | .04628 | .04615 | .04603 | .04591 | .04579 | .04567 | .04555 | .04543 |
| **3** | .04531 | .04519 | .04507 | .04495 | .04484 | .04472 | .04460 | .04449 | .04437 | .04426 |
| **4** | .04414 | .04403 | .04392 | .04381 | .04369 | .04358 | .04347 | .04336 | .04325 | .04314 |
| **5** | .04303 | .04292 | .04282 | .04271 | .04260 | .04250 | .04239 | .04229 | .04218 | .04208 |
| **6** | .04197 | .04187 | .04177 | .04166 | .04156 | .04146 | .04136 | .04126 | .04116 | .04106 |
| **7** | .04096 | .04086 | .04076 | .04066 | .04056 | .04047 | .04037 | .04027 | .04018 | .04008 |
| **8** | .03999 | .03989 | .03980 | .03971 | .03961 | .03952 | .03943 | .03933 | .03924 | .03915 |
| **9** | .03906 | .03897 | .03888 | .03879 | .03870 | .03861 | .03852 | .03843 | .03835 | .03826 |
| **10** | .03817 | .03809 | .03800 | .03791 | .03783 | .03774 | .03766 | .03757 | .03749 | .03741 |
| **11** | .03732 | .03724 | .03716 | .03707 | .03699 | .03691 | .03683 | .03675 | .03667 | .03659 |
| **12** | .03651 | .03643 | .03635 | .03627 | .03619 | .03611 | .03604 | .03596 | .03588 | .03581 |
| **13** | .03573 | .03565 | .03558 | .03550 | .03543 | .03535 | .03528 | .03520 | .03513 | .03505 |
| **14** | .03498 | .03491 | .03484 | .03476 | .03469 | .03462 | .03455 | .03448 | .03441 | .03433 |
| **15** | .03426 | .03419 | .03412 | .03406 | .03399 | .03392 | .03385 | .03378 | .03371 | .03364 |
| **16** | .03358 | .03351 | .03344 | .03338 | .03331 | .03324 | .03318 | .03311 | .03305 | .03298 |
| **17** | .03292 | .03285 | .03279 | .03272 | .03266 | .03260 | .03253 | .03247 | .03241 | .03235 |
| **18** | .03228 | .03222 | .03216 | .03210 | .03204 | .03198 | .03192 | .03186 | .03180 | .03174 |
| **19** | .03168 | .03162 | .03156 | .03150 | .03144 | .03138 | .03132 | .03126 | .03121 | .03115 |
| **20** | .03109 | .03103 | .03098 | .03092 | .03086 | .03081 | .03075 | .03070 | .03064 | .03059 |
| **21** | .03053 | .03048 | .03042 | .03037 | .03031 | .03026 | .03020 | .03015 | .03010 | .03004 |
| **22** | .02999 | .02994 | .02989 | .02983 | .02978 | .02973 | .02968 | .02963 | .02958 | .02952 |
| **23** | .02947 | .02942 | .02937 | .02932 | .02927 | .02922 | .02917 | .02912 | .02907 | .02902 |
| **24** | .02897 | .02893 | .02888 | .02883 | .02878 | .02873 | .02868 | .02864 | .02859 | .02854 |
| **25** | .02850 | .02845 | .02840 | .02835 | .02831 | .02826 | .02822 | .02817 | .02812 | .02808 |
| **26** | .02803 | .02799 | .02794 | .02790 | .02785 | .02781 | .02777 | .02772 | .02768 | .02763 |
| **27** | .02759 | .02755 | .02750 | .02746 | .02742 | .02737 | .02733 | .02729 | .02725 | .02720 |
| **28** | .02716 | .02712 | .02708 | .02704 | .02700 | .02695 | .02691 | .02687 | .02683 | .02679 |
| **29** | .02675 | .02671 | .02667 | .02663 | .02659 | .02655 | .02651 | .02647 | .02643 | .02639 |
| **30** | .02635 | .02632 | .02628 | .02624 | .02620 | .02616 | .02612 | .02609 | .02605 | .02601 |
| **31** | .02597 | .02594 | .02590 | .02586 | .02582 | .02579 | .02575 | .02571 | .02568 | .02564 |
| **32** | .02561 | .02557 | .02553 | .02550 | .02546 | .02543 | .02539 | .02536 | .02532 | .02529 |
| **33** | .02525 | .02522 | .02518 | .02515 | .02511 | .02508 | .02504 | .02501 | .02498 | .02494 |
| **34** | .02491 | .02488 | .02484 | .02481 | .02478 | .02474 | .02471 | .02468 | .02465 | .02461 |
| **35** | .02458 | .02455 | .02452 | .02448 | .02445 | .02442 | .02439 | .02436 | .02433 | .02429 |
| **36** | .02426 | .02423 | .02420 | .02417 | .02414 | .02411 | .02408 | .02405 | .02402 | .02399 |
| **37** | .02396 | .02393 | .02390 | .02387 | .02384 | .02381 | .02378 | .02375 | .02372 | .02369 |
| **38** | .02366 | .02363 | .02360 | .02358 | .02355 | .02352 | .02349 | .02346 | .02343 | .02341 |
| **39** | .02338 | .02335 | .02332 | .02329 | .02327 | .02324 | .02321 | .02318 | .02316 | .02313 |
| **40** | .02310 | .02308 | .02305 | .02302 | .02300 | .02297 | .02294 | .02292 | .02289 | .02286 |

*Source*: Freshwater, based on Benson and Krause (1980).

**Table 1.33** Bunsen Coefficient ($\beta$) for Nitrogen as a Function of Temperature (L real gas/(L atm))

| Temperature (°C) | $\Delta t$ (°C) | | | | | | | | | |
|---|---|---|---|---|---|---|---|---|---|---|
| | **0.0** | **0.1** | **0.2** | **0.3** | **0.4** | **0.5** | **0.6** | **0.7** | **0.8** | **0.9** |
| **0** | .02397 | .02391 | .02384 | .02378 | .02372 | .02366 | .02360 | .02354 | .02348 | .02342 |
| **1** | .02336 | .02330 | .02324 | .02318 | .02312 | .02306 | .02300 | .02295 | .02289 | .02283 |
| **2** | .02277 | .02272 | .02266 | .02261 | .02255 | .02249 | .02244 | .02238 | .02233 | .02227 |
| **3** | .02222 | .02216 | .02211 | .02206 | .02200 | .02195 | .02190 | .02185 | .02179 | .02174 |
| **4** | .02169 | .02164 | .02159 | .02153 | .02148 | .02143 | .02138 | .02133 | .02128 | .02123 |
| **5** | .02118 | .02113 | .02108 | .02103 | .02099 | .02094 | .02089 | .02084 | .02079 | .02075 |
| **6** | .02070 | .02065 | .02060 | .02056 | .02051 | .02046 | .02042 | .02037 | .02033 | .02028 |
| **7** | .02024 | .02019 | .02015 | .02010 | .02006 | .02001 | .01997 | .01992 | .01988 | .01984 |
| **8** | .01979 | .01975 | .01971 | .01966 | .01962 | .01958 | .01954 | .01950 | .01945 | .01941 |
| **9** | .01937 | .01933 | .01929 | .01925 | .01921 | .01917 | .01913 | .01909 | .01905 | .01901 |
| **10** | .01897 | .01893 | .01889 | .01885 | .01881 | .01877 | .01873 | .01869 | .01866 | .01862 |
| **11** | .01858 | .01854 | .01850 | .01847 | .01843 | .01839 | .01836 | .01832 | .01828 | .01825 |
| **12** | .01821 | .01817 | .01814 | .01810 | .01807 | .01803 | .01799 | .01796 | .01792 | .01789 |
| **13** | .01785 | .01782 | .01779 | .01775 | .01772 | .01768 | .01765 | .01762 | .01758 | .01755 |
| **14** | .01751 | .01748 | .01745 | .01742 | .01738 | .01735 | .01732 | .01729 | .01725 | .01722 |
| **15** | .01719 | .01716 | .01713 | .01709 | .01706 | .01703 | .01700 | .01697 | .01694 | .01691 |
| **16** | .01688 | .01685 | .01682 | .01679 | .01676 | .01673 | .01670 | .01667 | .01664 | .01661 |
| **17** | .01658 | .01655 | .01652 | .01649 | .01646 | .01644 | .01641 | .01638 | .01635 | .01632 |
| **18** | .01629 | .01627 | .01624 | .01621 | .01618 | .01616 | .01613 | .01610 | .01607 | .01605 |
| **19** | .01602 | .01599 | .01597 | .01594 | .01591 | .01589 | .01586 | .01583 | .01581 | .01578 |
| **20** | .01576 | .01573 | .01571 | .01568 | .01565 | .01563 | .01560 | .01558 | .01555 | .01553 |
| **21** | .01550 | .01548 | .01546 | .01543 | .01541 | .01538 | .01536 | .01533 | .01531 | .01529 |
| **22** | .01526 | .01524 | .01522 | .01519 | .01517 | .01515 | .01512 | .01510 | .01508 | .01505 |
| **23** | .01503 | .01501 | .01499 | .01496 | .01494 | .01492 | .01490 | .01487 | .01485 | .01483 |
| **24** | .01481 | .01479 | .01476 | .01474 | .01472 | .01470 | .01468 | .01466 | .01464 | .01462 |
| **25** | .01459 | .01457 | .01455 | .01453 | .01451 | .01449 | .01447 | .01445 | .01443 | .01441 |
| **26** | .01439 | .01437 | .01435 | .01433 | .01431 | .01429 | .01427 | .01425 | .01423 | .01421 |
| **27** | .01419 | .01418 | .01416 | .01414 | .01412 | .01410 | .01408 | .01406 | .01404 | .01402 |
| **28** | .01401 | .01399 | .01397 | .01395 | .01393 | .01392 | .01390 | .01388 | .01386 | .01384 |
| **29** | .01383 | .01381 | .01379 | .01377 | .01376 | .01374 | .01372 | .01370 | .01369 | .01367 |
| **30** | .01365 | .01364 | .01362 | .01360 | .01359 | .01357 | .01355 | .01354 | .01352 | .01350 |
| **31** | 0.01349 | .01347 | .01346 | .01344 | .01342 | .01341 | .01339 | .01338 | .01336 | .01334 |
| **32** | 0.01333 | .01331 | .01330 | .01328 | .01327 | .01325 | .01324 | .01322 | .01321 | .01319 |
| **33** | 0.01318 | .01316 | .01315 | .01313 | .01312 | .01310 | .01309 | .01307 | .01306 | .01305 |
| **34** | 0.01303 | .01302 | .01300 | .01299 | .01297 | .01296 | .01295 | .01293 | .01292 | .01291 |
| **35** | 0.01289 | .01288 | .01286 | .01285 | .01284 | .01282 | .01281 | .01280 | .01278 | .01277 |
| **36** | 0.01276 | .01275 | .01273 | .01272 | .01271 | .01269 | .01268 | .01267 | .01266 | .01264 |
| **37** | 0.01263 | .01262 | .01261 | .01259 | .01258 | .01257 | .01256 | .01255 | .01253 | .01252 |
| **38** | 0.01251 | .01250 | .01249 | .01247 | .01246 | .01245 | .01244 | .01243 | .01242 | .01240 |
| **39** | 0.01239 | .01238 | .01237 | .01236 | .01235 | .01234 | .01233 | .01232 | .01230 | .01229 |
| **40** | 0.01228 | .01227 | .01226 | .01225 | .01224 | .01223 | .01222 | .01221 | .01220 | .01219 |

*Source*: Freshwater, based on Hamme and Emerson (2004).

**Table 1.34** Bunsen Coefficient ($\beta$) for Argon as a Function of Temperature (L real gas/(L atm))

| Temperature (°C) | $\Delta t$ (°C) | | | | | | | | | |
|---|---|---|---|---|---|---|---|---|---|---|
| | 0.0 | 0.1 | 0.2 | 0.3 | 0.4 | 0.5 | 0.6 | 0.7 | 0.8 | 0.9 |
| 0 | .05378 | .05363 | .05349 | .05334 | .05319 | .05305 | .05290 | .05276 | .05262 | .05247 |
| 1 | .05233 | .05219 | .05205 | .05191 | .05177 | .05163 | .05149 | .05135 | .05122 | .05108 |
| 2 | .05095 | .05081 | .05068 | .05054 | .05041 | .05028 | .05015 | .05001 | .04988 | .04975 |
| 3 | .04962 | .04950 | .04937 | .04924 | .04911 | .04899 | .04886 | .04874 | .04861 | .04849 |
| 4 | .04836 | .04824 | .04812 | .04800 | .04787 | .04775 | .04763 | .04751 | .04739 | .04728 |
| 5 | .04716 | .04704 | .04692 | .04681 | .04669 | .04657 | .04646 | .04634 | .04623 | .04612 |
| 6 | .04600 | .04589 | .04578 | .04567 | .04556 | .04545 | .04534 | .04523 | .04512 | .04501 |
| 7 | .04490 | .04480 | .04469 | .04458 | .04448 | .04437 | .04427 | .04416 | .04406 | .04395 |
| 8 | .04385 | .04375 | .04364 | .04354 | .04344 | .04334 | .04324 | .04314 | .04304 | .04294 |
| 9 | .04284 | .04274 | .04265 | .04255 | .04245 | .04235 | .04226 | .04216 | .04207 | .04197 |
| 10 | .04188 | .04178 | .04169 | .04159 | .04150 | .04141 | .04132 | .04123 | .04113 | .04104 |
| 11 | .04095 | .04086 | .04077 | .04068 | .04059 | .04050 | .04042 | .04033 | .04024 | .04015 |
| 12 | .04007 | .03998 | .03989 | .03981 | .03972 | .03964 | .03955 | .03947 | .03939 | .03930 |
| 13 | .03922 | .03914 | .03905 | .03897 | .03889 | .03881 | .03873 | .03865 | .03856 | .03848 |
| 14 | .03840 | .03833 | .03825 | .03817 | .03809 | .03801 | .03793 | .03786 | .03778 | .03770 |
| 15 | .03762 | .03755 | .03747 | .03740 | .03732 | .03725 | .03717 | .03710 | .03702 | .03695 |
| 16 | .03688 | .03680 | .03673 | .03666 | .03659 | .03651 | .03644 | .03637 | .03630 | .03623 |
| 17 | .03616 | .03609 | .03602 | .03595 | .03588 | .03581 | .03574 | .03567 | .03560 | .03554 |
| 18 | .03547 | .03540 | .03533 | .03527 | .03520 | .03513 | .03507 | .03500 | .03494 | .03487 |
| 19 | .03481 | .03474 | .03468 | .03461 | .03455 | .03449 | .03442 | .03436 | .03430 | .03423 |
| 20 | .03417 | .03411 | .03405 | .03398 | .03392 | .03386 | .03380 | .03374 | .03368 | .03362 |
| 21 | .03356 | .03350 | .03344 | .03338 | .03332 | .03326 | .03320 | .03315 | .03309 | .03303 |
| 22 | .03297 | .03291 | .03286 | .03280 | .03274 | .03269 | .03263 | .03257 | .03252 | .03246 |
| 23 | .03241 | .03235 | .03230 | .03224 | .03219 | .03213 | .03208 | .03203 | .03197 | .03192 |
| 24 | .03187 | .03181 | .03176 | .03171 | .03165 | .03160 | .03155 | .03150 | .03145 | .03139 |
| 25 | .03134 | .03129 | .03124 | .03119 | .03114 | .03109 | .03104 | .03099 | .03094 | .03089 |
| 26 | .03084 | .03079 | .03074 | .03069 | .03065 | .03060 | .03055 | .03050 | .03045 | .03041 |
| 27 | .03036 | .03031 | .03026 | .03022 | .03017 | .03012 | .03008 | .03003 | .02999 | .02994 |
| 28 | .02989 | .02985 | .02980 | .02976 | .02971 | .02967 | .02962 | .02958 | .02953 | .02949 |
| 29 | .02945 | .02940 | .02936 | .02932 | .02927 | .02923 | .02919 | .02914 | .02910 | .02906 |
| 30 | .02902 | .02897 | .02893 | .02889 | .02885 | .02881 | .02877 | .02872 | .02868 | .02864 |
| 31 | 0.02860 | .02856 | .02852 | .02848 | .02844 | .02840 | .02836 | .02832 | .02828 | .02824 |
| 32 | 0.02820 | .02816 | .02813 | .02809 | .02805 | .02801 | .02797 | .02793 | .02790 | .02786 |
| 33 | 0.02782 | .02778 | .02774 | .02771 | .02767 | .02763 | .02760 | .02756 | .02752 | .02749 |
| 34 | 0.02745 | .02741 | .02738 | .02734 | .02731 | .02727 | .02724 | .02720 | .02717 | .02713 |
| 35 | 0.02710 | .02706 | .02703 | .02699 | .02696 | .02692 | .02689 | .02685 | .02682 | .02679 |
| 36 | 0.02675 | .02672 | .02669 | .02665 | .02662 | .02659 | .02655 | .02652 | .02649 | .02646 |
| 37 | 0.02642 | .02639 | .02636 | .02633 | .02630 | .02626 | .02623 | .02620 | .02617 | .02614 |
| 38 | 0.02611 | .02608 | .02605 | .02602 | .02598 | .02595 | .02592 | .02589 | .02586 | .02583 |
| 39 | 0.02580 | .02577 | .02574 | .02571 | .02569 | .02566 | .02563 | .02560 | .02557 | .02554 |
| 40 | 0.02551 | .02548 | .02545 | .02543 | .02540 | .02537 | .02534 | .02531 | .02529 | .02526 |

*Source*: Freshwater, based on Hamme and Emerson (2004).

**Table 1.35** Bunsen Coefficient ($\beta$) for Carbon Dioxide as a Function of Temperature (L real gas/(L atm))

| Temperature (°C) | Δt (°C) | | | | | | | | | |
|---|---|---|---|---|---|---|---|---|---|---|
| | 0.0 | 0.1 | 0.2 | 0.3 | 0.4 | 0.5 | 0.6 | 0.7 | 0.8 | 0.9 |
| 0 | 1.7272 | 1.7203 | 1.7135 | 1.7068 | 1.7000 | 1.6933 | 1.6867 | 1.6801 | 1.6735 | 1.6669 |
| 1 | 1.6604 | 1.6540 | 1.6475 | 1.6411 | 1.6347 | 1.6284 | 1.6221 | 1.6158 | 1.6096 | 1.6034 |
| 2 | 1.5972 | 1.5911 | 1.5850 | 1.5789 | 1.5729 | 1.5669 | 1.5609 | 1.5550 | 1.5490 | 1.5432 |
| 3 | 1.5373 | 1.5315 | 1.5257 | 1.5199 | 1.5142 | 1.5085 | 1.5029 | 1.4972 | 1.4916 | 1.4860 |
| 4 | 1.4805 | 1.4750 | 1.4695 | 1.4640 | 1.4586 | 1.4532 | 1.4478 | 1.4424 | 1.4371 | 1.4318 |
| 5 | 1.4265 | 1.4213 | 1.4161 | 1.4109 | 1.4057 | 1.4006 | 1.3955 | 1.3904 | 1.3854 | 1.3803 |
| 6 | 1.3753 | 1.3704 | 1.3654 | 1.3605 | 1.3556 | 1.3507 | 1.3458 | 1.3410 | 1.3362 | 1.3314 |
| 7 | 1.3267 | 1.3220 | 1.3173 | 1.3126 | 1.3079 | 1.3033 | 1.2987 | 1.2941 | 1.2895 | 1.2850 |
| 8 | 1.2805 | 1.2760 | 1.2715 | 1.2670 | 1.2626 | 1.2582 | 1.2538 | 1.2494 | 1.2451 | 1.2408 |
| 9 | 1.2365 | 1.2322 | 1.2280 | 1.2237 | 1.2195 | 1.2153 | 1.2111 | 1.2070 | 1.2029 | 1.1988 |
| 10 | 1.1947 | 1.1906 | 1.1865 | 1.1825 | 1.1785 | 1.1745 | 1.1705 | 1.1666 | 1.1627 | 1.1587 |
| 11 | 1.1548 | 1.1510 | 1.1471 | 1.1433 | 1.1395 | 1.1357 | 1.1319 | 1.1281 | 1.1244 | 1.1206 |
| 12 | 1.1169 | 1.1132 | 1.1096 | 1.1059 | 1.1023 | 1.0987 | 1.0951 | 1.0915 | 1.0879 | 1.0843 |
| 13 | 1.0808 | 1.0773 | 1.0738 | 1.0703 | 1.0668 | 1.0634 | 1.0600 | 1.0565 | 1.0531 | 1.0498 |
| 14 | 1.0464 | 1.0430 | 1.0397 | 1.0364 | 1.0331 | 1.0298 | 1.0265 | 1.0232 | 1.0200 | 1.0168 |
| 15 | 1.0135 | 1.0103 | 1.0072 | 1.0040 | 1.0008 | 0.9977 | 0.9946 | 0.9915 | 0.9884 | 0.9853 |
| 16 | 0.9822 | 0.9792 | 0.9761 | 0.9731 | 0.9701 | 0.9671 | 0.9641 | 0.9612 | 0.9582 | 0.9553 |
| 17 | 0.9523 | 0.9494 | 0.9465 | 0.9436 | 0.9408 | 0.9379 | 0.9350 | 0.9322 | 0.9294 | 0.9266 |
| 18 | 0.9238 | 0.9210 | 0.9182 | 0.9155 | 0.9127 | 0.9100 | 0.9073 | 0.9046 | 0.9019 | 0.8992 |
| 19 | 0.8965 | 0.8939 | 0.8912 | 0.8886 | 0.8859 | 0.8833 | 0.8807 | 0.8781 | 0.8756 | 0.8730 |
| 20 | 0.8705 | 0.8679 | 0.8654 | 0.8629 | 0.8604 | 0.8579 | 0.8554 | 0.8529 | 0.8504 | 0.8480 |
| 21 | 0.8455 | 0.8431 | 0.8407 | 0.8383 | 0.8359 | 0.8335 | 0.8311 | 0.8288 | 0.8264 | 0.8240 |
| 22 | 0.8217 | 0.8194 | 0.8171 | 0.8148 | 0.8125 | 0.8102 | 0.8079 | 0.8056 | 0.8034 | 0.8011 |
| 23 | 0.7989 | 0.7967 | 0.7945 | 0.7923 | 0.7901 | 0.7879 | 0.7857 | 0.7835 | 0.7814 | 0.7792 |
| 24 | 0.7771 | 0.7750 | 0.7728 | 0.7707 | 0.7686 | 0.7665 | 0.7644 | 0.7624 | 0.7603 | 0.7582 |
| 25 | 0.7562 | 0.7542 | 0.7521 | 0.7501 | 0.7481 | 0.7461 | 0.7441 | 0.7421 | 0.7401 | 0.7381 |
| 26 | 0.7362 | 0.7342 | 0.7323 | 0.7303 | 0.7284 | 0.7265 | 0.7246 | 0.7227 | 0.7208 | 0.7189 |
| 27 | 0.7170 | 0.7151 | 0.7133 | 0.7114 | 0.7096 | 0.7077 | 0.7059 | 0.7040 | 0.7022 | 0.7004 |
| 28 | 0.6986 | 0.6968 | 0.6950 | 0.6932 | 0.6915 | 0.6897 | 0.6879 | 0.6862 | 0.6845 | 0.6827 |
| 29 | 0.6810 | 0.6793 | 0.6775 | 0.6758 | 0.6741 | 0.6724 | 0.6708 | 0.6691 | 0.6674 | 0.6657 |
| 30 | 0.6641 | 0.6624 | 0.6608 | 0.6591 | 0.6575 | 0.6559 | 0.6543 | 0.6526 | 0.6510 | 0.6494 |
| 31 | 0.6478 | 0.6463 | 0.6447 | 0.6431 | 0.6415 | 0.6400 | 0.6384 | 0.6369 | 0.6353 | 0.6338 |
| 32 | 0.6323 | 0.6307 | 0.6292 | 0.6277 | 0.6262 | 0.6247 | 0.6232 | 0.6217 | 0.6203 | 0.6188 |
| 33 | 0.6173 | 0.6158 | 0.6144 | 0.6129 | 0.6115 | 0.6101 | 0.6086 | 0.6072 | 0.6058 | 0.6044 |
| 34 | 0.6029 | 0.6015 | 0.6001 | 0.5987 | 0.5974 | 0.5960 | 0.5946 | 0.5932 | 0.5919 | 0.5905 |
| 35 | 0.5891 | 0.5878 | 0.5864 | 0.5851 | 0.5838 | 0.5824 | 0.5811 | 0.5798 | 0.5785 | 0.5772 |
| 36 | 0.5759 | 0.5746 | 0.5733 | 0.5720 | 0.5707 | 0.5694 | 0.5682 | 0.5669 | 0.5656 | 0.5644 |
| 37 | 0.5631 | 0.5619 | 0.5606 | 0.5594 | 0.5582 | 0.5569 | 0.5557 | 0.5545 | 0.5533 | 0.5521 |
| 38 | 0.5509 | 0.5497 | 0.5485 | 0.5473 | 0.5461 | 0.5449 | 0.5437 | 0.5426 | 0.5414 | 0.5402 |
| 39 | 0.5391 | 0.5379 | 0.5368 | 0.5356 | 0.5345 | 0.5333 | 0.5322 | 0.5311 | 0.5299 | 0.5288 |
| 40 | 0.5277 | 0.5266 | 0.5255 | 0.5244 | 0.5233 | 0.5222 | 0.5211 | 0.5200 | 0.5189 | 0.5179 |

*Source*: Freshwater, based on Weiss (1974).

**Table 1.36** Bunsen Coefficient ($\beta'$) for Oxygen as a Function of Temperature (mg/(L mmHg))

| Temperature (°C) | Δt (°C) | | | | | | | | | |
|---|---|---|---|---|---|---|---|---|---|---|
| | **0.0** | **0.1** | **0.2** | **0.3** | **0.4** | **0.5** | **0.6** | **0.7** | **0.8** | **0.9** |
| **0** | .09240 | .09214 | .09189 | .09163 | .09138 | .09113 | .09088 | .09063 | .09038 | .09013 |
| **1** | .08988 | .08964 | .08939 | .08915 | .08891 | .08867 | .08843 | .08819 | .08795 | .08772 |
| **2** | .08748 | .08725 | .08701 | .08678 | .08655 | .08632 | .08609 | .08586 | .08564 | .08541 |
| **3** | .08519 | .08497 | .08474 | .08452 | .08430 | .08408 | .08386 | .08365 | .08343 | .08322 |
| **4** | .08300 | .08279 | .08258 | .08236 | .08215 | .08195 | .08174 | .08153 | .08132 | .08112 |
| **5** | .08091 | .08071 | .08051 | .08030 | .08010 | .07990 | .07970 | .07951 | .07931 | .07911 |
| **6** | .07892 | .07872 | .07853 | .07834 | .07814 | .07795 | .07776 | .07757 | .07739 | .07720 |
| **7** | .07701 | .07682 | .07664 | .07646 | .07627 | .07609 | .07591 | .07573 | .07555 | .07537 |
| **8** | .07519 | .07501 | .07483 | .07466 | .07448 | .07431 | .07413 | .07396 | .07379 | .07361 |
| **9** | .07344 | .07327 | .07310 | .07293 | .07277 | .07260 | .07243 | .07227 | .07210 | .07194 |
| **10** | .07177 | .07161 | .07145 | .07129 | .07113 | .07097 | .07081 | .07065 | .07049 | .07033 |
| **11** | .07018 | .07002 | .06986 | .06971 | .06956 | .06940 | .06925 | .06910 | .06895 | .06880 |
| **12** | .06865 | .06850 | .06835 | .06820 | .06805 | .06790 | .06776 | .06761 | .06747 | .06732 |
| **13** | .06718 | .06704 | .06689 | .06675 | .06661 | .06647 | .06633 | .06619 | .06605 | .06591 |
| **14** | .06577 | .06564 | .06550 | .06536 | .06523 | .06509 | .06496 | .06482 | .06469 | .06456 |
| **15** | .06443 | .06429 | .06416 | .06403 | .06390 | .06377 | .06364 | .06351 | .06339 | .06326 |
| **16** | .06313 | .06301 | .06288 | .06276 | .06263 | .06251 | .06238 | .06226 | .06214 | .06201 |
| **17** | .06189 | .06177 | .06165 | .06153 | .06141 | .06129 | .06117 | .06105 | .06094 | .06082 |
| **18** | .06070 | .06059 | .06047 | .06035 | .06024 | .06012 | .06001 | .05990 | .05978 | .05967 |
| **19** | .05956 | .05945 | .05934 | .05922 | .05911 | .05900 | .05889 | .05879 | .05868 | .05857 |
| **20** | .05846 | .05835 | .05825 | .05814 | .05803 | .05793 | .05782 | .05772 | .05761 | .05751 |
| **21** | .05741 | .05730 | .05720 | .05710 | .05699 | .05689 | .05679 | .05669 | .05659 | .05649 |
| **22** | .05639 | .05629 | .05619 | .05610 | .05600 | .05590 | .05580 | .05571 | .05561 | .05551 |
| **23** | .05542 | .05532 | .05523 | .05513 | .05504 | .05494 | .05485 | .05476 | .05466 | .05457 |
| **24** | .05448 | .05439 | .05430 | .05421 | .05411 | .05402 | .05393 | .05384 | .05376 | .05367 |
| **25** | .05358 | .05349 | .05340 | .05331 | .05323 | .05314 | .05305 | .05297 | .05288 | .05280 |
| **26** | .05271 | .05263 | .05254 | .05246 | .05237 | .05229 | .05221 | .05212 | .05204 | .05196 |
| **27** | .05188 | .05179 | .05171 | .05163 | .05155 | .05147 | .05139 | .05131 | .05123 | .05115 |
| **28** | .05107 | .05099 | .05092 | .05084 | .05076 | .05068 | .05060 | .05053 | .05045 | .05037 |
| **29** | .05030 | .05022 | .05015 | .05007 | .05000 | .04992 | .04985 | .04977 | .04970 | .04963 |
| **30** | .04955 | .04948 | .04941 | .04934 | .04926 | .04919 | .04912 | .04905 | .04898 | .04891 |
| **31** | .04884 | .04877 | .04870 | .04863 | .04856 | .04849 | .04842 | .04835 | .04828 | .04821 |
| **32** | .04814 | .04808 | .04801 | .04794 | .04788 | .04781 | .04774 | .04768 | .04761 | .04754 |
| **33** | .04748 | .04741 | .04735 | .04728 | .04722 | .04715 | .04709 | .04703 | .04696 | .04690 |
| **34** | .04684 | .04677 | .04671 | .04665 | .04659 | .04652 | .04646 | .04640 | .04634 | .04628 |
| **35** | .04622 | .04616 | .04610 | .04604 | .04598 | .04592 | .04586 | .04580 | .04574 | .04568 |
| **36** | .04562 | .04556 | .04550 | .04545 | .04539 | .04533 | .04527 | .04522 | .04516 | .04510 |
| **37** | .04505 | .04499 | .04493 | .04488 | .04482 | .04477 | .04471 | .04465 | .04460 | .04455 |
| **38** | .04449 | .04444 | .04438 | .04433 | .04427 | .04422 | .04417 | .04411 | .04406 | .04401 |
| **39** | .04396 | .04390 | .04385 | .04380 | .04375 | .04369 | .04364 | .04359 | .04354 | .04349 |
| **40** | .04344 | .04339 | .04334 | .04329 | .04324 | .04319 | .04314 | .04309 | .04304 | .04299 |

*Source*: Freshwater, based on Benson and Krause (1980) and Eq. (1.16).

**Table 1.37** Bunsen Coefficient ($\beta'$) for Nitrogen (mg/(L mmHg))

| Temperature (°C) | $\Delta t$ (°C) | | | | | | | | | |
|---|---|---|---|---|---|---|---|---|---|---|
| | **0.0** | **0.1** | **0.2** | **0.3** | **0.4** | **0.5** | **0.6** | **0.7** | **0.8** | **0.9** |
| **0** | .03943 | .03933 | .03923 | .03913 | .03903 | .03893 | .03883 | .03873 | .03863 | .03853 |
| **1** | .03843 | .03833 | .03823 | .03814 | .03804 | .03794 | .03785 | .03775 | .03766 | .03756 |
| **2** | .03747 | .03738 | .03728 | .03719 | .03710 | .03701 | .03692 | .03683 | .03674 | .03665 |
| **3** | .03656 | .03647 | .03638 | .03629 | .03620 | .03611 | .03603 | .03594 | .03585 | .03577 |
| **4** | .03568 | .03560 | .03551 | .03543 | .03535 | .03526 | .03518 | .03510 | .03501 | .03493 |
| **5** | .03485 | .03477 | .03469 | .03461 | .03453 | .03445 | .03437 | .03429 | .03421 | .03413 |
| **6** | .03405 | .03398 | .03390 | .03382 | .03374 | .03367 | .03359 | .03352 | .03344 | .03337 |
| **7** | .03329 | .03322 | .03314 | .03307 | .03300 | .03292 | .03285 | .03278 | .03271 | .03264 |
| **8** | .03257 | .03249 | .03242 | .03235 | .03228 | .03221 | .03214 | .03208 | .03201 | .03194 |
| **9** | .03187 | .03180 | .03173 | .03167 | .03160 | .03153 | .03147 | .03140 | .03134 | .03127 |
| **10** | .03120 | .03114 | .03107 | .03101 | .03095 | .03088 | .03082 | .03076 | .03069 | .03063 |
| **11** | .03057 | .03051 | .03044 | .03038 | .03032 | .03026 | .03020 | .03014 | .03008 | .03002 |
| **12** | .02996 | .02990 | .02984 | .02978 | .02972 | .02966 | .02961 | .02955 | .02949 | .02943 |
| **13** | .02938 | .02932 | .02926 | .02921 | .02915 | .02909 | .02904 | .02898 | .02893 | .02887 |
| **14** | .02882 | .02876 | .02871 | .02865 | .02860 | .02855 | .02849 | .02844 | .02839 | .02833 |
| **15** | .02828 | .02823 | .02818 | .02813 | .02807 | .02802 | .02797 | .02792 | .02787 | .02782 |
| **16** | .02777 | .02772 | .02767 | .02762 | .02757 | .02752 | .02747 | .02742 | .02737 | .02733 |
| **17** | .02728 | .02723 | .02718 | .02713 | .02709 | .02704 | .02699 | .02695 | .02690 | .02685 |
| **18** | .02681 | .02676 | .02672 | .02667 | .02662 | .02658 | .02653 | .02649 | .02644 | .02640 |
| **19** | .02636 | .02631 | .02627 | .02622 | .02618 | .02614 | .02609 | .02605 | .02601 | .02597 |
| **20** | .02592 | .02588 | .02584 | .02580 | .02576 | .02571 | .02567 | .02563 | .02559 | .02555 |
| **21** | .02551 | .02547 | .02543 | .02539 | .02535 | .02531 | .02527 | .02523 | .02519 | .02515 |
| **22** | .02511 | .02507 | .02503 | .02499 | .02496 | .02492 | .02488 | .02484 | .02480 | .02477 |
| **23** | .02473 | .02469 | .02465 | .02462 | .02458 | .02454 | .02451 | .02447 | .02444 | .02440 |
| **24** | .02436 | .02433 | .02429 | .02426 | .02422 | .02419 | .02415 | .02412 | .02408 | .02405 |
| **25** | .02401 | .02398 | .02394 | .02391 | .02388 | .02384 | .02381 | .02378 | .02374 | .02371 |
| **26** | .02368 | .02364 | .02361 | .02358 | .02355 | .02351 | .02348 | .02345 | .02342 | .02339 |
| **27** | .02335 | .02332 | .02329 | .02326 | .02323 | .02320 | .02317 | .02314 | .02310 | .02307 |
| **28** | .02304 | .02301 | .02298 | .02295 | .02292 | .02289 | .02286 | .02283 | .02281 | .02278 |
| **29** | .02275 | .02272 | .02269 | .02266 | .02263 | .02260 | .02258 | .02255 | .02252 | .02249 |
| **30** | .02246 | .02244 | .02241 | .02238 | .02235 | .02233 | .02230 | .02227 | .02224 | .02222 |
| **31** | .02219 | .02216 | .02214 | .02211 | .02208 | .02206 | .02203 | .02201 | .02198 | .02195 |
| **32** | .02193 | .02190 | .02188 | .02185 | .02183 | .02180 | .02178 | .02175 | .02173 | .02170 |
| **33** | .02168 | .02165 | .02163 | .02161 | .02158 | .02156 | .02153 | .02151 | .02149 | .02146 |
| **34** | .02144 | .02142 | .02139 | .02137 | .02135 | .02132 | .02130 | .02128 | .02126 | .02123 |
| **35** | .02121 | .02119 | .02117 | .02114 | .02112 | .02110 | .02108 | .02106 | .02103 | .02101 |
| **36** | .02099 | .02097 | .02095 | .02093 | .02091 | .02089 | .02086 | .02084 | .02082 | .02080 |
| **37** | .02078 | .02076 | .02074 | .02072 | .02070 | .02068 | .02066 | .02064 | .02062 | .02060 |
| **38** | .02058 | .02056 | .02054 | .02052 | .02050 | .02049 | .02047 | .02045 | .02043 | .02041 |
| **39** | .02039 | .02037 | .02035 | .02034 | .02032 | .02030 | .02028 | .02026 | .02024 | .02023 |
| **40** | .02021 | .02019 | .02017 | .02016 | .02014 | .02012 | .02010 | .02009 | .02007 | .02005 |

*Source*: Based on Hamme and Emerson (2004) and Eq. (1.16).

**Table 1.38** Bunsen Coefficient ($\beta'$) for Argon (mg/(L mmHg))

| Temperature (°C) | $\Delta t$ (°C) | | | | | | | | | |
|---|---|---|---|---|---|---|---|---|---|---|
| | **0.0** | **0.1** | **0.2** | **0.3** | **0.4** | **0.5** | **0.6** | **0.7** | **0.8** | **0.9** |
| **0** | .12624 | .12590 | .12555 | .12521 | .12486 | .12452 | .12418 | .12384 | .12351 | .12317 |
| **1** | .12284 | .12251 | .12217 | .12185 | .12152 | .12119 | .12087 | .12055 | .12022 | .11990 |
| **2** | .11959 | .11927 | .11895 | .11864 | .11833 | .11802 | .11771 | .11740 | .11709 | .11679 |
| **3** | .11648 | .11618 | .11588 | .11558 | .11528 | .11499 | .11469 | .11440 | .11410 | .11381 |
| **4** | .11352 | .11323 | .11295 | .11266 | .11237 | .11209 | .11181 | .11153 | .11125 | .11097 |
| **5** | .11069 | .11042 | .11014 | .10987 | .10960 | .10932 | .10905 | .10879 | .10852 | .10825 |
| **6** | .10799 | .10772 | .10746 | .10720 | .10694 | .10668 | .10642 | .10617 | .10591 | .10565 |
| **7** | .10540 | .10515 | .10490 | .10465 | .10440 | .10415 | .10390 | .10366 | .10341 | .10317 |
| **8** | .10293 | .10269 | .10245 | .10221 | .10197 | .10173 | .10150 | .10126 | .10103 | .10079 |
| **9** | .10056 | .10033 | .10010 | .09987 | .09964 | .09942 | .09919 | .09897 | .09874 | .09852 |
| **10** | .09830 | .09808 | .09786 | .09764 | .09742 | .09720 | .09698 | .09677 | .09655 | .09634 |
| **11** | .09613 | .09592 | .09570 | .09549 | .09529 | .09508 | .09487 | .09466 | .09446 | .09425 |
| **12** | .09405 | .09385 | .09364 | .09344 | .09324 | .09304 | .09284 | .09265 | .09245 | .09225 |
| **13** | .09206 | .09186 | .09167 | .09148 | .09128 | .09109 | .09090 | .09071 | .09052 | .09033 |
| **14** | .09015 | .08996 | .08978 | .08959 | .08941 | .08922 | .08904 | .08886 | .08868 | .08850 |
| **15** | .08832 | .08814 | .08796 | .08778 | .08760 | .08743 | .08725 | .08708 | .08691 | .08673 |
| **16** | .08656 | .08639 | .08622 | .08605 | .08588 | .08571 | .08554 | .08537 | .08521 | .08504 |
| **17** | .08487 | .08471 | .08454 | .08438 | .08422 | .08406 | .08389 | .08373 | .08357 | .08341 |
| **18** | .08326 | .08310 | .08294 | .08278 | .08263 | .08247 | .08232 | .08216 | .08201 | .08185 |
| **19** | .08170 | .08155 | .08140 | .08125 | .08110 | .08095 | .08080 | .08065 | .08050 | .08035 |
| **20** | .08021 | .08006 | .07992 | .07977 | .07963 | .07948 | .07934 | .07920 | .07906 | .07892 |
| **21** | .07877 | .07863 | .07849 | .07835 | .07822 | .07808 | .07794 | .07780 | .07767 | .07753 |
| **22** | .07740 | .07726 | .07713 | .07699 | .07686 | .07673 | .07659 | .07646 | .07633 | .07620 |
| **23** | .07607 | .07594 | .07581 | .07568 | .07556 | .07543 | .07530 | .07517 | .07505 | .07492 |
| **24** | .07480 | .07467 | .07455 | .07442 | .07430 | .07418 | .07406 | .07393 | .07381 | .07369 |
| **25** | .07357 | .07345 | .07333 | .07321 | .07310 | .07298 | .07286 | .07274 | .07263 | .07251 |
| **26** | .07239 | .07228 | .07216 | .07205 | .07194 | .07182 | .07171 | .07160 | .07148 | .07137 |
| **27** | .07126 | .07115 | .07104 | .07093 | .07082 | .07071 | .07060 | .07049 | .07038 | .07028 |
| **28** | .07017 | .07006 | .06996 | .06985 | .06974 | .06964 | .06954 | .06943 | .06933 | .06922 |
| **29** | .06912 | .06902 | .06891 | .06881 | .06871 | .06861 | .06851 | .06841 | .06831 | .06821 |
| **30** | .06811 | .06801 | .06791 | .06781 | .06772 | .06762 | .06752 | .06743 | .06733 | .06723 |
| **31** | 0.06714 | .06704 | .06695 | .06685 | .06676 | .06667 | .06657 | .06648 | .06639 | .06629 |
| **32** | 0.06620 | .06611 | .06602 | .06593 | .06584 | .06575 | .06566 | .06557 | .06548 | .06539 |
| **33** | 0.06530 | .06521 | .06513 | .06504 | .06495 | .06486 | .06478 | .06469 | .06461 | .06452 |
| **34** | 0.06444 | .06435 | .06427 | .06418 | .06410 | .06401 | .06393 | .06385 | .06377 | .06368 |
| **35** | 0.06360 | .06352 | .06344 | .06336 | .06328 | .06320 | .06312 | .06304 | .06296 | .06288 |
| **36** | 0.06280 | .06272 | .06264 | .06256 | .06249 | .06241 | .06233 | .06225 | .06218 | .06210 |
| **37** | 0.06203 | .06195 | .06188 | .06180 | .06173 | .06165 | .06158 | .06150 | .06143 | .06136 |
| **38** | 0.06128 | .06121 | .06114 | .06107 | .06099 | .06092 | .06085 | .06078 | .06071 | .06064 |
| **39** | 0.06057 | .06050 | .06043 | .06036 | .06029 | .06022 | .06015 | .06009 | .06002 | .05995 |
| **40** | 0.05988 | .05981 | .05975 | .05968 | .05961 | .05955 | .05948 | .05942 | .05935 | .05929 |

*Source*: Based on Hamme and Emerson (2004) and Eq. (1.16).

**Table 1.39** Bunsen Coefficient ($\beta'$) for Carbon Dioxide as a Function of Temperature (mg/(L mmHg))

| Temperature (°C) | Δt (°C) | | | | | | | | | |
|---|---|---|---|---|---|---|---|---|---|---|
| | 0.0 | 0.1 | 0.2 | 0.3 | 0.4 | 0.5 | 0.6 | 0.7 | 0.8 | 0.9 |
| 0 | 4.4924 | 4.4746 | 4.4569 | 4.4393 | 4.4218 | 4.4044 | 4.3871 | 4.3699 | 4.3528 | 4.3358 |
| 1 | 4.3188 | 4.3020 | 4.2853 | 4.2686 | 4.2520 | 4.2355 | 4.2191 | 4.2028 | 4.1866 | 4.1705 |
| 2 | 4.1544 | 4.1385 | 4.1226 | 4.1068 | 4.0911 | 4.0755 | 4.0599 | 4.0445 | 4.0291 | 4.0138 |
| 3 | 3.9986 | 3.9834 | 3.9684 | 3.9534 | 3.9385 | 3.9237 | 3.9090 | 3.8943 | 3.8797 | 3.8652 |
| 4 | 3.8508 | 3.8364 | 3.8221 | 3.8079 | 3.7938 | 3.7797 | 3.7657 | 3.7518 | 3.7379 | 3.7242 |
| 5 | 3.7105 | 3.6968 | 3.6833 | 3.6698 | 3.6564 | 3.6430 | 3.6297 | 3.6165 | 3.6034 | 3.5903 |
| 6 | 3.5773 | 3.5643 | 3.5515 | 3.5386 | 3.5259 | 3.5132 | 3.5006 | 3.4880 | 3.4755 | 3.4631 |
| 7 | 3.4508 | 3.4384 | 3.4262 | 3.4140 | 3.4019 | 3.3899 | 3.3779 | 3.3659 | 3.3541 | 3.3422 |
| 8 | 3.3305 | 3.3188 | 3.3072 | 3.2956 | 3.2841 | 3.2726 | 3.2612 | 3.2498 | 3.2386 | 3.2273 |
| 9 | 3.2161 | 3.2050 | 3.1939 | 3.1829 | 3.1720 | 3.1611 | 3.1502 | 3.1394 | 3.1287 | 3.1180 |
| 10 | 3.1073 | 3.0968 | 3.0862 | 3.0757 | 3.0653 | 3.0549 | 3.0446 | 3.0343 | 3.0241 | 3.0139 |
| 11 | 3.0038 | 2.9937 | 2.9837 | 2.9737 | 2.9638 | 2.9539 | 2.9440 | 2.9343 | 2.9245 | 2.9148 |
| 12 | 2.9052 | 2.8956 | 2.8860 | 2.8765 | 2.8670 | 2.8576 | 2.8483 | 2.8389 | 2.8296 | 2.8204 |
| 13 | 2.8112 | 2.8021 | 2.7930 | 2.7839 | 2.7749 | 2.7659 | 2.7570 | 2.7481 | 2.7392 | 2.7304 |
| 14 | 2.7217 | 2.7129 | 2.7043 | 2.6956 | 2.6870 | 2.6785 | 2.6699 | 2.6615 | 2.6530 | 2.6446 |
| 15 | 2.6363 | 2.6279 | 2.6197 | 2.6114 | 2.6032 | 2.5951 | 2.5869 | 2.5788 | 2.5708 | 2.5628 |
| 16 | 2.5548 | 2.5469 | 2.5390 | 2.5311 | 2.5233 | 2.5155 | 2.5077 | 2.5000 | 2.4923 | 2.4846 |
| 17 | 2.4770 | 2.4695 | 2.4619 | 2.4544 | 2.4469 | 2.4395 | 2.4321 | 2.4247 | 2.4174 | 2.4101 |
| 18 | 2.4028 | 2.3955 | 2.3883 | 2.3812 | 2.3740 | 2.3669 | 2.3598 | 2.3528 | 2.3458 | 2.3388 |
| 19 | 2.3319 | 2.3249 | 2.3181 | 2.3112 | 2.3044 | 2.2976 | 2.2908 | 2.2841 | 2.2774 | 2.2707 |
| 20 | 2.2641 | 2.2575 | 2.2509 | 2.2443 | 2.2378 | 2.2313 | 2.2248 | 2.2184 | 2.2120 | 2.2056 |
| 21 | 2.1993 | 2.1930 | 2.1867 | 2.1804 | 2.1742 | 2.1679 | 2.1618 | 2.1556 | 2.1495 | 2.1434 |
| 22 | 2.1373 | 2.1312 | 2.1252 | 2.1192 | 2.1133 | 2.1073 | 2.1014 | 2.0955 | 2.0896 | 2.0838 |
| 23 | 2.0780 | 2.0722 | 2.0664 | 2.0607 | 2.0550 | 2.0493 | 2.0436 | 2.0380 | 2.0324 | 2.0268 |
| 24 | 2.0212 | 2.0157 | 2.0102 | 2.0047 | 1.9992 | 1.9938 | 1.9883 | 1.9829 | 1.9776 | 1.9722 |
| 25 | 1.9669 | 1.9616 | 1.9563 | 1.9510 | 1.9458 | 1.9406 | 1.9354 | 1.9302 | 1.9251 | 1.9199 |
| 26 | 1.9148 | 1.9097 | 1.9047 | 1.8996 | 1.8946 | 1.8896 | 1.8846 | 1.8797 | 1.8747 | 1.8698 |
| 27 | 1.8649 | 1.8601 | 1.8552 | 1.8504 | 1.8456 | 1.8408 | 1.8360 | 1.8313 | 1.8265 | 1.8218 |
| 28 | 1.8171 | 1.8124 | 1.8078 | 1.8032 | 1.7985 | 1.7939 | 1.7894 | 1.7848 | 1.7803 | 1.7758 |
| 29 | 1.7713 | 1.7668 | 1.7623 | 1.7579 | 1.7534 | 1.7490 | 1.7446 | 1.7403 | 1.7359 | 1.7316 |
| 30 | 1.7273 | 1.7230 | 1.7187 | 1.7144 | 1.7102 | 1.7060 | 1.7017 | 1.6975 | 1.6934 | 1.6892 |
| 31 | 1.6851 | 1.6809 | 1.6768 | 1.6727 | 1.6687 | 1.6646 | 1.6606 | 1.6565 | 1.6525 | 1.6485 |
| 32 | 1.6445 | 1.6406 | 1.6366 | 1.6327 | 1.6288 | 1.6249 | 1.6210 | 1.6172 | 1.6133 | 1.6095 |
| 33 | 1.6056 | 1.6018 | 1.5981 | 1.5943 | 1.5905 | 1.5868 | 1.5830 | 1.5793 | 1.5756 | 1.5719 |
| 34 | 1.5683 | 1.5646 | 1.5610 | 1.5574 | 1.5537 | 1.5501 | 1.5466 | 1.5430 | 1.5394 | 1.5359 |
| 35 | 1.5324 | 1.5289 | 1.5254 | 1.5219 | 1.5184 | 1.5150 | 1.5115 | 1.5081 | 1.5047 | 1.5013 |
| 36 | 1.4979 | 1.4945 | 1.4911 | 1.4878 | 1.4844 | 1.4811 | 1.4778 | 1.4745 | 1.4712 | 1.4680 |
| 37 | 1.4647 | 1.4615 | 1.4582 | 1.4550 | 1.4518 | 1.4486 | 1.4454 | 1.4422 | 1.4391 | 1.4359 |
| 38 | 1.4328 | 1.4297 | 1.4266 | 1.4235 | 1.4204 | 1.4173 | 1.4142 | 1.4112 | 1.4082 | 1.4051 |
| 39 | 1.4021 | 1.3991 | 1.3961 | 1.3931 | 1.3902 | 1.3872 | 1.3843 | 1.3813 | 1.3784 | 1.3755 |
| 40 | 1.3726 | 1.3697 | 1.3668 | 1.3639 | 1.3611 | 1.3582 | 1.3554 | 1.3526 | 1.3498 | 1.3470 |

*Source*: Freshwater, based on Weiss (1974) and Eq. (1.16).

**Table 1.40** Bunsen Coefficient ($\beta''$) for Oxygen as a Function of Temperature (mg/(L kPa))

| Temperature (°C) | $\Delta t$ (°C) | | | | | | | | | |
|---|---|---|---|---|---|---|---|---|---|---|
| | **0.0** | **0.1** | **0.2** | **0.3** | **0.4** | **0.5** | **0.6** | **0.7** | **0.8** | **0.9** |
| **0** | .69307 | .69114 | .68922 | .68730 | .68540 | .68351 | .68162 | .67975 | .67788 | .67602 |
| **1** | .67417 | .67233 | .67050 | .66868 | .66686 | .66506 | .66326 | .66147 | .65969 | .65792 |
| **2** | .65615 | .65440 | .65265 | .65091 | .64918 | .64746 | .64574 | .64404 | .64234 | .64065 |
| **3** | .63897 | .63729 | .63562 | .63396 | .63231 | .63067 | .62903 | .62740 | .62578 | .62417 |
| **4** | .62256 | .62096 | .61937 | .61779 | .61621 | .61464 | .61308 | .61152 | .60997 | .60843 |
| **5** | .60690 | .60537 | .60385 | .60233 | .60083 | .59933 | .59784 | .59635 | .59487 | .59340 |
| **6** | .59193 | .59047 | .58902 | .58757 | .58613 | .58470 | .58327 | .58185 | .58044 | .57903 |
| **7** | .57763 | .57623 | .57484 | .57346 | .57208 | .57071 | .56935 | .56799 | .56664 | .56529 |
| **8** | .56395 | .56262 | .56129 | .55997 | .55865 | .55734 | .55603 | .55473 | .55344 | .55215 |
| **9** | .55087 | .54959 | .54832 | .54705 | .54579 | .54454 | .54329 | .54205 | .54081 | .53958 |
| **10** | .53835 | .53713 | .53591 | .53470 | .53349 | .53229 | .53109 | .52990 | .52872 | .52754 |
| **11** | .52636 | .52519 | .52403 | .52287 | .52171 | .52056 | .51942 | .51827 | .51714 | .51601 |
| **12** | .51488 | .51376 | .51264 | .51153 | .51043 | .50932 | .50823 | .50713 | .50605 | .50496 |
| **13** | .50388 | .50281 | .50174 | .50067 | .49961 | .49856 | .49750 | .49646 | .49541 | .49437 |
| **14** | .49334 | .49231 | .49128 | .49026 | .48925 | .48823 | .48722 | .48622 | .48522 | .48422 |
| **15** | .48323 | .48224 | .48126 | .48028 | .47930 | .47833 | .47736 | .47640 | .47544 | .47449 |
| **16** | .47353 | .47259 | .47164 | .47070 | .46977 | .46883 | .46791 | .46698 | .46606 | .46514 |
| **17** | .46423 | .46332 | .46241 | .46151 | .46061 | .45972 | .45883 | .45794 | .45706 | .45618 |
| **18** | .45530 | .45443 | .45356 | .45269 | .45183 | .45097 | .45011 | .44926 | .44841 | .44757 |
| **19** | .44672 | .44589 | .44505 | .44422 | .44339 | .44256 | .44174 | .44092 | .44011 | .43930 |
| **20** | .43849 | .43768 | .43688 | .43608 | .43529 | .43449 | .43370 | .43292 | .43213 | .43135 |
| **21** | .43058 | .42980 | .42903 | .42826 | .42750 | .42674 | .42598 | .42522 | .42447 | .42372 |
| **22** | .42297 | .42223 | .42149 | .42075 | .42001 | .41928 | .41855 | .41782 | .41710 | .41638 |
| **23** | .41566 | .41494 | .41423 | .41352 | .41282 | .41211 | .41141 | .41071 | .41001 | .40932 |
| **24** | .40863 | .40794 | .40726 | .40657 | .40589 | .40522 | .40454 | .40387 | .40320 | .40253 |
| **25** | .40187 | .40121 | .40055 | .39989 | .39924 | .39858 | .39793 | .39729 | .39664 | .39600 |
| **26** | .39536 | .39473 | .39409 | .39346 | .39283 | .39220 | .39158 | .39095 | .39033 | .38972 |
| **27** | .38910 | .38849 | .38788 | .38727 | .38666 | .38606 | .38546 | .38486 | .38426 | .38367 |
| **28** | .38307 | .38248 | .38190 | .38131 | .38073 | .38015 | .37957 | .37899 | .37842 | .37784 |
| **29** | .37727 | .37670 | .37614 | .37557 | .37501 | .37445 | .37389 | .37334 | .37279 | .37223 |
| **30** | .37168 | .37114 | .37059 | .37005 | .36951 | .36897 | .36843 | .36790 | .36736 | .36683 |
| **31** | .36630 | .36577 | .36525 | .36473 | .36420 | .36368 | .36317 | .36265 | .36214 | .36163 |
| **32** | .36112 | .36061 | .36010 | .35960 | .35909 | .35859 | .35810 | .35760 | .35710 | .35661 |
| **33** | .35612 | .35563 | .35514 | .35466 | .35417 | .35369 | .35321 | .35273 | .35225 | .35178 |
| **34** | .35130 | .35083 | .35036 | .34989 | .34943 | .34896 | .34850 | .34804 | .34758 | .34712 |
| **35** | .34666 | .34621 | .34575 | .34530 | .34485 | .34440 | .34395 | .34351 | .34307 | .34262 |
| **36** | .34218 | .34175 | .34131 | .34087 | .34044 | .34001 | .33958 | .33915 | .33872 | .33829 |
| **37** | .33787 | .33745 | .33702 | .33660 | .33619 | .33577 | .33535 | .33494 | .33453 | .33412 |
| **38** | .33371 | .33330 | .33289 | .33249 | .33208 | .33168 | .33128 | .33088 | .33048 | .33009 |
| **39** | .32969 | .32930 | .32891 | .32851 | .32813 | .32774 | .32735 | .32697 | .32658 | .32620 |
| **40** | .32582 | .32544 | .32506 | .32468 | .32431 | .32393 | .32356 | .32319 | .32282 | .32245 |

*Source*: Freshwater, based on Benson and Krause (1980) and Eq. (1.17).

**Table 1.41** Bunsen Coefficient ($\beta''$) for Nitrogen as a Function of Temperature (mg/(L kPa))

| Temperature (°C) | Δt (°C) | | | | | | | | | |
|---|---|---|---|---|---|---|---|---|---|---|
| | 0.0 | 0.1 | 0.2 | 0.3 | 0.4 | 0.5 | 0.6 | 0.7 | 0.8 | 0.9 |
| 0 | .29578 | .29501 | .29424 | .29348 | .29272 | .29197 | .29121 | .29047 | .28972 | .28898 |
| 1 | .28824 | .28751 | .28678 | .28605 | .28533 | .28460 | .28389 | .28317 | .28246 | .28176 |
| 2 | .28105 | .28035 | .27965 | .27896 | .27827 | .27758 | .27690 | .27622 | .27554 | .27486 |
| 3 | .27419 | .27352 | .27286 | .27220 | .27154 | .27088 | .27023 | .26958 | .26893 | .26829 |
| 4 | .26765 | .26701 | .26637 | .26574 | .26511 | .26448 | .26386 | .26324 | .26262 | .26201 |
| 5 | .26139 | .26078 | .26018 | .25957 | .25897 | .25837 | .25778 | .25718 | .25659 | .25601 |
| 6 | .25542 | .25484 | .25426 | .25368 | .25311 | .25254 | .25197 | .25140 | .25083 | .25027 |
| 7 | .24971 | .24916 | .24860 | .24805 | .24750 | .24696 | .24641 | .24587 | .24533 | .24479 |
| 8 | .24426 | .24373 | .24320 | .24267 | .24214 | .24162 | .24110 | .24058 | .24007 | .23955 |
| 9 | .23904 | .23853 | .23803 | .23752 | .23702 | .23652 | .23602 | .23553 | .23503 | .23454 |
| 10 | .23405 | .23356 | .23308 | .23260 | .23212 | .23164 | .23116 | .23069 | .23022 | .22975 |
| 11 | .22928 | .22881 | .22835 | .22789 | .22743 | .22697 | .22651 | .22606 | .22561 | .22516 |
| 12 | .22471 | .22426 | .22382 | .22338 | .22293 | .22250 | .22206 | .22163 | .22119 | .22076 |
| 13 | .22033 | .21991 | .21948 | .21906 | .21864 | .21822 | .21780 | .21738 | .21697 | .21655 |
| 14 | .21614 | .21573 | .21533 | .21492 | .21452 | .21412 | .21371 | .21332 | .21292 | .21252 |
| 15 | .21213 | .21174 | .21135 | .21096 | .21057 | .21019 | .20980 | .20942 | .20904 | .20866 |
| 16 | .20829 | .20791 | .20754 | .20716 | .20679 | .20642 | .20606 | .20569 | .20533 | .20496 |
| 17 | .20460 | .20424 | .20388 | .20353 | .20317 | .20282 | .20247 | .20211 | .20177 | .20142 |
| 18 | .20107 | .20073 | .20038 | .20004 | .19970 | .19936 | .19902 | .19869 | .19835 | .19802 |
| 19 | .19769 | .19736 | .19703 | .19670 | .19637 | .19605 | .19572 | .19540 | .19508 | .19476 |
| 20 | .19444 | .19412 | .19381 | .19349 | .19318 | .19287 | .19256 | .19225 | .19194 | .19164 |
| 21 | .19133 | .19103 | .19072 | .19042 | .19012 | .18982 | .18952 | .18923 | .18893 | .18864 |
| 22 | .18835 | .18805 | .18776 | .18748 | .18719 | .18690 | .18662 | .18633 | .18605 | .18577 |
| 23 | .18549 | .18521 | .18493 | .18465 | .18437 | .18410 | .18382 | .18355 | .18328 | .18301 |
| 24 | .18274 | .18247 | .18221 | .18194 | .18168 | .18141 | .18115 | .18089 | .18063 | .18037 |
| 25 | .18011 | .17985 | .17960 | .17934 | .17909 | .17883 | .17858 | .17833 | .17808 | .17783 |
| 26 | .17759 | .17734 | .17709 | .17685 | .17660 | .17636 | .17612 | .17588 | .17564 | .17540 |
| 27 | .17516 | .17493 | .17469 | .17446 | .17422 | .17399 | .17376 | .17353 | .17330 | .17307 |
| 28 | .17284 | .17262 | .17239 | .17217 | .17194 | .17172 | .17150 | .17128 | .17106 | .17084 |
| 29 | .17062 | .17040 | .17018 | .16997 | .16975 | .16954 | .16933 | .16912 | .16890 | .16869 |
| 30 | .16849 | .16828 | .16807 | .16786 | .16766 | .16745 | .16725 | .16705 | .16684 | .16664 |
| 31 | .16644 | .16624 | .16604 | .16584 | .16565 | .16545 | .16526 | .16506 | .16487 | .16467 |
| 32 | .16448 | .16429 | .16410 | .16391 | .16372 | .16353 | .16335 | .16316 | .16297 | .16279 |
| 33 | .16261 | .16242 | .16224 | .16206 | .16188 | .16170 | .16152 | .16134 | .16116 | .16099 |
| 34 | .16081 | .16063 | .16046 | .16029 | .16011 | .15994 | .15977 | .15960 | .15943 | .15926 |
| 35 | .15909 | .15892 | .15876 | .15859 | .15842 | .15826 | .15810 | .15793 | .15777 | .15761 |
| 36 | .15745 | .15729 | .15713 | .15697 | .15681 | .15665 | .15650 | .15634 | .15618 | .15603 |
| 37 | .15588 | .15572 | .15557 | .15542 | .15527 | .15512 | .15497 | .15482 | .15467 | .15452 |
| 38 | .15438 | .15423 | .15408 | .15394 | .15379 | .15365 | .15351 | .15337 | .15322 | .15308 |
| 39 | .15294 | .15280 | .15267 | .15253 | .15239 | .15225 | .15212 | .15198 | .15185 | .15171 |
| 40 | .15158 | .15145 | .15131 | .15118 | .15105 | .15092 | .15079 | .15066 | .15053 | .15041 |

*Source*: Freshwater, based on Hamme and Emerson (2004) and Eq. (1.17).

**Table 1.42** Bunsen Coefficient ($\beta''$) for Argon as a Function of Temperature (mg/(L kPa))

| Temperature (°C) | Δt (°C) | | | | | | | | | |
|---|---|---|---|---|---|---|---|---|---|---|
| | 0.0 | 0.1 | 0.2 | 0.3 | 0.4 | 0.5 | 0.6 | 0.7 | 0.8 | 0.9 |
| 0 | .94691 | .94430 | .94170 | .93912 | .93654 | .93398 | .93143 | .92889 | .92637 | .92386 |
| 1 | .92135 | .91886 | .91638 | .91392 | .91146 | .90902 | .90659 | .90417 | .90176 | .89936 |
| 2 | .89697 | .89459 | .89223 | .88988 | .88753 | .88520 | .88288 | .88057 | .87827 | .87598 |
| 3 | .87370 | .87143 | .86917 | .86693 | .86469 | .86246 | .86025 | .85804 | .85584 | .85366 |
| 4 | .85148 | .84931 | .84716 | .84501 | .84287 | .84075 | .83863 | .83652 | .83442 | .83233 |
| 5 | .83025 | .82818 | .82612 | .82407 | .82203 | .82000 | .81797 | .81596 | .81395 | .81196 |
| 6 | .80997 | .80799 | .80602 | .80406 | .80211 | .80016 | .79823 | .79630 | .79439 | .79248 |
| 7 | .79058 | .78868 | .78680 | .78492 | .78306 | .78120 | .77935 | .77751 | .77567 | .77384 |
| 8 | .77203 | .77022 | .76841 | .76662 | .76483 | .76305 | .76128 | .75952 | .75776 | .75602 |
| 9 | .75428 | .75254 | .75082 | .74910 | .74739 | .74569 | .74399 | .74230 | .74062 | .73895 |
| 10 | .73728 | .73563 | .73397 | .73233 | .73069 | .72906 | .72744 | .72582 | .72421 | .72261 |
| 11 | .72101 | .71942 | .71784 | .71627 | .71470 | .71313 | .71158 | .71003 | .70849 | .70695 |
| 12 | .70542 | .70390 | .70239 | .70088 | .69937 | .69788 | .69638 | .69490 | .69342 | .69195 |
| 13 | .69048 | .68903 | .68757 | .68612 | .68468 | .68325 | .68182 | .68040 | .67898 | .67757 |
| 14 | .67616 | .67476 | .67337 | .67198 | .67060 | .66922 | .66785 | .66649 | .66513 | .66377 |
| 15 | .66243 | .66108 | .65975 | .65842 | .65709 | .65577 | .65446 | .65315 | .65184 | .65054 |
| 16 | .64925 | .64796 | .64668 | .64540 | .64413 | .64286 | .64160 | .64034 | .63909 | .63785 |
| 17 | .63660 | .63537 | .63414 | .63291 | .63169 | .63047 | .62926 | .62806 | .62685 | .62566 |
| 18 | .62447 | .62328 | .62210 | .62092 | .61975 | .61858 | .61741 | .61626 | .61510 | .61395 |
| 19 | .61281 | .61167 | .61053 | .60940 | .60828 | .60715 | .60604 | .60492 | .60382 | .60271 |
| 20 | .60161 | .60052 | .59943 | .59834 | .59726 | .59618 | .59511 | .59404 | .59297 | .59191 |
| 21 | .59085 | .58980 | .58875 | .58771 | .58667 | .58563 | .58460 | .58358 | .58255 | .58153 |
| 22 | .58052 | .57951 | .57850 | .57749 | .57649 | .57550 | .57451 | .57352 | .57254 | .57156 |
| 23 | .57058 | .56961 | .56864 | .56767 | .56671 | .56576 | .56480 | .56385 | .56291 | .56196 |
| 24 | .56102 | .56009 | .55916 | .55823 | .55731 | .55639 | .55547 | .55456 | .55365 | .55274 |
| 25 | .55184 | .55094 | .55004 | .54915 | .54826 | .54737 | .54649 | .54561 | .54474 | .54387 |
| 26 | .54300 | .54213 | .54127 | .54041 | .53956 | .53871 | .53786 | .53701 | .53617 | .53533 |
| 27 | .53450 | .53366 | .53283 | .53201 | .53119 | .53037 | .52955 | .52874 | .52793 | .52712 |
| 28 | .52632 | .52551 | .52472 | .52392 | .52313 | .52234 | .52156 | .52077 | .51999 | .51922 |
| 29 | .51844 | .51767 | .51690 | .51614 | .51538 | .51462 | .51386 | .51311 | .51236 | .51161 |
| 30 | .51087 | .51013 | .50939 | .50865 | .50792 | .50719 | .50646 | .50573 | .50501 | .50429 |
| 31 | 0.50358 | .50286 | .50215 | .50144 | .50074 | .50003 | .49933 | .49864 | .49794 | .49725 |
| 32 | 0.49656 | .49587 | .49519 | .49451 | .49383 | .49315 | .49248 | .49180 | .49113 | .49047 |
| 33 | 0.48980 | .48914 | .48848 | .48783 | .48717 | .48652 | .48587 | .48523 | .48458 | .48394 |
| 34 | 0.48330 | .48267 | .48203 | .48140 | .48077 | .48015 | .47952 | .47890 | .47828 | .47766 |
| 35 | 0.47705 | .47643 | .47582 | .47522 | .47461 | .47401 | .47341 | .47281 | .47221 | .47162 |
| 36 | 0.47103 | .47044 | .46985 | .46927 | .46868 | .46810 | .46752 | .46695 | .46637 | .46580 |
| 37 | 0.46523 | .46467 | .46410 | .46354 | .46298 | .46242 | .46187 | .46131 | .46076 | .46021 |
| 38 | 0.45966 | .45912 | .45857 | .45803 | .45749 | .45696 | .45642 | .45589 | .45536 | .45483 |
| 39 | 0.45430 | .45378 | .45326 | .45274 | .45222 | .45170 | .45119 | .45068 | .45017 | .44966 |
| 40 | 0.44915 | .44865 | .44815 | .44765 | .44715 | .44665 | .44616 | .44567 | .44518 | .44469 |

*Source*: Freshwater, based on Hamme and Emerson (2004) and Eq. (1.17).

**Table 1.43** Bunsen Coefficient ($\beta''$) for Carbon Dioxide as a Function of Temperature (mg/(L kPa))

| Temperature (°C) | Δt (°C) | | | | | | | | | |
|---|---|---|---|---|---|---|---|---|---|---|
| | **0.0** | **0.1** | **0.2** | **0.3** | **0.4** | **0.5** | **0.6** | **0.7** | **0.8** | **0.9** |
| **0** | 33.696 | 33.562 | 33.430 | 33.298 | 33.166 | 33.036 | 32.906 | 32.777 | 32.649 | 32.521 |
| **1** | 32.394 | 32.268 | 32.142 | 32.017 | 31.893 | 31.769 | 31.646 | 31.524 | 31.402 | 31.281 |
| **2** | 31.161 | 31.041 | 30.922 | 30.804 | 30.686 | 30.569 | 30.452 | 30.336 | 30.221 | 30.106 |
| **3** | 29.992 | 29.878 | 29.765 | 29.653 | 29.541 | 29.430 | 29.320 | 29.210 | 29.100 | 28.991 |
| **4** | 28.883 | 28.775 | 28.668 | 28.562 | 28.456 | 28.350 | 28.245 | 28.141 | 28.037 | 27.934 |
| **5** | 27.831 | 27.729 | 27.627 | 27.526 | 27.425 | 27.325 | 27.225 | 27.126 | 27.028 | 26.929 |
| **6** | 26.832 | 26.735 | 26.638 | 26.542 | 26.446 | 26.351 | 26.257 | 26.162 | 26.069 | 25.975 |
| **7** | 25.883 | 25.790 | 25.699 | 25.607 | 25.516 | 25.426 | 25.336 | 25.247 | 25.158 | 25.069 |
| **8** | 24.981 | 24.893 | 24.806 | 24.719 | 24.632 | 24.547 | 24.461 | 24.376 | 24.291 | 24.207 |
| **9** | 24.123 | 24.040 | 23.957 | 23.874 | 23.792 | 23.710 | 23.629 | 23.548 | 23.467 | 23.387 |
| **10** | 23.307 | 23.228 | 23.149 | 23.070 | 22.992 | 22.914 | 22.836 | 22.759 | 22.683 | 22.606 |
| **11** | 22.530 | 22.455 | 22.379 | 22.305 | 22.230 | 22.156 | 22.082 | 22.009 | 21.936 | 21.863 |
| **12** | 21.791 | 21.719 | 21.647 | 21.576 | 21.505 | 21.434 | 21.364 | 21.294 | 21.224 | 21.155 |
| **13** | 21.086 | 21.017 | 20.949 | 20.881 | 20.813 | 20.746 | 20.679 | 20.612 | 20.546 | 20.480 |
| **14** | 20.414 | 20.349 | 20.284 | 20.219 | 20.154 | 20.090 | 20.026 | 19.963 | 19.899 | 19.836 |
| **15** | 19.774 | 19.711 | 19.649 | 19.587 | 19.526 | 19.465 | 19.404 | 19.343 | 19.282 | 19.222 |
| **16** | 19.163 | 19.103 | 19.044 | 18.985 | 18.926 | 18.868 | 18.809 | 18.751 | 18.694 | 18.636 |
| **17** | 18.579 | 18.522 | 18.466 | 18.410 | 18.353 | 18.298 | 18.242 | 18.187 | 18.132 | 18.077 |
| **18** | 18.022 | 17.968 | 17.914 | 17.860 | 17.807 | 17.753 | 17.700 | 17.647 | 17.595 | 17.542 |
| **19** | 17.490 | 17.438 | 17.387 | 17.335 | 17.284 | 17.233 | 17.183 | 17.132 | 17.082 | 17.032 |
| **20** | 16.982 | 16.932 | 16.883 | 16.834 | 16.785 | 16.736 | 16.688 | 16.639 | 16.591 | 16.544 |
| **21** | 16.496 | 16.448 | 16.401 | 16.354 | 16.307 | 16.261 | 16.215 | 16.168 | 16.122 | 16.077 |
| **22** | 16.031 | 15.986 | 15.941 | 15.896 | 15.851 | 15.806 | 15.762 | 15.718 | 15.674 | 15.630 |
| **23** | 15.586 | 15.543 | 15.500 | 15.457 | 15.414 | 15.371 | 15.329 | 15.286 | 15.244 | 15.202 |
| **24** | 15.161 | 15.119 | 15.078 | 15.036 | 14.995 | 14.954 | 14.914 | 14.873 | 14.833 | 14.793 |
| **25** | 14.753 | 14.713 | 14.673 | 14.634 | 14.595 | 14.556 | 14.517 | 14.478 | 14.439 | 14.401 |
| **26** | 14.362 | 14.324 | 14.286 | 14.248 | 14.211 | 14.173 | 14.136 | 14.099 | 14.062 | 14.025 |
| **27** | 13.988 | 13.952 | 13.915 | 13.879 | 13.843 | 13.807 | 13.771 | 13.736 | 13.700 | 13.665 |
| **28** | 13.629 | 13.594 | 13.560 | 13.525 | 13.490 | 13.456 | 13.421 | 13.387 | 13.353 | 13.319 |
| **29** | 13.286 | 13.252 | 13.218 | 13.185 | 13.152 | 13.119 | 13.086 | 13.053 | 13.021 | 12.988 |
| **30** | 12.956 | 12.923 | 12.891 | 12.859 | 12.827 | 12.796 | 12.764 | 12.733 | 12.701 | 12.670 |
| **31** | 12.639 | 12.608 | 12.577 | 12.547 | 12.516 | 12.486 | 12.455 | 12.425 | 12.395 | 12.365 |
| **32** | 12.335 | 12.305 | 12.276 | 12.246 | 12.217 | 12.188 | 12.159 | 12.130 | 12.101 | 12.072 |
| **33** | 12.043 | 12.015 | 11.986 | 11.958 | 11.930 | 11.902 | 11.874 | 11.846 | 11.818 | 11.791 |
| **34** | 11.763 | 11.736 | 11.708 | 11.681 | 11.654 | 11.627 | 11.600 | 11.573 | 11.547 | 11.520 |
| **35** | 11.494 | 11.467 | 11.441 | 11.415 | 11.389 | 11.363 | 11.337 | 11.312 | 11.286 | 11.260 |
| **36** | 11.235 | 11.210 | 11.184 | 11.159 | 11.134 | 11.109 | 11.084 | 11.060 | 11.035 | 11.011 |
| **37** | 10.986 | 10.962 | 10.938 | 10.913 | 10.889 | 10.865 | 10.841 | 10.818 | 10.794 | 10.770 |
| **38** | 10.747 | 10.723 | 10.700 | 10.677 | 10.654 | 10.631 | 10.608 | 10.585 | 10.562 | 10.539 |
| **39** | 10.517 | 10.494 | 10.472 | 10.449 | 10.427 | 10.405 | 10.383 | 10.361 | 10.339 | 10.317 |
| **40** | 10.295 | 10.274 | 10.252 | 10.230 | 10.209 | 10.188 | 10.166 | 10.145 | 10.124 | 10.103 |

*Source*: Freshwater, based on Weiss (1974) and Eq. (1.17).

**Table 1.44** Oxygen (mmHg per mg/L) as a Function of Temperature (Salinity = 0.0 g/kg)

| Temperature (°C) | Δt (°C) | | | | | | | | | |
|---|---|---|---|---|---|---|---|---|---|---|
| | **0.0** | **0.1** | **0.2** | **0.3** | **0.4** | **0.5** | **0.6** | **0.7** | **0.8** | **0.9** |
| **0** | 10.822 | 10.853 | 10.883 | 10.913 | 10.943 | 10.974 | 11.004 | 11.034 | 11.065 | 11.095 |
| **1** | 11.126 | 11.156 | 11.187 | 11.217 | 11.248 | 11.278 | 11.309 | 11.339 | 11.370 | 11.401 |
| **2** | 11.431 | 11.462 | 11.493 | 11.523 | 11.554 | 11.585 | 11.615 | 11.646 | 11.677 | 11.708 |
| **3** | 11.739 | 11.770 | 11.800 | 11.831 | 11.862 | 11.893 | 11.924 | 11.955 | 11.986 | 12.017 |
| **4** | 12.048 | 12.079 | 12.110 | 12.141 | 12.172 | 12.203 | 12.234 | 12.266 | 12.297 | 12.328 |
| **5** | 12.359 | 12.390 | 12.421 | 12.453 | 12.484 | 12.515 | 12.546 | 12.578 | 12.609 | 12.640 |
| **6** | 12.671 | 12.703 | 12.734 | 12.765 | 12.797 | 12.828 | 12.860 | 12.891 | 12.922 | 12.954 |
| **7** | 12.985 | 13.017 | 13.048 | 13.080 | 13.111 | 13.143 | 13.174 | 13.206 | 13.237 | 13.269 |
| **8** | 13.300 | 13.332 | 13.363 | 13.395 | 13.426 | 13.458 | 13.490 | 13.521 | 13.553 | 13.584 |
| **9** | 13.616 | 13.648 | 13.679 | 13.711 | 13.743 | 13.774 | 13.806 | 13.838 | 13.869 | 13.901 |
| **10** | 13.933 | 13.964 | 13.996 | 14.028 | 14.059 | 14.091 | 14.123 | 14.155 | 14.186 | 14.218 |
| **11** | 14.250 | 14.282 | 14.313 | 14.345 | 14.377 | 14.409 | 14.441 | 14.472 | 14.504 | 14.536 |
| **12** | 14.568 | 14.599 | 14.631 | 14.663 | 14.695 | 14.727 | 14.758 | 14.790 | 14.822 | 14.854 |
| **13** | 14.886 | 14.917 | 14.949 | 14.981 | 15.013 | 15.045 | 15.077 | 15.108 | 15.140 | 15.172 |
| **14** | 15.204 | 15.236 | 15.267 | 15.299 | 15.331 | 15.363 | 15.395 | 15.426 | 15.458 | 15.490 |
| **15** | 15.522 | 15.554 | 15.585 | 15.617 | 15.649 | 15.681 | 15.713 | 15.744 | 15.776 | 15.808 |
| **16** | 15.840 | 15.871 | 15.903 | 15.935 | 15.967 | 15.998 | 16.030 | 16.062 | 16.094 | 16.125 |
| **17** | 16.157 | 16.189 | 16.221 | 16.252 | 16.284 | 16.316 | 16.347 | 16.379 | 16.411 | 16.442 |
| **18** | 16.474 | 16.506 | 16.537 | 16.569 | 16.601 | 16.632 | 16.664 | 16.695 | 16.727 | 16.759 |
| **19** | 16.790 | 16.822 | 16.853 | 16.885 | 16.917 | 16.948 | 16.980 | 17.011 | 17.043 | 17.074 |
| **20** | 17.106 | 17.137 | 17.169 | 17.200 | 17.231 | 17.263 | 17.294 | 17.326 | 17.357 | 17.389 |
| **21** | 17.420 | 17.451 | 17.483 | 17.514 | 17.545 | 17.577 | 17.608 | 17.639 | 17.671 | 17.702 |
| **22** | 17.733 | 17.764 | 17.796 | 17.827 | 17.858 | 17.889 | 17.920 | 17.952 | 17.983 | 18.014 |
| **23** | 18.045 | 18.076 | 18.107 | 18.138 | 18.169 | 18.200 | 18.232 | 18.263 | 18.294 | 18.325 |
| **24** | 18.356 | 18.386 | 18.417 | 18.448 | 18.479 | 18.510 | 18.541 | 18.572 | 18.603 | 18.634 |
| **25** | 18.664 | 18.695 | 18.726 | 18.757 | 18.787 | 18.818 | 18.849 | 18.880 | 18.910 | 18.941 |
| **26** | 18.971 | 19.002 | 19.033 | 19.063 | 19.094 | 19.124 | 19.155 | 19.185 | 19.216 | 19.246 |
| **27** | 19.277 | 19.307 | 19.338 | 19.368 | 19.398 | 19.429 | 19.459 | 19.489 | 19.520 | 19.550 |
| **28** | 19.580 | 19.610 | 19.640 | 19.671 | 19.701 | 19.731 | 19.761 | 19.791 | 19.821 | 19.851 |
| **29** | 19.881 | 19.911 | 19.941 | 19.971 | 20.001 | 20.031 | 20.061 | 20.091 | 20.120 | 20.150 |
| **30** | 20.180 | 20.210 | 20.240 | 20.269 | 20.299 | 20.329 | 20.358 | 20.388 | 20.418 | 20.447 |
| **31** | 20.477 | 20.506 | 20.536 | 20.565 | 20.595 | 20.624 | 20.653 | 20.683 | 20.712 | 20.741 |
| **32** | 20.771 | 20.800 | 20.829 | 20.858 | 20.888 | 20.917 | 20.946 | 20.975 | 21.004 | 21.033 |
| **33** | 21.062 | 21.091 | 21.120 | 21.149 | 21.178 | 21.207 | 21.236 | 21.265 | 21.293 | 21.322 |
| **34** | 21.351 | 21.380 | 21.408 | 21.437 | 21.466 | 21.494 | 21.523 | 21.551 | 21.580 | 21.608 |
| **35** | 21.637 | 21.665 | 21.694 | 21.722 | 21.750 | 21.779 | 21.807 | 21.835 | 21.863 | 21.892 |
| **36** | 21.920 | 21.948 | 21.976 | 22.004 | 22.032 | 22.060 | 22.088 | 22.116 | 22.144 | 22.172 |
| **37** | 22.200 | 22.228 | 22.255 | 22.283 | 22.311 | 22.339 | 22.366 | 22.394 | 22.422 | 22.449 |
| **38** | 22.477 | 22.504 | 22.532 | 22.559 | 22.587 | 22.614 | 22.641 | 22.669 | 22.696 | 22.723 |
| **39** | 22.750 | 22.778 | 22.805 | 22.832 | 22.859 | 22.886 | 22.913 | 22.940 | 22.967 | 22.994 |
| **40** | 23.021 | 23.048 | 23.075 | 23.101 | 23.128 | 23.155 | 23.181 | 23.208 | 23.235 | 23.261 |

**Table 1.45** Nitrogen (mmHg per mg/L) as a Function of Temperature (Salinity = 0.0 g/kg)

| Temperature (°C) | Δt (°C) | | | | | | | | | |
|---|---|---|---|---|---|---|---|---|---|---|
| | 0.0 | 0.1 | 0.2 | 0.3 | 0.4 | 0.5 | 0.6 | 0.7 | 0.8 | 0.9 |
| 0 | 25.359 | 25.425 | 25.491 | 25.557 | 25.624 | 25.690 | 25.756 | 25.823 | 25.889 | 25.956 |
| 1 | 26.022 | 26.088 | 26.155 | 26.221 | 26.288 | 26.355 | 26.421 | 26.488 | 26.554 | 26.621 |
| 2 | 26.688 | 26.754 | 26.821 | 26.888 | 26.954 | 27.021 | 27.088 | 27.155 | 27.222 | 27.288 |
| 3 | 27.355 | 27.422 | 27.489 | 27.556 | 27.623 | 27.690 | 27.757 | 27.824 | 27.890 | 27.957 |
| 4 | 28.024 | 28.091 | 28.158 | 28.225 | 28.292 | 28.359 | 28.427 | 28.494 | 28.561 | 28.628 |
| 5 | 28.695 | 28.762 | 28.829 | 28.896 | 28.963 | 29.030 | 29.097 | 29.164 | 29.231 | 29.299 |
| 6 | 29.366 | 29.433 | 29.500 | 29.567 | 29.634 | 29.701 | 29.768 | 29.835 | 29.903 | 29.970 |
| 7 | 30.037 | 30.104 | 30.171 | 30.238 | 30.305 | 30.372 | 30.439 | 30.506 | 30.574 | 30.641 |
| 8 | 30.708 | 30.775 | 30.842 | 30.909 | 30.976 | 31.043 | 31.110 | 31.177 | 31.244 | 31.311 |
| 9 | 31.378 | 31.445 | 31.512 | 31.579 | 31.646 | 31.713 | 31.779 | 31.846 | 31.913 | 31.980 |
| 10 | 32.047 | 32.114 | 32.180 | 32.247 | 32.314 | 32.381 | 32.447 | 32.514 | 32.581 | 32.648 |
| 11 | 32.714 | 32.781 | 32.847 | 32.914 | 32.981 | 33.047 | 33.114 | 33.180 | 33.247 | 33.313 |
| 12 | 33.379 | 33.446 | 33.512 | 33.579 | 33.645 | 33.711 | 33.777 | 33.844 | 33.910 | 33.976 |
| 13 | 34.042 | 34.108 | 34.174 | 34.241 | 34.307 | 34.373 | 34.439 | 34.504 | 34.570 | 34.636 |
| 14 | 34.702 | 34.768 | 34.834 | 34.899 | 34.965 | 35.031 | 35.096 | 35.162 | 35.228 | 35.293 |
| 15 | 35.359 | 35.424 | 35.489 | 35.555 | 35.620 | 35.685 | 35.751 | 35.816 | 35.881 | 35.946 |
| 16 | 36.011 | 36.076 | 36.141 | 36.206 | 36.271 | 36.336 | 36.401 | 36.466 | 36.530 | 36.595 |
| 17 | 36.660 | 36.724 | 36.789 | 36.853 | 36.918 | 36.982 | 37.046 | 37.111 | 37.175 | 37.239 |
| 18 | 37.303 | 37.367 | 37.431 | 37.495 | 37.559 | 37.623 | 37.687 | 37.751 | 37.815 | 37.878 |
| 19 | 37.942 | 38.006 | 38.069 | 38.133 | 38.196 | 38.259 | 38.323 | 38.386 | 38.449 | 38.512 |
| 20 | 38.575 | 38.638 | 38.701 | 38.764 | 38.827 | 38.890 | 38.952 | 39.015 | 39.078 | 39.140 |
| 21 | 39.203 | 39.265 | 39.327 | 39.389 | 39.452 | 39.514 | 39.576 | 39.638 | 39.700 | 39.762 |
| 22 | 39.824 | 39.885 | 39.947 | 40.009 | 40.070 | 40.132 | 40.193 | 40.254 | 40.316 | 40.377 |
| 23 | 40.438 | 40.499 | 40.560 | 40.621 | 40.682 | 40.742 | 40.803 | 40.864 | 40.924 | 40.985 |
| 24 | 41.045 | 41.105 | 41.166 | 41.226 | 41.286 | 41.346 | 41.406 | 41.466 | 41.525 | 41.585 |
| 25 | 41.645 | 41.704 | 41.764 | 41.823 | 41.883 | 41.942 | 42.001 | 42.060 | 42.119 | 42.178 |
| 26 | 42.237 | 42.295 | 42.354 | 42.413 | 42.471 | 42.530 | 42.588 | 42.646 | 42.704 | 42.762 |
| 27 | 42.820 | 42.878 | 42.936 | 42.994 | 43.051 | 43.109 | 43.166 | 43.224 | 43.281 | 43.338 |
| 28 | 43.395 | 43.452 | 43.509 | 43.566 | 43.623 | 43.679 | 43.736 | 43.792 | 43.849 | 43.905 |
| 29 | 43.961 | 44.017 | 44.073 | 44.129 | 44.185 | 44.241 | 44.296 | 44.352 | 44.407 | 44.463 |
| 30 | 44.518 | 44.573 | 44.628 | 44.683 | 44.738 | 44.793 | 44.847 | 44.902 | 44.956 | 45.011 |
| 31 | 45.065 | 45.119 | 45.173 | 45.227 | 45.281 | 45.334 | 45.388 | 45.442 | 45.495 | 45.548 |
| 32 | 45.601 | 45.655 | 45.708 | 45.760 | 45.813 | 45.866 | 45.918 | 45.971 | 46.023 | 46.075 |
| 33 | 46.128 | 46.180 | 46.232 | 46.283 | 46.335 | 46.387 | 46.438 | 46.489 | 46.541 | 46.592 |
| 34 | 46.643 | 46.694 | 46.745 | 46.795 | 46.846 | 46.896 | 46.947 | 46.997 | 47.047 | 47.097 |
| 35 | 47.147 | 47.197 | 47.246 | 47.296 | 47.345 | 47.394 | 47.444 | 47.493 | 47.542 | 47.590 |
| 36 | 47.639 | 47.688 | 47.736 | 47.784 | 47.833 | 47.881 | 47.929 | 47.976 | 48.024 | 48.072 |
| 37 | 48.119 | 48.167 | 48.214 | 48.261 | 48.308 | 48.355 | 48.401 | 48.448 | 48.494 | 48.541 |
| 38 | 48.587 | 48.633 | 48.679 | 48.725 | 48.770 | 48.816 | 48.861 | 48.907 | 48.952 | 48.997 |
| 39 | 49.042 | 49.086 | 49.131 | 49.176 | 49.220 | 49.264 | 49.308 | 49.352 | 49.396 | 49.440 |
| 40 | 49.483 | 49.527 | 49.570 | 49.613 | 49.656 | 49.699 | 49.742 | 49.784 | 49.827 | 49.869 |

**Table 1.46** Argon (mmHg per mg/L) as a Function of Temperature (Salinity = 0.0 g/kg)

| Temperature (°C) | Δ*t* (°C) | | | | | | | | | |
|---|---|---|---|---|---|---|---|---|---|---|
| | **0.0** | **0.1** | **0.2** | **0.3** | **0.4** | **0.5** | **0.6** | **0.7** | **0.8** | **0.9** |
| **0** | 7.921 | 7.943 | 7.965 | 7.987 | 8.009 | 8.031 | 8.053 | 8.075 | 8.097 | 8.119 |
| **1** | 8.141 | 8.163 | 8.185 | 8.207 | 8.229 | 8.251 | 8.273 | 8.296 | 8.318 | 8.340 |
| **2** | 8.362 | 8.384 | 8.407 | 8.429 | 8.451 | 8.473 | 8.496 | 8.518 | 8.540 | 8.563 |
| **3** | 8.585 | 8.607 | 8.630 | 8.652 | 8.674 | 8.697 | 8.719 | 8.742 | 8.764 | 8.786 |
| **4** | 8.809 | 8.831 | 8.854 | 8.876 | 8.899 | 8.921 | 8.944 | 8.966 | 8.989 | 9.012 |
| **5** | 9.034 | 9.057 | 9.079 | 9.102 | 9.124 | 9.147 | 9.170 | 9.192 | 9.215 | 9.238 |
| **6** | 9.260 | 9.283 | 9.306 | 9.328 | 9.351 | 9.374 | 9.397 | 9.419 | 9.442 | 9.465 |
| **7** | 9.488 | 9.510 | 9.533 | 9.556 | 9.579 | 9.601 | 9.624 | 9.647 | 9.670 | 9.693 |
| **8** | 9.715 | 9.738 | 9.761 | 9.784 | 9.807 | 9.830 | 9.853 | 9.875 | 9.898 | 9.921 |
| **9** | 9.944 | 9.967 | 9.990 | 10.013 | 10.036 | 10.059 | 10.082 | 10.105 | 10.127 | 10.150 |
| **10** | 10.173 | 10.196 | 10.219 | 10.242 | 10.265 | 10.288 | 10.311 | 10.334 | 10.357 | 10.380 |
| **11** | 10.403 | 10.426 | 10.449 | 10.472 | 10.495 | 10.518 | 10.541 | 10.564 | 10.587 | 10.610 |
| **12** | 10.633 | 10.656 | 10.679 | 10.702 | 10.725 | 10.748 | 10.771 | 10.794 | 10.817 | 10.840 |
| **13** | 10.863 | 10.886 | 10.909 | 10.932 | 10.955 | 10.978 | 11.001 | 11.024 | 11.047 | 11.070 |
| **14** | 11.093 | 11.116 | 11.139 | 11.162 | 11.185 | 11.208 | 11.231 | 11.254 | 11.277 | 11.300 |
| **15** | 11.323 | 11.346 | 11.369 | 11.392 | 11.415 | 11.438 | 11.461 | 11.484 | 11.507 | 11.530 |
| **16** | 11.553 | 11.576 | 11.599 | 11.622 | 11.645 | 11.668 | 11.690 | 11.713 | 11.736 | 11.759 |
| **17** | 11.782 | 11.805 | 11.828 | 11.851 | 11.874 | 11.897 | 11.920 | 11.943 | 11.965 | 11.988 |
| **18** | 12.011 | 12.034 | 12.057 | 12.080 | 12.103 | 12.126 | 12.148 | 12.171 | 12.194 | 12.217 |
| **19** | 12.240 | 12.263 | 12.285 | 12.308 | 12.331 | 12.354 | 12.377 | 12.399 | 12.422 | 12.445 |
| **20** | 12.468 | 12.490 | 12.513 | 12.536 | 12.558 | 12.581 | 12.604 | 12.627 | 12.649 | 12.672 |
| **21** | 12.695 | 12.717 | 12.740 | 12.762 | 12.785 | 12.808 | 12.830 | 12.853 | 12.875 | 12.898 |
| **22** | 12.921 | 12.943 | 12.966 | 12.988 | 13.011 | 13.033 | 13.056 | 13.078 | 13.101 | 13.123 |
| **23** | 13.146 | 13.168 | 13.190 | 13.213 | 13.235 | 13.258 | 13.280 | 13.302 | 13.325 | 13.347 |
| **24** | 13.369 | 13.392 | 13.414 | 13.436 | 13.459 | 13.481 | 13.503 | 13.525 | 13.548 | 13.570 |
| **25** | 13.592 | 13.614 | 13.636 | 13.659 | 13.681 | 13.703 | 13.725 | 13.747 | 13.769 | 13.791 |
| **26** | 13.813 | 13.835 | 13.857 | 13.879 | 13.901 | 13.923 | 13.945 | 13.967 | 13.989 | 14.011 |
| **27** | 14.033 | 14.055 | 14.077 | 14.099 | 14.121 | 14.142 | 14.164 | 14.186 | 14.208 | 14.229 |
| **28** | 14.251 | 14.273 | 14.295 | 14.316 | 14.338 | 14.360 | 14.381 | 14.403 | 14.424 | 14.446 |
| **29** | 14.468 | 14.489 | 14.511 | 14.532 | 14.554 | 14.575 | 14.597 | 14.618 | 14.639 | 14.661 |
| **30** | 14.682 | 14.703 | 14.725 | 14.746 | 14.767 | 14.789 | 14.810 | 14.831 | 14.852 | 14.874 |
| **31** | 14.895 | 14.916 | 14.937 | 14.958 | 14.979 | 15.000 | 15.021 | 15.042 | 15.063 | 15.084 |
| **32** | 15.105 | 15.126 | 15.147 | 15.168 | 15.189 | 15.210 | 15.230 | 15.251 | 15.272 | 15.293 |
| **33** | 15.313 | 15.334 | 15.355 | 15.376 | 15.396 | 15.417 | 15.437 | 15.458 | 15.478 | 15.499 |
| **34** | 15.519 | 15.540 | 15.560 | 15.581 | 15.601 | 15.622 | 15.642 | 15.662 | 15.682 | 15.703 |
| **35** | 15.723 | 15.743 | 15.763 | 15.784 | 15.804 | 15.824 | 15.844 | 15.864 | 15.884 | 15.904 |
| **36** | 15.924 | 15.944 | 15.964 | 15.984 | 16.004 | 16.023 | 16.043 | 16.063 | 16.083 | 16.103 |
| **37** | 16.122 | 16.142 | 16.162 | 16.181 | 16.201 | 16.220 | 16.240 | 16.259 | 16.279 | 16.298 |
| **38** | 16.318 | 16.337 | 16.356 | 16.376 | 16.395 | 16.414 | 16.434 | 16.453 | 16.472 | 16.491 |
| **39** | 16.510 | 16.529 | 16.548 | 16.567 | 16.586 | 16.605 | 16.624 | 16.643 | 16.662 | 16.681 |
| **40** | 16.699 | 16.718 | 16.737 | 16.756 | 16.774 | 16.793 | 16.812 | 16.830 | 16.849 | 16.867 |

**Table 1.47** Nitrogen + Argon (mmHg per mg/L) as a Function of Temperature (Salinity = 0.0 g/kg)

| Temperature (°C) | Δt (°C) | | | | | | | | | |
|---|---|---|---|---|---|---|---|---|---|---|
| | **0.0** | **0.1** | **0.2** | **0.3** | **0.4** | **0.5** | **0.6** | **0.7** | **0.8** | **0.9** |
| **0** | 24.716 | 24.780 | 24.845 | 24.910 | 24.974 | 25.039 | 25.104 | 25.169 | 25.234 | 25.299 |
| **1** | 25.363 | 25.428 | 25.493 | 25.558 | 25.623 | 25.688 | 25.753 | 25.818 | 25.884 | 25.949 |
| **2** | 26.014 | 26.079 | 26.144 | 26.209 | 26.274 | 26.340 | 26.405 | 26.470 | 26.535 | 26.601 |
| **3** | 26.666 | 26.731 | 26.797 | 26.862 | 26.927 | 26.993 | 27.058 | 27.124 | 27.189 | 27.255 |
| **4** | 27.320 | 27.385 | 27.451 | 27.516 | 27.582 | 27.647 | 27.713 | 27.778 | 27.844 | 27.910 |
| **5** | 27.975 | 28.041 | 28.106 | 28.172 | 28.237 | 28.303 | 28.369 | 28.434 | 28.500 | 28.565 |
| **6** | 28.631 | 28.697 | 28.762 | 28.828 | 28.893 | 28.959 | 29.025 | 29.090 | 29.156 | 29.221 |
| **7** | 29.287 | 29.353 | 29.418 | 29.484 | 29.549 | 29.615 | 29.681 | 29.746 | 29.812 | 29.877 |
| **8** | 29.943 | 30.009 | 30.074 | 30.140 | 30.205 | 30.271 | 30.336 | 30.402 | 30.467 | 30.533 |
| **9** | 30.598 | 30.664 | 30.729 | 30.795 | 30.860 | 30.926 | 30.991 | 31.056 | 31.122 | 31.187 |
| **10** | 31.253 | 31.318 | 31.383 | 31.449 | 31.514 | 31.579 | 31.644 | 31.710 | 31.775 | 31.840 |
| **11** | 31.905 | 31.971 | 32.036 | 32.101 | 32.166 | 32.231 | 32.296 | 32.361 | 32.426 | 32.491 |
| **12** | 32.556 | 32.621 | 32.686 | 32.751 | 32.816 | 32.881 | 32.946 | 33.010 | 33.075 | 33.140 |
| **13** | 33.205 | 33.269 | 33.334 | 33.399 | 33.463 | 33.528 | 33.593 | 33.657 | 33.722 | 33.786 |
| **14** | 33.851 | 33.915 | 33.979 | 34.044 | 34.108 | 34.172 | 34.236 | 34.301 | 34.365 | 34.429 |
| **15** | 34.493 | 34.557 | 34.621 | 34.685 | 34.749 | 34.813 | 34.877 | 34.941 | 35.005 | 35.068 |
| **16** | 35.132 | 35.196 | 35.259 | 35.323 | 35.386 | 35.450 | 35.513 | 35.577 | 35.640 | 35.704 |
| **17** | 35.767 | 35.830 | 35.893 | 35.957 | 36.020 | 36.083 | 36.146 | 36.209 | 36.272 | 36.335 |
| **18** | 36.397 | 36.460 | 36.523 | 36.586 | 36.648 | 36.711 | 36.773 | 36.836 | 36.898 | 36.961 |
| **19** | 37.023 | 37.085 | 37.148 | 37.210 | 37.272 | 37.334 | 37.396 | 37.458 | 37.520 | 37.582 |
| **20** | 37.643 | 37.705 | 37.767 | 37.828 | 37.890 | 37.952 | 38.013 | 38.074 | 38.136 | 38.197 |
| **21** | 38.258 | 38.319 | 38.380 | 38.441 | 38.502 | 38.563 | 38.624 | 38.685 | 38.746 | 38.806 |
| **22** | 38.867 | 38.927 | 38.988 | 39.048 | 39.109 | 39.169 | 39.229 | 39.289 | 39.349 | 39.409 |
| **23** | 39.469 | 39.529 | 39.589 | 39.649 | 39.708 | 39.768 | 39.827 | 39.887 | 39.946 | 40.006 |
| **24** | 40.065 | 40.124 | 40.183 | 40.242 | 40.301 | 40.360 | 40.419 | 40.477 | 40.536 | 40.595 |
| **25** | 40.653 | 40.711 | 40.770 | 40.828 | 40.886 | 40.944 | 41.002 | 41.060 | 41.118 | 41.176 |
| **26** | 41.234 | 41.291 | 41.349 | 41.407 | 41.464 | 41.521 | 41.578 | 41.636 | 41.693 | 41.750 |
| **27** | 41.807 | 41.863 | 41.920 | 41.977 | 42.033 | 42.090 | 42.146 | 42.203 | 42.259 | 42.315 |
| **28** | 42.371 | 42.427 | 42.483 | 42.539 | 42.595 | 42.650 | 42.706 | 42.761 | 42.816 | 42.872 |
| **29** | 42.927 | 42.982 | 43.037 | 43.092 | 43.147 | 43.201 | 43.256 | 43.311 | 43.365 | 43.419 |
| **30** | 43.474 | 43.528 | 43.582 | 43.636 | 43.690 | 43.744 | 43.797 | 43.851 | 43.904 | 43.958 |
| **31** | 44.011 | 44.064 | 44.117 | 44.170 | 44.223 | 44.276 | 44.329 | 44.381 | 44.434 | 44.486 |
| **32** | 44.539 | 44.591 | 44.643 | 44.695 | 44.747 | 44.799 | 44.850 | 44.902 | 44.953 | 45.005 |
| **33** | 45.056 | 45.107 | 45.158 | 45.209 | 45.260 | 45.311 | 45.361 | 45.412 | 45.462 | 45.513 |
| **34** | 45.563 | 45.613 | 45.663 | 45.713 | 45.762 | 45.812 | 45.862 | 45.911 | 45.960 | 46.010 |
| **35** | 46.059 | 46.108 | 46.157 | 46.205 | 46.254 | 46.302 | 46.351 | 46.399 | 46.447 | 46.495 |
| **36** | 46.543 | 46.591 | 46.639 | 46.686 | 46.734 | 46.781 | 46.829 | 46.876 | 46.923 | 46.970 |
| **37** | 47.016 | 47.063 | 47.109 | 47.156 | 47.202 | 47.248 | 47.294 | 47.340 | 47.386 | 47.432 |
| **38** | 47.477 | 47.523 | 47.568 | 47.613 | 47.658 | 47.703 | 47.748 | 47.792 | 47.837 | 47.881 |
| **39** | 47.926 | 47.970 | 48.014 | 48.058 | 48.101 | 48.145 | 48.188 | 48.232 | 48.275 | 48.318 |
| **40** | 48.361 | 48.404 | 48.447 | 48.489 | 48.532 | 48.574 | 48.616 | 48.658 | 48.700 | 48.742 |

**Table 1.48** Carbon Dioxide (mmHg per mg/L) as a Function of Temperature (Salinity = 0.0 g/kg)

| Temperature (°C) | Δt (°C) | | | | | | | | | |
|---|---|---|---|---|---|---|---|---|---|---|
| | **0.0** | **0.1** | **0.2** | **0.3** | **0.4** | **0.5** | **0.6** | **0.7** | **0.8** | **0.9** |
| **0** | .22260 | .22348 | .22437 | .22526 | .22615 | .22704 | .22794 | .22884 | .22974 | .23064 |
| **1** | .23154 | .23245 | .23336 | .23427 | .23518 | .23610 | .23702 | .23793 | .23886 | .23978 |
| **2** | .24071 | .24164 | .24257 | .24350 | .24443 | .24537 | .24631 | .24725 | .24819 | .24914 |
| **3** | .25009 | .25104 | .25199 | .25295 | .25390 | .25486 | .25582 | .25679 | .25775 | .25872 |
| **4** | .25969 | .26066 | .26164 | .26261 | .26359 | .26457 | .26555 | .26654 | .26753 | .26852 |
| **5** | .26951 | .27050 | .27150 | .27250 | .27350 | .27450 | .27550 | .27651 | .27752 | .27853 |
| **6** | .27954 | .28056 | .28157 | .28259 | .28362 | .28464 | .28567 | .28669 | .28772 | .28876 |
| **7** | .28979 | .29083 | .29187 | .29291 | .29395 | .29500 | .29604 | .29709 | .29815 | .29920 |
| **8** | .30026 | .30131 | .30237 | .30344 | .30450 | .30557 | .30664 | .30771 | .30878 | .30985 |
| **9** | .31093 | .31201 | .31309 | .31418 | .31526 | .31635 | .31744 | .31853 | .31962 | .32072 |
| **10** | .32182 | .32292 | .32402 | .32513 | .32623 | .32734 | .32845 | .32956 | .33068 | .33179 |
| **11** | .33291 | .33403 | .33516 | .33628 | .33741 | .33854 | .33967 | .34080 | .34194 | .34307 |
| **12** | .34421 | .34536 | .34650 | .34764 | .34879 | .34994 | .35109 | .35225 | .35340 | .35456 |
| **13** | .35572 | .35688 | .35804 | .35921 | .36038 | .36154 | .36272 | .36389 | .36507 | .36624 |
| **14** | .36742 | .36860 | .36979 | .37097 | .37216 | .37335 | .37454 | .37573 | .37693 | .37813 |
| **15** | .37932 | .38053 | .38173 | .38293 | .38414 | .38535 | .38656 | .38777 | .38899 | .39020 |
| **16** | .39142 | .39264 | .39386 | .39509 | .39631 | .39754 | .39877 | .40000 | .40124 | .40247 |
| **17** | .40371 | .40495 | .40619 | .40743 | .40868 | .40992 | .41117 | .41242 | .41367 | .41493 |
| **18** | .41618 | .41744 | .41870 | .41996 | .42123 | .42249 | .42376 | .42503 | .42630 | .42757 |
| **19** | .42884 | .43012 | .43140 | .43268 | .43396 | .43524 | .43652 | .43781 | .43910 | .44039 |
| **20** | .44168 | .44298 | .44427 | .44557 | .44687 | .44817 | .44947 | .45077 | .45208 | .45339 |
| **21** | .45470 | .45601 | .45732 | .45863 | .45995 | .46127 | .46259 | .46391 | .46523 | .46655 |
| **22** | .46788 | .46921 | .47054 | .47187 | .47320 | .47454 | .47587 | .47721 | .47855 | .47989 |
| **23** | .48123 | .48258 | .48392 | .48527 | .48662 | .48797 | .48932 | .49068 | .49203 | .49339 |
| **24** | .49475 | .49611 | .49747 | .49883 | .50020 | .50156 | .50293 | .50430 | .50567 | .50704 |
| **25** | .50842 | .50979 | .51117 | .51255 | .51393 | .51531 | .51669 | .51808 | .51946 | .52085 |
| **26** | .52224 | .52363 | .52502 | .52642 | .52781 | .52921 | .53061 | .53201 | .53341 | .53481 |
| **27** | .53621 | .53762 | .53902 | .54043 | .54184 | .54325 | .54466 | .54607 | .54749 | .54891 |
| **28** | .55032 | .55174 | .55316 | .55458 | .55601 | .55743 | .55885 | .56028 | .56171 | .56314 |
| **29** | .56457 | .56600 | .56744 | .56887 | .57031 | .57174 | .57318 | .57462 | .57606 | .57750 |
| **30** | .57895 | .58039 | .58184 | .58329 | .58473 | .58618 | .58763 | .58909 | .59054 | .59199 |
| **31** | .59345 | .59491 | .59636 | .59782 | .59928 | .60075 | .60221 | .60367 | .60514 | .60660 |
| **32** | .60807 | .60954 | .61101 | .61248 | .61395 | .61542 | .61690 | .61837 | .61985 | .62132 |
| **33** | .62280 | .62428 | .62576 | .62724 | .62873 | .63021 | .63169 | .63318 | .63467 | .63615 |
| **34** | .63764 | .63913 | .64062 | .64211 | .64361 | .64510 | .64659 | .64809 | .64959 | .65108 |
| **35** | .65258 | .65408 | .65558 | .65708 | .65858 | .66009 | .66159 | .66310 | .66460 | .66611 |
| **36** | .66761 | .66912 | .67063 | .67214 | .67365 | .67516 | .67668 | .67819 | .67970 | .68122 |
| **37** | .68274 | .68425 | .68577 | .68729 | .68881 | .69033 | .69185 | .69337 | .69489 | .69641 |
| **38** | .69794 | .69946 | .70099 | .70251 | .70404 | .70556 | .70709 | .70862 | .71015 | .71168 |
| **39** | .71321 | .71474 | .71627 | .71781 | .71934 | .72087 | .72241 | .72394 | .72548 | .72702 |
| **40** | .72855 | .73009 | .73163 | .73317 | .73471 | .73625 | .73779 | .73933 | .74087 | .74241 |

# 2 Solubility of Atmospheric Gases in Brackish and Marine Waters

The solubilities of oxygen, nitrogen, argon, and carbon dioxide in terms of standard air solubility concentrations and Bunsen coefficients have been computed as functions of salinity and temperature from the equations presented in Appendix A. Values are given at coarse salinity intervals for 0–40 g/kg (which embraces the range likely to be encountered in most common nearshore conditions) and at finer intervals for 33–37 g/kg for open water conditions. Once the $C_o^\dagger$ or $C_o^*$ value has been determined for a given temperature and salinity, the solubility can be adjusted for depth, elevation, or barometric pressure by equations presented in Chapter 1. Conversion factors for other solubility units have been presented in Table 1.1{4}. The Bunsen coefficient can be used to compute the solubility of a gas with an arbitrary mole fraction. These tabular data may be found as follows.

## Standard Air Solubility Concentrations

| Gas (g/kg) | $C_o^\dagger$ (μmol/kg) | | $C_o^*$ (mg/L) | |
|---|---|---|---|---|
| | **Table** | **Page** | **Table** | **Page** |
| Oxygen | | | | |
| 0–40 | 2.1 | 78 | 2.11 | 88 |
| 33–37 | 2.2 | 79 | 2.12 | 89 |
| Nitrogen | | | | |
| 0–40 | 2.3 | 80 | 2.13 | 90 |
| 33–37 | 2.4 | 81 | 2.14 | 91 |
| Argon | | | | |
| 0–40 | 2.5 | 82 | 2.15 | 92 |
| 33–37 | 2.6 | 83 | 2.16 | 93 |
| Carbon dioxide—2010 | | | | |
| 0–40 | 2.7 | 84 | 2.17 | 94 |
| 33–37 | 2.8 | 85 | 2.18 | 95 |
| Carbon dioxide—2030 | | | | |
| 0–40 | 2.9 | 86 | 2.19 | 96 |
| 33–37 | 2.10 | 87 | 2.20 | 97 |

**Dissolved Gas Concentration in Water. DOI: 10.1016/B978-0-12-415916-7.00002-4**

Once the $C_o^{\dagger}$ or $C_o^{*}$ values have been determined for a given temperature and salinity, the solubility can be adjusted for depth, elevation, or barometric pressure by equations presented previously. Conversion factors for other solubility units have been presented in Table 1.1{4}.

## Standard Air Solubility Concentration for Carbon Dioxide as a Function of Mole Fraction—Seawater

The standard air solubility concentration factor ($F$) is presented below for mass and volume units for two ranges of salinity.

| Gas (g/kg) | $F^{\dagger}$ (μmol/kg) | | $F^{*}$ (μmol/L) | |
|---|---|---|---|---|
| | Table | Page | Table | Page |
| 0–40 | 2.21 | 98 | 2.23 | 100 |
| 33–37 | 2.22 | 99 | 2.24 | 101 |

## Bunsen Coefficients

| Gas (g/kg) | Bunsen coefficient ($\beta$) (L real gas/(L atm)) | | Bunsen coefficient ($\beta'$) (mg/(L mmHg)) | |
|---|---|---|---|---|
| | Table | Page | Table | Page |
| Oxygen | | | | |
| 0–40 | 2.25 | 102 | 2.33 | 110 |
| 33–37 | 2.26 | 103 | 2.34 | 111 |
| Nitrogen | | | | |
| 0–40 | 2.27 | 104 | 2.35 | 112 |
| 33–37 | 2.28 | 105 | 2.36 | 113 |
| Argon | | | | |
| 0–40 | 2.29 | 106 | 2.37 | 114 |
| 33–37 | 2.30 | 107 | 2.38 | 115 |
| Carbon dioxide | | | | |
| 0–40 | 2.31 | 108 | 2.39 | 116 |
| 33–37 | 2.32 | 109 | 2.40 | 117 |

## Vapor Pressure of Water ($P_{wv}$)—Seawater

The vapor pressure of seawater is presented in Table 2.41{118} (mmHg) and Table 2.42{119} (kPa) as a function of salinity and temperature.

## Hydrostatic Head of Water—Seawater

The hydrostatic head of water is presented in Table 2.43{120} (mmHg/m) and Table 2.44{121} (kPa/m) as a function of salinity and temperature.

## Computation of Gas Tension, mmHg—Seawater

The conversion factors between concentration in mg/L and mmHg are presented for two ranges of salinities in the tables below:

| Gas (g/kg) | mmHg/(mg/L) | |
|---|---|---|
| | Table | Page |
| Oxygen | | |
| 0–40 | 2.45 | 122 |
| 33–37 | 2.46 | 123 |
| Nitrogen | | |
| 0–40 | 2.47 | 124 |
| 33–37 | 2.48 | 125 |
| Argon | | |
| 0–40 | 2.49 | 126 |
| 33–37 | 2.50 | 127 |
| Nitrogen + Argon | | |
| 0–40 | 2.51 | 128 |
| 33–37 | 2.52 | 129 |
| Carbon dioxide | | |
| 0–40 | 2.53 | 130 |
| 33–37 | 2.54 | 131 |

### Example 2-1

Compute the standard air solubility concentration of the four major atmospheric gases in μmol/kg at 23°C and for salinities equal to 0, 5, and 40 g/kg. Use Tables 2.1, 2.3, 2.5, and 2.7.

| Gas | 0 g/kg | 5 g/kg | 40 g/kg | Source |
|---|---|---|---|---|
| $O_2$ | 268.73 | 260.12 | 207.05 | Table 2.1 |
| $N_2$ | 510.59 | 493.10 | 386.97 | Table 2.3 |
| Ar | 13.175 | 12.756 | 10.174 | Table 2.5 |
| $CO_2$ | 13.594 | 13.240 | 11.009 | Table 2.7 |
| Total | 806.089 | 779.216 | 615.203 | |

## Example 2-2

Compute the standard air solubility concentration of oxygen at 0°C and 40°C for freshwater and salinity = 40 g/kg in terms of mg/L and gas tension in mmHg. Compare the gas tension with the partial pressure in the air. Use Eqs (3.4) and (3.5).

### Data Needed

| Temperature (°C) | Salinity (g/kg) | $C_o^*$ (mg/L) | Source | $\beta$ (L/(L atm)) | Source |
|---|---|---|---|---|---|
| 0 | 0 | 14.621 | Table 1.9 | 0.04914 | Table 1.32 |
| 40 | 0 | 6.412 | Table 1.9 | 0.02310 | Table 1.32 |
| 0 | 40 | 11.050 | Table 2.11 | 0.03714 | Table 2.25 |
| 40 | 40 | 5.215 | Table 2.11 | 0.01876 | Table 2.25 |

| Temperature (°C) | Salinity (g/kg) | $P_{wv}$ (mmHg) | Source | Parameter | Value | Source |
|---|---|---|---|---|---|---|
| 0 | 0 | 4.581 | Table 2.41 | $A_{O_2}$ | 0.5318 | Table D-1 |
| 40 | 0 | 55.364 | Table 2.41 | $\chi_{O_2}$ | 0.20946 | Table D-1 |
| 0 | 40 | 4.482 | Table 2.41 | | | |
| 40 | 40 | 54.171 | Table 2.41 | | | |

### Reduction in Standard Air Solubility in Terms of mg/L

Using the solubility information presented above:

| Temperature (°C) | Freshwater | 40 g/kg | % Change |
|---|---|---|---|
| 0 | 14.621 | 11.050 | −24.42 |
| 40 | 6.412 | 5.215 | −18.57 |

Increasing the salinity from 0 to 40 g/kg significantly reduces the standard air solubility concentration of oxygen.

### Reduction in Standard Air Solubility in Terms of Gas Tension (mmHg)

From Eq. (3.4):

$$\text{Gas tension (mmHg)} = C\left[\frac{A_i}{\beta_i}\right]$$

For 0°C and salinity = 0 g/kg:

$$\text{Gas tension (mmHg)} = 14.621\left[\frac{0.5318}{0.04914}\right] = 158.23 \text{ mmHg}$$

| Temperature (°C) | Freshwater | 40 g/kg | % Change |
|---|---|---|---|
| 0 | **158.23** | 147.62 | −6.71 |
| 40 | 158.22 | 147.83 | −6.57 |

The gas tension of oxygen is reduced when the salinity is increased from 0 to 40 g/kg, but much less than when expressed as mg/L. Increasing temperature has a negligible impact on gas tension. In respiratory physiology, gas tension (gas tension gradients) can be more important than concentrations or concentration gradients.

**Compute the Partial Pressure of Oxygen in Air**

From Eq. (3.5):

$$P_i^g = \chi_i\,(BP - P_{wv})$$

For 0°C and salinity = 0 g/kg:

$$P_{O_2}^g = 0.20946(760.0 - 4.581) = 158.23 \text{ mmHg}$$

| Temperature (°C) | Salinity (g/kg) | Gas Tension (mmHg) | Partial Pressure (mmHg) |
|---|---|---|---|
| 0 | 0 | 158.23 | 158.23 |
| 40 | 0 | 147.62 | 147.59 |
| 0 | 40 | 158.22 | 158.25 |
| 40 | 40 | 147.83 | 147.84 |

The gas tension in the water is equal to the partial pressure in the air at saturation (within computational uncertainty).

**Table 2.1** Standard Air Saturation Concentration of Oxygen as a Function of Temperature and Salinity in μmol/kg, 0–40 g/kg (Seawater, 1 atm moist air)

| Temperature (°C) | Salinity (g/kg) | | | | | | | | |
|---|---|---|---|---|---|---|---|---|---|
| | **0.0** | **5.0** | **10.0** | **15.0** | **20.0** | **25.0** | **30.0** | **35.0** | **40.0** |
| **0** | 457.00 | 439.55 | 422.77 | 406.61 | 391.07 | 376.12 | 361.74 | 347.90 | 334.59 |
| **1** | 444.31 | 427.50 | 411.31 | 395.73 | 380.74 | 366.31 | 352.42 | 339.06 | 326.20 |
| **2** | 432.21 | 415.99 | 400.38 | 385.35 | 370.87 | 356.94 | 343.52 | 330.60 | 318.17 |
| **3** | 420.66 | 405.01 | 389.94 | 375.42 | 361.44 | 347.97 | 335.01 | 322.52 | 310.49 |
| **4** | 409.63 | 394.52 | 379.96 | 365.93 | 352.42 | 339.40 | 326.86 | 314.77 | 303.13 |
| **5** | 399.10 | 384.49 | 370.42 | 356.86 | 343.79 | 331.19 | 319.06 | 307.36 | 296.09 |
| **6** | 389.02 | 374.90 | 361.30 | 348.18 | 335.53 | 323.34 | 311.58 | 300.25 | 289.33 |
| **7** | 379.38 | 365.73 | 352.56 | 339.86 | 327.62 | 315.81 | 304.42 | 293.44 | 282.85 |
| **8** | 370.16 | 356.94 | 344.19 | 331.90 | 320.03 | 308.59 | 297.55 | 286.91 | 276.64 |
| **9** | 361.33 | 348.53 | 336.18 | 324.26 | 312.76 | 301.67 | 290.96 | 280.63 | 270.67 |
| **10** | 352.86 | 340.46 | 328.49 | 316.94 | 305.79 | 295.02 | 284.64 | 274.61 | 264.93 |
| **11** | 344.75 | 332.72 | 321.12 | 309.91 | 299.09 | 288.64 | 278.56 | 268.82 | 259.42 |
| **12** | 336.96 | 325.30 | 314.04 | 303.16 | 292.66 | 282.51 | 272.72 | 263.26 | 254.12 |
| **13** | 329.49 | 318.17 | 307.24 | 296.68 | 286.47 | 276.62 | 267.10 | 257.91 | 249.02 |
| **14** | 322.31 | 311.32 | 300.70 | 290.44 | 280.53 | 270.95 | 261.70 | 252.76 | 244.11 |
| **15** | 315.42 | 304.74 | 294.42 | 284.45 | 274.81 | 265.50 | 256.50 | 247.80 | 239.39 |
| **16** | 308.79 | 298.41 | 288.38 | 278.68 | 269.31 | 260.25 | 251.49 | 243.02 | 234.83 |
| **17** | 302.41 | 292.32 | 282.56 | 273.13 | 264.01 | 255.19 | 246.66 | 238.41 | 230.43 |
| **18** | 296.27 | 286.45 | 276.96 | 267.78 | 258.90 | 250.31 | 242.00 | 233.97 | 226.20 |
| **19** | 290.35 | 280.80 | 271.56 | 262.62 | 253.97 | 245.60 | 237.51 | 229.68 | 222.10 |
| **20** | 284.65 | 275.35 | 266.35 | 257.64 | 249.22 | 241.06 | 233.17 | 225.54 | 218.15 |
| **21** | 279.15 | 270.09 | 261.33 | 252.84 | 244.63 | 236.68 | 228.98 | 221.53 | 214.33 |
| **22** | 273.85 | 265.02 | 256.48 | 248.20 | 240.19 | 232.44 | 224.93 | 217.67 | 210.63 |
| **23** | 268.73 | 260.12 | 251.79 | 243.72 | 235.91 | 228.34 | 221.02 | 213.92 | 207.05 |
| **24** | 263.78 | 255.39 | 247.26 | 239.39 | 231.76 | 224.38 | 217.23 | 210.30 | 203.59 |
| **25** | 258.99 | 250.81 | 242.88 | 235.19 | 227.75 | 220.54 | 213.55 | 206.79 | 200.23 |
| **26** | 254.37 | 246.38 | 238.63 | 231.13 | 223.86 | 216.82 | 210.00 | 203.38 | 196.98 |
| **27** | 249.88 | 242.08 | 234.52 | 227.20 | 220.10 | 213.22 | 206.55 | 200.08 | 193.82 |
| **28** | 245.54 | 237.93 | 230.54 | 223.38 | 216.44 | 209.72 | 203.20 | 196.88 | 190.75 |
| **29** | 241.33 | 233.89 | 226.68 | 219.68 | 212.90 | 206.32 | 199.95 | 193.77 | 187.77 |
| **30** | 237.25 | 229.98 | 222.93 | 216.09 | 209.46 | 203.02 | 196.79 | 190.74 | 184.88 |
| **31** | 233.29 | 226.18 | 219.28 | 212.59 | 206.11 | 199.82 | 193.72 | 187.80 | 182.06 |
| **32** | 229.43 | 222.48 | 215.74 | 209.20 | 202.85 | 196.70 | 190.72 | 184.93 | 179.31 |
| **33** | 225.69 | 218.89 | 212.29 | 205.89 | 199.68 | 193.66 | 187.81 | 182.14 | 176.64 |
| **34** | 222.04 | 215.39 | 208.93 | 202.67 | 196.59 | 190.69 | 184.97 | 179.41 | 174.02 |
| **35** | 218.49 | 211.98 | 205.66 | 199.53 | 193.58 | 187.80 | 182.20 | 176.76 | 171.48 |
| **36** | 215.02 | 208.65 | 202.47 | 196.47 | 190.64 | 184.98 | 179.49 | 174.16 | 168.98 |
| **37** | 211.65 | 205.41 | 199.35 | 193.48 | 187.77 | 182.22 | 176.84 | 171.62 | 166.55 |
| **38** | 208.34 | 202.24 | 196.31 | 190.55 | 184.96 | 179.53 | 174.25 | 169.13 | 164.16 |
| **39** | 205.12 | 199.14 | 193.33 | 187.69 | 182.21 | 176.89 | 171.72 | 166.70 | 161.82 |
| **40** | 201.96 | 196.10 | 190.41 | 184.88 | 179.52 | 174.30 | 169.23 | 164.31 | 159.53 |

*Source*: Based on Eq. 22, Benson and Krause (1984).

**Table 2.2** Standard Air Saturation Concentration of Oxygen as a Function of Temperature and Salinity in μmol/kg, 33–37 g/kg (Seawater, 1 atm moist air)

| Temperature (°C) | Salinity (g/kg) | | | | | | | | |
|---|---|---|---|---|---|---|---|---|---|
| | **33.0** | **33.5** | **34.0** | **34.5** | **35.0** | **35.5** | **36.0** | **36.5** | **37.0** |
| **0** | 353.37 | 352.00 | 350.63 | 349.26 | 347.90 | 346.55 | 345.20 | 343.85 | 342.51 |
| **1** | 344.34 | 343.01 | 341.69 | 340.37 | 339.06 | 337.75 | 336.45 | 335.15 | 333.85 |
| **2** | 335.71 | 334.43 | 333.15 | 331.87 | 330.60 | 329.34 | 328.08 | 326.82 | 325.57 |
| **3** | 327.46 | 326.21 | 324.98 | 323.74 | 322.52 | 321.29 | 320.07 | 318.86 | 317.65 |
| **4** | 319.55 | 318.35 | 317.15 | 315.96 | 314.77 | 313.59 | 312.41 | 311.24 | 310.07 |
| **5** | 311.99 | 310.82 | 309.66 | 308.51 | 307.36 | 306.21 | 305.07 | 303.93 | 302.80 |
| **6** | 304.74 | 303.61 | 302.49 | 301.37 | 300.25 | 299.14 | 298.04 | 296.93 | 295.84 |
| **7** | 297.79 | 296.69 | 295.61 | 294.52 | 293.44 | 292.36 | 291.29 | 290.22 | 289.16 |
| **8** | 291.12 | 290.06 | 289.00 | 287.95 | 286.91 | 285.86 | 284.82 | 283.79 | 282.75 |
| **9** | 284.72 | 283.69 | 282.67 | 281.65 | 280.63 | 279.62 | 278.61 | 277.61 | 276.60 |
| **10** | 278.58 | 277.58 | 276.59 | 275.60 | 274.61 | 273.63 | 272.65 | 271.67 | 270.70 |
| **11** | 272.68 | 271.71 | 270.74 | 269.78 | 268.82 | 267.87 | 266.92 | 265.97 | 265.02 |
| **12** | 267.00 | 266.06 | 265.12 | 264.19 | 263.26 | 262.33 | 261.41 | 260.48 | 259.57 |
| **13** | 261.55 | 260.63 | 259.72 | 258.81 | 257.91 | 257.00 | 256.10 | 255.21 | 254.32 |
| **14** | 256.30 | 255.41 | 254.52 | 253.64 | 252.76 | 251.88 | 251.00 | 250.13 | 249.26 |
| **15** | 251.24 | 250.37 | 249.51 | 248.65 | 247.80 | 246.94 | 246.09 | 245.24 | 244.40 |
| **16** | 246.37 | 245.53 | 244.69 | 243.85 | 243.02 | 242.19 | 241.36 | 240.53 | 239.71 |
| **17** | 241.68 | 240.85 | 240.04 | 239.22 | 238.41 | 237.60 | 236.79 | 235.99 | 235.19 |
| **18** | 237.15 | 236.35 | 235.55 | 234.76 | 233.97 | 233.18 | 232.39 | 231.61 | 230.83 |
| **19** | 232.78 | 232.00 | 231.22 | 230.45 | 229.68 | 228.91 | 228.14 | 227.38 | 226.62 |
| **20** | 228.56 | 227.80 | 227.04 | 226.29 | 225.54 | 224.79 | 224.04 | 223.29 | 222.55 |
| **21** | 224.48 | 223.74 | 223.00 | 222.27 | 221.53 | 220.80 | 220.07 | 219.35 | 218.62 |
| **22** | 220.54 | 219.82 | 219.10 | 218.38 | 217.67 | 216.95 | 216.24 | 215.53 | 214.82 |
| **23** | 216.73 | 216.03 | 215.32 | 214.62 | 213.92 | 213.22 | 212.53 | 211.84 | 211.15 |
| **24** | 213.04 | 212.35 | 211.67 | 210.98 | 210.30 | 209.62 | 208.94 | 208.26 | 207.59 |
| **25** | 209.47 | 208.80 | 208.12 | 207.45 | 206.79 | 206.12 | 205.46 | 204.80 | 204.14 |
| **26** | 206.00 | 205.35 | 204.69 | 204.04 | 203.38 | 202.73 | 202.09 | 201.44 | 200.80 |
| **27** | 202.64 | 202.00 | 201.36 | 200.72 | 200.08 | 199.45 | 198.82 | 198.18 | 197.55 |
| **28** | 199.38 | 198.75 | 198.13 | 197.50 | 196.88 | 196.26 | 195.64 | 195.02 | 194.41 |
| **29** | 196.22 | 195.60 | 194.99 | 194.38 | 193.77 | 193.16 | 192.55 | 191.95 | 191.35 |
| **30** | 193.14 | 192.54 | 191.94 | 191.34 | 190.74 | 190.15 | 189.55 | 188.96 | 188.37 |
| **31** | 190.14 | 189.55 | 188.97 | 188.38 | 187.80 | 187.22 | 186.64 | 186.06 | 185.48 |
| **32** | 187.23 | 186.65 | 186.08 | 185.50 | 184.93 | 184.36 | 183.79 | 183.23 | 182.66 |
| **33** | 184.39 | 183.82 | 183.26 | 182.70 | 182.14 | 181.58 | 181.02 | 180.47 | 179.92 |
| **34** | 181.62 | 181.06 | 180.51 | 179.96 | 179.41 | 178.87 | 178.32 | 177.78 | 177.24 |
| **35** | 178.91 | 178.37 | 177.83 | 177.29 | 176.76 | 176.22 | 175.69 | 175.16 | 174.63 |
| **36** | 176.27 | 175.74 | 175.21 | 174.69 | 174.16 | 173.63 | 173.11 | 172.59 | 172.07 |
| **37** | 173.69 | 173.17 | 172.65 | 172.14 | 171.62 | 171.11 | 170.59 | 170.08 | 169.57 |
| **38** | 171.16 | 170.65 | 170.15 | 169.64 | 169.13 | 168.63 | 168.13 | 167.63 | 167.13 |
| **39** | 168.69 | 168.19 | 167.69 | 167.19 | 166.70 | 166.21 | 165.71 | 165.22 | 164.73 |
| **40** | 166.26 | 165.77 | 165.28 | 164.80 | 164.31 | 163.83 | 163.34 | 162.86 | 162.38 |

*Source*: Based on Eq. 22, Benson and Krause (1984).

**Table 2.3** Standard Air Saturation Concentration of Nitrogen as a Function of Temperature and Salinity, 0–40 g/kg in μmol/kg (Seawater, 1 atm moist air)

| Temperature (°C) | Salinity (g/kg) | | | | | | | | |
|---|---|---|---|---|---|---|---|---|---|
| | **0.0** | **5.0** | **10.0** | **15.0** | **20.0** | **25.0** | **30.0** | **35.0** | **40.0** |
| **0** | 830.45 | 796.87 | 764.64 | 733.71 | 704.04 | 675.57 | 648.25 | 622.03 | 596.87 |
| **1** | 808.87 | 776.45 | 745.32 | 715.44 | 686.76 | 659.22 | 632.80 | 607.43 | 583.08 |
| **2** | 788.28 | 756.95 | 726.87 | 697.98 | 670.24 | 643.60 | 618.02 | 593.46 | 569.88 |
| **3** | 768.63 | 738.34 | 709.25 | 681.30 | 654.45 | 628.67 | 603.89 | 580.10 | 557.24 |
| **4** | 749.85 | 720.56 | 692.40 | 665.35 | 639.35 | 614.37 | 590.37 | 567.30 | 545.14 |
| **5** | 731.91 | 703.55 | 676.29 | 650.09 | 624.90 | 600.69 | 577.42 | 555.04 | 533.54 |
| **6** | 714.76 | 687.29 | 660.88 | 635.49 | 611.07 | 587.59 | 565.01 | 543.29 | 522.42 |
| **7** | 698.35 | 671.73 | 646.13 | 621.50 | 597.81 | 575.03 | 553.11 | 532.03 | 511.75 |
| **8** | 682.65 | 656.84 | 632.00 | 608.11 | 585.11 | 562.99 | 541.70 | 521.22 | 501.51 |
| **9** | 667.61 | 642.57 | 618.47 | 595.27 | 572.94 | 551.44 | 530.76 | 510.85 | 491.68 |
| **10** | 653.21 | 628.90 | 605.49 | 582.95 | 561.25 | 540.36 | 520.25 | 500.89 | 482.24 |
| **11** | 639.40 | 615.78 | 593.04 | 571.13 | 550.04 | 529.72 | 510.16 | 491.31 | 473.17 |
| **12** | 626.16 | 603.21 | 581.10 | 559.79 | 539.27 | 519.50 | 500.46 | 482.11 | 464.44 |
| **13** | 613.46 | 591.14 | 569.63 | 548.90 | 528.93 | 509.68 | 491.13 | 473.26 | 456.04 |
| **14** | 601.27 | 579.55 | 558.61 | 538.43 | 518.98 | 500.23 | 482.16 | 464.74 | 447.96 |
| **15** | 589.56 | 568.41 | 548.03 | 528.37 | 509.42 | 491.15 | 473.53 | 456.55 | 440.17 |
| **16** | 578.31 | 557.71 | 537.85 | 518.69 | 500.21 | 482.40 | 465.22 | 448.65 | 432.67 |
| **17** | 567.49 | 547.42 | 528.05 | 509.37 | 491.36 | 473.97 | 457.21 | 441.03 | 425.43 |
| **18** | 557.09 | 537.52 | 518.63 | 500.40 | 482.82 | 465.86 | 449.49 | 433.69 | 418.45 |
| **19** | 547.08 | 527.98 | 509.55 | 491.76 | 474.60 | 458.03 | 442.04 | 426.61 | 411.72 |
| **20** | 537.44 | 518.80 | 500.81 | 483.44 | 466.67 | 450.48 | 434.86 | 419.77 | 405.21 |
| **21** | 528.16 | 509.95 | 492.38 | 475.40 | 459.02 | 443.20 | 427.92 | 413.17 | 398.93 |
| **22** | 519.21 | 501.42 | 484.25 | 467.65 | 451.63 | 436.16 | 421.22 | 406.79 | 392.85 |
| **23** | 510.59 | 493.20 | 476.40 | 460.17 | 444.50 | 429.36 | 414.74 | 400.61 | 386.97 |
| **24** | 502.26 | 485.25 | 468.82 | 452.95 | 437.61 | 422.79 | 408.47 | 394.64 | 381.28 |
| **25** | 494.23 | 477.58 | 461.50 | 445.96 | 430.94 | 416.43 | 402.41 | 388.86 | 375.76 |
| **26** | 486.47 | 470.17 | 454.42 | 439.20 | 424.49 | 410.28 | 396.53 | 383.25 | 370.42 |
| **27** | 478.97 | 463.01 | 447.58 | 432.67 | 418.25 | 404.32 | 390.84 | 377.82 | 365.23 |
| **28** | 471.71 | 456.07 | 440.96 | 426.34 | 412.20 | 398.54 | 385.33 | 372.55 | 360.20 |
| **29** | 464.70 | 449.36 | 434.54 | 420.20 | 406.34 | 392.94 | 379.97 | 367.44 | 355.32 |
| **30** | 457.90 | 442.87 | 428.32 | 414.26 | 400.66 | 387.50 | 374.77 | 362.47 | 350.57 |
| **31** | 451.32 | 436.57 | 422.29 | 408.49 | 395.14 | 382.22 | 369.72 | 357.64 | 345.95 |
| **32** | 444.94 | 430.46 | 416.45 | 402.89 | 389.78 | 377.09 | 364.81 | 352.94 | 341.45 |
| **33** | 438.75 | 424.53 | 410.77 | 397.45 | 384.57 | 372.10 | 360.04 | 348.37 | 337.07 |
| **34** | 432.74 | 418.77 | 405.25 | 392.16 | 379.50 | 367.24 | 355.39 | 343.91 | 332.81 |
| **35** | 426.91 | 413.18 | 399.88 | 387.02 | 374.57 | 362.52 | 350.85 | 339.57 | 328.64 |
| **36** | 421.24 | 407.73 | 394.66 | 382.01 | 369.76 | 357.91 | 346.43 | 335.33 | 324.58 |
| **37** | 415.72 | 402.44 | 389.58 | 377.13 | 365.08 | 353.41 | 342.12 | 331.19 | 320.61 |
| **38** | 410.35 | 397.28 | 384.63 | 372.37 | 360.51 | 349.03 | 337.91 | 327.14 | 316.72 |
| **39** | 405.12 | 392.25 | 379.79 | 367.73 | 356.05 | 344.74 | 333.79 | 323.19 | 312.92 |
| **40** | 400.02 | 387.35 | 375.08 | 363.20 | 351.69 | 340.55 | 329.76 | 319.31 | 309.20 |

*Source*: Based on Eq. 1, Hamme and Emerson (2004).

**Table 2.4** Standard Air Saturation Concentration of Nitrogen as a Function of Temperature and Salinity, 33–37 g/kg in μmol/kg (Seawater, 1 atm moist air)

| Temperature (°C) | Salinity (g/kg) | | | | | | | | |
|---|---|---|---|---|---|---|---|---|---|
| | **33.0** | **33.5** | **34.0** | **34.5** | **35.0** | **35.5** | **36.0** | **36.5** | **37.0** |
| **0** | 632.39 | 629.78 | 627.19 | 624.60 | 622.03 | 619.47 | 616.91 | 614.37 | 611.84 |
| **1** | 617.45 | 614.93 | 612.42 | 609.92 | 607.43 | 604.95 | 602.48 | 600.02 | 597.57 |
| **2** | 603.17 | 600.73 | 598.30 | 595.87 | 593.46 | 591.06 | 588.67 | 586.29 | 583.91 |
| **3** | 589.50 | 587.14 | 584.78 | 582.43 | 580.10 | 577.77 | 575.45 | 573.14 | 570.84 |
| **4** | 576.42 | 574.12 | 571.84 | 569.57 | 567.30 | 565.04 | 562.80 | 560.56 | 558.33 |
| **5** | 563.89 | 561.66 | 559.45 | 557.24 | 555.04 | 552.85 | 550.67 | 548.50 | 546.34 |
| **6** | 551.88 | 549.72 | 547.57 | 545.43 | 543.29 | 541.17 | 539.05 | 536.95 | 534.85 |
| **7** | 540.36 | 538.27 | 536.18 | 534.10 | 532.03 | 529.97 | 527.91 | 525.86 | 523.82 |
| **8** | 529.32 | 527.28 | 525.25 | 523.23 | 521.22 | 519.22 | 517.22 | 515.23 | 513.25 |
| **9** | 518.72 | 516.74 | 514.77 | 512.80 | 510.85 | 508.90 | 506.96 | 505.02 | 503.09 |
| **10** | 508.54 | 506.62 | 504.70 | 502.79 | 500.89 | 498.99 | 497.10 | 495.22 | 493.34 |
| **11** | 498.77 | 496.89 | 495.03 | 493.17 | 491.31 | 489.47 | 487.63 | 485.80 | 483.97 |
| **12** | 489.37 | 487.54 | 485.73 | 483.92 | 482.11 | 480.31 | 478.52 | 476.74 | 474.96 |
| **13** | 480.33 | 478.55 | 476.78 | 475.02 | 473.26 | 471.51 | 469.77 | 468.03 | 466.30 |
| **14** | 471.64 | 469.90 | 468.18 | 466.46 | 464.74 | 463.04 | 461.34 | 459.64 | 457.96 |
| **15** | 463.26 | 461.58 | 459.89 | 458.22 | 456.55 | 454.88 | 453.22 | 451.57 | 449.92 |
| **16** | 455.20 | 453.56 | 451.91 | 450.28 | 448.65 | 447.02 | 445.40 | 443.79 | 442.19 |
| **17** | 447.43 | 445.83 | 444.22 | 442.63 | 441.03 | 439.45 | 437.87 | 436.30 | 434.73 |
| **18** | 439.94 | 438.37 | 436.81 | 435.25 | 433.69 | 432.15 | 430.60 | 429.06 | 427.53 |
| **19** | 432.72 | 431.18 | 429.65 | 428.13 | 426.61 | 425.10 | 423.59 | 422.09 | 420.59 |
| **20** | 425.74 | 424.24 | 422.75 | 421.26 | 419.77 | 418.29 | 416.82 | 415.35 | 413.89 |
| **21** | 419.01 | 417.54 | 416.08 | 414.62 | 413.17 | 411.72 | 410.28 | 408.84 | 407.41 |
| **22** | 412.50 | 411.06 | 409.63 | 408.21 | 406.79 | 405.37 | 403.96 | 402.55 | 401.15 |
| **23** | 406.20 | 404.80 | 403.40 | 402.00 | 400.61 | 399.23 | 397.85 | 396.47 | 395.10 |
| **24** | 400.12 | 398.74 | 397.37 | 396.00 | 394.64 | 393.28 | 391.93 | 390.58 | 389.24 |
| **25** | 394.22 | 392.87 | 391.53 | 390.19 | 388.86 | 387.53 | 386.20 | 384.88 | 383.56 |
| **26** | 388.51 | 387.19 | 385.87 | 384.56 | 383.25 | 381.95 | 380.65 | 379.36 | 378.07 |
| **27** | 382.98 | 381.68 | 380.39 | 379.10 | 377.82 | 376.54 | 375.27 | 374.00 | 372.73 |
| **28** | 377.61 | 376.34 | 375.07 | 373.81 | 372.55 | 371.30 | 370.05 | 368.80 | 367.56 |
| **29** | 372.40 | 371.15 | 369.91 | 368.67 | 367.44 | 366.21 | 364.98 | 363.76 | 362.54 |
| **30** | 367.34 | 366.12 | 364.90 | 363.68 | 362.47 | 361.26 | 360.06 | 358.86 | 357.66 |
| **31** | 362.42 | 361.22 | 360.02 | 358.83 | 357.64 | 356.45 | 355.27 | 354.09 | 352.92 |
| **32** | 357.64 | 356.46 | 355.28 | 354.11 | 352.94 | 351.77 | 350.61 | 349.45 | 348.30 |
| **33** | 352.99 | 351.83 | 350.67 | 349.52 | 348.37 | 347.22 | 346.08 | 344.94 | 343.80 |
| **34** | 348.46 | 347.31 | 346.18 | 345.04 | 343.91 | 342.78 | 341.66 | 340.54 | 339.42 |
| **35** | 344.04 | 342.91 | 341.79 | 340.68 | 339.57 | 338.46 | 337.35 | 336.25 | 335.15 |
| **36** | 339.73 | 338.62 | 337.52 | 336.42 | 335.33 | 334.24 | 333.15 | 332.07 | 330.99 |
| **37** | 335.52 | 334.43 | 333.35 | 332.27 | 331.19 | 330.11 | 329.04 | 327.98 | 326.91 |
| **38** | 331.41 | 330.34 | 329.27 | 328.20 | 327.14 | 326.09 | 325.03 | 323.98 | 322.93 |
| **39** | 327.39 | 326.33 | 325.28 | 324.23 | 323.19 | 322.14 | 321.11 | 320.07 | 319.04 |
| **40** | 323.45 | 322.41 | 321.37 | 320.34 | 319.31 | 318.29 | 317.26 | 316.24 | 315.23 |

*Source*: Based on Eq. 1, Hamme and Emerson (2004).

**Table 2.5** Standard Air Saturation Concentration of Argon as a Function of Temperature and Salinity, 0–40 g/kg in μmol/kg (Seawater, 1 atm moist air)

| Temperature (°C) | Salinity (g/kg) | | | | | | | | |
|---|---|---|---|---|---|---|---|---|---|
| | **0.0** | **5.0** | **10.0** | **15.0** | **20.0** | **25.0** | **30.0** | **35.0** | **40.0** |
| **0** | 22.301 | 21.456 | 20.643 | 19.861 | 19.109 | 18.385 | 17.689 | 17.019 | 16.374 |
| **1** | 21.688 | 20.873 | 20.089 | 19.335 | 18.609 | 17.910 | 17.237 | 16.590 | 15.967 |
| **2** | 21.103 | 20.317 | 19.560 | 18.832 | 18.131 | 17.456 | 16.806 | 16.180 | 15.577 |
| **3** | 20.544 | 19.786 | 19.055 | 18.351 | 17.674 | 17.021 | 16.393 | 15.788 | 15.205 |
| **4** | 20.010 | 19.278 | 18.572 | 17.892 | 17.237 | 16.606 | 15.998 | 15.412 | 14.848 |
| **5** | 19.500 | 18.792 | 18.110 | 17.452 | 16.819 | 16.208 | 15.619 | 15.052 | 14.506 |
| **6** | 19.012 | 18.328 | 17.668 | 17.031 | 16.418 | 15.827 | 15.257 | 14.707 | 14.178 |
| **7** | 18.546 | 17.883 | 17.244 | 16.628 | 16.034 | 15.462 | 14.909 | 14.377 | 13.863 |
| **8** | 18.099 | 17.457 | 16.839 | 16.242 | 15.666 | 15.111 | 14.576 | 14.059 | 13.561 |
| **9** | 17.670 | 17.049 | 16.450 | 15.872 | 15.314 | 14.775 | 14.256 | 13.755 | 13.271 |
| **10** | 17.260 | 16.658 | 16.077 | 15.516 | 14.975 | 14.453 | 13.949 | 13.462 | 12.993 |
| **11** | 16.866 | 16.283 | 15.719 | 15.175 | 14.650 | 14.143 | 13.653 | 13.181 | 12.725 |
| **12** | 16.489 | 15.922 | 15.376 | 14.847 | 14.338 | 13.845 | 13.370 | 12.911 | 12.467 |
| **13** | 16.126 | 15.576 | 15.045 | 14.533 | 14.037 | 13.559 | 13.097 | 12.650 | 12.219 |
| **14** | 15.778 | 15.244 | 14.728 | 14.230 | 13.749 | 13.283 | 12.834 | 12.400 | 11.980 |
| **15** | 15.443 | 14.924 | 14.423 | 13.939 | 13.471 | 13.018 | 12.581 | 12.159 | 11.750 |
| **16** | 15.121 | 14.617 | 14.129 | 13.658 | 13.203 | 12.763 | 12.337 | 11.926 | 11.528 |
| **17** | 14.811 | 14.321 | 13.847 | 13.388 | 12.945 | 12.517 | 12.102 | 11.702 | 11.314 |
| **18** | 14.513 | 14.036 | 13.574 | 13.128 | 12.697 | 12.279 | 11.876 | 11.485 | 11.108 |
| **19** | 14.225 | 13.761 | 13.312 | 12.877 | 12.457 | 12.050 | 11.657 | 11.276 | 10.908 |
| **20** | 13.948 | 13.496 | 13.059 | 12.635 | 12.226 | 11.829 | 11.446 | 11.075 | 10.715 |
| **21** | 13.681 | 13.241 | 12.814 | 12.402 | 12.002 | 11.616 | 11.241 | 10.879 | 10.529 |
| **22** | 13.424 | 12.994 | 12.578 | 12.176 | 11.786 | 11.409 | 11.044 | 10.691 | 10.349 |
| **23** | 13.175 | 12.756 | 12.350 | 11.958 | 11.577 | 11.209 | 10.853 | 10.508 | 10.174 |
| **24** | 12.934 | 12.526 | 12.130 | 11.747 | 11.376 | 11.016 | 10.668 | 10.331 | 10.005 |
| **25** | 12.702 | 12.303 | 11.917 | 11.542 | 11.180 | 10.829 | 10.489 | 10.160 | 9.841 |
| **26** | 12.477 | 12.088 | 11.710 | 11.345 | 10.991 | 10.648 | 10.315 | 9.993 | 9.681 |
| **27** | 12.259 | 11.879 | 11.510 | 11.153 | 10.807 | 10.472 | 10.147 | 9.832 | 9.527 |
| **28** | 12.049 | 11.677 | 11.317 | 10.967 | 10.629 | 10.301 | 9.983 | 9.675 | 9.377 |
| **29** | 11.844 | 11.481 | 11.129 | 10.787 | 10.456 | 10.135 | 9.825 | 9.523 | 9.231 |
| **30** | 11.646 | 11.291 | 10.946 | 10.612 | 10.288 | 9.975 | 9.670 | 9.375 | 9.089 |
| **31** | 11.454 | 11.106 | 10.769 | 10.442 | 10.125 | 9.818 | 9.520 | 9.231 | 8.951 |
| **32** | 11.267 | 10.927 | 10.597 | 10.277 | 9.967 | 9.666 | 9.374 | 9.091 | 8.816 |
| **33** | 11.086 | 10.753 | 10.430 | 10.116 | 9.812 | 9.517 | 9.231 | 8.954 | 8.685 |
| **34** | 10.909 | 10.583 | 10.267 | 9.960 | 9.662 | 9.373 | 9.093 | 8.821 | 8.557 |
| **35** | 10.738 | 10.418 | 10.108 | 9.807 | 9.515 | 9.232 | 8.957 | 8.691 | 8.432 |
| **36** | 10.571 | 10.257 | 9.954 | 9.659 | 9.372 | 9.095 | 8.825 | 8.564 | 8.310 |
| **37** | 10.408 | 10.101 | 9.803 | 9.514 | 9.233 | 8.960 | 8.696 | 8.440 | 8.191 |
| **38** | 10.249 | 9.948 | 9.656 | 9.372 | 9.097 | 8.829 | 8.570 | 8.318 | 8.074 |
| **39** | 10.094 | 9.799 | 9.512 | 9.234 | 8.963 | 8.701 | 8.446 | 8.199 | 7.959 |
| **40** | 9.943 | 9.653 | 9.371 | 9.098 | 8.833 | 8.575 | 8.325 | 8.083 | 7.847 |

*Source*: Based on Eq. 1, Hamme and Emerson (2004).

**Table 2.6** Standard Air Saturation Concentration of Argon as a Function of Temperature and Salinity, 33–37 g/kg in μmol/kg (Seawater, 1 atm moist air)

| Temperature (°C) | Salinity (g/kg) | | | | | | | | |
|---|---|---|---|---|---|---|---|---|---|
| | **33.0** | **33.5** | **34.0** | **34.5** | **35.0** | **35.5** | **36.0** | **36.5** | **37.0** |
| **0** | 17.284 | 17.217 | 17.151 | 17.084 | 17.019 | 16.953 | 16.888 | 16.823 | 16.758 |
| **1** | 16.846 | 16.781 | 16.717 | 16.654 | 16.590 | 16.526 | 16.463 | 16.400 | 16.338 |
| **2** | 16.427 | 16.365 | 16.303 | 16.241 | 16.180 | 16.119 | 16.058 | 15.997 | 15.936 |
| **3** | 16.027 | 15.967 | 15.907 | 15.847 | 15.788 | 15.728 | 15.669 | 15.610 | 15.552 |
| **4** | 15.644 | 15.585 | 15.527 | 15.470 | 15.412 | 15.355 | 15.297 | 15.240 | 15.184 |
| **5** | 15.277 | 15.220 | 15.164 | 15.108 | 15.052 | 14.997 | 14.941 | 14.886 | 14.831 |
| **6** | 14.925 | 14.870 | 14.816 | 14.761 | 14.707 | 14.654 | 14.600 | 14.546 | 14.493 |
| **7** | 14.587 | 14.534 | 14.482 | 14.429 | 14.377 | 14.324 | 14.272 | 14.221 | 14.169 |
| **8** | 14.264 | 14.212 | 14.161 | 14.110 | 14.059 | 14.009 | 13.958 | 13.908 | 13.858 |
| **9** | 13.953 | 13.903 | 13.854 | 13.804 | 13.755 | 13.706 | 13.657 | 13.608 | 13.559 |
| **10** | 13.655 | 13.606 | 13.558 | 13.510 | 13.462 | 13.414 | 13.367 | 13.320 | 13.272 |
| **11** | 13.368 | 13.321 | 13.274 | 13.227 | 13.181 | 13.135 | 13.088 | 13.042 | 12.997 |
| **12** | 13.092 | 13.047 | 13.001 | 12.956 | 12.911 | 12.865 | 12.821 | 12.776 | 12.731 |
| **13** | 12.827 | 12.783 | 12.738 | 12.694 | 12.650 | 12.607 | 12.563 | 12.519 | 12.476 |
| **14** | 12.572 | 12.529 | 12.485 | 12.443 | 12.400 | 12.357 | 12.315 | 12.272 | 12.230 |
| **15** | 12.326 | 12.284 | 12.242 | 12.200 | 12.159 | 12.117 | 12.076 | 12.035 | 11.994 |
| **16** | 12.089 | 12.048 | 12.007 | 11.967 | 11.926 | 11.886 | 11.845 | 11.805 | 11.765 |
| **17** | 11.860 | 11.821 | 11.781 | 11.741 | 11.702 | 11.662 | 11.623 | 11.584 | 11.545 |
| **18** | 11.640 | 11.601 | 11.562 | 11.524 | 11.485 | 11.447 | 11.409 | 11.371 | 11.333 |
| **19** | 11.427 | 11.389 | 11.351 | 11.314 | 11.276 | 11.239 | 11.202 | 11.165 | 11.128 |
| **20** | 11.222 | 11.185 | 11.148 | 11.111 | 11.075 | 11.038 | 11.002 | 10.966 | 10.930 |
| **21** | 11.023 | 10.987 | 10.951 | 10.915 | 10.879 | 10.844 | 10.808 | 10.773 | 10.738 |
| **22** | 10.831 | 10.795 | 10.760 | 10.726 | 10.691 | 10.656 | 10.621 | 10.587 | 10.553 |
| **23** | 10.645 | 10.610 | 10.576 | 10.542 | 10.508 | 10.474 | 10.440 | 10.407 | 10.373 |
| **24** | 10.465 | 10.431 | 10.398 | 10.364 | 10.331 | 10.298 | 10.265 | 10.232 | 10.199 |
| **25** | 10.290 | 10.257 | 10.225 | 10.192 | 10.160 | 10.127 | 10.095 | 10.063 | 10.031 |
| **26** | 10.121 | 10.089 | 10.057 | 10.025 | 9.993 | 9.962 | 9.930 | 9.899 | 9.867 |
| **27** | 9.957 | 9.925 | 9.894 | 9.863 | 9.832 | 9.801 | 9.770 | 9.739 | 9.709 |
| **28** | 9.797 | 9.767 | 9.736 | 9.706 | 9.675 | 9.645 | 9.615 | 9.585 | 9.555 |
| **29** | 9.643 | 9.613 | 9.583 | 9.553 | 9.523 | 9.493 | 9.464 | 9.435 | 9.405 |
| **30** | 9.492 | 9.463 | 9.433 | 9.404 | 9.375 | 9.346 | 9.317 | 9.288 | 9.260 |
| **31** | 9.346 | 9.317 | 9.288 | 9.260 | 9.231 | 9.203 | 9.174 | 9.146 | 9.118 |
| **32** | 9.203 | 9.175 | 9.147 | 9.119 | 9.091 | 9.063 | 9.035 | 9.008 | 8.980 |
| **33** | 9.064 | 9.036 | 9.009 | 8.981 | 8.954 | 8.927 | 8.900 | 8.873 | 8.845 |
| **34** | 8.929 | 8.902 | 8.875 | 8.848 | 8.821 | 8.794 | 8.767 | 8.741 | 8.714 |
| **35** | 8.796 | 8.770 | 8.743 | 8.717 | 8.691 | 8.665 | 8.638 | 8.612 | 8.586 |
| **36** | 8.667 | 8.641 | 8.615 | 8.589 | 8.564 | 8.538 | 8.512 | 8.487 | 8.461 |
| **37** | 8.541 | 8.516 | 8.490 | 8.465 | 8.440 | 8.414 | 8.389 | 8.364 | 8.339 |
| **38** | 8.418 | 8.393 | 8.368 | 8.343 | 8.318 | 8.293 | 8.269 | 8.244 | 8.219 |
| **39** | 8.297 | 8.273 | 8.248 | 8.224 | 8.199 | 8.175 | 8.151 | 8.126 | 8.102 |
| **40** | 8.179 | 8.155 | 8.131 | 8.107 | 8.083 | 8.059 | 8.035 | 8.011 | 7.988 |

*Source*: Based on Eq. 1, Hamme and Emerson (2004).

**Table 2.7** Standard Air Saturation Concentration of Carbon Dioxide as a Function of Temperature and Salinity, 0–40 g/kg in μmol/kg (Seawater, 1 atm moist air; mole fraction = 390 μatm)

| Temperature (°C) | Salinity (g/kg) | | | | | | | | |
|---|---|---|---|---|---|---|---|---|---|
| | 0.0 | 5.0 | 10.0 | 15.0 | 20.0 | 25.0 | 30.0 | 35.0 | 40.0 |
| **0** | 29.940 | 29.055 | 28.196 | 27.362 | 26.554 | 25.768 | 25.007 | 24.267 | 23.550 |
| **1** | 28.771 | 27.923 | 27.101 | 26.302 | 25.527 | 24.775 | 24.045 | 23.337 | 22.649 |
| **2** | 27.663 | 26.851 | 26.063 | 25.298 | 24.555 | 23.834 | 23.135 | 22.456 | 21.796 |
| **3** | 26.613 | 25.835 | 25.079 | 24.346 | 23.634 | 22.943 | 22.272 | 21.620 | 20.988 |
| **4** | 25.616 | 24.870 | 24.146 | 23.443 | 22.760 | 22.097 | 21.453 | 20.828 | 20.222 |
| **5** | 24.670 | 23.955 | 23.260 | 22.585 | 21.930 | 21.294 | 20.676 | 20.077 | 19.494 |
| **6** | 23.772 | 23.085 | 22.419 | 21.771 | 21.142 | 20.532 | 19.939 | 19.363 | 18.804 |
| **7** | 22.918 | 22.259 | 21.619 | 20.998 | 20.394 | 19.808 | 19.238 | 18.685 | 18.148 |
| **8** | 22.106 | 21.474 | 20.859 | 20.262 | 19.682 | 19.119 | 18.572 | 18.041 | 17.524 |
| **9** | 21.334 | 20.726 | 20.136 | 19.563 | 19.006 | 18.464 | 17.939 | 17.428 | 16.932 |
| **10** | 20.598 | 20.015 | 19.448 | 18.897 | 18.361 | 17.841 | 17.336 | 16.845 | 16.367 |
| **11** | 19.898 | 19.337 | 18.792 | 18.262 | 17.748 | 17.248 | 16.762 | 16.289 | 15.830 |
| **12** | 19.230 | 18.691 | 18.167 | 17.658 | 17.163 | 16.682 | 16.215 | 15.760 | 15.319 |
| **13** | 18.594 | 18.075 | 17.572 | 17.082 | 16.606 | 16.143 | 15.693 | 15.256 | 14.831 |
| **14** | 17.986 | 17.488 | 17.004 | 16.532 | 16.074 | 15.629 | 15.196 | 14.775 | 14.366 |
| **15** | 17.407 | 16.927 | 16.461 | 16.008 | 15.567 | 15.138 | 14.722 | 14.316 | 13.922 |
| **16** | 16.853 | 16.392 | 15.943 | 15.507 | 15.083 | 14.670 | 14.269 | 13.878 | 13.498 |
| **17** | 16.324 | 15.880 | 15.448 | 15.028 | 14.620 | 14.222 | 13.836 | 13.460 | 13.094 |
| **18** | 15.818 | 15.391 | 14.975 | 14.571 | 14.177 | 13.795 | 13.422 | 13.060 | 12.707 |
| **19** | 15.334 | 14.923 | 14.523 | 14.133 | 13.754 | 13.385 | 13.026 | 12.677 | 12.337 |
| **20** | 14.871 | 14.475 | 14.089 | 13.714 | 13.349 | 12.994 | 12.648 | 12.311 | 11.983 |
| **21** | 14.427 | 14.046 | 13.674 | 13.313 | 12.961 | 12.619 | 12.285 | 11.961 | 11.645 |
| **22** | 14.002 | 13.635 | 13.277 | 12.929 | 12.590 | 12.260 | 11.938 | 11.625 | 11.320 |
| **23** | 13.594 | 13.240 | 12.896 | 12.560 | 12.234 | 11.915 | 11.605 | 11.303 | 11.009 |
| **24** | 13.203 | 12.862 | 12.530 | 12.207 | 11.892 | 11.585 | 11.286 | 10.995 | 10.711 |
| **25** | 12.828 | 12.500 | 12.180 | 11.868 | 11.564 | 11.268 | 10.980 | 10.699 | 10.425 |
| **26** | 12.467 | 12.151 | 11.843 | 11.542 | 11.250 | 10.964 | 10.686 | 10.415 | 10.151 |
| **27** | 12.121 | 11.816 | 11.519 | 11.230 | 10.947 | 10.672 | 10.404 | 10.142 | 9.887 |
| **28** | 11.788 | 11.494 | 11.208 | 10.929 | 10.657 | 10.391 | 10.132 | 9.880 | 9.634 |
| **29** | 11.468 | 11.185 | 10.909 | 10.640 | 10.377 | 10.121 | 9.872 | 9.628 | 9.390 |
| **30** | 11.159 | 10.887 | 10.621 | 10.361 | 10.108 | 9.861 | 9.620 | 9.385 | 9.156 |
| **31** | 10.862 | 10.600 | 10.343 | 10.093 | 9.849 | 9.611 | 9.379 | 9.152 | 8.931 |
| **32** | 10.576 | 10.323 | 10.076 | 9.835 | 9.600 | 9.370 | 9.146 | 8.927 | 8.713 |
| **33** | 10.300 | 10.057 | 9.819 | 9.586 | 9.359 | 9.138 | 8.921 | 8.710 | 8.504 |
| **34** | 10.034 | 9.799 | 9.570 | 9.346 | 9.127 | 8.913 | 8.705 | 8.501 | 8.302 |
| **35** | 9.777 | 9.551 | 9.330 | 9.114 | 8.903 | 8.697 | 8.496 | 8.299 | 8.107 |
| **36** | 9.529 | 9.311 | 9.098 | 8.890 | 8.687 | 8.488 | 8.294 | 8.104 | 7.919 |
| **37** | 9.289 | 9.079 | 8.874 | 8.674 | 8.478 | 8.286 | 8.099 | 7.916 | 7.737 |
| **38** | 9.057 | 8.855 | 8.657 | 8.464 | 8.275 | 8.091 | 7.910 | 7.734 | 7.561 |
| **39** | 8.832 | 8.638 | 8.448 | 8.262 | 8.080 | 7.902 | 7.728 | 7.557 | 7.391 |
| **40** | 8.615 | 8.428 | 8.244 | 8.065 | 7.890 | 7.719 | 7.551 | 7.387 | 7.226 |

*Source*: Based on Eqs. 5 and 9, Weiss (1974); mole fraction estimated from Mauna Loa data for 2010.

**Table 2.8** Standard Air Saturation Concentration of Carbon Dioxide as a Function of Temperature and Salinity, 33–37 g/kg in μmol/kg (Seawater, 1 atm moist air; mole fraction = 390 μatm)

| Temperature (°C) | Salinity (g/kg) | | | | | | | | |
|---|---|---|---|---|---|---|---|---|---|
| | **33.0** | **33.5** | **34.0** | **34.5** | **35.0** | **35.5** | **36.0** | **36.5** | **37.0** |
| **0** | 24.560 | 24.487 | 24.413 | 24.340 | 24.267 | 24.195 | 24.122 | 24.050 | 23.978 |
| **1** | 23.618 | 23.547 | 23.477 | 23.407 | 23.337 | 23.267 | 23.198 | 23.129 | 23.059 |
| **2** | 22.725 | 22.657 | 22.590 | 22.523 | 22.456 | 22.389 | 22.322 | 22.256 | 22.190 |
| **3** | 21.879 | 21.814 | 21.749 | 21.685 | 21.620 | 21.556 | 21.492 | 21.429 | 21.365 |
| **4** | 21.076 | 21.014 | 20.952 | 20.890 | 20.828 | 20.767 | 20.706 | 20.644 | 20.583 |
| **5** | 20.314 | 20.255 | 20.195 | 20.136 | 20.077 | 20.018 | 19.959 | 19.900 | 19.842 |
| **6** | 19.591 | 19.534 | 19.477 | 19.420 | 19.363 | 19.306 | 19.250 | 19.194 | 19.137 |
| **7** | 18.904 | 18.849 | 18.794 | 18.740 | 18.685 | 18.631 | 18.576 | 18.522 | 18.468 |
| **8** | 18.251 | 18.199 | 18.146 | 18.093 | 18.041 | 17.988 | 17.936 | 17.884 | 17.832 |
| **9** | 17.630 | 17.580 | 17.529 | 17.478 | 17.428 | 17.378 | 17.327 | 17.277 | 17.228 |
| **10** | 17.039 | 16.990 | 16.942 | 16.893 | 16.845 | 16.796 | 16.748 | 16.700 | 16.652 |
| **11** | 16.477 | 16.430 | 16.383 | 16.336 | 16.289 | 16.243 | 16.196 | 16.150 | 16.104 |
| **12** | 15.940 | 15.895 | 15.850 | 15.805 | 15.760 | 15.716 | 15.671 | 15.626 | 15.582 |
| **13** | 15.429 | 15.386 | 15.343 | 15.299 | 15.256 | 15.213 | 15.170 | 15.127 | 15.085 |
| **14** | 14.942 | 14.900 | 14.858 | 14.817 | 14.775 | 14.734 | 14.692 | 14.651 | 14.610 |
| **15** | 14.477 | 14.437 | 14.396 | 14.356 | 14.316 | 14.276 | 14.237 | 14.197 | 14.157 |
| **16** | 14.033 | 13.994 | 13.955 | 13.917 | 13.878 | 13.840 | 13.801 | 13.763 | 13.725 |
| **17** | 13.609 | 13.571 | 13.534 | 13.497 | 13.460 | 13.423 | 13.386 | 13.349 | 13.312 |
| **18** | 13.203 | 13.167 | 13.131 | 13.095 | 13.060 | 13.024 | 12.988 | 12.953 | 12.917 |
| **19** | 12.816 | 12.781 | 12.746 | 12.712 | 12.677 | 12.643 | 12.608 | 12.574 | 12.540 |
| **20** | 12.445 | 12.411 | 12.378 | 12.344 | 12.311 | 12.278 | 12.245 | 12.212 | 12.179 |
| **21** | 12.090 | 12.057 | 12.025 | 11.993 | 11.961 | 11.929 | 11.897 | 11.865 | 11.833 |
| **22** | 11.749 | 11.718 | 11.687 | 11.656 | 11.625 | 11.594 | 11.563 | 11.533 | 11.502 |
| **23** | 11.423 | 11.393 | 11.363 | 11.333 | 11.303 | 11.274 | 11.244 | 11.214 | 11.185 |
| **24** | 11.111 | 11.082 | 11.053 | 11.024 | 10.995 | 10.966 | 10.938 | 10.909 | 10.881 |
| **25** | 10.811 | 10.783 | 10.755 | 10.727 | 10.699 | 10.671 | 10.644 | 10.616 | 10.589 |
| **26** | 10.523 | 10.496 | 10.469 | 10.442 | 10.415 | 10.388 | 10.362 | 10.335 | 10.309 |
| **27** | 10.246 | 10.220 | 10.194 | 10.168 | 10.142 | 10.117 | 10.091 | 10.065 | 10.040 |
| **28** | 9.980 | 9.955 | 9.930 | 9.905 | 9.880 | 9.855 | 9.830 | 9.806 | 9.781 |
| **29** | 9.725 | 9.700 | 9.676 | 9.652 | 9.628 | 9.604 | 9.580 | 9.556 | 9.532 |
| **30** | 9.479 | 9.455 | 9.432 | 9.409 | 9.385 | 9.362 | 9.339 | 9.316 | 9.293 |
| **31** | 9.242 | 9.219 | 9.197 | 9.174 | 9.152 | 9.129 | 9.107 | 9.085 | 9.063 |
| **32** | 9.014 | 8.992 | 8.970 | 8.949 | 8.927 | 8.905 | 8.884 | 8.862 | 8.841 |
| **33** | 8.794 | 8.773 | 8.752 | 8.731 | 8.710 | 8.689 | 8.668 | 8.648 | 8.627 |
| **34** | 8.582 | 8.562 | 8.541 | 8.521 | 8.501 | 8.481 | 8.461 | 8.441 | 8.421 |
| **35** | 8.377 | 8.358 | 8.338 | 8.318 | 8.299 | 8.280 | 8.260 | 8.241 | 8.222 |
| **36** | 8.179 | 8.161 | 8.142 | 8.123 | 8.104 | 8.085 | 8.067 | 8.048 | 8.029 |
| **37** | 7.988 | 7.970 | 7.952 | 7.934 | 7.916 | 7.898 | 7.880 | 7.862 | 7.844 |
| **38** | 7.804 | 7.786 | 7.769 | 7.751 | 7.734 | 7.716 | 7.699 | 7.681 | 7.664 |
| **39** | 7.625 | 7.608 | 7.591 | 7.574 | 7.557 | 7.541 | 7.524 | 7.507 | 7.490 |
| **40** | 7.452 | 7.436 | 7.419 | 7.403 | 7.387 | 7.370 | 7.354 | 7.338 | 7.322 |

*Source*: Based on Eqs. 5 and 9, Weiss (1974); mole fraction estimated from Mauna Loa data for 2010.

**Table 2.9** Standard Air Saturation Concentration of Carbon Dioxide as a Function of Temperature and Salinity, 0–40 g/kg in μmol/kg (Seawater, 1 atm moist air; mole fraction = 440 μatm)

| Temperature (°C) | Salinity (g/kg) | | | | | | | | |
|---|---|---|---|---|---|---|---|---|---|
| | 0.0 | 5.0 | 10.0 | 15.0 | 20.0 | 25.0 | 30.0 | 35.0 | 40.0 |
| 0 | 33.779 | 32.780 | 31.811 | 30.871 | 29.958 | 29.072 | 28.213 | 27.379 | 26.569 |
| 1 | 32.460 | 31.503 | 30.575 | 29.674 | 28.800 | 27.952 | 27.128 | 26.329 | 25.553 |
| 2 | 31.210 | 30.294 | 29.404 | 28.541 | 27.703 | 26.890 | 26.101 | 25.335 | 24.591 |
| 3 | 30.025 | 29.147 | 28.294 | 27.467 | 26.664 | 25.884 | 25.127 | 24.392 | 23.679 |
| 4 | 28.900 | 28.059 | 27.241 | 26.448 | 25.678 | 24.930 | 24.204 | 23.499 | 22.814 |
| 5 | 27.833 | 27.026 | 26.242 | 25.481 | 24.742 | 24.024 | 23.327 | 22.651 | 21.994 |
| 6 | 26.820 | 26.045 | 25.293 | 24.562 | 23.853 | 23.164 | 22.495 | 21.845 | 21.215 |
| 7 | 25.856 | 25.113 | 24.391 | 23.690 | 23.009 | 22.347 | 21.705 | 21.081 | 20.475 |
| 8 | 24.940 | 24.227 | 23.533 | 22.860 | 22.206 | 21.570 | 20.953 | 20.354 | 19.771 |
| 9 | 24.069 | 23.383 | 22.717 | 22.071 | 21.442 | 20.832 | 20.238 | 19.662 | 19.102 |
| 10 | 23.239 | 22.581 | 21.941 | 21.319 | 20.715 | 20.128 | 19.558 | 19.004 | 18.466 |
| 11 | 22.449 | 21.816 | 21.201 | 20.604 | 20.023 | 19.459 | 18.911 | 18.378 | 17.860 |
| 12 | 21.696 | 21.087 | 20.496 | 19.922 | 19.364 | 18.821 | 18.294 | 17.781 | 17.282 |
| 13 | 20.977 | 20.393 | 19.825 | 19.272 | 18.735 | 18.213 | 17.705 | 17.212 | 16.732 |
| 14 | 20.292 | 19.730 | 19.183 | 18.652 | 18.135 | 17.633 | 17.144 | 16.669 | 16.208 |
| 15 | 19.638 | 19.097 | 18.572 | 18.060 | 17.563 | 17.079 | 16.609 | 16.152 | 15.707 |
| 16 | 19.013 | 18.493 | 17.987 | 17.495 | 17.016 | 16.551 | 16.098 | 15.657 | 15.229 |
| 17 | 18.417 | 17.916 | 17.429 | 16.955 | 16.494 | 16.046 | 15.610 | 15.185 | 14.772 |
| 18 | 17.846 | 17.364 | 16.895 | 16.439 | 15.995 | 15.563 | 15.143 | 14.734 | 14.336 |
| 19 | 17.300 | 16.836 | 16.384 | 15.945 | 15.517 | 15.101 | 14.696 | 14.302 | 13.919 |
| 20 | 16.777 | 16.330 | 15.896 | 15.472 | 15.061 | 14.660 | 14.269 | 13.889 | 13.520 |
| 21 | 16.276 | 15.846 | 15.428 | 15.020 | 14.623 | 14.237 | 13.860 | 13.494 | 13.138 |
| 22 | 15.797 | 15.383 | 14.979 | 14.586 | 14.204 | 13.831 | 13.469 | 13.116 | 12.772 |
| 23 | 15.337 | 14.938 | 14.549 | 14.171 | 13.802 | 13.443 | 13.093 | 12.753 | 12.421 |
| 24 | 14.896 | 14.511 | 14.137 | 13.772 | 13.417 | 13.071 | 12.733 | 12.405 | 12.085 |
| 25 | 14.472 | 14.102 | 13.741 | 13.390 | 13.047 | 12.713 | 12.388 | 12.071 | 11.762 |
| 26 | 14.066 | 13.709 | 13.361 | 13.022 | 12.692 | 12.370 | 12.056 | 11.750 | 11.452 |
| 27 | 13.675 | 13.331 | 12.996 | 12.669 | 12.351 | 12.040 | 11.738 | 11.443 | 11.155 |
| 28 | 13.299 | 12.968 | 12.645 | 12.330 | 12.023 | 11.724 | 11.432 | 11.147 | 10.869 |
| 29 | 12.938 | 12.619 | 12.307 | 12.004 | 11.708 | 11.419 | 11.137 | 10.862 | 10.594 |
| 30 | 12.590 | 12.283 | 11.983 | 11.690 | 11.404 | 11.126 | 10.854 | 10.589 | 10.330 |
| 31 | 12.255 | 11.959 | 11.670 | 11.387 | 11.112 | 10.843 | 10.581 | 10.325 | 10.075 |
| 32 | 11.932 | 11.647 | 11.368 | 11.096 | 10.830 | 10.571 | 10.318 | 10.071 | 9.830 |
| 33 | 11.621 | 11.346 | 11.077 | 10.815 | 10.559 | 10.309 | 10.065 | 9.827 | 9.594 |
| 34 | 11.321 | 11.056 | 10.797 | 10.544 | 10.297 | 10.056 | 9.821 | 9.591 | 9.366 |
| 35 | 11.031 | 10.776 | 10.526 | 10.282 | 10.044 | 9.812 | 9.585 | 9.363 | 9.146 |
| 36 | 10.751 | 10.505 | 10.265 | 10.030 | 9.800 | 9.576 | 9.357 | 9.143 | 8.934 |
| 37 | 10.480 | 10.243 | 10.012 | 9.786 | 9.564 | 9.348 | 9.137 | 8.931 | 8.729 |
| 38 | 10.218 | 9.990 | 9.767 | 9.549 | 9.336 | 9.128 | 8.924 | 8.725 | 8.530 |
| 39 | 9.965 | 9.745 | 9.531 | 9.321 | 9.115 | 8.915 | 8.718 | 8.526 | 8.338 |
| 40 | 9.719 | 9.508 | 9.301 | 9.099 | 8.902 | 8.708 | 8.519 | 8.334 | 8.153 |

*Source*: Based on Eqs. 5 and 9, Weiss (1974); mole fraction estimated from Mauna Loa data for 2030.

**Table 2.10** Standard Air Saturation Concentration of Carbon Dioxide as a Function of Temperature, 33–37 g/kg in μmol/kg (Seawater, 1 atm moist air; mole fraction = 440 μatm)

| Temperature (°C) | Salinity (g/kg) | | | | | | | | |
|---|---|---|---|---|---|---|---|---|---|
| | 33.0 | 33.5 | 34.0 | 34.5 | 35.0 | 35.5 | 36.0 | 36.5 | 37.0 |
| 0 | 27.709 | 27.626 | 27.543 | 27.461 | 27.379 | 27.296 | 27.215 | 27.133 | 27.052 |
| 1 | 26.646 | 26.566 | 26.487 | 26.408 | 26.329 | 26.250 | 26.172 | 26.094 | 26.016 |
| 2 | 25.638 | 25.562 | 25.486 | 25.410 | 25.335 | 25.259 | 25.184 | 25.109 | 25.034 |
| 3 | 24.684 | 24.610 | 24.538 | 24.465 | 24.392 | 24.320 | 24.248 | 24.176 | 24.104 |
| 4 | 23.778 | 23.708 | 23.638 | 23.568 | 23.499 | 23.429 | 23.360 | 23.291 | 23.222 |
| 5 | 22.919 | 22.852 | 22.784 | 22.717 | 22.651 | 22.584 | 22.518 | 22.451 | 22.385 |
| 6 | 22.103 | 22.038 | 21.974 | 21.910 | 21.845 | 21.782 | 21.718 | 21.654 | 21.591 |
| 7 | 21.328 | 21.266 | 21.204 | 21.142 | 21.081 | 21.019 | 20.958 | 20.897 | 20.836 |
| 8 | 20.591 | 20.532 | 20.472 | 20.413 | 20.354 | 20.295 | 20.236 | 20.177 | 20.119 |
| 9 | 19.891 | 19.833 | 19.776 | 19.719 | 19.662 | 19.605 | 19.549 | 19.492 | 19.436 |
| 10 | 19.224 | 19.169 | 19.114 | 19.059 | 19.004 | 18.950 | 18.895 | 18.841 | 18.787 |
| 11 | 18.589 | 18.536 | 18.483 | 18.430 | 18.378 | 18.325 | 18.273 | 18.221 | 18.169 |
| 12 | 17.984 | 17.933 | 17.882 | 17.831 | 17.781 | 17.730 | 17.680 | 17.630 | 17.580 |
| 13 | 17.408 | 17.358 | 17.309 | 17.261 | 17.212 | 17.163 | 17.115 | 17.067 | 17.018 |
| 14 | 16.858 | 16.810 | 16.763 | 16.716 | 16.669 | 16.623 | 16.576 | 16.529 | 16.483 |
| 15 | 16.333 | 16.288 | 16.242 | 16.197 | 16.152 | 16.107 | 16.062 | 16.017 | 15.972 |
| 16 | 15.832 | 15.788 | 15.744 | 15.701 | 15.657 | 15.614 | 15.571 | 15.528 | 15.485 |
| 17 | 15.354 | 15.311 | 15.269 | 15.227 | 15.185 | 15.143 | 15.102 | 15.060 | 15.019 |
| 18 | 14.896 | 14.855 | 14.815 | 14.774 | 14.734 | 14.694 | 14.653 | 14.613 | 14.573 |
| 19 | 14.459 | 14.419 | 14.380 | 14.341 | 14.302 | 14.264 | 14.225 | 14.186 | 14.148 |
| 20 | 14.040 | 14.002 | 13.965 | 13.927 | 13.889 | 13.852 | 13.815 | 13.777 | 13.740 |
| 21 | 13.639 | 13.603 | 13.567 | 13.530 | 13.494 | 13.458 | 13.422 | 13.386 | 13.350 |
| 22 | 13.256 | 13.220 | 13.185 | 13.150 | 13.116 | 13.081 | 13.046 | 13.011 | 12.977 |
| 23 | 12.888 | 12.854 | 12.820 | 12.786 | 12.753 | 12.719 | 12.686 | 12.652 | 12.619 |
| 24 | 12.535 | 12.502 | 12.470 | 12.437 | 12.405 | 12.372 | 12.340 | 12.308 | 12.276 |
| 25 | 12.197 | 12.165 | 12.134 | 12.102 | 12.071 | 12.040 | 12.008 | 11.977 | 11.946 |
| 26 | 11.872 | 11.841 | 11.811 | 11.781 | 11.750 | 11.720 | 11.690 | 11.660 | 11.630 |
| 27 | 11.560 | 11.530 | 11.501 | 11.472 | 11.443 | 11.413 | 11.384 | 11.356 | 11.327 |
| 28 | 11.260 | 11.231 | 11.203 | 11.175 | 11.147 | 11.119 | 11.091 | 11.063 | 11.035 |
| 29 | 10.971 | 10.944 | 10.917 | 10.889 | 10.862 | 10.835 | 10.808 | 10.781 | 10.754 |
| 30 | 10.694 | 10.667 | 10.641 | 10.615 | 10.589 | 10.562 | 10.536 | 10.510 | 10.484 |
| 31 | 10.427 | 10.401 | 10.376 | 10.350 | 10.325 | 10.300 | 10.275 | 10.250 | 10.225 |
| 32 | 10.169 | 10.145 | 10.120 | 10.096 | 10.071 | 10.047 | 10.023 | 9.998 | 9.974 |
| 33 | 9.921 | 9.898 | 9.874 | 9.850 | 9.827 | 9.803 | 9.780 | 9.756 | 9.733 |
| 34 | 9.682 | 9.659 | 9.636 | 9.613 | 9.591 | 9.568 | 9.545 | 9.523 | 9.500 |
| 35 | 9.451 | 9.429 | 9.407 | 9.385 | 9.363 | 9.341 | 9.319 | 9.297 | 9.276 |
| 36 | 9.228 | 9.207 | 9.186 | 9.164 | 9.143 | 9.122 | 9.101 | 9.080 | 9.059 |
| 37 | 9.013 | 8.992 | 8.972 | 8.951 | 8.931 | 8.910 | 8.890 | 8.870 | 8.849 |
| 38 | 8.804 | 8.784 | 8.765 | 8.745 | 8.725 | 8.705 | 8.686 | 8.666 | 8.647 |
| 39 | 8.603 | 8.583 | 8.564 | 8.545 | 8.526 | 8.507 | 8.488 | 8.469 | 8.451 |
| 40 | 8.407 | 8.389 | 8.370 | 8.352 | 8.334 | 8.315 | 8.297 | 8.279 | 8.261 |

*Source*: Based on Eqs. 5 and 9, Weiss (1974); mole fraction estimated from Mauna Loa data for 2030.

**Table 2.11** Standard Air Saturation Concentration of Oxygen as a Function of Temperature and Salinity in mg/L, 0–40 g/kg (Seawater, 1 atm moist air)

| Temperature (°C) | Salinity (g/kg) | | | | | | | | |
|---|---|---|---|---|---|---|---|---|---|
| | **0.0** | **5.0** | **10.0** | **15.0** | **20.0** | **25.0** | **30.0** | **35.0** | **40.0** |
| **0** | 14.621 | 14.120 | 13.635 | 13.167 | 12.714 | 12.276 | 11.854 | 11.445 | 11.050 |
| **1** | 14.216 | 13.733 | 13.266 | 12.815 | 12.378 | 11.956 | 11.548 | 11.153 | 10.772 |
| **2** | 13.829 | 13.364 | 12.914 | 12.478 | 12.057 | 11.649 | 11.255 | 10.875 | 10.506 |
| **3** | 13.460 | 13.011 | 12.577 | 12.156 | 11.750 | 11.356 | 10.976 | 10.608 | 10.252 |
| **4** | 13.107 | 12.674 | 12.255 | 11.849 | 11.456 | 11.076 | 10.708 | 10.352 | 10.008 |
| **5** | 12.770 | 12.352 | 11.946 | 11.554 | 11.174 | 10.807 | 10.451 | 10.107 | 9.774 |
| **6** | 12.447 | 12.043 | 11.652 | 11.272 | 10.905 | 10.550 | 10.205 | 9.872 | 9.550 |
| **7** | 12.138 | 11.748 | 11.369 | 11.002 | 10.647 | 10.303 | 9.970 | 9.647 | 9.335 |
| **8** | 11.843 | 11.465 | 11.098 | 10.743 | 10.399 | 10.066 | 9.743 | 9.431 | 9.128 |
| **9** | 11.559 | 11.194 | 10.839 | 10.495 | 10.162 | 9.839 | 9.526 | 9.223 | 8.930 |
| **10** | 11.288 | 10.933 | 10.590 | 10.257 | 9.934 | 9.621 | 9.318 | 9.024 | 8.739 |
| **11** | 11.027 | 10.684 | 10.351 | 10.028 | 9.715 | 9.411 | 9.117 | 8.832 | 8.556 |
| **12** | 10.777 | 10.444 | 10.121 | 9.808 | 9.505 | 9.210 | 8.925 | 8.648 | 8.379 |
| **13** | 10.536 | 10.214 | 9.901 | 9.597 | 9.302 | 9.016 | 8.739 | 8.470 | 8.209 |
| **14** | 10.306 | 9.993 | 9.689 | 9.394 | 9.108 | 8.830 | 8.561 | 8.299 | 8.046 |
| **15** | 10.084 | 9.780 | 9.485 | 9.198 | 8.920 | 8.651 | 8.389 | 8.135 | 7.888 |
| **16** | 9.870 | 9.575 | 9.289 | 9.010 | 8.740 | 8.478 | 8.223 | 7.976 | 7.736 |
| **17** | 9.665 | 9.378 | 9.099 | 8.829 | 8.566 | 8.311 | 8.064 | 7.823 | 7.590 |
| **18** | 9.467 | 9.188 | 8.917 | 8.654 | 8.399 | 8.151 | 7.910 | 7.676 | 7.448 |
| **19** | 9.276 | 9.005 | 8.742 | 8.486 | 8.237 | 7.996 | 7.761 | 7.533 | 7.312 |
| **20** | 9.092 | 8.828 | 8.572 | 8.323 | 8.081 | 7.846 | 7.617 | 7.395 | 7.180 |
| **21** | 8.914 | 8.658 | 8.408 | 8.166 | 7.930 | 7.701 | 7.479 | 7.262 | 7.052 |
| **22** | 8.743 | 8.493 | 8.250 | 8.014 | 7.785 | 7.561 | 7.344 | 7.134 | 6.929 |
| **23** | 8.578 | 8.334 | 8.098 | 7.868 | 7.644 | 7.426 | 7.215 | 7.009 | 6.809 |
| **24** | 8.418 | 8.181 | 7.950 | 7.726 | 7.507 | 7.295 | 7.089 | 6.888 | 6.693 |
| **25** | 8.263 | 8.032 | 7.807 | 7.588 | 7.375 | 7.168 | 6.967 | 6.771 | 6.581 |
| **26** | 8.113 | 7.888 | 7.668 | 7.455 | 7.247 | 7.045 | 6.849 | 6.658 | 6.472 |
| **27** | 7.968 | 7.748 | 7.534 | 7.326 | 7.123 | 6.926 | 6.734 | 6.548 | 6.366 |
| **28** | 7.827 | 7.613 | 7.404 | 7.201 | 7.003 | 6.811 | 6.623 | 6.441 | 6.263 |
| **29** | 7.691 | 7.482 | 7.278 | 7.079 | 6.886 | 6.698 | 6.515 | 6.337 | 6.164 |
| **30** | 7.559 | 7.354 | 7.155 | 6.961 | 6.773 | 6.589 | 6.410 | 6.236 | 6.066 |
| **31** | 7.430 | 7.230 | 7.036 | 6.847 | 6.662 | 6.483 | 6.308 | 6.138 | 5.972 |
| **32** | 7.305 | 7.110 | 6.920 | 6.735 | 6.555 | 6.379 | 6.208 | 6.042 | 5.880 |
| **33** | 7.183 | 6.993 | 6.807 | 6.626 | 6.450 | 6.279 | 6.111 | 5.949 | 5.790 |
| **34** | 7.065 | 6.879 | 6.697 | 6.521 | 6.348 | 6.180 | 6.017 | 5.857 | 5.702 |
| **35** | 6.949 | 6.768 | 6.590 | 6.417 | 6.249 | 6.085 | 5.925 | 5.769 | 5.617 |
| **36** | 6.837 | 6.659 | 6.486 | 6.316 | 6.152 | 5.991 | 5.834 | 5.682 | 5.533 |
| **37** | 6.727 | 6.553 | 6.383 | 6.218 | 6.057 | 5.899 | 5.746 | 5.597 | 5.451 |
| **38** | 6.620 | 6.450 | 6.284 | 6.122 | 5.964 | 5.810 | 5.660 | 5.514 | 5.371 |
| **39** | 6.515 | 6.348 | 6.186 | 6.027 | 5.873 | 5.722 | 5.575 | 5.432 | 5.292 |
| **40** | 6.412 | 6.249 | 6.090 | 5.935 | 5.784 | 5.636 | 5.492 | 5.352 | 5.215 |

*Source*: Based on Eq. 22, Benson and Krause (1984).

**Table 2.12** Standard Air Saturation Concentration of Oxygen as a Function of Temperature and Salinity in mg/L, 33–37 g/kg (Seawater, 1 atm moist air)

| Temperature (°C) | Salinity (g/kg) | | | | | | | | |
|---|---|---|---|---|---|---|---|---|---|
| | **33.0** | **33.5** | **34.0** | **34.5** | **35.0** | **35.5** | **36.0** | **36.5** | **37.0** |
| **0** | 11.607 | 11.566 | 11.526 | 11.485 | 11.445 | 11.405 | 11.365 | 11.325 | 11.285 |
| **1** | 11.310 | 11.270 | 11.231 | 11.192 | 11.153 | 11.115 | 11.076 | 11.038 | 10.999 |
| **2** | 11.025 | 10.987 | 10.950 | 10.912 | 10.875 | 10.837 | 10.800 | 10.763 | 10.726 |
| **3** | 10.753 | 10.717 | 10.680 | 10.644 | 10.608 | 10.572 | 10.536 | 10.500 | 10.464 |
| **4** | 10.493 | 10.458 | 10.422 | 10.387 | 10.352 | 10.317 | 10.282 | 10.248 | 10.213 |
| **5** | 10.243 | 10.209 | 10.175 | 10.141 | 10.107 | 10.073 | 10.040 | 10.006 | 9.973 |
| **6** | 10.004 | 9.971 | 9.938 | 9.905 | 9.872 | 9.840 | 9.807 | 9.774 | 9.742 |
| **7** | 9.775 | 9.743 | 9.711 | 9.679 | 9.647 | 9.615 | 9.584 | 9.552 | 9.521 |
| **8** | 9.555 | 9.524 | 9.493 | 9.462 | 9.431 | 9.400 | 9.369 | 9.339 | 9.309 |
| **9** | 9.343 | 9.313 | 9.283 | 9.253 | 9.223 | 9.193 | 9.164 | 9.134 | 9.105 |
| **10** | 9.140 | 9.111 | 9.082 | 9.053 | 9.024 | 8.995 | 8.966 | 8.937 | 8.909 |
| **11** | 8.945 | 8.917 | 8.888 | 8.860 | 8.832 | 8.804 | 8.776 | 8.748 | 8.720 |
| **12** | 8.757 | 8.730 | 8.702 | 8.675 | 8.648 | 8.620 | 8.593 | 8.566 | 8.539 |
| **13** | 8.577 | 8.550 | 8.523 | 8.497 | 8.470 | 8.444 | 8.417 | 8.391 | 8.365 |
| **14** | 8.403 | 8.377 | 8.351 | 8.325 | 8.299 | 8.274 | 8.248 | 8.223 | 8.197 |
| **15** | 8.236 | 8.210 | 8.185 | 8.160 | 8.135 | 8.110 | 8.085 | 8.060 | 8.035 |
| **16** | 8.074 | 8.050 | 8.025 | 8.001 | 7.976 | 7.952 | 7.928 | 7.904 | 7.879 |
| **17** | 7.919 | 7.895 | 7.871 | 7.847 | 7.823 | 7.800 | 7.776 | 7.753 | 7.729 |
| **18** | 7.768 | 7.745 | 7.722 | 7.699 | 7.676 | 7.653 | 7.630 | 7.607 | 7.584 |
| **19** | 7.623 | 7.601 | 7.578 | 7.556 | 7.533 | 7.511 | 7.488 | 7.466 | 7.444 |
| **20** | 7.483 | 7.461 | 7.439 | 7.417 | 7.395 | 7.374 | 7.352 | 7.330 | 7.308 |
| **21** | 7.348 | 7.327 | 7.305 | 7.284 | 7.262 | 7.241 | 7.220 | 7.199 | 7.178 |
| **22** | 7.217 | 7.196 | 7.175 | 7.154 | 7.134 | 7.113 | 7.092 | 7.071 | 7.051 |
| **23** | 7.090 | 7.070 | 7.050 | 7.029 | 7.009 | 6.989 | 6.968 | 6.948 | 6.928 |
| **24** | 6.968 | 6.948 | 6.928 | 6.908 | 6.888 | 6.868 | 6.849 | 6.829 | 6.810 |
| **25** | 6.849 | 6.829 | 6.810 | 6.791 | 6.771 | 6.752 | 6.733 | 6.714 | 6.694 |
| **26** | 6.734 | 6.715 | 6.696 | 6.677 | 6.658 | 6.639 | 6.620 | 6.602 | 6.583 |
| **27** | 6.622 | 6.603 | 6.585 | 6.566 | 6.548 | 6.529 | 6.511 | 6.493 | 6.475 |
| **28** | 6.513 | 6.495 | 6.477 | 6.459 | 6.441 | 6.423 | 6.405 | 6.387 | 6.369 |
| **29** | 6.408 | 6.390 | 6.372 | 6.355 | 6.337 | 6.319 | 6.302 | 6.284 | 6.267 |
| **30** | 6.305 | 6.288 | 6.270 | 6.253 | 6.236 | 6.219 | 6.202 | 6.185 | 6.168 |
| **31** | 6.205 | 6.188 | 6.171 | 6.154 | 6.138 | 6.121 | 6.104 | 6.087 | 6.071 |
| **32** | 6.108 | 6.091 | 6.075 | 6.058 | 6.042 | 6.025 | 6.009 | 5.993 | 5.977 |
| **33** | 6.013 | 5.997 | 5.981 | 5.965 | 5.949 | 5.933 | 5.916 | 5.901 | 5.885 |
| **34** | 5.921 | 5.905 | 5.889 | 5.873 | 5.857 | 5.842 | 5.826 | 5.811 | 5.795 |
| **35** | 5.830 | 5.815 | 5.799 | 5.784 | 5.769 | 5.753 | 5.738 | 5.723 | 5.707 |
| **36** | 5.742 | 5.727 | 5.712 | 5.697 | 5.682 | 5.667 | 5.652 | 5.637 | 5.622 |
| **37** | 5.656 | 5.641 | 5.626 | 5.612 | 5.597 | 5.582 | 5.567 | 5.553 | 5.538 |
| **38** | 5.572 | 5.557 | 5.543 | 5.528 | 5.514 | 5.499 | 5.485 | 5.470 | 5.456 |
| **39** | 5.489 | 5.475 | 5.460 | 5.446 | 5.432 | 5.418 | 5.404 | 5.390 | 5.376 |
| **40** | 5.408 | 5.394 | 5.380 | 5.366 | 5.352 | 5.338 | 5.324 | 5.311 | 5.297 |

*Source*: Based on Eq. 22, Benson and Krause (1984).

**Table 2.13** Standard Air Saturation Concentration of Nitrogen as a Function of Temperature and Salinity, 0–40 g/kg in mg/L (Seawater, 1 atm moist air)

| Temperature (°C) | Salinity (g/kg) | | | | | | | | |
|---|---|---|---|---|---|---|---|---|---|
| | **0.0** | **5.0** | **10.0** | **15.0** | **20.0** | **25.0** | **30.0** | **35.0** | **40.0** |
| **0** | 23.261 | 22.411 | 21.591 | 20.801 | 20.039 | 19.305 | 18.597 | 17.915 | 17.258 |
| **1** | 22.658 | 21.837 | 21.046 | 20.283 | 19.547 | 18.837 | 18.153 | 17.494 | 16.858 |
| **2** | 22.082 | 21.290 | 20.525 | 19.787 | 19.076 | 18.390 | 17.728 | 17.090 | 16.475 |
| **3** | 21.532 | 20.766 | 20.027 | 19.314 | 18.626 | 17.962 | 17.322 | 16.704 | 16.108 |
| **4** | 21.006 | 20.266 | 19.551 | 18.861 | 18.195 | 17.553 | 16.932 | 16.334 | 15.757 |
| **5** | 20.503 | 19.787 | 19.096 | 18.428 | 17.783 | 17.160 | 16.559 | 15.979 | 15.419 |
| **6** | 20.022 | 19.329 | 18.659 | 18.012 | 17.388 | 16.784 | 16.202 | 15.639 | 15.096 |
| **7** | 19.562 | 18.891 | 18.242 | 17.615 | 17.009 | 16.424 | 15.859 | 15.313 | 14.786 |
| **8** | 19.121 | 18.470 | 17.841 | 17.233 | 16.646 | 16.078 | 15.530 | 15.000 | 14.488 |
| **9** | 18.698 | 18.068 | 17.458 | 16.868 | 16.297 | 15.746 | 15.214 | 14.699 | 14.202 |
| **10** | 18.294 | 17.682 | 17.089 | 16.517 | 15.963 | 15.428 | 14.910 | 14.410 | 13.926 |
| **11** | 17.905 | 17.311 | 16.736 | 16.180 | 15.642 | 15.122 | 14.619 | 14.132 | 13.662 |
| **12** | 17.533 | 16.955 | 16.397 | 15.856 | 15.333 | 14.827 | 14.338 | 13.865 | 13.407 |
| **13** | 17.175 | 16.614 | 16.071 | 15.545 | 15.037 | 14.545 | 14.068 | 13.608 | 13.162 |
| **14** | 16.831 | 16.286 | 15.758 | 15.246 | 14.751 | 14.272 | 13.809 | 13.360 | 12.926 |
| **15** | 16.501 | 15.971 | 15.457 | 14.959 | 14.477 | 14.010 | 13.559 | 13.122 | 12.699 |
| **16** | 16.184 | 15.667 | 15.167 | 14.682 | 14.213 | 13.758 | 13.318 | 12.892 | 12.480 |
| **17** | 15.878 | 15.375 | 14.888 | 14.416 | 13.958 | 13.515 | 13.086 | 12.670 | 12.268 |
| **18** | 15.584 | 15.094 | 14.619 | 14.159 | 13.713 | 13.281 | 12.862 | 12.457 | 12.064 |
| **19** | 15.302 | 14.824 | 14.360 | 13.911 | 13.476 | 13.055 | 12.646 | 12.250 | 11.867 |
| **20** | 15.029 | 14.563 | 14.111 | 13.673 | 13.248 | 12.836 | 12.437 | 12.051 | 11.676 |
| **21** | 14.766 | 14.311 | 13.870 | 13.442 | 13.028 | 12.626 | 12.236 | 11.858 | 11.492 |
| **22** | 14.513 | 14.069 | 13.638 | 13.220 | 12.815 | 12.422 | 12.041 | 11.672 | 11.314 |
| **23** | 14.268 | 13.835 | 13.414 | 13.005 | 12.609 | 12.225 | 11.853 | 11.491 | 11.141 |
| **24** | 14.032 | 13.608 | 13.197 | 12.798 | 12.410 | 12.035 | 11.670 | 11.317 | 10.974 |
| **25** | 13.804 | 13.390 | 12.987 | 12.597 | 12.218 | 11.850 | 11.494 | 11.148 | 10.812 |
| **26** | 13.584 | 13.179 | 12.785 | 12.403 | 12.032 | 11.672 | 11.323 | 10.984 | 10.655 |
| **27** | 13.371 | 12.974 | 12.589 | 12.214 | 11.851 | 11.499 | 11.157 | 10.825 | 10.503 |
| **28** | 13.165 | 12.776 | 12.399 | 12.032 | 11.676 | 11.331 | 10.996 | 10.670 | 10.355 |
| **29** | 12.965 | 12.585 | 12.215 | 11.855 | 11.507 | 11.168 | 10.840 | 10.521 | 10.211 |
| **30** | 12.772 | 12.399 | 12.036 | 11.684 | 11.342 | 11.010 | 10.688 | 10.375 | 10.071 |
| **31** | 12.584 | 12.219 | 11.863 | 11.518 | 11.182 | 10.857 | 10.540 | 10.233 | 9.935 |
| **32** | 12.403 | 12.044 | 11.695 | 11.356 | 11.027 | 10.707 | 10.397 | 10.095 | 9.802 |
| **33** | 12.226 | 11.874 | 11.532 | 11.199 | 10.876 | 10.562 | 10.257 | 9.961 | 9.673 |
| **34** | 12.055 | 11.709 | 11.373 | 11.046 | 10.729 | 10.420 | 10.121 | 9.830 | 9.547 |
| **35** | 11.888 | 11.549 | 11.218 | 10.897 | 10.586 | 10.283 | 9.988 | 9.702 | 9.424 |
| **36** | 11.726 | 11.392 | 11.068 | 10.753 | 10.446 | 10.148 | 9.859 | 9.578 | 9.304 |
| **37** | 11.568 | 11.240 | 10.921 | 10.611 | 10.310 | 10.017 | 9.732 | 9.456 | 9.187 |
| **38** | 11.415 | 11.092 | 10.779 | 10.474 | 10.177 | 9.889 | 9.609 | 9.337 | 9.072 |
| **39** | 11.265 | 10.948 | 10.639 | 10.339 | 10.047 | 9.764 | 9.488 | 9.220 | 8.960 |
| **40** | 11.119 | 10.807 | 10.503 | 10.208 | 9.920 | 9.641 | 9.370 | 9.106 | 8.850 |

*Source*: Based on Eq. 1, Hamme and Emerson (2004).

**Table 2.14** Standard Air Saturation Concentration of Nitrogen as a Function of Temperature and Salinity, 33–37 g/kg in mg/L (Seawater, 1 atm moist air)

| Temperature (°C) | Salinity (g/kg) | | | | | | | | |
|---|---|---|---|---|---|---|---|---|---|
| | **33.0** | **33.5** | **34.0** | **34.5** | **35.0** | **35.5** | **36.0** | **36.5** | **37.0** |
| **0** | 18.185 | 18.117 | 18.050 | 17.982 | 17.915 | 17.848 | 17.782 | 17.716 | 17.650 |
| **1** | 17.755 | 17.689 | 17.624 | 17.559 | 17.494 | 17.429 | 17.365 | 17.301 | 17.237 |
| **2** | 17.343 | 17.279 | 17.216 | 17.153 | 17.090 | 17.028 | 16.965 | 16.903 | 16.842 |
| **3** | 16.948 | 16.887 | 16.826 | 16.765 | 16.704 | 16.643 | 16.583 | 16.523 | 16.463 |
| **4** | 16.571 | 16.511 | 16.452 | 16.393 | 16.334 | 16.275 | 16.217 | 16.159 | 16.101 |
| **5** | 16.209 | 16.151 | 16.094 | 16.036 | 15.979 | 15.922 | 15.866 | 15.809 | 15.753 |
| **6** | 15.862 | 15.806 | 15.750 | 15.695 | 15.639 | 15.584 | 15.529 | 15.474 | 15.420 |
| **7** | 15.529 | 15.475 | 15.421 | 15.367 | 15.313 | 15.259 | 15.206 | 15.153 | 15.100 |
| **8** | 15.210 | 15.157 | 15.104 | 15.052 | 15.000 | 14.948 | 14.896 | 14.844 | 14.793 |
| **9** | 14.903 | 14.852 | 14.801 | 14.750 | 14.699 | 14.648 | 14.598 | 14.548 | 14.498 |
| **10** | 14.608 | 14.558 | 14.509 | 14.459 | 14.410 | 14.361 | 14.312 | 14.263 | 14.215 |
| **11** | 14.325 | 14.276 | 14.228 | 14.180 | 14.132 | 14.084 | 14.037 | 13.989 | 13.942 |
| **12** | 14.052 | 14.005 | 13.958 | 13.912 | 13.865 | 13.819 | 13.772 | 13.726 | 13.680 |
| **13** | 13.790 | 13.744 | 13.699 | 13.653 | 13.608 | 13.563 | 13.518 | 13.473 | 13.428 |
| **14** | 13.538 | 13.493 | 13.449 | 13.405 | 13.360 | 13.316 | 13.272 | 13.229 | 13.185 |
| **15** | 13.295 | 13.251 | 13.208 | 13.165 | 13.122 | 13.079 | 13.036 | 12.993 | 12.951 |
| **16** | 13.061 | 13.018 | 12.976 | 12.934 | 12.892 | 12.850 | 12.808 | 12.767 | 12.725 |
| **17** | 12.835 | 12.794 | 12.752 | 12.711 | 12.670 | 12.630 | 12.589 | 12.548 | 12.508 |
| **18** | 12.617 | 12.577 | 12.537 | 12.497 | 12.457 | 12.417 | 12.377 | 12.337 | 12.298 |
| **19** | 12.407 | 12.368 | 12.328 | 12.289 | 12.250 | 12.211 | 12.172 | 12.134 | 12.095 |
| **20** | 12.204 | 12.165 | 12.127 | 12.089 | 12.051 | 12.013 | 11.975 | 11.937 | 11.899 |
| **21** | 12.008 | 11.970 | 11.933 | 11.895 | 11.858 | 11.821 | 11.784 | 11.747 | 11.710 |
| **22** | 11.818 | 11.781 | 11.745 | 11.708 | 11.672 | 11.635 | 11.599 | 11.563 | 11.527 |
| **23** | 11.635 | 11.599 | 11.563 | 11.527 | 11.491 | 11.456 | 11.420 | 11.385 | 11.350 |
| **24** | 11.457 | 11.422 | 11.387 | 11.352 | 11.317 | 11.282 | 11.247 | 11.213 | 11.178 |
| **25** | 11.285 | 11.250 | 11.216 | 11.182 | 11.148 | 11.114 | 11.080 | 11.046 | 11.012 |
| **26** | 11.118 | 11.084 | 11.051 | 11.017 | 10.984 | 10.950 | 10.917 | 10.884 | 10.851 |
| **27** | 10.956 | 10.923 | 10.890 | 10.858 | 10.825 | 10.792 | 10.760 | 10.727 | 10.695 |
| **28** | 10.799 | 10.767 | 10.735 | 10.703 | 10.670 | 10.638 | 10.607 | 10.575 | 10.543 |
| **29** | 10.647 | 10.615 | 10.584 | 10.552 | 10.521 | 10.489 | 10.458 | 10.427 | 10.396 |
| **30** | 10.499 | 10.468 | 10.437 | 10.406 | 10.375 | 10.344 | 10.313 | 10.283 | 10.252 |
| **31** | 10.355 | 10.324 | 10.294 | 10.263 | 10.233 | 10.203 | 10.173 | 10.143 | 10.113 |
| **32** | 10.215 | 10.185 | 10.155 | 10.125 | 10.095 | 10.066 | 10.036 | 10.006 | 9.977 |
| **33** | 10.078 | 10.049 | 10.019 | 9.990 | 9.961 | 9.932 | 9.903 | 9.874 | 9.845 |
| **34** | 9.945 | 9.916 | 9.887 | 9.859 | 9.830 | 9.801 | 9.773 | 9.744 | 9.716 |
| **35** | 9.816 | 9.787 | 9.759 | 9.730 | 9.702 | 9.674 | 9.646 | 9.618 | 9.590 |
| **36** | 9.689 | 9.661 | 9.633 | 9.605 | 9.578 | 9.550 | 9.522 | 9.495 | 9.467 |
| **37** | 9.565 | 9.538 | 9.510 | 9.483 | 9.456 | 9.429 | 9.401 | 9.374 | 9.347 |
| **38** | 9.445 | 9.417 | 9.390 | 9.364 | 9.337 | 9.310 | 9.283 | 9.257 | 9.230 |
| **39** | 9.326 | 9.300 | 9.273 | 9.247 | 9.220 | 9.194 | 9.167 | 9.141 | 9.115 |
| **40** | 9.211 | 9.184 | 9.158 | 9.132 | 9.106 | 9.080 | 9.054 | 9.028 | 9.003 |

*Source*: Based on Eq. 1, Hamme and Emerson (2004).

**Table 2.15** Standard Air Saturation Concentration of Argon as a Function of Temperature and Salinity, 0–40 g/kg in mg/L (Seawater, 1 atm moist air)

| Temperature (°C) | Salinity (g/kg) | | | | | | | | |
|---|---|---|---|---|---|---|---|---|---|
| | 0.0 | 5.0 | 10.0 | 15.0 | 20.0 | 25.0 | 30.0 | 35.0 | 40.0 |
| 0 | 0.8907 | 0.8605 | 0.8312 | 0.8029 | 0.7756 | 0.7492 | 0.7236 | 0.6990 | 0.6751 |
| 1 | 0.8663 | 0.8371 | 0.8089 | 0.7817 | 0.7553 | 0.7298 | 0.7051 | 0.6813 | 0.6583 |
| 2 | 0.8430 | 0.8148 | 0.7876 | 0.7613 | 0.7359 | 0.7112 | 0.6874 | 0.6644 | 0.6422 |
| 3 | 0.8207 | 0.7935 | 0.7673 | 0.7419 | 0.7173 | 0.6935 | 0.6705 | 0.6483 | 0.6268 |
| 4 | 0.7994 | 0.7732 | 0.7478 | 0.7233 | 0.6995 | 0.6765 | 0.6543 | 0.6328 | 0.6120 |
| 5 | 0.7790 | 0.7537 | 0.7292 | 0.7055 | 0.6825 | 0.6603 | 0.6388 | 0.6179 | 0.5978 |
| 6 | 0.7595 | 0.7350 | 0.7113 | 0.6884 | 0.6662 | 0.6447 | 0.6239 | 0.6037 | 0.5842 |
| 7 | 0.7408 | 0.7171 | 0.6942 | 0.6720 | 0.6505 | 0.6297 | 0.6096 | 0.5901 | 0.5712 |
| 8 | 0.7229 | 0.7000 | 0.6779 | 0.6564 | 0.6356 | 0.6154 | 0.5959 | 0.5770 | 0.5586 |
| 9 | 0.7057 | 0.6836 | 0.6621 | 0.6413 | 0.6212 | 0.6016 | 0.5827 | 0.5644 | 0.5466 |
| 10 | 0.6893 | 0.6679 | 0.6471 | 0.6269 | 0.6074 | 0.5884 | 0.5701 | 0.5523 | 0.5350 |
| 11 | 0.6735 | 0.6527 | 0.6326 | 0.6130 | 0.5941 | 0.5757 | 0.5579 | 0.5407 | 0.5239 |
| 12 | 0.6584 | 0.6382 | 0.6187 | 0.5997 | 0.5813 | 0.5635 | 0.5462 | 0.5295 | 0.5132 |
| 13 | 0.6438 | 0.6243 | 0.6053 | 0.5869 | 0.5691 | 0.5518 | 0.5350 | 0.5187 | 0.5029 |
| 14 | 0.6298 | 0.6109 | 0.5925 | 0.5746 | 0.5573 | 0.5404 | 0.5241 | 0.5083 | 0.4930 |
| 15 | 0.6164 | 0.5980 | 0.5801 | 0.5627 | 0.5459 | 0.5296 | 0.5137 | 0.4983 | 0.4834 |
| 16 | 0.6034 | 0.5855 | 0.5682 | 0.5513 | 0.5349 | 0.5191 | 0.5036 | 0.4887 | 0.4742 |
| 17 | 0.5909 | 0.5736 | 0.5567 | 0.5403 | 0.5244 | 0.5089 | 0.4939 | 0.4794 | 0.4653 |
| 18 | 0.5789 | 0.5621 | 0.5456 | 0.5297 | 0.5142 | 0.4992 | 0.4846 | 0.4704 | 0.4567 |
| 19 | 0.5674 | 0.5509 | 0.5350 | 0.5195 | 0.5044 | 0.4898 | 0.4755 | 0.4617 | 0.4483 |
| 20 | 0.5562 | 0.5402 | 0.5247 | 0.5096 | 0.4949 | 0.4807 | 0.4668 | 0.4534 | 0.4403 |
| 21 | 0.5454 | 0.5299 | 0.5148 | 0.5000 | 0.4858 | 0.4719 | 0.4584 | 0.4453 | 0.4325 |
| 22 | 0.5350 | 0.5199 | 0.5052 | 0.4908 | 0.4769 | 0.4634 | 0.4502 | 0.4374 | 0.4250 |
| 23 | 0.5250 | 0.5102 | 0.4959 | 0.4819 | 0.4683 | 0.4551 | 0.4423 | 0.4298 | 0.4177 |
| 24 | 0.5153 | 0.5009 | 0.4869 | 0.4733 | 0.4600 | 0.4472 | 0.4346 | 0.4225 | 0.4106 |
| 25 | 0.5059 | 0.4919 | 0.4782 | 0.4649 | 0.4520 | 0.4394 | 0.4272 | 0.4153 | 0.4038 |
| 26 | 0.4968 | 0.4831 | 0.4698 | 0.4568 | 0.4442 | 0.4319 | 0.4200 | 0.4084 | 0.3971 |
| 27 | 0.4880 | 0.4747 | 0.4617 | 0.4490 | 0.4367 | 0.4247 | 0.4130 | 0.4017 | 0.3907 |
| 28 | 0.4795 | 0.4665 | 0.4537 | 0.4414 | 0.4293 | 0.4176 | 0.4062 | 0.3952 | 0.3844 |
| 29 | 0.4712 | 0.4585 | 0.4461 | 0.4340 | 0.4222 | 0.4108 | 0.3997 | 0.3888 | 0.3783 |
| 30 | 0.4632 | 0.4508 | 0.4386 | 0.4268 | 0.4153 | 0.4041 | 0.3933 | 0.3827 | 0.3723 |
| 31 | 0.4554 | 0.4433 | 0.4314 | 0.4199 | 0.4086 | 0.3977 | 0.3870 | 0.3766 | 0.3666 |
| 32 | 0.4479 | 0.4360 | 0.4244 | 0.4131 | 0.4021 | 0.3914 | 0.3809 | 0.3708 | 0.3609 |
| 33 | 0.4405 | 0.4289 | 0.4175 | 0.4065 | 0.3957 | 0.3852 | 0.3750 | 0.3651 | 0.3554 |
| 34 | 0.4334 | 0.4220 | 0.4109 | 0.4000 | 0.3895 | 0.3793 | 0.3693 | 0.3595 | 0.3501 |
| 35 | 0.4264 | 0.4152 | 0.4044 | 0.3938 | 0.3835 | 0.3734 | 0.3636 | 0.3541 | 0.3448 |
| 36 | 0.4196 | 0.4087 | 0.3981 | 0.3877 | 0.3776 | 0.3677 | 0.3581 | 0.3488 | 0.3397 |
| 37 | 0.4130 | 0.4023 | 0.3919 | 0.3817 | 0.3718 | 0.3622 | 0.3528 | 0.3436 | 0.3347 |
| 38 | 0.4066 | 0.3961 | 0.3859 | 0.3759 | 0.3662 | 0.3567 | 0.3475 | 0.3385 | 0.3298 |
| 39 | 0.4003 | 0.3900 | 0.3800 | 0.3702 | 0.3607 | 0.3514 | 0.3424 | 0.3336 | 0.3250 |
| 40 | 0.3941 | 0.3840 | 0.3742 | 0.3646 | 0.3553 | 0.3462 | 0.3373 | 0.3287 | 0.3203 |

*Source*: Based on Eq. 1, Hamme and Emerson (2004).

**Table 2.16** Standard Air Saturation Concentration of Argon as a Function of Temperature and Salinity, 33–37 g/kg in mg/L (Seawater, 1 atm moist air)

| Temperature (°C) | Salinity (g/kg) | | | | | | | | |
|---|---|---|---|---|---|---|---|---|---|
| | **33.0** | **33.5** | **34.0** | **34.5** | **35.0** | **35.5** | **36.0** | **36.5** | **37.0** |
| **0** | 0.7087 | 0.7063 | 0.7038 | 0.7014 | 0.6990 | 0.6965 | 0.6941 | 0.6917 | 0.6893 |
| **1** | 0.6907 | 0.6884 | 0.6860 | 0.6837 | 0.6813 | 0.6790 | 0.6767 | 0.6743 | 0.6720 |
| **2** | 0.6735 | 0.6713 | 0.6690 | 0.6667 | 0.6644 | 0.6622 | 0.6599 | 0.6577 | 0.6554 |
| **3** | 0.6571 | 0.6549 | 0.6527 | 0.6505 | 0.6483 | 0.6461 | 0.6439 | 0.6417 | 0.6396 |
| **4** | 0.6413 | 0.6392 | 0.6370 | 0.6349 | 0.6328 | 0.6307 | 0.6286 | 0.6265 | 0.6244 |
| **5** | 0.6262 | 0.6241 | 0.6221 | 0.6200 | 0.6179 | 0.6159 | 0.6139 | 0.6118 | 0.6098 |
| **6** | 0.6117 | 0.6097 | 0.6077 | 0.6057 | 0.6037 | 0.6017 | 0.5998 | 0.5978 | 0.5958 |
| **7** | 0.5978 | 0.5959 | 0.5939 | 0.5920 | 0.5901 | 0.5881 | 0.5862 | 0.5843 | 0.5824 |
| **8** | 0.5845 | 0.5826 | 0.5807 | 0.5788 | 0.5770 | 0.5751 | 0.5733 | 0.5714 | 0.5696 |
| **9** | 0.5716 | 0.5698 | 0.5680 | 0.5662 | 0.5644 | 0.5626 | 0.5608 | 0.5590 | 0.5572 |
| **10** | 0.5593 | 0.5576 | 0.5558 | 0.5540 | 0.5523 | 0.5505 | 0.5488 | 0.5471 | 0.5453 |
| **11** | 0.5475 | 0.5458 | 0.5441 | 0.5424 | 0.5407 | 0.5390 | 0.5373 | 0.5356 | 0.5339 |
| **12** | 0.5361 | 0.5344 | 0.5328 | 0.5311 | 0.5295 | 0.5278 | 0.5262 | 0.5245 | 0.5229 |
| **13** | 0.5251 | 0.5235 | 0.5219 | 0.5203 | 0.5187 | 0.5171 | 0.5155 | 0.5139 | 0.5123 |
| **14** | 0.5146 | 0.5130 | 0.5114 | 0.5099 | 0.5083 | 0.5068 | 0.5052 | 0.5037 | 0.5021 |
| **15** | 0.5044 | 0.5029 | 0.5014 | 0.4998 | 0.4983 | 0.4968 | 0.4953 | 0.4938 | 0.4923 |
| **16** | 0.4946 | 0.4931 | 0.4916 | 0.4902 | 0.4887 | 0.4872 | 0.4857 | 0.4843 | 0.4828 |
| **17** | 0.4852 | 0.4837 | 0.4823 | 0.4808 | 0.4794 | 0.4780 | 0.4765 | 0.4751 | 0.4737 |
| **18** | 0.4760 | 0.4746 | 0.4732 | 0.4718 | 0.4704 | 0.4690 | 0.4676 | 0.4662 | 0.4649 |
| **19** | 0.4672 | 0.4658 | 0.4645 | 0.4631 | 0.4617 | 0.4604 | 0.4590 | 0.4577 | 0.4563 |
| **20** | 0.4587 | 0.4574 | 0.4560 | 0.4547 | 0.4534 | 0.4520 | 0.4507 | 0.4494 | 0.4481 |
| **21** | 0.4505 | 0.4492 | 0.4478 | 0.4466 | 0.4453 | 0.4440 | 0.4427 | 0.4414 | 0.4401 |
| **22** | 0.4425 | 0.4412 | 0.4399 | 0.4387 | 0.4374 | 0.4362 | 0.4349 | 0.4337 | 0.4324 |
| **23** | 0.4348 | 0.4335 | 0.4323 | 0.4311 | 0.4298 | 0.4286 | 0.4274 | 0.4261 | 0.4249 |
| **24** | 0.4273 | 0.4261 | 0.4249 | 0.4237 | 0.4225 | 0.4213 | 0.4201 | 0.4189 | 0.4177 |
| **25** | 0.4200 | 0.4189 | 0.4177 | 0.4165 | 0.4153 | 0.4142 | 0.4130 | 0.4118 | 0.4107 |
| **26** | 0.4130 | 0.4119 | 0.4107 | 0.4096 | 0.4084 | 0.4073 | 0.4061 | 0.4050 | 0.4039 |
| **27** | 0.4062 | 0.4051 | 0.4039 | 0.4028 | 0.4017 | 0.4006 | 0.3995 | 0.3984 | 0.3972 |
| **28** | 0.3996 | 0.3985 | 0.3974 | 0.3963 | 0.3952 | 0.3941 | 0.3930 | 0.3919 | 0.3908 |
| **29** | 0.3931 | 0.3920 | 0.3910 | 0.3899 | 0.3888 | 0.3878 | 0.3867 | 0.3856 | 0.3846 |
| **30** | 0.3869 | 0.3858 | 0.3848 | 0.3837 | 0.3827 | 0.3816 | 0.3806 | 0.3795 | 0.3785 |
| **31** | 0.3808 | 0.3797 | 0.3787 | 0.3777 | 0.3766 | 0.3756 | 0.3746 | 0.3736 | 0.3726 |
| **32** | 0.3748 | 0.3738 | 0.3728 | 0.3718 | 0.3708 | 0.3698 | 0.3688 | 0.3678 | 0.3668 |
| **33** | 0.3690 | 0.3680 | 0.3671 | 0.3661 | 0.3651 | 0.3641 | 0.3631 | 0.3622 | 0.3612 |
| **34** | 0.3634 | 0.3624 | 0.3615 | 0.3605 | 0.3595 | 0.3586 | 0.3576 | 0.3567 | 0.3557 |
| **35** | 0.3579 | 0.3569 | 0.3560 | 0.3550 | 0.3541 | 0.3532 | 0.3522 | 0.3513 | 0.3504 |
| **36** | 0.3525 | 0.3516 | 0.3506 | 0.3497 | 0.3488 | 0.3479 | 0.3470 | 0.3460 | 0.3451 |
| **37** | 0.3472 | 0.3463 | 0.3454 | 0.3445 | 0.3436 | 0.3427 | 0.3418 | 0.3409 | 0.3400 |
| **38** | 0.3421 | 0.3412 | 0.3403 | 0.3394 | 0.3385 | 0.3376 | 0.3368 | 0.3359 | 0.3350 |
| **39** | 0.3371 | 0.3362 | 0.3353 | 0.3344 | 0.3336 | 0.3327 | 0.3318 | 0.3310 | 0.3301 |
| **40** | 0.3321 | 0.3313 | 0.3304 | 0.3295 | 0.3287 | 0.3278 | 0.3270 | 0.3261 | 0.3253 |

*Source*: Based on Eq. 1, Hamme and Emerson (2004).

**Table 2.17** Standard Air Saturation Concentration of Carbon Dioxide as a Function of Temperature and Salinity, 0−40 g/kg in mg/L (Seawater, 1 atm moist air; mole fraction = 390 μatm)

| Temperature (°C) | Salinity (g/kg) | | | | | | | | |
|---|---|---|---|---|---|---|---|---|---|
| | **0.0** | **5.0** | **10.0** | **15.0** | **20.0** | **25.0** | **30.0** | **35.0** | **40.0** |
| **0** | 1.3174 | 1.2837 | 1.2508 | 1.2186 | 1.1873 | 1.1568 | 1.1270 | 1.0980 | 1.0697 |
| **1** | 1.2661 | 1.2337 | 1.2022 | 1.1714 | 1.1414 | 1.1122 | 1.0836 | 1.0558 | 1.0287 |
| **2** | 1.2174 | 1.1864 | 1.1562 | 1.1267 | 1.0979 | 1.0699 | 1.0425 | 1.0159 | 0.9899 |
| **3** | 1.1712 | 1.1415 | 1.1125 | 1.0842 | 1.0567 | 1.0298 | 1.0036 | 0.9780 | 0.9531 |
| **4** | 1.1273 | 1.0989 | 1.0711 | 1.0440 | 1.0175 | 0.9918 | 0.9666 | 0.9421 | 0.9182 |
| **5** | 1.0857 | 1.0584 | 1.0317 | 1.0057 | 0.9804 | 0.9556 | 0.9315 | 0.9080 | 0.8851 |
| **6** | 1.0461 | 1.0199 | 0.9944 | 0.9694 | 0.9451 | 0.9214 | 0.8982 | 0.8756 | 0.8536 |
| **7** | 1.0085 | 0.9834 | 0.9588 | 0.9349 | 0.9115 | 0.8888 | 0.8665 | 0.8449 | 0.8237 |
| **8** | 0.9727 | 0.9486 | 0.9251 | 0.9021 | 0.8796 | 0.8578 | 0.8364 | 0.8156 | 0.7953 |
| **9** | 0.9387 | 0.9155 | 0.8929 | 0.8708 | 0.8493 | 0.8283 | 0.8078 | 0.7878 | 0.7683 |
| **10** | 0.9062 | 0.8840 | 0.8623 | 0.8411 | 0.8204 | 0.8002 | 0.7805 | 0.7613 | 0.7425 |
| **11** | 0.8753 | 0.8540 | 0.8331 | 0.8128 | 0.7929 | 0.7735 | 0.7545 | 0.7361 | 0.7180 |
| **12** | 0.8459 | 0.8254 | 0.8053 | 0.7858 | 0.7666 | 0.7480 | 0.7298 | 0.7120 | 0.6947 |
| **13** | 0.8178 | 0.7981 | 0.7788 | 0.7600 | 0.7416 | 0.7237 | 0.7062 | 0.6891 | 0.6725 |
| **14** | 0.7910 | 0.7720 | 0.7535 | 0.7354 | 0.7178 | 0.7005 | 0.6837 | 0.6673 | 0.6512 |
| **15** | 0.7654 | 0.7472 | 0.7294 | 0.7120 | 0.6950 | 0.6784 | 0.6622 | 0.6464 | 0.6310 |
| **16** | 0.7409 | 0.7234 | 0.7063 | 0.6896 | 0.6732 | 0.6573 | 0.6417 | 0.6265 | 0.6116 |
| **17** | 0.7175 | 0.7007 | 0.6842 | 0.6681 | 0.6524 | 0.6371 | 0.6221 | 0.6075 | 0.5932 |
| **18** | 0.6951 | 0.6790 | 0.6631 | 0.6477 | 0.6326 | 0.6178 | 0.6034 | 0.5893 | 0.5755 |
| **19** | 0.6738 | 0.6582 | 0.6430 | 0.6281 | 0.6135 | 0.5993 | 0.5854 | 0.5719 | 0.5586 |
| **20** | 0.6533 | 0.6383 | 0.6236 | 0.6093 | 0.5953 | 0.5817 | 0.5683 | 0.5552 | 0.5424 |
| **21** | 0.6336 | 0.6192 | 0.6051 | 0.5914 | 0.5779 | 0.5647 | 0.5519 | 0.5393 | 0.5270 |
| **22** | 0.6148 | 0.6010 | 0.5874 | 0.5742 | 0.5612 | 0.5485 | 0.5361 | 0.5240 | 0.5122 |
| **23** | 0.5968 | 0.5835 | 0.5704 | 0.5577 | 0.5452 | 0.5330 | 0.5210 | 0.5094 | 0.4979 |
| **24** | 0.5795 | 0.5667 | 0.5541 | 0.5418 | 0.5298 | 0.5181 | 0.5066 | 0.4953 | 0.4843 |
| **25** | 0.5629 | 0.5505 | 0.5385 | 0.5266 | 0.5151 | 0.5038 | 0.4927 | 0.4819 | 0.4713 |
| **26** | 0.5469 | 0.5350 | 0.5234 | 0.5120 | 0.5009 | 0.4900 | 0.4794 | 0.4689 | 0.4587 |
| **27** | 0.5316 | 0.5202 | 0.5090 | 0.4980 | 0.4873 | 0.4768 | 0.4665 | 0.4565 | 0.4467 |
| **28** | 0.5168 | 0.5058 | 0.4951 | 0.4845 | 0.4742 | 0.4641 | 0.4542 | 0.4446 | 0.4351 |
| **29** | 0.5026 | 0.4921 | 0.4817 | 0.4716 | 0.4616 | 0.4519 | 0.4424 | 0.4331 | 0.4239 |
| **30** | 0.4890 | 0.4788 | 0.4689 | 0.4591 | 0.4495 | 0.4402 | 0.4310 | 0.4220 | 0.4132 |
| **31** | 0.4758 | 0.4661 | 0.4565 | 0.4471 | 0.4379 | 0.4289 | 0.4200 | 0.4114 | 0.4029 |
| **32** | 0.4631 | 0.4537 | 0.4445 | 0.4355 | 0.4266 | 0.4180 | 0.4095 | 0.4011 | 0.3930 |
| **33** | 0.4509 | 0.4419 | 0.4330 | 0.4243 | 0.4158 | 0.4075 | 0.3993 | 0.3912 | 0.3834 |
| **34** | 0.4391 | 0.4304 | 0.4219 | 0.4136 | 0.4054 | 0.3973 | 0.3894 | 0.3817 | 0.3741 |
| **35** | 0.4277 | 0.4194 | 0.4112 | 0.4032 | 0.3953 | 0.3875 | 0.3799 | 0.3725 | 0.3652 |
| **36** | 0.4167 | 0.4087 | 0.4008 | 0.3931 | 0.3855 | 0.3781 | 0.3708 | 0.3636 | 0.3566 |
| **37** | 0.4061 | 0.3984 | 0.3908 | 0.3834 | 0.3761 | 0.3689 | 0.3619 | 0.3550 | 0.3483 |
| **38** | 0.3958 | 0.3884 | 0.3811 | 0.3740 | 0.3670 | 0.3601 | 0.3534 | 0.3467 | 0.3402 |
| **39** | 0.3858 | 0.3787 | 0.3718 | 0.3649 | 0.3582 | 0.3516 | 0.3451 | 0.3387 | 0.3324 |
| **40** | 0.3762 | 0.3694 | 0.3627 | 0.3561 | 0.3496 | 0.3433 | 0.3370 | 0.3309 | 0.3249 |

*Source*: Based on Eqs. 5 and 9, Weiss (1974); mole fraction estimated from Mauna Loa data for 2010.

**Table 2.18** Standard Air Saturation Concentration of Carbon Dioxide as a Function of Temperature, 33–37 g/kg in mg/L (Seawater, 1 atm moist air; mole fraction = 390 μatm)

| Temperature (°C) | Salinity (g/kg) | | | | | | | | |
|---|---|---|---|---|---|---|---|---|---|
| | 33.0 | 33.5 | 34.0 | 34.5 | 35.0 | 35.5 | 36.0 | 36.5 | 37.0 |
| 0 | 1.1095 | 1.1066 | 1.1037 | 1.1009 | 1.0980 | 1.0951 | 1.0923 | 1.0894 | 1.0866 |
| 1 | 1.0669 | 1.0641 | 1.0613 | 1.0586 | 1.0558 | 1.0531 | 1.0504 | 1.0476 | 1.0449 |
| 2 | 1.0265 | 1.0238 | 1.0212 | 1.0185 | 1.0159 | 1.0133 | 1.0106 | 1.0080 | 1.0054 |
| 3 | 0.9882 | 0.9856 | 0.9831 | 0.9806 | 0.9780 | 0.9755 | 0.9730 | 0.9705 | 0.9680 |
| 4 | 0.9518 | 0.9494 | 0.9470 | 0.9445 | 0.9421 | 0.9397 | 0.9373 | 0.9349 | 0.9325 |
| 5 | 0.9173 | 0.9150 | 0.9127 | 0.9103 | 0.9080 | 0.9057 | 0.9034 | 0.9011 | 0.8988 |
| 6 | 0.8846 | 0.8823 | 0.8801 | 0.8779 | 0.8756 | 0.8734 | 0.8712 | 0.8690 | 0.8668 |
| 7 | 0.8535 | 0.8513 | 0.8492 | 0.8470 | 0.8449 | 0.8427 | 0.8406 | 0.8385 | 0.8363 |
| 8 | 0.8239 | 0.8218 | 0.8197 | 0.8177 | 0.8156 | 0.8136 | 0.8115 | 0.8095 | 0.8074 |
| 9 | 0.7957 | 0.7937 | 0.7917 | 0.7898 | 0.7878 | 0.7858 | 0.7838 | 0.7819 | 0.7799 |
| 10 | 0.7689 | 0.7670 | 0.7651 | 0.7632 | 0.7613 | 0.7594 | 0.7575 | 0.7556 | 0.7537 |
| 11 | 0.7434 | 0.7416 | 0.7397 | 0.7379 | 0.7361 | 0.7342 | 0.7324 | 0.7306 | 0.7288 |
| 12 | 0.7191 | 0.7173 | 0.7156 | 0.7138 | 0.7120 | 0.7103 | 0.7085 | 0.7068 | 0.7051 |
| 13 | 0.6959 | 0.6942 | 0.6925 | 0.6908 | 0.6891 | 0.6874 | 0.6858 | 0.6841 | 0.6824 |
| 14 | 0.6738 | 0.6722 | 0.6705 | 0.6689 | 0.6673 | 0.6656 | 0.6640 | 0.6624 | 0.6608 |
| 15 | 0.6527 | 0.6511 | 0.6495 | 0.6480 | 0.6464 | 0.6448 | 0.6433 | 0.6417 | 0.6402 |
| 16 | 0.6325 | 0.6310 | 0.6295 | 0.6280 | 0.6265 | 0.6250 | 0.6235 | 0.6220 | 0.6205 |
| 17 | 0.6133 | 0.6118 | 0.6104 | 0.6089 | 0.6075 | 0.6060 | 0.6046 | 0.6031 | 0.6017 |
| 18 | 0.5949 | 0.5935 | 0.5921 | 0.5907 | 0.5893 | 0.5879 | 0.5865 | 0.5851 | 0.5837 |
| 19 | 0.5773 | 0.5759 | 0.5746 | 0.5732 | 0.5719 | 0.5705 | 0.5692 | 0.5679 | 0.5665 |
| 20 | 0.5604 | 0.5591 | 0.5578 | 0.5565 | 0.5552 | 0.5539 | 0.5526 | 0.5514 | 0.5501 |
| 21 | 0.5443 | 0.5430 | 0.5418 | 0.5405 | 0.5393 | 0.5380 | 0.5368 | 0.5356 | 0.5343 |
| 22 | 0.5288 | 0.5276 | 0.5264 | 0.5252 | 0.5240 | 0.5228 | 0.5216 | 0.5204 | 0.5192 |
| 23 | 0.5140 | 0.5128 | 0.5117 | 0.5105 | 0.5094 | 0.5082 | 0.5071 | 0.5059 | 0.5048 |
| 24 | 0.4998 | 0.4987 | 0.4976 | 0.4964 | 0.4953 | 0.4942 | 0.4931 | 0.4920 | 0.4909 |
| 25 | 0.4862 | 0.4851 | 0.4840 | 0.4829 | 0.4819 | 0.4808 | 0.4797 | 0.4786 | 0.4776 |
| 26 | 0.4731 | 0.4720 | 0.4710 | 0.4700 | 0.4689 | 0.4679 | 0.4669 | 0.4658 | 0.4648 |
| 27 | 0.4605 | 0.4595 | 0.4585 | 0.4575 | 0.4565 | 0.4555 | 0.4545 | 0.4535 | 0.4525 |
| 28 | 0.4484 | 0.4474 | 0.4465 | 0.4455 | 0.4446 | 0.4436 | 0.4426 | 0.4417 | 0.4407 |
| 29 | 0.4368 | 0.4358 | 0.4349 | 0.4340 | 0.4331 | 0.4321 | 0.4312 | 0.4303 | 0.4294 |
| 30 | 0.4256 | 0.4247 | 0.4238 | 0.4229 | 0.4220 | 0.4211 | 0.4202 | 0.4194 | 0.4185 |
| 31 | 0.4148 | 0.4140 | 0.4131 | 0.4122 | 0.4114 | 0.4105 | 0.4097 | 0.4088 | 0.4080 |
| 32 | 0.4044 | 0.4036 | 0.4028 | 0.4020 | 0.4011 | 0.4003 | 0.3995 | 0.3987 | 0.3978 |
| 33 | 0.3944 | 0.3936 | 0.3928 | 0.3920 | 0.3912 | 0.3904 | 0.3897 | 0.3889 | 0.3881 |
| 34 | 0.3848 | 0.3840 | 0.3832 | 0.3825 | 0.3817 | 0.3809 | 0.3802 | 0.3794 | 0.3787 |
| 35 | 0.3755 | 0.3747 | 0.3740 | 0.3732 | 0.3725 | 0.3718 | 0.3710 | 0.3703 | 0.3696 |
| 36 | 0.3665 | 0.3658 | 0.3650 | 0.3643 | 0.3636 | 0.3629 | 0.3622 | 0.3615 | 0.3608 |
| 37 | 0.3578 | 0.3571 | 0.3564 | 0.3557 | 0.3550 | 0.3544 | 0.3537 | 0.3530 | 0.3523 |
| 38 | 0.3494 | 0.3487 | 0.3481 | 0.3474 | 0.3467 | 0.3461 | 0.3454 | 0.3448 | 0.3441 |
| 39 | 0.3412 | 0.3406 | 0.3400 | 0.3393 | 0.3387 | 0.3381 | 0.3374 | 0.3368 | 0.3362 |
| 40 | 0.3334 | 0.3327 | 0.3321 | 0.3315 | 0.3309 | 0.3303 | 0.3297 | 0.3291 | 0.3285 |

*Source*: Based on Eqs. 5 and 9, Weiss (1974); mole fraction estimated from Mauna Loa data for 2010.

**Table 2.19** Standard Air Saturation Concentration of Carbon Dioxide as a Function of Temperature and Salinity, 0–40 g/kg in mg/L (Seawater, 1 atm moist air; mole fraction = 440 μatm)

| Temperature (°C) | Salinity (g/kg) | | | | | | | | |
|---|---|---|---|---|---|---|---|---|---|
| | **0.0** | **5.0** | **10.0** | **15.0** | **20.0** | **25.0** | **30.0** | **35.0** | **40.0** |
| **0** | 1.4863 | 1.4483 | 1.4111 | 1.3749 | 1.3395 | 1.3051 | 1.2715 | 1.2388 | 1.2069 |
| **1** | 1.4284 | 1.3919 | 1.3563 | 1.3216 | 1.2877 | 1.2547 | 1.2226 | 1.1912 | 1.1606 |
| **2** | 1.3734 | 1.3385 | 1.3044 | 1.2711 | 1.2387 | 1.2070 | 1.1762 | 1.1461 | 1.1168 |
| **3** | 1.3213 | 1.2878 | 1.2551 | 1.2232 | 1.1921 | 1.1618 | 1.1322 | 1.1034 | 1.0753 |
| **4** | 1.2718 | 1.2397 | 1.2084 | 1.1778 | 1.1480 | 1.1189 | 1.0905 | 1.0629 | 1.0359 |
| **5** | 1.2249 | 1.1941 | 1.1640 | 1.1347 | 1.1061 | 1.0782 | 1.0510 | 1.0244 | 0.9985 |
| **6** | 1.1802 | 1.1507 | 1.1219 | 1.0937 | 1.0663 | 1.0395 | 1.0134 | 0.9879 | 0.9631 |
| **7** | 1.1378 | 1.1095 | 1.0818 | 1.0548 | 1.0284 | 1.0027 | 0.9776 | 0.9532 | 0.9293 |
| **8** | 1.0974 | 1.0702 | 1.0437 | 1.0177 | 0.9924 | 0.9677 | 0.9437 | 0.9202 | 0.8973 |
| **9** | 1.0590 | 1.0329 | 1.0074 | 0.9825 | 0.9582 | 0.9345 | 0.9113 | 0.8888 | 0.8668 |
| **10** | 1.0224 | 0.9973 | 0.9728 | 0.9489 | 0.9256 | 0.9028 | 0.8806 | 0.8589 | 0.8377 |
| **11** | 0.9876 | 0.9635 | 0.9399 | 0.9170 | 0.8945 | 0.8726 | 0.8513 | 0.8304 | 0.8101 |
| **12** | 0.9543 | 0.9312 | 0.9086 | 0.8865 | 0.8649 | 0.8439 | 0.8234 | 0.8033 | 0.7838 |
| **13** | 0.9226 | 0.9004 | 0.8787 | 0.8574 | 0.8367 | 0.8165 | 0.7967 | 0.7775 | 0.7587 |
| **14** | 0.8924 | 0.8710 | 0.8501 | 0.8297 | 0.8098 | 0.7903 | 0.7714 | 0.7528 | 0.7347 |
| **15** | 0.8635 | 0.8429 | 0.8229 | 0.8032 | 0.7841 | 0.7654 | 0.7471 | 0.7293 | 0.7119 |
| **16** | 0.8359 | 0.8161 | 0.7968 | 0.7780 | 0.7595 | 0.7415 | 0.7240 | 0.7068 | 0.6900 |
| **17** | 0.8095 | 0.7905 | 0.7719 | 0.7538 | 0.7361 | 0.7188 | 0.7019 | 0.6853 | 0.6692 |
| **18** | 0.7843 | 0.7660 | 0.7482 | 0.7307 | 0.7137 | 0.6970 | 0.6807 | 0.6648 | 0.6493 |
| **19** | 0.7601 | 0.7426 | 0.7254 | 0.7086 | 0.6922 | 0.6762 | 0.6605 | 0.6452 | 0.6302 |
| **20** | 0.7370 | 0.7201 | 0.7036 | 0.6874 | 0.6717 | 0.6562 | 0.6411 | 0.6264 | 0.6120 |
| **21** | 0.7149 | 0.6986 | 0.6827 | 0.6672 | 0.6520 | 0.6371 | 0.6226 | 0.6084 | 0.5945 |
| **22** | 0.6937 | 0.6780 | 0.6627 | 0.6478 | 0.6331 | 0.6188 | 0.6049 | 0.5912 | 0.5778 |
| **23** | 0.6733 | 0.6583 | 0.6436 | 0.6292 | 0.6151 | 0.6013 | 0.5878 | 0.5747 | 0.5618 |
| **24** | 0.6538 | 0.6393 | 0.6252 | 0.6113 | 0.5977 | 0.5845 | 0.5715 | 0.5588 | 0.5464 |
| **25** | 0.6350 | 0.6211 | 0.6075 | 0.5942 | 0.5811 | 0.5683 | 0.5558 | 0.5436 | 0.5317 |
| **26** | 0.6170 | 0.6036 | 0.5905 | 0.5777 | 0.5651 | 0.5528 | 0.5408 | 0.5290 | 0.5175 |
| **27** | 0.5997 | 0.5868 | 0.5742 | 0.5619 | 0.5498 | 0.5379 | 0.5264 | 0.5150 | 0.5039 |
| **28** | 0.5831 | 0.5707 | 0.5586 | 0.5467 | 0.5350 | 0.5236 | 0.5125 | 0.5015 | 0.4909 |
| **29** | 0.5671 | 0.5552 | 0.5435 | 0.5320 | 0.5208 | 0.5099 | 0.4991 | 0.4886 | 0.4783 |
| **30** | 0.5517 | 0.5402 | 0.5290 | 0.5180 | 0.5072 | 0.4966 | 0.4863 | 0.4761 | 0.4662 |
| **31** | 0.5368 | 0.5258 | 0.5150 | 0.5044 | 0.4940 | 0.4838 | 0.4739 | 0.4641 | 0.4546 |
| **32** | 0.5225 | 0.5119 | 0.5015 | 0.4913 | 0.4813 | 0.4715 | 0.4620 | 0.4526 | 0.4433 |
| **33** | 0.5087 | 0.4985 | 0.4885 | 0.4787 | 0.4691 | 0.4597 | 0.4505 | 0.4414 | 0.4325 |
| **34** | 0.4954 | 0.4856 | 0.4760 | 0.4666 | 0.4573 | 0.4483 | 0.4394 | 0.4306 | 0.4221 |
| **35** | 0.4826 | 0.4731 | 0.4639 | 0.4548 | 0.4459 | 0.4372 | 0.4287 | 0.4203 | 0.4120 |
| **36** | 0.4701 | 0.4611 | 0.4522 | 0.4435 | 0.4349 | 0.4266 | 0.4183 | 0.4102 | 0.4023 |
| **37** | 0.4581 | 0.4495 | 0.4409 | 0.4325 | 0.4243 | 0.4162 | 0.4083 | 0.4006 | 0.3929 |
| **38** | 0.4465 | 0.4382 | 0.4300 | 0.4219 | 0.4140 | 0.4063 | 0.3987 | 0.3912 | 0.3839 |
| **39** | 0.4353 | 0.4273 | 0.4194 | 0.4117 | 0.4041 | 0.3966 | 0.3893 | 0.3821 | 0.3751 |
| **40** | 0.4244 | 0.4167 | 0.4092 | 0.4018 | 0.3945 | 0.3873 | 0.3803 | 0.3733 | 0.3666 |

*Source*: Based on Eqs. 5 and 9, Weiss (1974); mole fraction estimated from Mauna Loa data for 2030.

**Table 2.20** Standard Air Saturation Concentration of Carbon Dioxide as a Function of Temperature and Salinity, 33−37 g/kg in mg/L (Seawater, 1 atm moist air; mole fraction = 440 μatm)

| Temperature (°C) | Salinity (g/kg) | | | | | | | | |
|---|---|---|---|---|---|---|---|---|---|
| | **33.0** | **33.5** | **34.0** | **34.5** | **35.0** | **35.5** | **36.0** | **36.5** | **37.0** |
| **0** | 1.2518 | 1.2485 | 1.2452 | 1.2420 | 1.2388 | 1.2355 | 1.2323 | 1.2291 | 1.2259 |
| **1** | 1.2037 | 1.2005 | 1.1974 | 1.1943 | 1.1912 | 1.1881 | 1.1850 | 1.1820 | 1.1789 |
| **2** | 1.1581 | 1.1551 | 1.1521 | 1.1491 | 1.1461 | 1.1432 | 1.1402 | 1.1373 | 1.1343 |
| **3** | 1.1149 | 1.1120 | 1.1091 | 1.1063 | 1.1034 | 1.1006 | 1.0977 | 1.0949 | 1.0921 |
| **4** | 1.0739 | 1.0711 | 1.0684 | 1.0656 | 1.0629 | 1.0602 | 1.0574 | 1.0547 | 1.0520 |
| **5** | 1.0350 | 1.0323 | 1.0297 | 1.0270 | 1.0244 | 1.0218 | 1.0192 | 1.0166 | 1.0140 |
| **6** | 0.9980 | 0.9955 | 0.9929 | 0.9904 | 0.9879 | 0.9854 | 0.9829 | 0.9804 | 0.9779 |
| **7** | 0.9629 | 0.9604 | 0.9580 | 0.9556 | 0.9532 | 0.9508 | 0.9484 | 0.9460 | 0.9436 |
| **8** | 0.9295 | 0.9272 | 0.9248 | 0.9225 | 0.9202 | 0.9179 | 0.9155 | 0.9132 | 0.9109 |
| **9** | 0.8977 | 0.8955 | 0.8932 | 0.8910 | 0.8888 | 0.8866 | 0.8843 | 0.8821 | 0.8799 |
| **10** | 0.8675 | 0.8653 | 0.8632 | 0.8610 | 0.8589 | 0.8568 | 0.8546 | 0.8525 | 0.8504 |
| **11** | 0.8387 | 0.8366 | 0.8346 | 0.8325 | 0.8304 | 0.8284 | 0.8263 | 0.8243 | 0.8222 |
| **12** | 0.8113 | 0.8093 | 0.8073 | 0.8053 | 0.8033 | 0.8013 | 0.7994 | 0.7974 | 0.7954 |
| **13** | 0.7851 | 0.7832 | 0.7813 | 0.7794 | 0.7775 | 0.7756 | 0.7737 | 0.7718 | 0.7699 |
| **14** | 0.7602 | 0.7583 | 0.7565 | 0.7546 | 0.7528 | 0.7510 | 0.7492 | 0.7473 | 0.7455 |
| **15** | 0.7364 | 0.7346 | 0.7328 | 0.7310 | 0.7293 | 0.7275 | 0.7258 | 0.7240 | 0.7223 |
| **16** | 0.7136 | 0.7119 | 0.7102 | 0.7085 | 0.7068 | 0.7051 | 0.7034 | 0.7017 | 0.7001 |
| **17** | 0.6919 | 0.6903 | 0.6886 | 0.6870 | 0.6853 | 0.6837 | 0.6821 | 0.6805 | 0.6788 |
| **18** | 0.6711 | 0.6695 | 0.6680 | 0.6664 | 0.6648 | 0.6632 | 0.6617 | 0.6601 | 0.6586 |
| **19** | 0.6513 | 0.6497 | 0.6482 | 0.6467 | 0.6452 | 0.6437 | 0.6422 | 0.6407 | 0.6392 |
| **20** | 0.6323 | 0.6308 | 0.6293 | 0.6279 | 0.6264 | 0.6249 | 0.6235 | 0.6220 | 0.6206 |
| **21** | 0.6140 | 0.6126 | 0.6112 | 0.6098 | 0.6084 | 0.6070 | 0.6056 | 0.6042 | 0.6028 |
| **22** | 0.5966 | 0.5952 | 0.5939 | 0.5925 | 0.5912 | 0.5898 | 0.5885 | 0.5871 | 0.5858 |
| **23** | 0.5799 | 0.5786 | 0.5773 | 0.5760 | 0.5747 | 0.5734 | 0.5721 | 0.5708 | 0.5695 |
| **24** | 0.5639 | 0.5626 | 0.5613 | 0.5601 | 0.5588 | 0.5576 | 0.5563 | 0.5551 | 0.5538 |
| **25** | 0.5485 | 0.5473 | 0.5460 | 0.5448 | 0.5436 | 0.5424 | 0.5412 | 0.5400 | 0.5388 |
| **26** | 0.5337 | 0.5325 | 0.5314 | 0.5302 | 0.5290 | 0.5279 | 0.5267 | 0.5256 | 0.5244 |
| **27** | 0.5195 | 0.5184 | 0.5173 | 0.5161 | 0.5150 | 0.5139 | 0.5128 | 0.5117 | 0.5106 |
| **28** | 0.5059 | 0.5048 | 0.5037 | 0.5026 | 0.5015 | 0.5005 | 0.4994 | 0.4983 | 0.4972 |
| **29** | 0.4928 | 0.4917 | 0.4907 | 0.4896 | 0.4886 | 0.4875 | 0.4865 | 0.4855 | 0.4844 |
| **30** | 0.4801 | 0.4791 | 0.4781 | 0.4771 | 0.4761 | 0.4751 | 0.4741 | 0.4731 | 0.4721 |
| **31** | 0.4680 | 0.4670 | 0.4661 | 0.4651 | 0.4641 | 0.4632 | 0.4622 | 0.4612 | 0.4603 |
| **32** | 0.4563 | 0.4554 | 0.4544 | 0.4535 | 0.4526 | 0.4516 | 0.4507 | 0.4498 | 0.4488 |
| **33** | 0.4450 | 0.4441 | 0.4432 | 0.4423 | 0.4414 | 0.4405 | 0.4396 | 0.4387 | 0.4378 |
| **34** | 0.4341 | 0.4332 | 0.4324 | 0.4315 | 0.4306 | 0.4298 | 0.4289 | 0.4281 | 0.4272 |
| **35** | 0.4236 | 0.4228 | 0.4219 | 0.4211 | 0.4203 | 0.4194 | 0.4186 | 0.4178 | 0.4170 |
| **36** | 0.4135 | 0.4127 | 0.4118 | 0.4110 | 0.4102 | 0.4094 | 0.4087 | 0.4079 | 0.4071 |
| **37** | 0.4037 | 0.4029 | 0.4021 | 0.4013 | 0.4006 | 0.3998 | 0.3990 | 0.3983 | 0.3975 |
| **38** | 0.3942 | 0.3934 | 0.3927 | 0.3919 | 0.3912 | 0.3905 | 0.3897 | 0.3890 | 0.3882 |
| **39** | 0.3850 | 0.3843 | 0.3836 | 0.3828 | 0.3821 | 0.3814 | 0.3807 | 0.3800 | 0.3793 |
| **40** | 0.3761 | 0.3754 | 0.3747 | 0.3740 | 0.3733 | 0.3727 | 0.3720 | 0.3713 | 0.3706 |

*Source*: Based on Eqs. 5 and 9, Weiss (1974); mole fraction estimated from Mauna Loa data for 2030.

**Table 2.21** $F^{\dagger} \times 10^{+2}$ for Carbon Dioxide as a Function of Temperature and Salinity, 0–40 g/kg in mol/(kg atm) (At 10°C and 35 g/kg, the table value is 4.3191; the $F^{\dagger}$ value is 4.3191/100 or 0.043191 mol/(kg atm))

| Temperature (°C) | Salinity (g/kg) | | | | | | | | |
|---|---|---|---|---|---|---|---|---|---|
| | 0.0 | 5.0 | 10.0 | 15.0 | 20.0 | 25.0 | 30.0 | 35.0 | 40.0 |
| 0 | 7.6770 | 7.4500 | 7.2298 | 7.0160 | 6.8086 | 6.6073 | 6.4120 | 6.2224 | 6.0384 |
| 1 | 7.3772 | 7.1599 | 6.9489 | 6.7442 | 6.5455 | 6.3526 | 6.1655 | 5.9838 | 5.8075 |
| 2 | 7.0931 | 6.8849 | 6.6828 | 6.4866 | 6.2962 | 6.1114 | 5.9320 | 5.7579 | 5.5888 |
| 3 | 6.8238 | 6.6243 | 6.4305 | 6.2425 | 6.0599 | 5.8827 | 5.7107 | 5.5437 | 5.3816 |
| 4 | 6.5683 | 6.3770 | 6.1912 | 6.0109 | 5.8358 | 5.6658 | 5.5008 | 5.3406 | 5.1850 |
| 5 | 6.3257 | 6.1423 | 5.9641 | 5.7911 | 5.6231 | 5.4600 | 5.3016 | 5.1479 | 4.9985 |
| 6 | 6.0954 | 5.9193 | 5.7484 | 5.5824 | 5.4211 | 5.2646 | 5.1125 | 4.9649 | 4.8215 |
| 7 | 5.8764 | 5.7075 | 5.5434 | 5.3840 | 5.2292 | 5.0789 | 4.9329 | 4.7911 | 4.6533 |
| 8 | 5.6682 | 5.5060 | 5.3485 | 5.1954 | 5.0468 | 4.9024 | 4.7621 | 4.6258 | 4.4935 |
| 9 | 5.4701 | 5.3144 | 5.1631 | 5.0160 | 4.8732 | 4.7345 | 4.5996 | 4.4687 | 4.3414 |
| 10 | 5.2816 | 5.1319 | 4.9866 | 4.8453 | 4.7080 | 4.5747 | 4.4451 | 4.3191 | 4.1968 |
| 11 | 5.1020 | 4.9582 | 4.8185 | 4.6827 | 4.5507 | 4.4225 | 4.2979 | 4.1767 | 4.0590 |
| 12 | 4.9308 | 4.7926 | 4.6583 | 4.5277 | 4.4008 | 4.2775 | 4.1576 | 4.0411 | 3.9278 |
| 13 | 4.7676 | 4.6347 | 4.5056 | 4.3800 | 4.2580 | 4.1393 | 4.0239 | 3.9118 | 3.8028 |
| 14 | 4.6119 | 4.4841 | 4.3599 | 4.2391 | 4.1217 | 4.0075 | 3.8964 | 3.7885 | 3.6835 |
| 15 | 4.4632 | 4.3403 | 4.2208 | 4.1046 | 3.9916 | 3.8817 | 3.7748 | 3.6708 | 3.5697 |
| 16 | 4.3212 | 4.2030 | 4.0880 | 3.9761 | 3.8673 | 3.7615 | 3.6586 | 3.5585 | 3.4611 |
| 17 | 4.1856 | 4.0718 | 3.9611 | 3.8534 | 3.7487 | 3.6468 | 3.5476 | 3.4512 | 3.3574 |
| 18 | 4.0558 | 3.9463 | 3.8398 | 3.7361 | 3.6352 | 3.5371 | 3.4415 | 3.3486 | 3.2582 |
| 19 | 3.9317 | 3.8263 | 3.7237 | 3.6239 | 3.5267 | 3.4321 | 3.3401 | 3.2505 | 3.1634 |
| 20 | 3.8130 | 3.7114 | 3.6126 | 3.5165 | 3.4229 | 3.3317 | 3.2430 | 3.1567 | 3.0727 |
| 21 | 3.6992 | 3.6014 | 3.5063 | 3.4136 | 3.3234 | 3.2356 | 3.1501 | 3.0669 | 2.9858 |
| 22 | 3.5902 | 3.4960 | 3.4044 | 3.3151 | 3.2282 | 3.1435 | 3.0611 | 2.9808 | 2.9026 |
| 23 | 3.4857 | 3.3950 | 3.3067 | 3.2206 | 3.1368 | 3.0552 | 2.9757 | 2.8983 | 2.8229 |
| 24 | 3.3854 | 3.2980 | 3.2129 | 3.1300 | 3.0493 | 2.9706 | 2.8939 | 2.8192 | 2.7465 |
| 25 | 3.2892 | 3.2050 | 3.1230 | 3.0431 | 2.9652 | 2.8893 | 2.8154 | 2.7434 | 2.6732 |
| 26 | 3.1967 | 3.1157 | 3.0366 | 2.9596 | 2.8845 | 2.8114 | 2.7400 | 2.6705 | 2.6028 |
| 27 | 3.1079 | 3.0298 | 2.9536 | 2.8794 | 2.8070 | 2.7364 | 2.6677 | 2.6006 | 2.5352 |
| 28 | 3.0226 | 2.9473 | 2.8739 | 2.8023 | 2.7325 | 2.6644 | 2.5981 | 2.5334 | 2.4703 |
| 29 | 2.9404 | 2.8679 | 2.7971 | 2.7281 | 2.6608 | 2.5952 | 2.5312 | 2.4687 | 2.4078 |
| 30 | 2.8614 | 2.7915 | 2.7233 | 2.6568 | 2.5919 | 2.5285 | 2.4668 | 2.4065 | 2.3477 |
| 31 | 2.7853 | 2.7179 | 2.6522 | 2.5880 | 2.5255 | 2.4644 | 2.4048 | 2.3466 | 2.2899 |
| 32 | 2.7119 | 2.6470 | 2.5836 | 2.5218 | 2.4615 | 2.4026 | 2.3451 | 2.2889 | 2.2342 |
| 33 | 2.6411 | 2.5786 | 2.5176 | 2.4580 | 2.3998 | 2.3430 | 2.2875 | 2.2333 | 2.1805 |
| 34 | 2.5729 | 2.5127 | 2.4538 | 2.3964 | 2.3403 | 2.2855 | 2.2320 | 2.1797 | 2.1287 |
| 35 | 2.5070 | 2.4490 | 2.3923 | 2.3369 | 2.2828 | 2.2300 | 2.1784 | 2.1280 | 2.0787 |
| 36 | 2.4434 | 2.3875 | 2.3329 | 2.2795 | 2.2274 | 2.1764 | 2.1266 | 2.0780 | 2.0304 |
| 37 | 2.3818 | 2.3280 | 2.2754 | 2.2240 | 2.1737 | 2.1246 | 2.0766 | 2.0297 | 1.9838 |
| 38 | 2.3223 | 2.2705 | 2.2198 | 2.1703 | 2.1219 | 2.0745 | 2.0282 | 1.9830 | 1.9387 |
| 39 | 2.2647 | 2.2148 | 2.1661 | 2.1184 | 2.0717 | 2.0261 | 1.9814 | 1.9378 | 1.8951 |
| 40 | 2.2089 | 2.1609 | 2.1140 | 2.0680 | 2.0231 | 1.9791 | 1.9361 | 1.8940 | 1.8528 |

*Source*: Based on Weiss (1974), Weiss and Price (1980).

**Table 2.22** $F^{\dagger} \times 10^{+2}$ for Carbon Dioxide as a Function of Temperature and Salinity, 33−37 g/kg in mol/(kg atm) (At 10°C and 35 g/kg, the table value is 4.3191; the $F^{\dagger}$ value is 4.3191/100 or 0.043191 mol/(kg atm))

| Temperature (°C) | Salinity (g/kg) | | | | | | | | |
|---|---|---|---|---|---|---|---|---|---|
| | **33.0** | **33.5** | **34.0** | **34.5** | **35.0** | **35.5** | **36.0** | **36.5** | **37.0** |
| **0** | 6.2975 | 6.2787 | 6.2599 | 6.2411 | 6.2224 | 6.2037 | 6.1852 | 6.1666 | 6.1481 |
| **1** | 6.0558 | 6.0378 | 6.0197 | 6.0018 | 5.9838 | 5.9660 | 5.9482 | 5.9304 | 5.9127 |
| **2** | 5.8269 | 5.8096 | 5.7923 | 5.7750 | 5.7579 | 5.7407 | 5.7237 | 5.7066 | 5.6897 |
| **3** | 5.6099 | 5.5933 | 5.5767 | 5.5602 | 5.5437 | 5.5273 | 5.5109 | 5.4946 | 5.4783 |
| **4** | 5.4041 | 5.3882 | 5.3723 | 5.3564 | 5.3406 | 5.3248 | 5.3091 | 5.2934 | 5.2778 |
| **5** | 5.2088 | 5.1935 | 5.1783 | 5.1630 | 5.1479 | 5.1327 | 5.1177 | 5.1026 | 5.0876 |
| **6** | 5.0234 | 5.0087 | 4.9941 | 4.9795 | 4.9649 | 4.9504 | 4.9359 | 4.9214 | 4.9070 |
| **7** | 4.8473 | 4.8332 | 4.8191 | 4.8050 | 4.7911 | 4.7771 | 4.7632 | 4.7493 | 4.7355 |
| **8** | 4.6798 | 4.6663 | 4.6528 | 4.6393 | 4.6258 | 4.6124 | 4.5990 | 4.5857 | 4.5724 |
| **9** | 4.5206 | 4.5076 | 4.4946 | 4.4816 | 4.4687 | 4.4558 | 4.4429 | 4.4301 | 4.4173 |
| **10** | 4.3691 | 4.3565 | 4.3440 | 4.3316 | 4.3191 | 4.3067 | 4.2944 | 4.2820 | 4.2698 |
| **11** | 4.2248 | 4.2127 | 4.2007 | 4.1887 | 4.1767 | 4.1648 | 4.1529 | 4.1411 | 4.1293 |
| **12** | 4.0873 | 4.0757 | 4.0641 | 4.0526 | 4.0411 | 4.0296 | 4.0182 | 4.0068 | 3.9954 |
| **13** | 3.9563 | 3.9451 | 3.9340 | 3.9229 | 3.9118 | 3.9008 | 3.8898 | 3.8788 | 3.8678 |
| **14** | 3.8313 | 3.8206 | 3.8098 | 3.7992 | 3.7885 | 3.7779 | 3.7673 | 3.7567 | 3.7462 |
| **15** | 3.7121 | 3.7017 | 3.6914 | 3.6811 | 3.6708 | 3.6606 | 3.6504 | 3.6402 | 3.6301 |
| **16** | 3.5982 | 3.5882 | 3.5783 | 3.5684 | 3.5585 | 3.5486 | 3.5388 | 3.5290 | 3.5192 |
| **17** | 3.4894 | 3.4798 | 3.4703 | 3.4607 | 3.4512 | 3.4417 | 3.4322 | 3.4228 | 3.4133 |
| **18** | 3.3855 | 3.3762 | 3.3670 | 3.3578 | 3.3486 | 3.3395 | 3.3303 | 3.3212 | 3.3122 |
| **19** | 3.2861 | 3.2772 | 3.2683 | 3.2594 | 3.2505 | 3.2417 | 3.2329 | 3.2241 | 3.2154 |
| **20** | 3.1909 | 3.1823 | 3.1738 | 3.1652 | 3.1567 | 3.1482 | 3.1397 | 3.1312 | 3.1228 |
| **21** | 3.0999 | 3.0916 | 3.0833 | 3.0751 | 3.0669 | 3.0586 | 3.0505 | 3.0423 | 3.0342 |
| **22** | 3.0127 | 3.0047 | 2.9967 | 2.9887 | 2.9808 | 2.9729 | 2.9650 | 2.9571 | 2.9493 |
| **23** | 2.9291 | 2.9213 | 2.9136 | 2.9060 | 2.8983 | 2.8907 | 2.8831 | 2.8755 | 2.8679 |
| **24** | 2.8489 | 2.8414 | 2.8340 | 2.8266 | 2.8192 | 2.8119 | 2.8045 | 2.7972 | 2.7899 |
| **25** | 2.7720 | 2.7648 | 2.7576 | 2.7505 | 2.7434 | 2.7363 | 2.7292 | 2.7221 | 2.7151 |
| **26** | 2.6981 | 2.6912 | 2.6843 | 2.6774 | 2.6705 | 2.6637 | 2.6569 | 2.6500 | 2.6432 |
| **27** | 2.6272 | 2.6205 | 2.6139 | 2.6072 | 2.6006 | 2.5940 | 2.5874 | 2.5808 | 2.5742 |
| **28** | 2.5590 | 2.5526 | 2.5462 | 2.5398 | 2.5334 | 2.5270 | 2.5206 | 2.5143 | 2.5079 |
| **29** | 2.4935 | 2.4873 | 2.4811 | 2.4749 | 2.4687 | 2.4625 | 2.4564 | 2.4503 | 2.4442 |
| **30** | 2.4304 | 2.4244 | 2.4184 | 2.4125 | 2.4065 | 2.4006 | 2.3946 | 2.3887 | 2.3828 |
| **31** | 2.3697 | 2.3639 | 2.3581 | 2.3524 | 2.3466 | 2.3409 | 2.3352 | 2.3295 | 2.3238 |
| **32** | 2.3112 | 2.3056 | 2.3001 | 2.2945 | 2.2889 | 2.2834 | 2.2779 | 2.2724 | 2.2669 |
| **33** | 2.2548 | 2.2495 | 2.2441 | 2.2387 | 2.2333 | 2.2280 | 2.2227 | 2.2173 | 2.2120 |
| **34** | 2.2005 | 2.1953 | 2.1901 | 2.1849 | 2.1797 | 2.1746 | 2.1694 | 2.1643 | 2.1592 |
| **35** | 2.1480 | 2.1430 | 2.1379 | 2.1329 | 2.1280 | 2.1230 | 2.1180 | 2.1131 | 2.1081 |
| **36** | 2.0973 | 2.0925 | 2.0876 | 2.0828 | 2.0780 | 2.0732 | 2.0684 | 2.0636 | 2.0588 |
| **37** | 2.0483 | 2.0436 | 2.0390 | 2.0343 | 2.0297 | 2.0250 | 2.0204 | 2.0158 | 2.0112 |
| **38** | 2.0010 | 1.9964 | 1.9919 | 1.9875 | 1.9830 | 1.9785 | 1.9740 | 1.9696 | 1.9652 |
| **39** | 1.9551 | 1.9508 | 1.9464 | 1.9421 | 1.9378 | 1.9335 | 1.9292 | 1.9249 | 1.9206 |
| **40** | 1.9107 | 1.9065 | 1.9024 | 1.8982 | 1.8940 | 1.8899 | 1.8857 | 1.8816 | 1.8774 |

*Source*: Based on Weiss (1974), Weiss and Price (1980).

**Table 2.23** $F^* \times 10^{+2}$ for Carbon Dioxide as a Function of Temperature and Salinity, 0–40 g/kg in mol/(L atm) (At 10°C and 35 g/kg, the table value is 4.4355; the $F^*$ value is 4.4355/100 or 0.044355 mol/(L atm))

| Temperature (°C) | Salinity (g/kg) | | | | | | | | |
|---|---|---|---|---|---|---|---|---|---|
| | 0.0 | 5.0 | 10.0 | 15.0 | 20.0 | 25.0 | 30.0 | 35.0 | 40.0 |
| 0 | 7.6758 | 7.4792 | 7.2873 | 7.1001 | 6.9176 | 6.7397 | 6.5663 | 6.3973 | 6.2325 |
| 1 | 7.3765 | 7.1882 | 7.0044 | 6.8251 | 6.6502 | 6.4798 | 6.3136 | 6.1517 | 5.9938 |
| 2 | 7.0927 | 6.9123 | 6.7362 | 6.5644 | 6.3968 | 6.2334 | 6.0742 | 5.9189 | 5.7676 |
| 3 | 6.8236 | 6.6506 | 6.4818 | 6.3171 | 6.1565 | 5.9999 | 5.8472 | 5.6983 | 5.5532 |
| 4 | 6.5681 | 6.4023 | 6.2404 | 6.0825 | 5.9285 | 5.7783 | 5.6318 | 5.4890 | 5.3498 |
| 5 | 6.3255 | 6.1665 | 6.0113 | 5.8598 | 5.7120 | 5.5679 | 5.4274 | 5.2903 | 5.1567 |
| 6 | 6.0950 | 5.9425 | 5.7935 | 5.6482 | 5.5064 | 5.3681 | 5.2332 | 5.1017 | 4.9734 |
| 7 | 5.8758 | 5.7295 | 5.5865 | 5.4470 | 5.3110 | 5.1782 | 5.0487 | 4.9224 | 4.7993 |
| 8 | 5.6674 | 5.5269 | 5.3897 | 5.2558 | 5.1251 | 4.9976 | 4.8733 | 4.7520 | 4.6337 |
| 9 | 5.4690 | 5.3341 | 5.2023 | 5.0738 | 4.9483 | 4.8258 | 4.7064 | 4.5899 | 4.4762 |
| 10 | 5.2800 | 5.1505 | 5.0240 | 4.9005 | 4.7799 | 4.6623 | 4.5475 | 4.4355 | 4.3263 |
| 11 | 5.1000 | 4.9756 | 4.8540 | 4.7354 | 4.6196 | 4.5065 | 4.3962 | 4.2886 | 4.1835 |
| 12 | 4.9283 | 4.8088 | 4.6921 | 4.5780 | 4.4667 | 4.3581 | 4.2520 | 4.1485 | 4.0476 |
| 13 | 4.7646 | 4.6498 | 4.5376 | 4.4280 | 4.3210 | 4.2165 | 4.1146 | 4.0150 | 3.9179 |
| 14 | 4.6084 | 4.4980 | 4.3902 | 4.2848 | 4.1820 | 4.0815 | 3.9834 | 3.8877 | 3.7943 |
| 15 | 4.4592 | 4.3531 | 4.2494 | 4.1482 | 4.0492 | 3.9526 | 3.8583 | 3.7662 | 3.6762 |
| 16 | 4.3167 | 4.2147 | 4.1150 | 4.0176 | 3.9224 | 3.8295 | 3.7387 | 3.6501 | 3.5636 |
| 17 | 4.1805 | 4.0824 | 3.9865 | 3.8928 | 3.8013 | 3.7119 | 3.6245 | 3.5392 | 3.4559 |
| 18 | 4.0502 | 3.9559 | 3.8637 | 3.7735 | 3.6855 | 3.5994 | 3.5154 | 3.4332 | 3.3530 |
| 19 | 3.9255 | 3.8348 | 3.7461 | 3.6594 | 3.5747 | 3.4918 | 3.4109 | 3.3319 | 3.2546 |
| 20 | 3.8061 | 3.7189 | 3.6336 | 3.5501 | 3.4686 | 3.3889 | 3.3110 | 3.2349 | 3.1605 |
| 21 | 3.6918 | 3.6079 | 3.5258 | 3.4455 | 3.3670 | 3.2903 | 3.2153 | 3.1420 | 3.0703 |
| 22 | 3.5822 | 3.5015 | 3.4225 | 3.3452 | 3.2697 | 3.1958 | 3.1236 | 3.0530 | 2.9840 |
| 23 | 3.4771 | 3.3994 | 3.3234 | 3.2491 | 3.1764 | 3.1052 | 3.0357 | 2.9677 | 2.9012 |
| 24 | 3.3763 | 3.3016 | 3.2284 | 3.1569 | 3.0869 | 3.0184 | 2.9514 | 2.8859 | 2.8218 |
| 25 | 3.2795 | 3.2076 | 3.1372 | 3.0683 | 3.0010 | 2.9350 | 2.8705 | 2.8074 | 2.7457 |
| 26 | 3.1865 | 3.1173 | 3.0496 | 2.9833 | 2.9185 | 2.8550 | 2.7928 | 2.7321 | 2.6726 |
| 27 | 3.0971 | 3.0306 | 2.9655 | 2.9017 | 2.8392 | 2.7781 | 2.7182 | 2.6597 | 2.6024 |
| 28 | 3.0112 | 2.9472 | 2.8845 | 2.8231 | 2.7630 | 2.7041 | 2.6465 | 2.5901 | 2.5349 |
| 29 | 2.9285 | 2.8670 | 2.8067 | 2.7476 | 2.6897 | 2.6330 | 2.5775 | 2.5232 | 2.4700 |
| 30 | 2.8489 | 2.7898 | 2.7317 | 2.6749 | 2.6191 | 2.5646 | 2.5111 | 2.4588 | 2.4075 |
| 31 | 2.7723 | 2.7154 | 2.6595 | 2.6048 | 2.5512 | 2.4987 | 2.4472 | 2.3968 | 2.3474 |
| 32 | 2.6984 | 2.6437 | 2.5900 | 2.5373 | 2.4857 | 2.4352 | 2.3856 | 2.3371 | 2.2895 |
| 33 | 2.6272 | 2.5746 | 2.5229 | 2.4723 | 2.4226 | 2.3740 | 2.3263 | 2.2795 | 2.2337 |
| 34 | 2.5584 | 2.5078 | 2.4582 | 2.4095 | 2.3617 | 2.3149 | 2.2690 | 2.2240 | 2.1798 |
| 35 | 2.4921 | 2.4434 | 2.3957 | 2.3489 | 2.3030 | 2.2579 | 2.2137 | 2.1704 | 2.1279 |
| 36 | 2.4279 | 2.3812 | 2.3354 | 2.2903 | 2.2462 | 2.2028 | 2.1603 | 2.1186 | 2.0777 |
| 37 | 2.3660 | 2.3211 | 2.2770 | 2.2338 | 2.1913 | 2.1496 | 2.1087 | 2.0686 | 2.0292 |
| 38 | 2.3060 | 2.2629 | 2.2206 | 2.1790 | 2.1382 | 2.0981 | 2.0588 | 2.0202 | 1.9823 |
| 39 | 2.2480 | 2.2066 | 2.1660 | 2.1260 | 2.0868 | 2.0483 | 2.0105 | 1.9734 | 1.9370 |
| 40 | 2.1918 | 2.1521 | 2.1131 | 2.0747 | 2.0371 | 2.0001 | 1.9637 | 1.9281 | 1.8930 |

*Source*: Based on Weiss (1974), Weiss and Price (1980).

**Table 2.24** $F^* \times 10^{+2}$ for Carbon Dioxide as a Function of Temperature and Salinity, 33–37 g/kg in mol/(L atm) (At 10°C and 35 g/kg, the table value is 4.4355; the $F^*$ value is 4.4355/100 or 0.044355 mol/(L atm))

| Temperature (°C) | Salinity (g/kg) | | | | | | | | |
|---|---|---|---|---|---|---|---|---|---|
| | **33.0** | **33.5** | **34.0** | **34.5** | **35.0** | **35.5** | **36.0** | **36.5** | **37.0** |
| **0** | 6.4644 | 6.4475 | 6.4307 | 6.4140 | 6.3973 | 6.3806 | 6.3640 | 6.3474 | 6.3309 |
| **1** | 6.2159 | 6.1998 | 6.1837 | 6.1677 | 6.1517 | 6.1357 | 6.1198 | 6.1039 | 6.0880 |
| **2** | 5.9806 | 5.9651 | 5.9497 | 5.9343 | 5.9189 | 5.9036 | 5.8883 | 5.8731 | 5.8579 |
| **3** | 5.7574 | 5.7426 | 5.7278 | 5.7130 | 5.6983 | 5.6836 | 5.6690 | 5.6544 | 5.6398 |
| **4** | 5.5457 | 5.5315 | 5.5173 | 5.5031 | 5.4890 | 5.4749 | 5.4609 | 5.4468 | 5.4329 |
| **5** | 5.3447 | 5.3311 | 5.3175 | 5.3039 | 5.2903 | 5.2768 | 5.2633 | 5.2499 | 5.2365 |
| **6** | 5.1539 | 5.1408 | 5.1277 | 5.1147 | 5.1017 | 5.0887 | 5.0758 | 5.0629 | 5.0500 |
| **7** | 4.9726 | 4.9600 | 4.9474 | 4.9349 | 4.9224 | 4.9100 | 4.8975 | 4.8851 | 4.8728 |
| **8** | 4.8001 | 4.7881 | 4.7760 | 4.7640 | 4.7520 | 4.7400 | 4.7281 | 4.7162 | 4.7043 |
| **9** | 4.6361 | 4.6245 | 4.6129 | 4.6014 | 4.5899 | 4.5784 | 4.5669 | 4.5555 | 4.5440 |
| **10** | 4.4800 | 4.4688 | 4.4577 | 4.4466 | 4.4355 | 4.4245 | 4.4135 | 4.4025 | 4.3915 |
| **11** | 4.3313 | 4.3206 | 4.3099 | 4.2992 | 4.2886 | 4.2780 | 4.2674 | 4.2568 | 4.2463 |
| **12** | 4.1896 | 4.1793 | 4.1690 | 4.1588 | 4.1485 | 4.1383 | 4.1281 | 4.1180 | 4.1079 |
| **13** | 4.0546 | 4.0447 | 4.0348 | 4.0249 | 4.0150 | 4.0052 | 3.9954 | 3.9857 | 3.9759 |
| **14** | 3.9257 | 3.9162 | 3.9067 | 3.8972 | 3.8877 | 3.8783 | 3.8688 | 3.8594 | 3.8501 |
| **15** | 3.8027 | 3.7936 | 3.7844 | 3.7753 | 3.7662 | 3.7571 | 3.7480 | 3.7390 | 3.7299 |
| **16** | 3.6853 | 3.6765 | 3.6677 | 3.6589 | 3.6501 | 3.6414 | 3.6326 | 3.6239 | 3.6153 |
| **17** | 3.5731 | 3.5646 | 3.5561 | 3.5477 | 3.5392 | 3.5308 | 3.5224 | 3.5140 | 3.5057 |
| **18** | 3.4659 | 3.4577 | 3.4495 | 3.4414 | 3.4332 | 3.4251 | 3.4171 | 3.4090 | 3.4009 |
| **19** | 3.3633 | 3.3554 | 3.3475 | 3.3397 | 3.3319 | 3.3241 | 3.3163 | 3.3085 | 3.3008 |
| **20** | 3.2651 | 3.2575 | 3.2499 | 3.2424 | 3.2349 | 3.2273 | 3.2198 | 3.2124 | 3.2049 |
| **21** | 3.1711 | 3.1638 | 3.1565 | 3.1492 | 3.1420 | 3.1347 | 3.1275 | 3.1203 | 3.1131 |
| **22** | 3.0810 | 3.0740 | 3.0670 | 3.0600 | 3.0530 | 3.0460 | 3.0391 | 3.0321 | 3.0252 |
| **23** | 2.9947 | 2.9879 | 2.9812 | 2.9744 | 2.9677 | 2.9610 | 2.9543 | 2.9476 | 2.9409 |
| **24** | 2.9119 | 2.9054 | 2.8989 | 2.8924 | 2.8859 | 2.8794 | 2.8730 | 2.8665 | 2.8601 |
| **25** | 2.8325 | 2.8262 | 2.8199 | 2.8137 | 2.8074 | 2.8012 | 2.7950 | 2.7888 | 2.7826 |
| **26** | 2.7562 | 2.7502 | 2.7441 | 2.7381 | 2.7321 | 2.7261 | 2.7201 | 2.7141 | 2.7081 |
| **27** | 2.6829 | 2.6771 | 2.6713 | 2.6655 | 2.6597 | 2.6539 | 2.6481 | 2.6424 | 2.6366 |
| **28** | 2.6125 | 2.6069 | 2.6013 | 2.5957 | 2.5901 | 2.5845 | 2.5790 | 2.5734 | 2.5679 |
| **29** | 2.5448 | 2.5394 | 2.5340 | 2.5286 | 2.5232 | 2.5178 | 2.5125 | 2.5071 | 2.5018 |
| **30** | 2.4796 | 2.4744 | 2.4692 | 2.4640 | 2.4588 | 2.4536 | 2.4485 | 2.4433 | 2.4382 |
| **31** | 2.4168 | 2.4118 | 2.4068 | 2.4018 | 2.3968 | 2.3918 | 2.3868 | 2.3819 | 2.3769 |
| **32** | 2.3564 | 2.3515 | 2.3467 | 2.3419 | 2.3371 | 2.3323 | 2.3275 | 2.3227 | 2.3179 |
| **33** | 2.2981 | 2.2934 | 2.2888 | 2.2841 | 2.2795 | 2.2749 | 2.2703 | 2.2657 | 2.2611 |
| **34** | 2.2419 | 2.2374 | 2.2329 | 2.2284 | 2.2240 | 2.2195 | 2.2151 | 2.2106 | 2.2062 |
| **35** | 2.1876 | 2.1833 | 2.1790 | 2.1747 | 2.1704 | 2.1661 | 2.1618 | 2.1575 | 2.1533 |
| **36** | 2.1352 | 2.1310 | 2.1269 | 2.1227 | 2.1186 | 2.1145 | 2.1104 | 2.1063 | 2.1022 |
| **37** | 2.0845 | 2.0805 | 2.0766 | 2.0726 | 2.0686 | 2.0646 | 2.0607 | 2.0567 | 2.0527 |
| **38** | 2.0356 | 2.0317 | 2.0279 | 2.0240 | 2.0202 | 2.0164 | 2.0126 | 2.0088 | 2.0050 |
| **39** | 1.9882 | 1.9845 | 1.9808 | 1.9771 | 1.9734 | 1.9697 | 1.9661 | 1.9624 | 1.9587 |
| **40** | 1.9423 | 1.9387 | 1.9351 | 1.9316 | 1.9281 | 1.9245 | 1.9210 | 1.9175 | 1.9140 |

*Source*: Based on Weiss (1974), Weiss and Price (1980).

**Table 2.25** Bunsen Coefficient ($\beta$) for Oxygen as a Function of Temperature and Salinity, 0–40 g/kg (L real gas/(L atm))

| Temperature (°C) | Salinity (g/kg) | | | | | | | | |
|---|---|---|---|---|---|---|---|---|---|
| | **0.0** | **5.0** | **10.0** | **15.0** | **20.0** | **25.0** | **30.0** | **35.0** | **40.0** |
| **0** | 0.04914 | 0.04746 | 0.04583 | 0.04425 | 0.04273 | 0.04126 | 0.03984 | 0.03846 | 0.03714 |
| **1** | 0.04780 | 0.04618 | 0.04461 | 0.04309 | 0.04162 | 0.04020 | 0.03883 | 0.03750 | 0.03622 |
| **2** | 0.04653 | 0.04496 | 0.04344 | 0.04198 | 0.04056 | 0.03919 | 0.03786 | 0.03658 | 0.03534 |
| **3** | 0.04531 | 0.04380 | 0.04233 | 0.04092 | 0.03955 | 0.03822 | 0.03694 | 0.03570 | 0.03450 |
| **4** | 0.04414 | 0.04268 | 0.04127 | 0.03990 | 0.03858 | 0.03730 | 0.03606 | 0.03486 | 0.03370 |
| **5** | 0.04303 | 0.04162 | 0.04026 | 0.03893 | 0.03765 | 0.03641 | 0.03521 | 0.03405 | 0.03293 |
| **6** | 0.04197 | 0.04061 | 0.03929 | 0.03801 | 0.03677 | 0.03557 | 0.03441 | 0.03328 | 0.03220 |
| **7** | 0.04096 | 0.03964 | 0.03836 | 0.03712 | 0.03592 | 0.03476 | 0.03363 | 0.03255 | 0.03149 |
| **8** | 0.03999 | 0.03871 | 0.03747 | 0.03627 | 0.03511 | 0.03398 | 0.03289 | 0.03184 | 0.03081 |
| **9** | 0.03906 | 0.03782 | 0.03662 | 0.03546 | 0.03433 | 0.03324 | 0.03218 | 0.03116 | 0.03017 |
| **10** | 0.03817 | 0.03697 | 0.03581 | 0.03468 | 0.03359 | 0.03253 | 0.03150 | 0.03051 | 0.02955 |
| **11** | 0.03732 | 0.03616 | 0.03503 | 0.03394 | 0.03288 | 0.03185 | 0.03085 | 0.02989 | 0.02895 |
| **12** | 0.03651 | 0.03538 | 0.03429 | 0.03322 | 0.03219 | 0.03120 | 0.03023 | 0.02929 | 0.02838 |
| **13** | 0.03573 | 0.03463 | 0.03357 | 0.03254 | 0.03154 | 0.03057 | 0.02963 | 0.02871 | 0.02783 |
| **14** | 0.03498 | 0.03392 | 0.03288 | 0.03188 | 0.03091 | 0.02997 | 0.02905 | 0.02816 | 0.02730 |
| **15** | 0.03426 | 0.03323 | 0.03223 | 0.03125 | 0.03031 | 0.02939 | 0.02850 | 0.02763 | 0.02680 |
| **16** | 0.03358 | 0.03257 | 0.03160 | 0.03065 | 0.02973 | 0.02883 | 0.02797 | 0.02712 | 0.02631 |
| **17** | 0.03292 | 0.03194 | 0.03099 | 0.03007 | 0.02917 | 0.02830 | 0.02746 | 0.02664 | 0.02584 |
| **18** | 0.03228 | 0.03133 | 0.03041 | 0.02951 | 0.02864 | 0.02779 | 0.02697 | 0.02617 | 0.02539 |
| **19** | 0.03168 | 0.03075 | 0.02985 | 0.02897 | 0.02812 | 0.02730 | 0.02649 | 0.02571 | 0.02496 |
| **20** | 0.03109 | 0.03019 | 0.02931 | 0.02846 | 0.02763 | 0.02682 | 0.02604 | 0.02528 | 0.02454 |
| **21** | 0.03053 | 0.02965 | 0.02879 | 0.02796 | 0.02715 | 0.02637 | 0.02560 | 0.02486 | 0.02414 |
| **22** | 0.02999 | 0.02913 | 0.02830 | 0.02749 | 0.02670 | 0.02593 | 0.02518 | 0.02446 | 0.02375 |
| **23** | 0.02947 | 0.02864 | 0.02782 | 0.02703 | 0.02626 | 0.02551 | 0.02478 | 0.02407 | 0.02338 |
| **24** | 0.02897 | 0.02816 | 0.02736 | 0.02659 | 0.02583 | 0.02510 | 0.02439 | 0.02370 | 0.02302 |
| **25** | 0.02850 | 0.02770 | 0.02692 | 0.02616 | 0.02543 | 0.02471 | 0.02401 | 0.02334 | 0.02268 |
| **26** | 0.02803 | 0.02725 | 0.02649 | 0.02575 | 0.02503 | 0.02433 | 0.02365 | 0.02299 | 0.02235 |
| **27** | 0.02759 | 0.02683 | 0.02608 | 0.02536 | 0.02466 | 0.02397 | 0.02330 | 0.02266 | 0.02203 |
| **28** | 0.02716 | 0.02642 | 0.02569 | 0.02498 | 0.02429 | 0.02362 | 0.02297 | 0.02233 | 0.02172 |
| **29** | 0.02675 | 0.02602 | 0.02531 | 0.02462 | 0.02394 | 0.02329 | 0.02265 | 0.02202 | 0.02142 |
| **30** | 0.02635 | 0.02564 | 0.02494 | 0.02426 | 0.02360 | 0.02296 | 0.02233 | 0.02173 | 0.02113 |
| **31** | 0.02597 | 0.02527 | 0.02459 | 0.02393 | 0.02328 | 0.02265 | 0.02203 | 0.02144 | 0.02086 |
| **32** | 0.02561 | 0.02492 | 0.02425 | 0.02360 | 0.02296 | 0.02235 | 0.02174 | 0.02116 | 0.02059 |
| **33** | 0.02525 | 0.02458 | 0.02392 | 0.02328 | 0.02266 | 0.02206 | 0.02147 | 0.02089 | 0.02033 |
| **34** | 0.02491 | 0.02425 | 0.02361 | 0.02298 | 0.02237 | 0.02177 | 0.02120 | 0.02063 | 0.02008 |
| **35** | 0.02458 | 0.02393 | 0.02330 | 0.02269 | 0.02209 | 0.02150 | 0.02094 | 0.02038 | 0.01984 |
| **36** | 0.02426 | 0.02363 | 0.02301 | 0.02240 | 0.02182 | 0.02124 | 0.02068 | 0.02014 | 0.01961 |
| **37** | 0.02396 | 0.02333 | 0.02273 | 0.02213 | 0.02155 | 0.02099 | 0.02044 | 0.01991 | 0.01939 |
| **38** | 0.02366 | 0.02305 | 0.02245 | 0.02187 | 0.02130 | 0.02075 | 0.02021 | 0.01968 | 0.01917 |
| **39** | 0.02338 | 0.02278 | 0.02219 | 0.02162 | 0.02106 | 0.02051 | 0.01998 | 0.01947 | 0.01896 |
| **40** | 0.02310 | 0.02251 | 0.02193 | 0.02137 | 0.02082 | 0.02029 | 0.01976 | 0.01926 | 0.01876 |

*Source*: Based on Benson and Krause (1980) and Eq. (1.16).

**Table 2.26** Bunsen Coefficient ($\beta$) for Oxygen as a Function of Temperature and Salinity, 33−37 g/kg (L real gas/(L atm))

| Temperature (°C) | Salinity (g/kg) | | | | | | | | |
|---|---|---|---|---|---|---|---|---|---|
| | **33.0** | **33.5** | **34.0** | **34.5** | **35.0** | **35.5** | **36.0** | **36.5** | **37.0** |
| **0** | 0.03901 | 0.03887 | 0.03874 | 0.03860 | 0.03846 | 0.03833 | 0.03820 | 0.03806 | 0.03793 |
| **1** | 0.03803 | 0.03789 | 0.03776 | 0.03763 | 0.03750 | 0.03737 | 0.03724 | 0.03711 | 0.03698 |
| **2** | 0.03709 | 0.03696 | 0.03683 | 0.03671 | 0.03658 | 0.03646 | 0.03633 | 0.03620 | 0.03608 |
| **3** | 0.03619 | 0.03607 | 0.03595 | 0.03582 | 0.03570 | 0.03558 | 0.03546 | 0.03534 | 0.03522 |
| **4** | 0.03533 | 0.03521 | 0.03510 | 0.03498 | 0.03486 | 0.03474 | 0.03462 | 0.03451 | 0.03439 |
| **5** | 0.03451 | 0.03440 | 0.03428 | 0.03417 | 0.03405 | 0.03394 | 0.03383 | 0.03371 | 0.03360 |
| **6** | 0.03373 | 0.03362 | 0.03351 | 0.03339 | 0.03328 | 0.03317 | 0.03306 | 0.03295 | 0.03284 |
| **7** | 0.03298 | 0.03287 | 0.03276 | 0.03265 | 0.03255 | 0.03244 | 0.03233 | 0.03223 | 0.03212 |
| **8** | 0.03226 | 0.03215 | 0.03205 | 0.03194 | 0.03184 | 0.03173 | 0.03163 | 0.03153 | 0.03142 |
| **9** | 0.03157 | 0.03146 | 0.03136 | 0.03126 | 0.03116 | 0.03106 | 0.03096 | 0.03086 | 0.03076 |
| **10** | 0.03090 | 0.03081 | 0.03071 | 0.03061 | 0.03051 | 0.03041 | 0.03031 | 0.03022 | 0.03012 |
| **11** | 0.03027 | 0.03017 | 0.03008 | 0.02998 | 0.02989 | 0.02979 | 0.02970 | 0.02960 | 0.02951 |
| **12** | 0.02966 | 0.02957 | 0.02947 | 0.02938 | 0.02929 | 0.02920 | 0.02910 | 0.02901 | 0.02892 |
| **13** | 0.02908 | 0.02899 | 0.02889 | 0.02880 | 0.02871 | 0.02862 | 0.02854 | 0.02845 | 0.02836 |
| **14** | 0.02852 | 0.02843 | 0.02834 | 0.02825 | 0.02816 | 0.02808 | 0.02799 | 0.02790 | 0.02782 |
| **15** | 0.02798 | 0.02789 | 0.02780 | 0.02772 | 0.02763 | 0.02755 | 0.02746 | 0.02738 | 0.02730 |
| **16** | 0.02746 | 0.02737 | 0.02729 | 0.02721 | 0.02712 | 0.02704 | 0.02696 | 0.02688 | 0.02680 |
| **17** | 0.02696 | 0.02688 | 0.02680 | 0.02672 | 0.02664 | 0.02656 | 0.02647 | 0.02639 | 0.02631 |
| **18** | 0.02648 | 0.02640 | 0.02632 | 0.02624 | 0.02617 | 0.02609 | 0.02601 | 0.02593 | 0.02585 |
| **19** | 0.02602 | 0.02595 | 0.02587 | 0.02579 | 0.02571 | 0.02564 | 0.02556 | 0.02548 | 0.02541 |
| **20** | 0.02558 | 0.02550 | 0.02543 | 0.02535 | 0.02528 | 0.02520 | 0.02513 | 0.02506 | 0.02498 |
| **21** | 0.02515 | 0.02508 | 0.02501 | 0.02493 | 0.02486 | 0.02479 | 0.02471 | 0.02464 | 0.02457 |
| **22** | 0.02475 | 0.02467 | 0.02460 | 0.02453 | 0.02446 | 0.02439 | 0.02432 | 0.02424 | 0.02417 |
| **23** | 0.02435 | 0.02428 | 0.02421 | 0.02414 | 0.02407 | 0.02400 | 0.02393 | 0.02386 | 0.02379 |
| **24** | 0.02397 | 0.02390 | 0.02383 | 0.02377 | 0.02370 | 0.02363 | 0.02356 | 0.02349 | 0.02343 |
| **25** | 0.02361 | 0.02354 | 0.02347 | 0.02340 | 0.02334 | 0.02327 | 0.02320 | 0.02314 | 0.02307 |
| **26** | 0.02325 | 0.02319 | 0.02312 | 0.02306 | 0.02299 | 0.02293 | 0.02286 | 0.02280 | 0.02273 |
| **27** | 0.02291 | 0.02285 | 0.02279 | 0.02272 | 0.02266 | 0.02259 | 0.02253 | 0.02247 | 0.02240 |
| **28** | 0.02259 | 0.02252 | 0.02246 | 0.02240 | 0.02233 | 0.02227 | 0.02221 | 0.02215 | 0.02209 |
| **29** | 0.02227 | 0.02221 | 0.02215 | 0.02209 | 0.02202 | 0.02196 | 0.02190 | 0.02184 | 0.02178 |
| **30** | 0.02197 | 0.02191 | 0.02185 | 0.02179 | 0.02173 | 0.02167 | 0.02161 | 0.02155 | 0.02149 |
| **31** | 0.02167 | 0.02161 | 0.02156 | 0.02150 | 0.02144 | 0.02138 | 0.02132 | 0.02126 | 0.02120 |
| **32** | 0.02139 | 0.02133 | 0.02127 | 0.02122 | 0.02116 | 0.02110 | 0.02104 | 0.02099 | 0.02093 |
| **33** | 0.02112 | 0.02106 | 0.02100 | 0.02095 | 0.02089 | 0.02083 | 0.02078 | 0.02072 | 0.02066 |
| **34** | 0.02086 | 0.02080 | 0.02074 | 0.02069 | 0.02063 | 0.02058 | 0.02052 | 0.02046 | 0.02041 |
| **35** | 0.02060 | 0.02055 | 0.02049 | 0.02044 | 0.02038 | 0.02033 | 0.02027 | 0.02022 | 0.02016 |
| **36** | 0.02036 | 0.02030 | 0.02025 | 0.02019 | 0.02014 | 0.02009 | 0.02003 | 0.01998 | 0.01993 |
| **37** | 0.02012 | 0.02007 | 0.02001 | 0.01996 | 0.01991 | 0.01985 | 0.01980 | 0.01975 | 0.01970 |
| **38** | 0.01989 | 0.01984 | 0.01979 | 0.01973 | 0.01968 | 0.01963 | 0.01958 | 0.01953 | 0.01948 |
| **39** | 0.01967 | 0.01962 | 0.01957 | 0.01952 | 0.01947 | 0.01941 | 0.01936 | 0.01931 | 0.01926 |
| **40** | 0.01946 | 0.01941 | 0.01936 | 0.01931 | 0.01926 | 0.01921 | 0.01916 | 0.01911 | 0.01906 |

*Source*: Based on Benson and Krause (1980) and Eq. (1.16).

**Table 2.27** Bunsen Coefficient ($\beta$) for Nitrogen as a Function of Temperature and Salinity, 0–40 g/kg (L real gas/(L atm))

| Temperature (°C) | Salinity (g/kg) | | | | | | | | |
|---|---|---|---|---|---|---|---|---|---|
| | **0.0** | **5.0** | **10.0** | **15.0** | **20.0** | **25.0** | **30.0** | **35.0** | **40.0** |
| **0** | 0.02397 | 0.02309 | 0.02225 | 0.02143 | 0.02065 | 0.01989 | 0.01916 | 0.01846 | 0.01778 |
| **1** | 0.02336 | 0.02251 | 0.02170 | 0.02091 | 0.02015 | 0.01942 | 0.01871 | 0.01803 | 0.01738 |
| **2** | 0.02277 | 0.02196 | 0.02117 | 0.02041 | 0.01967 | 0.01897 | 0.01828 | 0.01762 | 0.01699 |
| **3** | 0.02222 | 0.02143 | 0.02067 | 0.01993 | 0.01922 | 0.01853 | 0.01787 | 0.01723 | 0.01662 |
| **4** | 0.02169 | 0.02092 | 0.02019 | 0.01947 | 0.01878 | 0.01812 | 0.01748 | 0.01686 | 0.01627 |
| **5** | 0.02118 | 0.02044 | 0.01973 | 0.01904 | 0.01837 | 0.01773 | 0.01711 | 0.01651 | 0.01593 |
| **6** | 0.02070 | 0.01998 | 0.01929 | 0.01862 | 0.01797 | 0.01735 | 0.01675 | 0.01616 | 0.01560 |
| **7** | 0.02024 | 0.01954 | 0.01887 | 0.01822 | 0.01759 | 0.01699 | 0.01640 | 0.01584 | 0.01529 |
| **8** | 0.01979 | 0.01912 | 0.01847 | 0.01784 | 0.01723 | 0.01664 | 0.01607 | 0.01552 | 0.01499 |
| **9** | 0.01937 | 0.01872 | 0.01808 | 0.01747 | 0.01688 | 0.01631 | 0.01576 | 0.01522 | 0.01471 |
| **10** | 0.01897 | 0.01833 | 0.01772 | 0.01712 | 0.01655 | 0.01599 | 0.01546 | 0.01494 | 0.01443 |
| **11** | 0.01858 | 0.01796 | 0.01736 | 0.01679 | 0.01623 | 0.01569 | 0.01517 | 0.01466 | 0.01417 |
| **12** | 0.01821 | 0.01761 | 0.01703 | 0.01647 | 0.01592 | 0.01540 | 0.01489 | 0.01440 | 0.01392 |
| **13** | 0.01785 | 0.01727 | 0.01671 | 0.01616 | 0.01563 | 0.01512 | 0.01462 | 0.01414 | 0.01368 |
| **14** | 0.01751 | 0.01695 | 0.01640 | 0.01586 | 0.01535 | 0.01485 | 0.01437 | 0.01390 | 0.01345 |
| **15** | 0.01719 | 0.01664 | 0.01610 | 0.01558 | 0.01508 | 0.01459 | 0.01412 | 0.01367 | 0.01322 |
| **16** | 0.01688 | 0.01634 | 0.01582 | 0.01531 | 0.01482 | 0.01435 | 0.01389 | 0.01344 | 0.01301 |
| **17** | 0.01658 | 0.01605 | 0.01554 | 0.01505 | 0.01457 | 0.01411 | 0.01366 | 0.01323 | 0.01280 |
| **18** | 0.01629 | 0.01578 | 0.01528 | 0.01480 | 0.01433 | 0.01388 | 0.01344 | 0.01302 | 0.01261 |
| **19** | 0.01602 | 0.01552 | 0.01503 | 0.01456 | 0.01411 | 0.01366 | 0.01323 | 0.01282 | 0.01242 |
| **20** | 0.01576 | 0.01527 | 0.01479 | 0.01433 | 0.01389 | 0.01345 | 0.01303 | 0.01263 | 0.01224 |
| **21** | 0.01550 | 0.01503 | 0.01456 | 0.01411 | 0.01368 | 0.01325 | 0.01284 | 0.01244 | 0.01206 |
| **22** | 0.01526 | 0.01479 | 0.01434 | 0.01390 | 0.01347 | 0.01306 | 0.01266 | 0.01227 | 0.01189 |
| **23** | 0.01503 | 0.01457 | 0.01413 | 0.01370 | 0.01328 | 0.01287 | 0.01248 | 0.01210 | 0.01173 |
| **24** | 0.01481 | 0.01436 | 0.01392 | 0.01350 | 0.01309 | 0.01269 | 0.01231 | 0.01194 | 0.01157 |
| **25** | 0.01459 | 0.01416 | 0.01373 | 0.01331 | 0.01291 | 0.01252 | 0.01215 | 0.01178 | 0.01142 |
| **26** | 0.01439 | 0.01396 | 0.01354 | 0.01314 | 0.01274 | 0.01236 | 0.01199 | 0.01163 | 0.01128 |
| **27** | 0.01419 | 0.01377 | 0.01336 | 0.01296 | 0.01258 | 0.01220 | 0.01184 | 0.01148 | 0.01114 |
| **28** | 0.01401 | 0.01359 | 0.01319 | 0.01280 | 0.01242 | 0.01205 | 0.01169 | 0.01134 | 0.01101 |
| **29** | 0.01383 | 0.01342 | 0.01302 | 0.01264 | 0.01227 | 0.01190 | 0.01155 | 0.01121 | 0.01088 |
| **30** | 0.01365 | 0.01325 | 0.01286 | 0.01249 | 0.01212 | 0.01176 | 0.01142 | 0.01108 | 0.01076 |
| **31** | 0.01349 | 0.01309 | 0.01271 | 0.01234 | 0.01198 | 0.01163 | 0.01129 | 0.01096 | 0.01064 |
| **32** | 0.01333 | 0.01294 | 0.01256 | 0.01220 | 0.01184 | 0.01150 | 0.01116 | 0.01084 | 0.01052 |
| **33** | 0.01318 | 0.01280 | 0.01242 | 0.01206 | 0.01171 | 0.01138 | 0.01105 | 0.01072 | 0.01041 |
| **34** | 0.01303 | 0.01266 | 0.01229 | 0.01194 | 0.01159 | 0.01126 | 0.01093 | 0.01062 | 0.01031 |
| **35** | 0.01289 | 0.01252 | 0.01216 | 0.01181 | 0.01147 | 0.01114 | 0.01082 | 0.01051 | 0.01021 |
| **36** | 0.01276 | 0.01239 | 0.01204 | 0.01169 | 0.01136 | 0.01103 | 0.01072 | 0.01041 | 0.01011 |
| **37** | 0.01263 | 0.01227 | 0.01192 | 0.01158 | 0.01125 | 0.01093 | 0.01062 | 0.01031 | 0.01002 |
| **38** | 0.01251 | 0.01215 | 0.01181 | 0.01147 | 0.01114 | 0.01083 | 0.01052 | 0.01022 | 0.00993 |
| **39** | 0.01239 | 0.01204 | 0.01170 | 0.01137 | 0.01105 | 0.01073 | 0.01043 | 0.01013 | 0.00984 |
| **40** | 0.01228 | 0.01194 | 0.01160 | 0.01127 | 0.01095 | 0.01064 | 0.01034 | 0.01004 | 0.00976 |

*Source*: Based on Hamme and Emerson (2004) and Eq. (1.16).

**Table 2.28** Bunsen Coefficient ($\beta$) for Nitrogen and Salinity, 33–37 g/kg (L real gas/(L atm))

| Temperature (°C) | Salinity (g/kg) | | | | | | | | |
|---|---|---|---|---|---|---|---|---|---|
| | **33.0** | **33.5** | **34.0** | **34.5** | **35.0** | **35.5** | **36.0** | **36.5** | **37.0** |
| **0** | 0.01874 | 0.01867 | 0.01860 | 0.01853 | 0.01846 | 0.01839 | 0.01832 | 0.01825 | 0.01818 |
| **1** | 0.01830 | 0.01823 | 0.01817 | 0.01810 | 0.01803 | 0.01797 | 0.01790 | 0.01783 | 0.01777 |
| **2** | 0.01788 | 0.01782 | 0.01775 | 0.01769 | 0.01762 | 0.01756 | 0.01750 | 0.01743 | 0.01737 |
| **3** | 0.01749 | 0.01742 | 0.01736 | 0.01730 | 0.01723 | 0.01717 | 0.01711 | 0.01705 | 0.01699 |
| **4** | 0.01711 | 0.01705 | 0.01698 | 0.01692 | 0.01686 | 0.01680 | 0.01674 | 0.01668 | 0.01662 |
| **5** | 0.01674 | 0.01668 | 0.01662 | 0.01656 | 0.01651 | 0.01645 | 0.01639 | 0.01633 | 0.01627 |
| **6** | 0.01639 | 0.01634 | 0.01628 | 0.01622 | 0.01616 | 0.01611 | 0.01605 | 0.01599 | 0.01594 |
| **7** | 0.01606 | 0.01600 | 0.01595 | 0.01589 | 0.01584 | 0.01578 | 0.01573 | 0.01567 | 0.01562 |
| **8** | 0.01574 | 0.01569 | 0.01563 | 0.01558 | 0.01552 | 0.01547 | 0.01542 | 0.01536 | 0.01531 |
| **9** | 0.01544 | 0.01538 | 0.01533 | 0.01528 | 0.01522 | 0.01517 | 0.01512 | 0.01507 | 0.01502 |
| **10** | 0.01514 | 0.01509 | 0.01504 | 0.01499 | 0.01494 | 0.01489 | 0.01483 | 0.01478 | 0.01473 |
| **11** | 0.01486 | 0.01481 | 0.01476 | 0.01471 | 0.01466 | 0.01461 | 0.01456 | 0.01451 | 0.01446 |
| **12** | 0.01459 | 0.01454 | 0.01449 | 0.01444 | 0.01440 | 0.01435 | 0.01430 | 0.01425 | 0.01420 |
| **13** | 0.01433 | 0.01428 | 0.01424 | 0.01419 | 0.01414 | 0.01410 | 0.01405 | 0.01400 | 0.01396 |
| **14** | 0.01408 | 0.01404 | 0.01399 | 0.01394 | 0.01390 | 0.01385 | 0.01381 | 0.01376 | 0.01372 |
| **15** | 0.01385 | 0.01380 | 0.01376 | 0.01371 | 0.01367 | 0.01362 | 0.01358 | 0.01353 | 0.01349 |
| **16** | 0.01362 | 0.01357 | 0.01353 | 0.01348 | 0.01344 | 0.01340 | 0.01335 | 0.01331 | 0.01327 |
| **17** | 0.01340 | 0.01335 | 0.01331 | 0.01327 | 0.01323 | 0.01318 | 0.01314 | 0.01310 | 0.01306 |
| **18** | 0.01319 | 0.01314 | 0.01310 | 0.01306 | 0.01302 | 0.01298 | 0.01294 | 0.01289 | 0.01285 |
| **19** | 0.01298 | 0.01294 | 0.01290 | 0.01286 | 0.01282 | 0.01278 | 0.01274 | 0.01270 | 0.01266 |
| **20** | 0.01279 | 0.01275 | 0.01271 | 0.01267 | 0.01263 | 0.01259 | 0.01255 | 0.01251 | 0.01247 |
| **21** | 0.01260 | 0.01256 | 0.01252 | 0.01248 | 0.01244 | 0.01241 | 0.01237 | 0.01233 | 0.01229 |
| **22** | 0.01242 | 0.01238 | 0.01235 | 0.01231 | 0.01227 | 0.01223 | 0.01219 | 0.01215 | 0.01212 |
| **23** | 0.01225 | 0.01221 | 0.01217 | 0.01214 | 0.01210 | 0.01206 | 0.01202 | 0.01199 | 0.01195 |
| **24** | 0.01208 | 0.01205 | 0.01201 | 0.01197 | 0.01194 | 0.01190 | 0.01186 | 0.01183 | 0.01179 |
| **25** | 0.01192 | 0.01189 | 0.01185 | 0.01182 | 0.01178 | 0.01174 | 0.01171 | 0.01167 | 0.01164 |
| **26** | 0.01177 | 0.01174 | 0.01170 | 0.01166 | 0.01163 | 0.01159 | 0.01156 | 0.01152 | 0.01149 |
| **27** | 0.01162 | 0.01159 | 0.01155 | 0.01152 | 0.01148 | 0.01145 | 0.01141 | 0.01138 | 0.01135 |
| **28** | 0.01148 | 0.01145 | 0.01141 | 0.01138 | 0.01134 | 0.01131 | 0.01128 | 0.01124 | 0.01121 |
| **29** | 0.01135 | 0.01131 | 0.01128 | 0.01124 | 0.01121 | 0.01118 | 0.01114 | 0.01111 | 0.01108 |
| **30** | 0.01121 | 0.01118 | 0.01115 | 0.01111 | 0.01108 | 0.01105 | 0.01102 | 0.01098 | 0.01095 |
| **31** | 0.01109 | 0.01106 | 0.01102 | 0.01099 | 0.01096 | 0.01093 | 0.01089 | 0.01086 | 0.01083 |
| **32** | 0.01097 | 0.01094 | 0.01090 | 0.01087 | 0.01084 | 0.01081 | 0.01078 | 0.01074 | 0.01071 |
| **33** | 0.01085 | 0.01082 | 0.01079 | 0.01076 | 0.01072 | 0.01069 | 0.01066 | 0.01063 | 0.01060 |
| **34** | 0.01074 | 0.01071 | 0.01068 | 0.01065 | 0.01062 | 0.01058 | 0.01055 | 0.01052 | 0.01049 |
| **35** | 0.01063 | 0.01060 | 0.01057 | 0.01054 | 0.01051 | 0.01048 | 0.01045 | 0.01042 | 0.01039 |
| **36** | 0.01053 | 0.01050 | 0.01047 | 0.01044 | 0.01041 | 0.01038 | 0.01035 | 0.01032 | 0.01029 |
| **37** | 0.01043 | 0.01040 | 0.01037 | 0.01034 | 0.01031 | 0.01028 | 0.01025 | 0.01022 | 0.01019 |
| **38** | 0.01034 | 0.01031 | 0.01028 | 0.01025 | 0.01022 | 0.01019 | 0.01016 | 0.01013 | 0.01010 |
| **39** | 0.01025 | 0.01022 | 0.01019 | 0.01016 | 0.01013 | 0.01010 | 0.01007 | 0.01004 | 0.01001 |
| **40** | 0.01016 | 0.01013 | 0.01010 | 0.01007 | 0.01004 | 0.01002 | 0.00999 | 0.00996 | 0.00993 |

*Source*: Based on Hamme and Emerson (2004) and Eq. (1.16).

**Table 2.29** Bunsen Coefficient ($\beta$) for Argon as a Function of Temperature and Salinity, 0–40 g/kg (L real gas/(L atm))

| Temperature (°C) | Salinity (g/kg) | | | | | | | | |
|---|---|---|---|---|---|---|---|---|---|
| | **0.0** | **5.0** | **10.0** | **15.0** | **20.0** | **25.0** | **30.0** | **35.0** | **40.0** |
| **0** | 0.05378 | 0.05196 | 0.05019 | 0.04848 | 0.04683 | 0.04523 | 0.04369 | 0.04220 | 0.04076 |
| **1** | 0.05233 | 0.05057 | 0.04886 | 0.04722 | 0.04562 | 0.04408 | 0.04259 | 0.04115 | 0.03976 |
| **2** | 0.05095 | 0.04925 | 0.04760 | 0.04601 | 0.04447 | 0.04298 | 0.04154 | 0.04015 | 0.03881 |
| **3** | 0.04962 | 0.04798 | 0.04639 | 0.04486 | 0.04337 | 0.04193 | 0.04054 | 0.03919 | 0.03789 |
| **4** | 0.04836 | 0.04678 | 0.04524 | 0.04376 | 0.04232 | 0.04093 | 0.03958 | 0.03828 | 0.03702 |
| **5** | 0.04716 | 0.04562 | 0.04414 | 0.04270 | 0.04131 | 0.03997 | 0.03866 | 0.03740 | 0.03618 |
| **6** | 0.04600 | 0.04452 | 0.04309 | 0.04170 | 0.04035 | 0.03905 | 0.03779 | 0.03656 | 0.03538 |
| **7** | 0.04490 | 0.04347 | 0.04208 | 0.04073 | 0.03943 | 0.03817 | 0.03694 | 0.03576 | 0.03461 |
| **8** | 0.04385 | 0.04246 | 0.04112 | 0.03981 | 0.03855 | 0.03732 | 0.03614 | 0.03499 | 0.03388 |
| **9** | 0.04284 | 0.04150 | 0.04019 | 0.03893 | 0.03770 | 0.03652 | 0.03537 | 0.03425 | 0.03317 |
| **10** | 0.04188 | 0.04057 | 0.03931 | 0.03808 | 0.03689 | 0.03574 | 0.03463 | 0.03354 | 0.03250 |
| **11** | 0.04095 | 0.03969 | 0.03846 | 0.03727 | 0.03612 | 0.03500 | 0.03392 | 0.03287 | 0.03185 |
| **12** | 0.04007 | 0.03884 | 0.03765 | 0.03649 | 0.03537 | 0.03429 | 0.03323 | 0.03221 | 0.03122 |
| **13** | 0.03922 | 0.03803 | 0.03687 | 0.03575 | 0.03466 | 0.03360 | 0.03258 | 0.03159 | 0.03063 |
| **14** | 0.03840 | 0.03725 | 0.03612 | 0.03503 | 0.03398 | 0.03295 | 0.03195 | 0.03099 | 0.03005 |
| **15** | 0.03762 | 0.03650 | 0.03541 | 0.03435 | 0.03332 | 0.03232 | 0.03135 | 0.03041 | 0.02950 |
| **16** | 0.03688 | 0.03578 | 0.03472 | 0.03369 | 0.03269 | 0.03171 | 0.03077 | 0.02985 | 0.02897 |
| **17** | 0.03616 | 0.03509 | 0.03406 | 0.03305 | 0.03208 | 0.03113 | 0.03021 | 0.02932 | 0.02846 |
| **18** | 0.03547 | 0.03443 | 0.03342 | 0.03245 | 0.03150 | 0.03057 | 0.02968 | 0.02881 | 0.02796 |
| **19** | 0.03481 | 0.03380 | 0.03282 | 0.03186 | 0.03094 | 0.03004 | 0.02916 | 0.02831 | 0.02749 |
| **20** | 0.03417 | 0.03319 | 0.03223 | 0.03130 | 0.03040 | 0.02952 | 0.02867 | 0.02784 | 0.02704 |
| **21** | 0.03356 | 0.03260 | 0.03167 | 0.03076 | 0.02988 | 0.02902 | 0.02819 | 0.02738 | 0.02660 |
| **22** | 0.03297 | 0.03204 | 0.03113 | 0.03024 | 0.02938 | 0.02854 | 0.02773 | 0.02694 | 0.02617 |
| **23** | 0.03241 | 0.03149 | 0.03060 | 0.02974 | 0.02890 | 0.02808 | 0.02729 | 0.02652 | 0.02577 |
| **24** | 0.03187 | 0.03097 | 0.03010 | 0.02926 | 0.02844 | 0.02764 | 0.02686 | 0.02611 | 0.02538 |
| **25** | 0.03134 | 0.03047 | 0.02962 | 0.02880 | 0.02799 | 0.02721 | 0.02645 | 0.02572 | 0.02500 |
| **26** | 0.03084 | 0.02999 | 0.02916 | 0.02835 | 0.02757 | 0.02680 | 0.02606 | 0.02534 | 0.02463 |
| **27** | 0.03036 | 0.02952 | 0.02871 | 0.02792 | 0.02715 | 0.02641 | 0.02568 | 0.02497 | 0.02428 |
| **28** | 0.02989 | 0.02908 | 0.02828 | 0.02751 | 0.02676 | 0.02602 | 0.02531 | 0.02462 | 0.02394 |
| **29** | 0.02945 | 0.02865 | 0.02787 | 0.02711 | 0.02637 | 0.02566 | 0.02496 | 0.02428 | 0.02362 |
| **30** | 0.02902 | 0.02823 | 0.02747 | 0.02673 | 0.02600 | 0.02530 | 0.02462 | 0.02395 | 0.02330 |
| **31** | 0.02860 | 0.02783 | 0.02709 | 0.02636 | 0.02565 | 0.02496 | 0.02429 | 0.02363 | 0.02300 |
| **32** | 0.02820 | 0.02745 | 0.02672 | 0.02600 | 0.02531 | 0.02463 | 0.02397 | 0.02333 | 0.02270 |
| **33** | 0.02782 | 0.02708 | 0.02636 | 0.02566 | 0.02498 | 0.02431 | 0.02366 | 0.02303 | 0.02242 |
| **34** | 0.02745 | 0.02673 | 0.02602 | 0.02533 | 0.02466 | 0.02401 | 0.02337 | 0.02275 | 0.02215 |
| **35** | 0.02710 | 0.02638 | 0.02569 | 0.02501 | 0.02435 | 0.02371 | 0.02309 | 0.02248 | 0.02188 |
| **36** | 0.02675 | 0.02605 | 0.02537 | 0.02471 | 0.02406 | 0.02343 | 0.02281 | 0.02221 | 0.02163 |
| **37** | 0.02642 | 0.02574 | 0.02506 | 0.02441 | 0.02377 | 0.02315 | 0.02255 | 0.02196 | 0.02138 |
| **38** | 0.02611 | 0.02543 | 0.02477 | 0.02413 | 0.02350 | 0.02289 | 0.02229 | 0.02171 | 0.02115 |
| **39** | 0.02580 | 0.02514 | 0.02449 | 0.02385 | 0.02323 | 0.02263 | 0.02205 | 0.02147 | 0.02092 |
| **40** | 0.02551 | 0.02485 | 0.02421 | 0.02359 | 0.02298 | 0.02239 | 0.02181 | 0.02125 | 0.02070 |

*Source*: Based on Hamme and Emerson (2004) and Eq. (1.16).

**Table 2.30** Bunsen Coefficient ($\beta$) for Argon as a Function of Temperature and Salinity, 33–37 g/kg (L real gas/(L atm))

| Temperature (°C) | Salinity (g/kg) | | | | | | | | |
|---|---|---|---|---|---|---|---|---|---|
| | **33.0** | **33.5** | **34.0** | **34.5** | **35.0** | **35.5** | **36.0** | **36.5** | **37.0** |
| **0** | 0.04279 | 0.04264 | 0.04249 | 0.04235 | 0.04220 | 0.04205 | 0.04191 | 0.04176 | 0.04162 |
| **1** | 0.04172 | 0.04158 | 0.04144 | 0.04129 | 0.04115 | 0.04101 | 0.04087 | 0.04073 | 0.04059 |
| **2** | 0.04070 | 0.04056 | 0.04043 | 0.04029 | 0.04015 | 0.04001 | 0.03988 | 0.03974 | 0.03961 |
| **3** | 0.03973 | 0.03959 | 0.03946 | 0.03933 | 0.03919 | 0.03906 | 0.03893 | 0.03880 | 0.03867 |
| **4** | 0.03879 | 0.03867 | 0.03854 | 0.03841 | 0.03828 | 0.03815 | 0.03802 | 0.03790 | 0.03777 |
| **5** | 0.03790 | 0.03778 | 0.03765 | 0.03753 | 0.03740 | 0.03728 | 0.03716 | 0.03703 | 0.03691 |
| **6** | 0.03705 | 0.03693 | 0.03681 | 0.03668 | 0.03656 | 0.03644 | 0.03632 | 0.03621 | 0.03609 |
| **7** | 0.03623 | 0.03611 | 0.03599 | 0.03588 | 0.03576 | 0.03564 | 0.03553 | 0.03541 | 0.03530 |
| **8** | 0.03545 | 0.03533 | 0.03522 | 0.03510 | 0.03499 | 0.03488 | 0.03477 | 0.03465 | 0.03454 |
| **9** | 0.03469 | 0.03458 | 0.03447 | 0.03436 | 0.03425 | 0.03414 | 0.03403 | 0.03393 | 0.03382 |
| **10** | 0.03397 | 0.03387 | 0.03376 | 0.03365 | 0.03354 | 0.03344 | 0.03333 | 0.03323 | 0.03312 |
| **11** | 0.03328 | 0.03318 | 0.03307 | 0.03297 | 0.03287 | 0.03276 | 0.03266 | 0.03256 | 0.03245 |
| **12** | 0.03262 | 0.03252 | 0.03242 | 0.03231 | 0.03221 | 0.03211 | 0.03201 | 0.03191 | 0.03181 |
| **13** | 0.03198 | 0.03188 | 0.03178 | 0.03169 | 0.03159 | 0.03149 | 0.03139 | 0.03130 | 0.03120 |
| **14** | 0.03137 | 0.03127 | 0.03118 | 0.03108 | 0.03099 | 0.03089 | 0.03080 | 0.03070 | 0.03061 |
| **15** | 0.03078 | 0.03069 | 0.03060 | 0.03050 | 0.03041 | 0.03032 | 0.03023 | 0.03013 | 0.03004 |
| **16** | 0.03022 | 0.03013 | 0.03004 | 0.02995 | 0.02985 | 0.02976 | 0.02968 | 0.02959 | 0.02950 |
| **17** | 0.02968 | 0.02959 | 0.02950 | 0.02941 | 0.02932 | 0.02923 | 0.02915 | 0.02906 | 0.02897 |
| **18** | 0.02915 | 0.02907 | 0.02898 | 0.02889 | 0.02881 | 0.02872 | 0.02864 | 0.02855 | 0.02847 |
| **19** | 0.02865 | 0.02857 | 0.02848 | 0.02840 | 0.02831 | 0.02823 | 0.02815 | 0.02806 | 0.02798 |
| **20** | 0.02817 | 0.02809 | 0.02800 | 0.02792 | 0.02784 | 0.02776 | 0.02768 | 0.02760 | 0.02752 |
| **21** | 0.02770 | 0.02762 | 0.02754 | 0.02746 | 0.02738 | 0.02730 | 0.02722 | 0.02714 | 0.02707 |
| **22** | 0.02726 | 0.02718 | 0.02710 | 0.02702 | 0.02694 | 0.02686 | 0.02679 | 0.02671 | 0.02663 |
| **23** | 0.02682 | 0.02675 | 0.02667 | 0.02659 | 0.02652 | 0.02644 | 0.02637 | 0.02629 | 0.02622 |
| **24** | 0.02641 | 0.02633 | 0.02626 | 0.02618 | 0.02611 | 0.02604 | 0.02596 | 0.02589 | 0.02581 |
| **25** | 0.02601 | 0.02594 | 0.02586 | 0.02579 | 0.02572 | 0.02564 | 0.02557 | 0.02550 | 0.02543 |
| **26** | 0.02562 | 0.02555 | 0.02548 | 0.02541 | 0.02534 | 0.02527 | 0.02519 | 0.02512 | 0.02505 |
| **27** | 0.02525 | 0.02518 | 0.02511 | 0.02504 | 0.02497 | 0.02490 | 0.02483 | 0.02476 | 0.02469 |
| **28** | 0.02489 | 0.02482 | 0.02476 | 0.02469 | 0.02462 | 0.02455 | 0.02448 | 0.02441 | 0.02435 |
| **29** | 0.02455 | 0.02448 | 0.02441 | 0.02435 | 0.02428 | 0.02421 | 0.02414 | 0.02408 | 0.02401 |
| **30** | 0.02421 | 0.02415 | 0.02408 | 0.02402 | 0.02395 | 0.02388 | 0.02382 | 0.02375 | 0.02369 |
| **31** | 0.02389 | 0.02383 | 0.02376 | 0.02370 | 0.02363 | 0.02357 | 0.02351 | 0.02344 | 0.02338 |
| **32** | 0.02358 | 0.02352 | 0.02346 | 0.02339 | 0.02333 | 0.02327 | 0.02320 | 0.02314 | 0.02308 |
| **33** | 0.02328 | 0.02322 | 0.02316 | 0.02310 | 0.02303 | 0.02297 | 0.02291 | 0.02285 | 0.02279 |
| **34** | 0.02300 | 0.02293 | 0.02287 | 0.02281 | 0.02275 | 0.02269 | 0.02263 | 0.02257 | 0.02251 |
| **35** | 0.02272 | 0.02266 | 0.02260 | 0.02254 | 0.02248 | 0.02242 | 0.02236 | 0.02230 | 0.02224 |
| **36** | 0.02245 | 0.02239 | 0.02233 | 0.02227 | 0.02221 | 0.02215 | 0.02209 | 0.02204 | 0.02198 |
| **37** | 0.02219 | 0.02213 | 0.02207 | 0.02202 | 0.02196 | 0.02190 | 0.02184 | 0.02178 | 0.02173 |
| **38** | 0.02194 | 0.02188 | 0.02183 | 0.02177 | 0.02171 | 0.02165 | 0.02160 | 0.02154 | 0.02148 |
| **39** | 0.02170 | 0.02164 | 0.02159 | 0.02153 | 0.02147 | 0.02142 | 0.02136 | 0.02131 | 0.02125 |
| **40** | 0.02147 | 0.02141 | 0.02136 | 0.02130 | 0.02125 | 0.02119 | 0.02113 | 0.02108 | 0.02102 |

*Source*: Based on Hamme and Emerson (2004) and Eq. (1.16).

**Table 2.31** Bunsen Coefficient ($\beta$) for Carbon Dioxide as a Function of Temperature and Salinity, 0–40 g/kg (L real gas/(L atm))

| Temperature (°C) | Salinity (g/kg) | | | | | | | | |
|---|---|---|---|---|---|---|---|---|---|
| | **0.0** | **5.0** | **10.0** | **15.0** | **20.0** | **25.0** | **30.0** | **35.0** | **40.0** |
| **0** | 1.7272 | 1.6827 | 1.6395 | 1.5973 | 1.5562 | 1.5162 | 1.4772 | 1.4392 | 1.4022 |
| **1** | 1.6604 | 1.6179 | 1.5764 | 1.5360 | 1.4967 | 1.4583 | 1.4209 | 1.3845 | 1.3490 |
| **2** | 1.5972 | 1.5564 | 1.5167 | 1.4780 | 1.4402 | 1.4034 | 1.3676 | 1.3327 | 1.2987 |
| **3** | 1.5373 | 1.4982 | 1.4601 | 1.4229 | 1.3867 | 1.3515 | 1.3171 | 1.2836 | 1.2509 |
| **4** | 1.4805 | 1.4430 | 1.4064 | 1.3708 | 1.3360 | 1.3022 | 1.2692 | 1.2370 | 1.2057 |
| **5** | 1.4265 | 1.3906 | 1.3555 | 1.3213 | 1.2879 | 1.2554 | 1.2238 | 1.1929 | 1.1628 |
| **6** | 1.3753 | 1.3408 | 1.3071 | 1.2743 | 1.2423 | 1.2111 | 1.1807 | 1.1510 | 1.1221 |
| **7** | 1.3267 | 1.2935 | 1.2612 | 1.2297 | 1.1989 | 1.1689 | 1.1397 | 1.1112 | 1.0834 |
| **8** | 1.2805 | 1.2486 | 1.2175 | 1.1872 | 1.1577 | 1.1289 | 1.1008 | 1.0734 | 1.0467 |
| **9** | 1.2365 | 1.2059 | 1.1760 | 1.1469 | 1.1185 | 1.0909 | 1.0639 | 1.0375 | 1.0118 |
| **10** | 1.1947 | 1.1652 | 1.1366 | 1.1086 | 1.0813 | 1.0547 | 1.0287 | 1.0034 | 0.9787 |
| **11** | 1.1548 | 1.1266 | 1.0990 | 1.0721 | 1.0458 | 1.0202 | 0.9953 | 0.9709 | 0.9471 |
| **12** | 1.1169 | 1.0897 | 1.0632 | 1.0373 | 1.0121 | 0.9875 | 0.9634 | 0.9400 | 0.9171 |
| **13** | 1.0808 | 1.0547 | 1.0291 | 1.0042 | 0.9799 | 0.9562 | 0.9331 | 0.9105 | 0.8885 |
| **14** | 1.0464 | 1.0212 | 0.9967 | 0.9727 | 0.9493 | 0.9265 | 0.9042 | 0.8825 | 0.8613 |
| **15** | 1.0135 | 0.9893 | 0.9657 | 0.9426 | 0.9201 | 0.8981 | 0.8767 | 0.8558 | 0.8353 |
| **16** | 0.9822 | 0.9589 | 0.9362 | 0.9140 | 0.8923 | 0.8711 | 0.8504 | 0.8303 | 0.8106 |
| **17** | 0.9523 | 0.9299 | 0.9080 | 0.8866 | 0.8657 | 0.8453 | 0.8254 | 0.8060 | 0.7870 |
| **18** | 0.9238 | 0.9022 | 0.8811 | 0.8605 | 0.8404 | 0.8207 | 0.8015 | 0.7828 | 0.7645 |
| **19** | 0.8965 | 0.8757 | 0.8554 | 0.8355 | 0.8161 | 0.7972 | 0.7787 | 0.7606 | 0.7430 |
| **20** | 0.8705 | 0.8504 | 0.8308 | 0.8117 | 0.7930 | 0.7747 | 0.7569 | 0.7395 | 0.7224 |
| **21** | 0.8455 | 0.8262 | 0.8074 | 0.7889 | 0.7709 | 0.7533 | 0.7361 | 0.7193 | 0.7028 |
| **22** | 0.8217 | 0.8031 | 0.7849 | 0.7671 | 0.7498 | 0.7328 | 0.7162 | 0.7000 | 0.6841 |
| **23** | 0.7989 | 0.7810 | 0.7634 | 0.7463 | 0.7295 | 0.7132 | 0.6971 | 0.6815 | 0.6662 |
| **24** | 0.7771 | 0.7598 | 0.7429 | 0.7264 | 0.7102 | 0.6944 | 0.6789 | 0.6638 | 0.6491 |
| **25** | 0.7562 | 0.7395 | 0.7232 | 0.7073 | 0.6917 | 0.6764 | 0.6615 | 0.6469 | 0.6327 |
| **26** | 0.7362 | 0.7201 | 0.7044 | 0.6890 | 0.6740 | 0.6592 | 0.6448 | 0.6308 | 0.6170 |
| **27** | 0.7170 | 0.7015 | 0.6863 | 0.6715 | 0.6570 | 0.6428 | 0.6289 | 0.6153 | 0.6020 |
| **28** | 0.6986 | 0.6837 | 0.6690 | 0.6547 | 0.6407 | 0.6270 | 0.6136 | 0.6005 | 0.5876 |
| **29** | 0.6810 | 0.6666 | 0.6525 | 0.6386 | 0.6251 | 0.6119 | 0.5989 | 0.5863 | 0.5738 |
| **30** | 0.6641 | 0.6502 | 0.6366 | 0.6232 | 0.6102 | 0.5974 | 0.5849 | 0.5726 | 0.5607 |
| **31** | 0.6478 | 0.6344 | 0.6213 | 0.6084 | 0.5958 | 0.5835 | 0.5714 | 0.5596 | 0.5480 |
| **32** | 0.6323 | 0.6193 | 0.6067 | 0.5942 | 0.5821 | 0.5702 | 0.5585 | 0.5471 | 0.5359 |
| **33** | 0.6173 | 0.6048 | 0.5926 | 0.5806 | 0.5689 | 0.5574 | 0.5461 | 0.5351 | 0.5243 |
| **34** | 0.6029 | 0.5909 | 0.5791 | 0.5676 | 0.5562 | 0.5451 | 0.5342 | 0.5236 | 0.5131 |
| **35** | 0.5891 | 0.5775 | 0.5662 | 0.5550 | 0.5441 | 0.5333 | 0.5228 | 0.5125 | 0.5024 |
| **36** | 0.5759 | 0.5647 | 0.5537 | 0.5429 | 0.5324 | 0.5220 | 0.5119 | 0.5019 | 0.4922 |
| **37** | 0.5631 | 0.5523 | 0.5417 | 0.5313 | 0.5211 | 0.5112 | 0.5013 | 0.4917 | 0.4823 |
| **38** | 0.5509 | 0.5404 | 0.5302 | 0.5202 | 0.5104 | 0.5007 | 0.4912 | 0.4820 | 0.4728 |
| **39** | 0.5391 | 0.5290 | 0.5192 | 0.5095 | 0.5000 | 0.4907 | 0.4815 | 0.4726 | 0.4638 |
| **40** | 0.5277 | 0.5180 | 0.5085 | 0.4992 | 0.4900 | 0.4810 | 0.4722 | 0.4635 | 0.4550 |

*Source*: Based on Weiss (1974).

**Table 2.32** Bunsen Coefficient ($\beta$) for Carbon Dioxide as a Function of Temperature and Salinity, 33–37 g/kg (L real gas/(L atm))

| Temperature (°C) | Salinity (g/kg) | | | | | | | | |
|---|---|---|---|---|---|---|---|---|---|
| | 33.0 | 33.5 | 34.0 | 34.5 | 35.0 | 35.5 | 36.0 | 36.5 | 37.0 |
| 0 | 1.4543 | 1.4505 | 1.4468 | 1.4430 | 1.4392 | 1.4355 | 1.4318 | 1.4280 | 1.4243 |
| 1 | 1.3990 | 1.3953 | 1.3917 | 1.3881 | 1.3845 | 1.3809 | 1.3773 | 1.3738 | 1.3702 |
| 2 | 1.3465 | 1.3431 | 1.3396 | 1.3361 | 1.3327 | 1.3292 | 1.3258 | 1.3224 | 1.3190 |
| 3 | 1.2969 | 1.2935 | 1.2902 | 1.2869 | 1.2836 | 1.2803 | 1.2770 | 1.2737 | 1.2704 |
| 4 | 1.2498 | 1.2466 | 1.2434 | 1.2402 | 1.2370 | 1.2339 | 1.2307 | 1.2276 | 1.2244 |
| 5 | 1.2052 | 1.2021 | 1.1990 | 1.1959 | 1.1929 | 1.1899 | 1.1868 | 1.1838 | 1.1808 |
| 6 | 1.1628 | 1.1598 | 1.1569 | 1.1539 | 1.1510 | 1.1481 | 1.1452 | 1.1422 | 1.1393 |
| 7 | 1.1225 | 1.1197 | 1.1169 | 1.1140 | 1.1112 | 1.1084 | 1.1056 | 1.1028 | 1.1000 |
| 8 | 1.0843 | 1.0816 | 1.0789 | 1.0761 | 1.0734 | 1.0707 | 1.0680 | 1.0654 | 1.0627 |
| 9 | 1.0480 | 1.0454 | 1.0427 | 1.0401 | 1.0375 | 1.0349 | 1.0323 | 1.0298 | 1.0272 |
| 10 | 1.0134 | 1.0109 | 1.0084 | 1.0059 | 1.0034 | 1.0009 | 0.9984 | 0.9959 | 0.9934 |
| 11 | 0.9806 | 0.9781 | 0.9757 | 0.9733 | 0.9709 | 0.9685 | 0.9661 | 0.9637 | 0.9613 |
| 12 | 0.9493 | 0.9469 | 0.9446 | 0.9423 | 0.9400 | 0.9377 | 0.9353 | 0.9330 | 0.9308 |
| 13 | 0.9195 | 0.9172 | 0.9150 | 0.9128 | 0.9105 | 0.9083 | 0.9061 | 0.9039 | 0.9017 |
| 14 | 0.8911 | 0.8889 | 0.8868 | 0.8846 | 0.8825 | 0.8803 | 0.8782 | 0.8761 | 0.8739 |
| 15 | 0.8641 | 0.8620 | 0.8599 | 0.8578 | 0.8558 | 0.8537 | 0.8516 | 0.8496 | 0.8475 |
| 16 | 0.8383 | 0.8363 | 0.8343 | 0.8323 | 0.8303 | 0.8283 | 0.8263 | 0.8243 | 0.8223 |
| 17 | 0.8137 | 0.8117 | 0.8098 | 0.8079 | 0.8060 | 0.8040 | 0.8021 | 0.8002 | 0.7983 |
| 18 | 0.7902 | 0.7883 | 0.7865 | 0.7846 | 0.7828 | 0.7809 | 0.7791 | 0.7772 | 0.7754 |
| 19 | 0.7678 | 0.7660 | 0.7642 | 0.7624 | 0.7606 | 0.7588 | 0.7571 | 0.7553 | 0.7535 |
| 20 | 0.7464 | 0.7447 | 0.7429 | 0.7412 | 0.7395 | 0.7377 | 0.7360 | 0.7343 | 0.7326 |
| 21 | 0.7259 | 0.7243 | 0.7226 | 0.7209 | 0.7193 | 0.7176 | 0.7159 | 0.7143 | 0.7126 |
| 22 | 0.7064 | 0.7048 | 0.7032 | 0.7016 | 0.7000 | 0.6984 | 0.6968 | 0.6952 | 0.6936 |
| 23 | 0.6877 | 0.6862 | 0.6846 | 0.6830 | 0.6815 | 0.6799 | 0.6784 | 0.6769 | 0.6753 |
| 24 | 0.6698 | 0.6683 | 0.6668 | 0.6653 | 0.6638 | 0.6623 | 0.6609 | 0.6594 | 0.6579 |
| 25 | 0.6527 | 0.6513 | 0.6498 | 0.6484 | 0.6469 | 0.6455 | 0.6441 | 0.6426 | 0.6412 |
| 26 | 0.6364 | 0.6350 | 0.6336 | 0.6322 | 0.6308 | 0.6294 | 0.6280 | 0.6266 | 0.6252 |
| 27 | 0.6207 | 0.6193 | 0.6180 | 0.6166 | 0.6153 | 0.6139 | 0.6126 | 0.6113 | 0.6099 |
| 28 | 0.6057 | 0.6044 | 0.6031 | 0.6018 | 0.6005 | 0.5992 | 0.5979 | 0.5966 | 0.5953 |
| 29 | 0.5913 | 0.5900 | 0.5888 | 0.5875 | 0.5863 | 0.5850 | 0.5838 | 0.5825 | 0.5813 |
| 30 | 0.5775 | 0.5763 | 0.5751 | 0.5739 | 0.5726 | 0.5714 | 0.5702 | 0.5690 | 0.5678 |
| 31 | 0.5643 | 0.5631 | 0.5619 | 0.5608 | 0.5596 | 0.5584 | 0.5573 | 0.5561 | 0.5549 |
| 32 | 0.5516 | 0.5505 | 0.5493 | 0.5482 | 0.5471 | 0.5460 | 0.5448 | 0.5437 | 0.5426 |
| 33 | 0.5395 | 0.5384 | 0.5373 | 0.5362 | 0.5351 | 0.5340 | 0.5329 | 0.5318 | 0.5307 |
| 34 | 0.5278 | 0.5267 | 0.5257 | 0.5246 | 0.5236 | 0.5225 | 0.5215 | 0.5204 | 0.5194 |
| 35 | 0.5166 | 0.5156 | 0.5146 | 0.5135 | 0.5125 | 0.5115 | 0.5105 | 0.5095 | 0.5085 |
| 36 | 0.5059 | 0.5049 | 0.5039 | 0.5029 | 0.5019 | 0.5009 | 0.4999 | 0.4990 | 0.4980 |
| 37 | 0.4956 | 0.4946 | 0.4936 | 0.4927 | 0.4917 | 0.4908 | 0.4898 | 0.4889 | 0.4879 |
| 38 | 0.4857 | 0.4847 | 0.4838 | 0.4829 | 0.4820 | 0.4810 | 0.4801 | 0.4792 | 0.4783 |
| 39 | 0.4761 | 0.4752 | 0.4743 | 0.4735 | 0.4726 | 0.4717 | 0.4708 | 0.4699 | 0.4690 |
| 40 | 0.4670 | 0.4661 | 0.4653 | 0.4644 | 0.4635 | 0.4627 | 0.4618 | 0.4610 | 0.4601 |

*Source*: Based on Weiss (1974).

**Table 2.33** Bunsen Coefficient ($\beta'$) for Oxygen as a Function of Temperature and Salinity, 0−40 g/kg (mg/(L mmHg))

| Temperature (°C) | Salinity (g/kg) | | | | | | | | |
|---|---|---|---|---|---|---|---|---|---|
| | **0.0** | **5.0** | **10.0** | **15.0** | **20.0** | **25.0** | **30.0** | **35.0** | **40.0** |
| **0** | 0.09240 | 0.08923 | 0.08617 | 0.08321 | 0.08035 | 0.07758 | 0.07491 | 0.07232 | 0.06983 |
| **1** | 0.08988 | 0.08683 | 0.08388 | 0.08102 | 0.07826 | 0.07559 | 0.07301 | 0.07051 | 0.06810 |
| **2** | 0.08748 | 0.08454 | 0.08169 | 0.07893 | 0.07626 | 0.07368 | 0.07119 | 0.06878 | 0.06645 |
| **3** | 0.08519 | 0.08235 | 0.07960 | 0.07693 | 0.07436 | 0.07187 | 0.06946 | 0.06713 | 0.06487 |
| **4** | 0.08300 | 0.08026 | 0.07760 | 0.07503 | 0.07254 | 0.07013 | 0.06780 | 0.06554 | 0.06336 |
| **5** | 0.08091 | 0.07826 | 0.07569 | 0.07321 | 0.07080 | 0.06847 | 0.06621 | 0.06403 | 0.06192 |
| **6** | 0.07892 | 0.07635 | 0.07387 | 0.07146 | 0.06913 | 0.06688 | 0.06469 | 0.06258 | 0.06054 |
| **7** | 0.07701 | 0.07453 | 0.07213 | 0.06980 | 0.06754 | 0.06536 | 0.06324 | 0.06119 | 0.05921 |
| **8** | 0.07519 | 0.07279 | 0.07046 | 0.06820 | 0.06602 | 0.06390 | 0.06185 | 0.05986 | 0.05794 |
| **9** | 0.07344 | 0.07112 | 0.06886 | 0.06668 | 0.06456 | 0.06250 | 0.06052 | 0.05859 | 0.05672 |
| **10** | 0.07177 | 0.06952 | 0.06733 | 0.06521 | 0.06316 | 0.06117 | 0.05924 | 0.05737 | 0.05555 |
| **11** | 0.07018 | 0.06799 | 0.06587 | 0.06381 | 0.06182 | 0.05989 | 0.05801 | 0.05619 | 0.05443 |
| **12** | 0.06865 | 0.06652 | 0.06447 | 0.06247 | 0.06053 | 0.05866 | 0.05683 | 0.05507 | 0.05336 |
| **13** | 0.06718 | 0.06512 | 0.06312 | 0.06118 | 0.05930 | 0.05748 | 0.05571 | 0.05399 | 0.05233 |
| **14** | 0.06577 | 0.06377 | 0.06183 | 0.05995 | 0.05812 | 0.05634 | 0.05462 | 0.05295 | 0.05133 |
| **15** | 0.06443 | 0.06248 | 0.06059 | 0.05876 | 0.05698 | 0.05526 | 0.05358 | 0.05196 | 0.05038 |
| **16** | 0.06313 | 0.06124 | 0.05941 | 0.05762 | 0.05589 | 0.05421 | 0.05258 | 0.05100 | 0.04947 |
| **17** | 0.06189 | 0.06005 | 0.05827 | 0.05653 | 0.05485 | 0.05321 | 0.05162 | 0.05008 | 0.04859 |
| **18** | 0.06070 | 0.05891 | 0.05717 | 0.05548 | 0.05384 | 0.05225 | 0.05070 | 0.04920 | 0.04774 |
| **19** | 0.05956 | 0.05782 | 0.05612 | 0.05448 | 0.05288 | 0.05132 | 0.04981 | 0.04835 | 0.04693 |
| **20** | 0.05846 | 0.05676 | 0.05511 | 0.05351 | 0.05195 | 0.05043 | 0.04896 | 0.04753 | 0.04614 |
| **21** | 0.05741 | 0.05575 | 0.05414 | 0.05257 | 0.05105 | 0.04958 | 0.04814 | 0.04674 | 0.04539 |
| **22** | 0.05639 | 0.05478 | 0.05321 | 0.05168 | 0.05019 | 0.04875 | 0.04735 | 0.04599 | 0.04466 |
| **23** | 0.05542 | 0.05384 | 0.05231 | 0.05082 | 0.04937 | 0.04796 | 0.04659 | 0.04526 | 0.04396 |
| **24** | 0.05448 | 0.05294 | 0.05144 | 0.04999 | 0.04857 | 0.04720 | 0.04586 | 0.04456 | 0.04329 |
| **25** | 0.05358 | 0.05208 | 0.05061 | 0.04919 | 0.04781 | 0.04646 | 0.04515 | 0.04388 | 0.04264 |
| **26** | 0.05271 | 0.05124 | 0.04981 | 0.04842 | 0.04707 | 0.04575 | 0.04447 | 0.04323 | 0.04202 |
| **27** | 0.05188 | 0.05044 | 0.04904 | 0.04768 | 0.04636 | 0.04507 | 0.04382 | 0.04260 | 0.04142 |
| **28** | 0.05107 | 0.04967 | 0.04830 | 0.04697 | 0.04568 | 0.04442 | 0.04319 | 0.04200 | 0.04083 |
| **29** | 0.05030 | 0.04893 | 0.04759 | 0.04628 | 0.04502 | 0.04378 | 0.04258 | 0.04141 | 0.04027 |
| **30** | 0.04955 | 0.04821 | 0.04690 | 0.04562 | 0.04438 | 0.04317 | 0.04200 | 0.04085 | 0.03973 |
| **31** | 0.04884 | 0.04752 | 0.04624 | 0.04499 | 0.04377 | 0.04258 | 0.04143 | 0.04031 | 0.03921 |
| **32** | 0.04814 | 0.04685 | 0.04560 | 0.04437 | 0.04318 | 0.04202 | 0.04089 | 0.03978 | 0.03871 |
| **33** | 0.04748 | 0.04621 | 0.04498 | 0.04378 | 0.04261 | 0.04147 | 0.04036 | 0.03928 | 0.03823 |
| **34** | 0.04684 | 0.04560 | 0.04439 | 0.04321 | 0.04206 | 0.04094 | 0.03985 | 0.03879 | 0.03776 |
| **35** | 0.04622 | 0.04500 | 0.04381 | 0.04266 | 0.04153 | 0.04043 | 0.03936 | 0.03832 | 0.03731 |
| **36** | 0.04562 | 0.04443 | 0.04326 | 0.04213 | 0.04102 | 0.03994 | 0.03889 | 0.03787 | 0.03687 |
| **37** | 0.04505 | 0.04387 | 0.04273 | 0.04161 | 0.04053 | 0.03947 | 0.03844 | 0.03743 | 0.03645 |
| **38** | 0.04449 | 0.04334 | 0.04222 | 0.04112 | 0.04005 | 0.03901 | 0.03800 | 0.03701 | 0.03604 |
| **39** | 0.04396 | 0.04282 | 0.04172 | 0.04064 | 0.03959 | 0.03857 | 0.03757 | 0.03660 | 0.03565 |
| **40** | 0.04344 | 0.04233 | 0.04124 | 0.04018 | 0.03915 | 0.03814 | 0.03716 | 0.03621 | 0.03527 |

*Source*: Based on Benson and Krause (1980) and Eq. (1.16).

**Table 2.34** Bunsen Coefficient ($\beta'$) for Oxygen as a Function of Temperature and Salinity, 33–37 g/kg (mg/(L mmHg))

| Temperature (°C) | Salinity (g/kg) | | | | | | | | |
|---|---|---|---|---|---|---|---|---|---|
| | **33.0** | **33.5** | **34.0** | **34.5** | **35.0** | **35.5** | **36.0** | **36.5** | **37.0** |
| **0** | 0.07334 | 0.07309 | 0.07283 | 0.07258 | 0.07232 | 0.07207 | 0.07182 | 0.07156 | 0.07131 |
| **1** | 0.07150 | 0.07125 | 0.07100 | 0.07076 | 0.07051 | 0.07027 | 0.07002 | 0.06978 | 0.06954 |
| **2** | 0.06974 | 0.06950 | 0.06926 | 0.06902 | 0.06878 | 0.06854 | 0.06831 | 0.06807 | 0.06784 |
| **3** | 0.06805 | 0.06782 | 0.06759 | 0.06736 | 0.06713 | 0.06690 | 0.06667 | 0.06644 | 0.06622 |
| **4** | 0.06644 | 0.06621 | 0.06599 | 0.06577 | 0.06554 | 0.06532 | 0.06510 | 0.06488 | 0.06466 |
| **5** | 0.06489 | 0.06468 | 0.06446 | 0.06425 | 0.06403 | 0.06382 | 0.06360 | 0.06339 | 0.06318 |
| **6** | 0.06342 | 0.06321 | 0.06300 | 0.06279 | 0.06258 | 0.06237 | 0.06217 | 0.06196 | 0.06175 |
| **7** | 0.06200 | 0.06180 | 0.06160 | 0.06139 | 0.06119 | 0.06099 | 0.06079 | 0.06059 | 0.06039 |
| **8** | 0.06065 | 0.06045 | 0.06026 | 0.06006 | 0.05986 | 0.05967 | 0.05947 | 0.05928 | 0.05909 |
| **9** | 0.05935 | 0.05916 | 0.05897 | 0.05878 | 0.05859 | 0.05840 | 0.05821 | 0.05802 | 0.05783 |
| **10** | 0.05811 | 0.05792 | 0.05774 | 0.05755 | 0.05737 | 0.05718 | 0.05700 | 0.05682 | 0.05663 |
| **11** | 0.05691 | 0.05673 | 0.05655 | 0.05637 | 0.05619 | 0.05602 | 0.05584 | 0.05566 | 0.05548 |
| **12** | 0.05577 | 0.05559 | 0.05542 | 0.05524 | 0.05507 | 0.05490 | 0.05472 | 0.05455 | 0.05438 |
| **13** | 0.05467 | 0.05450 | 0.05433 | 0.05416 | 0.05399 | 0.05382 | 0.05365 | 0.05349 | 0.05332 |
| **14** | 0.05362 | 0.05345 | 0.05328 | 0.05312 | 0.05295 | 0.05279 | 0.05263 | 0.05246 | 0.05230 |
| **15** | 0.05260 | 0.05244 | 0.05228 | 0.05212 | 0.05196 | 0.05180 | 0.05164 | 0.05148 | 0.05132 |
| **16** | 0.05163 | 0.05147 | 0.05131 | 0.05116 | 0.05100 | 0.05085 | 0.05069 | 0.05054 | 0.05038 |
| **17** | 0.05069 | 0.05054 | 0.05039 | 0.05023 | 0.05008 | 0.04993 | 0.04978 | 0.04963 | 0.04948 |
| **18** | 0.04979 | 0.04964 | 0.04950 | 0.04935 | 0.04920 | 0.04905 | 0.04890 | 0.04876 | 0.04861 |
| **19** | 0.04893 | 0.04878 | 0.04864 | 0.04849 | 0.04835 | 0.04820 | 0.04806 | 0.04792 | 0.04777 |
| **20** | 0.04810 | 0.04796 | 0.04781 | 0.04767 | 0.04753 | 0.04739 | 0.04725 | 0.04711 | 0.04697 |
| **21** | 0.04730 | 0.04716 | 0.04702 | 0.04688 | 0.04674 | 0.04661 | 0.04647 | 0.04633 | 0.04620 |
| **22** | 0.04653 | 0.04639 | 0.04626 | 0.04612 | 0.04599 | 0.04585 | 0.04572 | 0.04559 | 0.04545 |
| **23** | 0.04579 | 0.04565 | 0.04552 | 0.04539 | 0.04526 | 0.04513 | 0.04500 | 0.04487 | 0.04474 |
| **24** | 0.04507 | 0.04494 | 0.04481 | 0.04468 | 0.04456 | 0.04443 | 0.04430 | 0.04417 | 0.04405 |
| **25** | 0.04438 | 0.04426 | 0.04413 | 0.04401 | 0.04388 | 0.04375 | 0.04363 | 0.04350 | 0.04338 |
| **26** | 0.04372 | 0.04360 | 0.04347 | 0.04335 | 0.04323 | 0.04311 | 0.04298 | 0.04286 | 0.04274 |
| **27** | 0.04308 | 0.04296 | 0.04284 | 0.04272 | 0.04260 | 0.04248 | 0.04236 | 0.04224 | 0.04212 |
| **28** | 0.04247 | 0.04235 | 0.04223 | 0.04211 | 0.04200 | 0.04188 | 0.04176 | 0.04164 | 0.04153 |
| **29** | 0.04188 | 0.04176 | 0.04164 | 0.04153 | 0.04141 | 0.04130 | 0.04118 | 0.04107 | 0.04095 |
| **30** | 0.04130 | 0.04119 | 0.04108 | 0.04096 | 0.04085 | 0.04074 | 0.04062 | 0.04051 | 0.04040 |
| **31** | 0.04075 | 0.04064 | 0.04053 | 0.04042 | 0.04031 | 0.04020 | 0.04009 | 0.03998 | 0.03987 |
| **32** | 0.04022 | 0.04011 | 0.04000 | 0.03989 | 0.03978 | 0.03968 | 0.03957 | 0.03946 | 0.03935 |
| **33** | 0.03971 | 0.03960 | 0.03949 | 0.03939 | 0.03928 | 0.03917 | 0.03907 | 0.03896 | 0.03885 |
| **34** | 0.03921 | 0.03911 | 0.03900 | 0.03890 | 0.03879 | 0.03869 | 0.03858 | 0.03848 | 0.03838 |
| **35** | 0.03874 | 0.03863 | 0.03853 | 0.03842 | 0.03832 | 0.03822 | 0.03812 | 0.03801 | 0.03791 |
| **36** | 0.03827 | 0.03817 | 0.03807 | 0.03797 | 0.03787 | 0.03777 | 0.03767 | 0.03757 | 0.03747 |
| **37** | 0.03783 | 0.03773 | 0.03763 | 0.03753 | 0.03743 | 0.03733 | 0.03723 | 0.03713 | 0.03704 |
| **38** | 0.03740 | 0.03730 | 0.03720 | 0.03711 | 0.03701 | 0.03691 | 0.03681 | 0.03672 | 0.03662 |
| **39** | 0.03699 | 0.03689 | 0.03679 | 0.03670 | 0.03660 | 0.03650 | 0.03641 | 0.03631 | 0.03622 |
| **40** | 0.03658 | 0.03649 | 0.03639 | 0.03630 | 0.03621 | 0.03611 | 0.03602 | 0.03592 | 0.03583 |

*Source*: Based on Benson and Krause (1980) and Eq. (1.16).

**Table 2.35** Bunsen Coefficient ($\beta'$) for Nitrogen as a Function of Temperature and Salinity, 0–40 g/kg (mg/(L mmHg))

| Temperature (°C) | Salinity (g/kg) | | | | | | | | |
|---|---|---|---|---|---|---|---|---|---|
| | **0.0** | **5.0** | **10.0** | **15.0** | **20.0** | **25.0** | **30.0** | **35.0** | **40.0** |
| **0** | 0.03943 | 0.03799 | 0.03660 | 0.03526 | 0.03397 | 0.03272 | 0.03152 | 0.03037 | 0.02925 |
| **1** | 0.03843 | 0.03704 | 0.03569 | 0.03440 | 0.03315 | 0.03195 | 0.03079 | 0.02967 | 0.02859 |
| **2** | 0.03747 | 0.03613 | 0.03483 | 0.03358 | 0.03237 | 0.03120 | 0.03008 | 0.02900 | 0.02795 |
| **3** | 0.03656 | 0.03526 | 0.03400 | 0.03279 | 0.03162 | 0.03049 | 0.02941 | 0.02836 | 0.02734 |
| **4** | 0.03568 | 0.03443 | 0.03321 | 0.03204 | 0.03091 | 0.02981 | 0.02876 | 0.02774 | 0.02676 |
| **5** | 0.03485 | 0.03363 | 0.03246 | 0.03132 | 0.03022 | 0.02916 | 0.02814 | 0.02716 | 0.02620 |
| **6** | 0.03405 | 0.03287 | 0.03173 | 0.03063 | 0.02957 | 0.02854 | 0.02755 | 0.02659 | 0.02567 |
| **7** | 0.03329 | 0.03215 | 0.03104 | 0.02998 | 0.02894 | 0.02795 | 0.02699 | 0.02606 | 0.02516 |
| **8** | 0.03257 | 0.03146 | 0.03038 | 0.02935 | 0.02835 | 0.02738 | 0.02644 | 0.02554 | 0.02467 |
| **9** | 0.03187 | 0.03079 | 0.02975 | 0.02875 | 0.02777 | 0.02683 | 0.02593 | 0.02505 | 0.02420 |
| **10** | 0.03120 | 0.03016 | 0.02915 | 0.02817 | 0.02723 | 0.02631 | 0.02543 | 0.02457 | 0.02375 |
| **11** | 0.03057 | 0.02955 | 0.02857 | 0.02762 | 0.02670 | 0.02581 | 0.02495 | 0.02412 | 0.02332 |
| **12** | 0.02996 | 0.02897 | 0.02802 | 0.02709 | 0.02620 | 0.02533 | 0.02449 | 0.02369 | 0.02290 |
| **13** | 0.02938 | 0.02841 | 0.02749 | 0.02659 | 0.02571 | 0.02487 | 0.02406 | 0.02327 | 0.02251 |
| **14** | 0.02882 | 0.02788 | 0.02698 | 0.02610 | 0.02525 | 0.02443 | 0.02364 | 0.02287 | 0.02212 |
| **15** | 0.02828 | 0.02737 | 0.02649 | 0.02563 | 0.02481 | 0.02401 | 0.02323 | 0.02248 | 0.02176 |
| **16** | 0.02777 | 0.02688 | 0.02602 | 0.02519 | 0.02438 | 0.02360 | 0.02285 | 0.02211 | 0.02140 |
| **17** | 0.02728 | 0.02641 | 0.02557 | 0.02476 | 0.02397 | 0.02321 | 0.02247 | 0.02176 | 0.02107 |
| **18** | 0.02681 | 0.02596 | 0.02514 | 0.02435 | 0.02358 | 0.02284 | 0.02212 | 0.02142 | 0.02074 |
| **19** | 0.02636 | 0.02553 | 0.02473 | 0.02396 | 0.02321 | 0.02248 | 0.02177 | 0.02109 | 0.02043 |
| **20** | 0.02592 | 0.02512 | 0.02434 | 0.02358 | 0.02285 | 0.02213 | 0.02145 | 0.02078 | 0.02013 |
| **21** | 0.02551 | 0.02472 | 0.02396 | 0.02322 | 0.02250 | 0.02180 | 0.02113 | 0.02048 | 0.01984 |
| **22** | 0.02511 | 0.02434 | 0.02359 | 0.02287 | 0.02217 | 0.02149 | 0.02082 | 0.02018 | 0.01956 |
| **23** | 0.02473 | 0.02398 | 0.02324 | 0.02254 | 0.02185 | 0.02118 | 0.02053 | 0.01991 | 0.01930 |
| **24** | 0.02436 | 0.02363 | 0.02291 | 0.02221 | 0.02154 | 0.02089 | 0.02025 | 0.01964 | 0.01904 |
| **25** | 0.02401 | 0.02329 | 0.02259 | 0.02191 | 0.02125 | 0.02060 | 0.01998 | 0.01938 | 0.01879 |
| **26** | 0.02368 | 0.02297 | 0.02228 | 0.02161 | 0.02096 | 0.02033 | 0.01972 | 0.01913 | 0.01856 |
| **27** | 0.02335 | 0.02266 | 0.02198 | 0.02133 | 0.02069 | 0.02007 | 0.01947 | 0.01889 | 0.01833 |
| **28** | 0.02304 | 0.02236 | 0.02170 | 0.02105 | 0.02043 | 0.01982 | 0.01924 | 0.01866 | 0.01811 |
| **29** | 0.02275 | 0.02208 | 0.02143 | 0.02079 | 0.02018 | 0.01958 | 0.01901 | 0.01844 | 0.01790 |
| **30** | 0.02246 | 0.02180 | 0.02116 | 0.02054 | 0.01994 | 0.01935 | 0.01878 | 0.01823 | 0.01770 |
| **31** | 0.02219 | 0.02154 | 0.02091 | 0.02030 | 0.01971 | 0.01913 | 0.01857 | 0.01803 | 0.01750 |
| **32** | 0.02193 | 0.02129 | 0.02067 | 0.02007 | 0.01949 | 0.01892 | 0.01837 | 0.01783 | 0.01731 |
| **33** | 0.02168 | 0.02105 | 0.02044 | 0.01985 | 0.01927 | 0.01872 | 0.01817 | 0.01765 | 0.01713 |
| **34** | 0.02144 | 0.02082 | 0.02022 | 0.01964 | 0.01907 | 0.01852 | 0.01798 | 0.01746 | 0.01696 |
| **35** | 0.02121 | 0.02060 | 0.02001 | 0.01943 | 0.01887 | 0.01833 | 0.01780 | 0.01729 | 0.01679 |
| **36** | 0.02099 | 0.02039 | 0.01981 | 0.01924 | 0.01869 | 0.01815 | 0.01763 | 0.01713 | 0.01663 |
| **37** | 0.02078 | 0.02019 | 0.01961 | 0.01905 | 0.01851 | 0.01798 | 0.01747 | 0.01697 | 0.01648 |
| **38** | 0.02058 | 0.02000 | 0.01943 | 0.01887 | 0.01834 | 0.01781 | 0.01731 | 0.01681 | 0.01633 |
| **39** | 0.02039 | 0.01981 | 0.01925 | 0.01870 | 0.01817 | 0.01766 | 0.01715 | 0.01667 | 0.01619 |
| **40** | 0.02021 | 0.01964 | 0.01908 | 0.01854 | 0.01802 | 0.01750 | 0.01701 | 0.01653 | 0.01606 |

*Source*: Based on Hamme and Emerson (2004) and Eq. (1.16).

**Table 2.36** Bunsen Coefficient ($\beta'$) for Nitrogen as a Function of Temperature and Salinity, 33–37 g/kg (mg/(L mmHg))

| Temperature (°C) | Salinity (g/kg) | | | | | | | | |
|---|---|---|---|---|---|---|---|---|---|
| | 33.0 | 33.5 | 34.0 | 34.5 | 35.0 | 35.5 | 36.0 | 36.5 | 37.0 |
| 0 | 0.03083 | 0.03071 | 0.03060 | 0.03048 | 0.03037 | 0.03026 | 0.03014 | 0.03003 | 0.02992 |
| 1 | 0.03011 | 0.03000 | 0.02989 | 0.02978 | 0.02967 | 0.02956 | 0.02945 | 0.02934 | 0.02923 |
| 2 | 0.02943 | 0.02932 | 0.02921 | 0.02910 | 0.02900 | 0.02889 | 0.02878 | 0.02868 | 0.02857 |
| 3 | 0.02877 | 0.02867 | 0.02856 | 0.02846 | 0.02836 | 0.02825 | 0.02815 | 0.02805 | 0.02795 |
| 4 | 0.02815 | 0.02804 | 0.02794 | 0.02784 | 0.02774 | 0.02764 | 0.02754 | 0.02744 | 0.02735 |
| 5 | 0.02755 | 0.02745 | 0.02735 | 0.02725 | 0.02716 | 0.02706 | 0.02696 | 0.02687 | 0.02677 |
| 6 | 0.02697 | 0.02688 | 0.02678 | 0.02669 | 0.02659 | 0.02650 | 0.02641 | 0.02631 | 0.02622 |
| 7 | 0.02642 | 0.02633 | 0.02624 | 0.02615 | 0.02606 | 0.02597 | 0.02587 | 0.02578 | 0.02569 |
| 8 | 0.02590 | 0.02581 | 0.02572 | 0.02563 | 0.02554 | 0.02545 | 0.02536 | 0.02528 | 0.02519 |
| 9 | 0.02540 | 0.02531 | 0.02522 | 0.02513 | 0.02505 | 0.02496 | 0.02488 | 0.02479 | 0.02470 |
| 10 | 0.02491 | 0.02483 | 0.02474 | 0.02466 | 0.02457 | 0.02449 | 0.02441 | 0.02432 | 0.02424 |
| 11 | 0.02445 | 0.02437 | 0.02428 | 0.02420 | 0.02412 | 0.02404 | 0.02396 | 0.02388 | 0.02380 |
| 12 | 0.02401 | 0.02393 | 0.02385 | 0.02377 | 0.02369 | 0.02361 | 0.02353 | 0.02345 | 0.02337 |
| 13 | 0.02358 | 0.02350 | 0.02342 | 0.02335 | 0.02327 | 0.02319 | 0.02311 | 0.02304 | 0.02296 |
| 14 | 0.02317 | 0.02310 | 0.02302 | 0.02294 | 0.02287 | 0.02279 | 0.02272 | 0.02264 | 0.02257 |
| 15 | 0.02278 | 0.02271 | 0.02263 | 0.02256 | 0.02248 | 0.02241 | 0.02234 | 0.02226 | 0.02219 |
| 16 | 0.02240 | 0.02233 | 0.02226 | 0.02219 | 0.02211 | 0.02204 | 0.02197 | 0.02190 | 0.02183 |
| 17 | 0.02204 | 0.02197 | 0.02190 | 0.02183 | 0.02176 | 0.02169 | 0.02162 | 0.02155 | 0.02148 |
| 18 | 0.02170 | 0.02163 | 0.02156 | 0.02149 | 0.02142 | 0.02135 | 0.02128 | 0.02121 | 0.02115 |
| 19 | 0.02136 | 0.02129 | 0.02123 | 0.02116 | 0.02109 | 0.02102 | 0.02096 | 0.02089 | 0.02082 |
| 20 | 0.02104 | 0.02098 | 0.02091 | 0.02084 | 0.02078 | 0.02071 | 0.02065 | 0.02058 | 0.02052 |
| 21 | 0.02073 | 0.02067 | 0.02060 | 0.02054 | 0.02048 | 0.02041 | 0.02035 | 0.02028 | 0.02022 |
| 22 | 0.02044 | 0.02037 | 0.02031 | 0.02025 | 0.02018 | 0.02012 | 0.02006 | 0.02000 | 0.01993 |
| 23 | 0.02015 | 0.02009 | 0.02003 | 0.01997 | 0.01991 | 0.01984 | 0.01978 | 0.01972 | 0.01966 |
| 24 | 0.01988 | 0.01982 | 0.01976 | 0.01970 | 0.01964 | 0.01958 | 0.01952 | 0.01946 | 0.01940 |
| 25 | 0.01962 | 0.01956 | 0.01950 | 0.01944 | 0.01938 | 0.01932 | 0.01926 | 0.01920 | 0.01914 |
| 26 | 0.01937 | 0.01931 | 0.01925 | 0.01919 | 0.01913 | 0.01907 | 0.01902 | 0.01896 | 0.01890 |
| 27 | 0.01912 | 0.01907 | 0.01901 | 0.01895 | 0.01889 | 0.01884 | 0.01878 | 0.01872 | 0.01867 |
| 28 | 0.01889 | 0.01883 | 0.01878 | 0.01872 | 0.01866 | 0.01861 | 0.01855 | 0.01850 | 0.01844 |
| 29 | 0.01867 | 0.01861 | 0.01855 | 0.01850 | 0.01844 | 0.01839 | 0.01833 | 0.01828 | 0.01822 |
| 30 | 0.01845 | 0.01840 | 0.01834 | 0.01829 | 0.01823 | 0.01818 | 0.01812 | 0.01807 | 0.01802 |
| 31 | 0.01824 | 0.01819 | 0.01814 | 0.01808 | 0.01803 | 0.01798 | 0.01792 | 0.01787 | 0.01782 |
| 32 | 0.01805 | 0.01799 | 0.01794 | 0.01789 | 0.01783 | 0.01778 | 0.01773 | 0.01768 | 0.01762 |
| 33 | 0.01785 | 0.01780 | 0.01775 | 0.01770 | 0.01765 | 0.01759 | 0.01754 | 0.01749 | 0.01744 |
| 34 | 0.01767 | 0.01762 | 0.01757 | 0.01752 | 0.01746 | 0.01741 | 0.01736 | 0.01731 | 0.01726 |
| 35 | 0.01749 | 0.01744 | 0.01739 | 0.01734 | 0.01729 | 0.01724 | 0.01719 | 0.01714 | 0.01709 |
| 36 | 0.01733 | 0.01728 | 0.01723 | 0.01717 | 0.01713 | 0.01708 | 0.01703 | 0.01698 | 0.01693 |
| 37 | 0.01716 | 0.01711 | 0.01706 | 0.01701 | 0.01697 | 0.01692 | 0.01687 | 0.01682 | 0.01677 |
| 38 | 0.01701 | 0.01696 | 0.01691 | 0.01686 | 0.01681 | 0.01676 | 0.01672 | 0.01667 | 0.01662 |
| 39 | 0.01686 | 0.01681 | 0.01676 | 0.01671 | 0.01667 | 0.01662 | 0.01657 | 0.01652 | 0.01647 |
| 40 | 0.01672 | 0.01667 | 0.01662 | 0.01657 | 0.01653 | 0.01648 | 0.01643 | 0.01638 | 0.01634 |

*Source*: Based on Hamme and Emerson (2004) and Eq. (1.16).

**Table 2.37** Bunsen Coefficient ($\beta'$) for Argon as a Function of Temperature and Salinity, 0–40 g/kg (mg/(L mmHg))

| Temperature (°C) | Salinity (g/kg) | | | | | | | | |
|---|---|---|---|---|---|---|---|---|---|
| | 0.0 | 5.0 | 10.0 | 15.0 | 20.0 | 25.0 | 30.0 | 35.0 | 40.0 |
| 0 | 0.12624 | 0.12196 | 0.11781 | 0.11379 | 0.10992 | 0.10617 | 0.10255 | 0.09905 | 0.09567 |
| 1 | 0.12284 | 0.11870 | 0.11470 | 0.11083 | 0.10709 | 0.10347 | 0.09998 | 0.09660 | 0.09333 |
| 2 | 0.11959 | 0.11560 | 0.11173 | 0.10800 | 0.10438 | 0.10089 | 0.09751 | 0.09425 | 0.09109 |
| 3 | 0.11648 | 0.11263 | 0.10890 | 0.10529 | 0.10180 | 0.09843 | 0.09516 | 0.09200 | 0.08895 |
| 4 | 0.11352 | 0.10980 | 0.10620 | 0.10271 | 0.09933 | 0.09607 | 0.09291 | 0.08985 | 0.08690 |
| 5 | 0.11069 | 0.10709 | 0.10361 | 0.10024 | 0.09697 | 0.09381 | 0.09075 | 0.08780 | 0.08493 |
| 6 | 0.10799 | 0.10451 | 0.10114 | 0.09787 | 0.09471 | 0.09165 | 0.08869 | 0.08583 | 0.08305 |
| 7 | 0.10540 | 0.10204 | 0.09877 | 0.09561 | 0.09255 | 0.08959 | 0.08672 | 0.08394 | 0.08125 |
| 8 | 0.10293 | 0.09967 | 0.09651 | 0.09345 | 0.09048 | 0.08761 | 0.08483 | 0.08213 | 0.07952 |
| 9 | 0.10056 | 0.09740 | 0.09434 | 0.09138 | 0.08850 | 0.08571 | 0.08302 | 0.08040 | 0.07787 |
| 10 | 0.09830 | 0.09524 | 0.09227 | 0.08939 | 0.08660 | 0.08390 | 0.08128 | 0.07874 | 0.07628 |
| 11 | 0.09613 | 0.09316 | 0.09028 | 0.08749 | 0.08478 | 0.08216 | 0.07961 | 0.07715 | 0.07476 |
| 12 | 0.09405 | 0.09117 | 0.08837 | 0.08566 | 0.08303 | 0.08048 | 0.07801 | 0.07562 | 0.07329 |
| 13 | 0.09206 | 0.08926 | 0.08655 | 0.08391 | 0.08136 | 0.07888 | 0.07648 | 0.07415 | 0.07189 |
| 14 | 0.09015 | 0.08743 | 0.08479 | 0.08223 | 0.07975 | 0.07734 | 0.07500 | 0.07274 | 0.07054 |
| 15 | 0.08832 | 0.08568 | 0.08311 | 0.08062 | 0.07821 | 0.07586 | 0.07359 | 0.07138 | 0.06924 |
| 16 | 0.08656 | 0.08399 | 0.08150 | 0.07907 | 0.07672 | 0.07444 | 0.07223 | 0.07008 | 0.06799 |
| 17 | 0.08487 | 0.08237 | 0.07995 | 0.07759 | 0.07530 | 0.07308 | 0.07092 | 0.06883 | 0.06679 |
| 18 | 0.08326 | 0.08082 | 0.07846 | 0.07616 | 0.07393 | 0.07177 | 0.06966 | 0.06762 | 0.06564 |
| 19 | 0.08170 | 0.07933 | 0.07703 | 0.07479 | 0.07262 | 0.07050 | 0.06845 | 0.06646 | 0.06453 |
| 20 | 0.08021 | 0.07790 | 0.07565 | 0.07347 | 0.07135 | 0.06929 | 0.06729 | 0.06535 | 0.06346 |
| 21 | 0.07877 | 0.07652 | 0.07433 | 0.07220 | 0.07013 | 0.06812 | 0.06617 | 0.06427 | 0.06243 |
| 22 | 0.07740 | 0.07520 | 0.07306 | 0.07098 | 0.06896 | 0.06700 | 0.06509 | 0.06324 | 0.06144 |
| 23 | 0.07607 | 0.07393 | 0.07184 | 0.06981 | 0.06784 | 0.06592 | 0.06406 | 0.06225 | 0.06049 |
| 24 | 0.07480 | 0.07270 | 0.07066 | 0.06868 | 0.06675 | 0.06488 | 0.06306 | 0.06129 | 0.05957 |
| 25 | 0.07357 | 0.07152 | 0.06953 | 0.06759 | 0.06571 | 0.06388 | 0.06210 | 0.06036 | 0.05868 |
| 26 | 0.07239 | 0.07039 | 0.06844 | 0.06655 | 0.06470 | 0.06291 | 0.06117 | 0.05947 | 0.05782 |
| 27 | 0.07126 | 0.06930 | 0.06740 | 0.06554 | 0.06374 | 0.06198 | 0.06027 | 0.05861 | 0.05700 |
| 28 | 0.07017 | 0.06825 | 0.06639 | 0.06457 | 0.06280 | 0.06109 | 0.05941 | 0.05779 | 0.05620 |
| 29 | 0.06912 | 0.06724 | 0.06542 | 0.06364 | 0.06191 | 0.06022 | 0.05858 | 0.05699 | 0.05544 |
| 30 | 0.06811 | 0.06627 | 0.06448 | 0.06274 | 0.06104 | 0.05939 | 0.05778 | 0.05622 | 0.05470 |
| 31 | 0.06714 | 0.06534 | 0.06358 | 0.06187 | 0.06021 | 0.05859 | 0.05701 | 0.05548 | 0.05398 |
| 32 | 0.06620 | 0.06443 | 0.06271 | 0.06104 | 0.05940 | 0.05781 | 0.05627 | 0.05476 | 0.05329 |
| 33 | 0.06530 | 0.06357 | 0.06188 | 0.06023 | 0.05863 | 0.05707 | 0.05555 | 0.05407 | 0.05263 |
| 34 | 0.06444 | 0.06273 | 0.06107 | 0.05946 | 0.05788 | 0.05635 | 0.05486 | 0.05340 | 0.05199 |
| 35 | 0.06360 | 0.06193 | 0.06030 | 0.05871 | 0.05716 | 0.05566 | 0.05419 | 0.05276 | 0.05137 |
| 36 | 0.06280 | 0.06115 | 0.05955 | 0.05799 | 0.05647 | 0.05499 | 0.05354 | 0.05214 | 0.05077 |
| 37 | 0.06203 | 0.06041 | 0.05883 | 0.05730 | 0.05580 | 0.05434 | 0.05292 | 0.05154 | 0.05019 |
| 38 | 0.06128 | 0.05969 | 0.05814 | 0.05663 | 0.05516 | 0.05372 | 0.05233 | 0.05096 | 0.04964 |
| 39 | 0.06057 | 0.05900 | 0.05748 | 0.05599 | 0.05454 | 0.05313 | 0.05175 | 0.05041 | 0.04910 |
| 40 | 0.05988 | 0.05834 | 0.05684 | 0.05537 | 0.05394 | 0.05255 | 0.05119 | 0.04987 | 0.04858 |

*Source*: Based on Hamme and Emerson (2004) and Eq. (1.16).

**Table 2.38** Bunsen Coefficient ($\beta'$) for Argon as a Function of Temperature and Salinity, 33−37 g/kg (mg/(L mmHg))

| Temperature (°C) | Salinity (g/kg) | | | | | | | | |
|---|---|---|---|---|---|---|---|---|---|
| | **33.0** | **33.5** | **34.0** | **34.5** | **35.0** | **35.5** | **36.0** | **36.5** | **37.0** |
| **0** | 0.10044 | 0.10009 | 0.09974 | 0.09940 | 0.09905 | 0.09871 | 0.09837 | 0.09803 | 0.09769 |
| **1** | 0.09793 | 0.09760 | 0.09726 | 0.09693 | 0.09660 | 0.09626 | 0.09593 | 0.09560 | 0.09528 |
| **2** | 0.09554 | 0.09522 | 0.09489 | 0.09457 | 0.09425 | 0.09393 | 0.09361 | 0.09329 | 0.09297 |
| **3** | 0.09325 | 0.09294 | 0.09262 | 0.09231 | 0.09200 | 0.09169 | 0.09138 | 0.09107 | 0.09077 |
| **4** | 0.09106 | 0.09076 | 0.09046 | 0.09015 | 0.08985 | 0.08955 | 0.08925 | 0.08895 | 0.08866 |
| **5** | 0.08897 | 0.08867 | 0.08838 | 0.08809 | 0.08780 | 0.08751 | 0.08722 | 0.08693 | 0.08664 |
| **6** | 0.08696 | 0.08668 | 0.08639 | 0.08611 | 0.08583 | 0.08555 | 0.08526 | 0.08498 | 0.08471 |
| **7** | 0.08504 | 0.08477 | 0.08449 | 0.08421 | 0.08394 | 0.08367 | 0.08340 | 0.08312 | 0.08285 |
| **8** | 0.08320 | 0.08293 | 0.08267 | 0.08240 | 0.08213 | 0.08187 | 0.08161 | 0.08134 | 0.08108 |
| **9** | 0.08144 | 0.08118 | 0.08092 | 0.08066 | 0.08040 | 0.08014 | 0.07989 | 0.07963 | 0.07938 |
| **10** | 0.07975 | 0.07949 | 0.07924 | 0.07899 | 0.07874 | 0.07849 | 0.07824 | 0.07799 | 0.07775 |
| **11** | 0.07812 | 0.07788 | 0.07763 | 0.07739 | 0.07715 | 0.07690 | 0.07666 | 0.07642 | 0.07618 |
| **12** | 0.07657 | 0.07633 | 0.07609 | 0.07585 | 0.07562 | 0.07538 | 0.07515 | 0.07491 | 0.07468 |
| **13** | 0.07507 | 0.07484 | 0.07461 | 0.07438 | 0.07415 | 0.07392 | 0.07369 | 0.07346 | 0.07324 |
| **14** | 0.07364 | 0.07341 | 0.07318 | 0.07296 | 0.07274 | 0.07251 | 0.07229 | 0.07207 | 0.07185 |
| **15** | 0.07226 | 0.07204 | 0.07182 | 0.07160 | 0.07138 | 0.07116 | 0.07095 | 0.07073 | 0.07052 |
| **16** | 0.07093 | 0.07072 | 0.07050 | 0.07029 | 0.07008 | 0.06987 | 0.06966 | 0.06945 | 0.06924 |
| **17** | 0.06966 | 0.06945 | 0.06924 | 0.06903 | 0.06883 | 0.06862 | 0.06841 | 0.06821 | 0.06801 |
| **18** | 0.06843 | 0.06823 | 0.06803 | 0.06782 | 0.06762 | 0.06742 | 0.06722 | 0.06702 | 0.06682 |
| **19** | 0.06725 | 0.06705 | 0.06686 | 0.06666 | 0.06646 | 0.06627 | 0.06607 | 0.06588 | 0.06568 |
| **20** | 0.06612 | 0.06592 | 0.06573 | 0.06554 | 0.06535 | 0.06516 | 0.06497 | 0.06478 | 0.06459 |
| **21** | 0.06503 | 0.06484 | 0.06465 | 0.06446 | 0.06427 | 0.06409 | 0.06390 | 0.06372 | 0.06353 |
| **22** | 0.06398 | 0.06379 | 0.06361 | 0.06342 | 0.06324 | 0.06306 | 0.06288 | 0.06270 | 0.06251 |
| **23** | 0.06296 | 0.06278 | 0.06260 | 0.06242 | 0.06225 | 0.06207 | 0.06189 | 0.06171 | 0.06154 |
| **24** | 0.06199 | 0.06181 | 0.06164 | 0.06146 | 0.06129 | 0.06111 | 0.06094 | 0.06077 | 0.06059 |
| **25** | 0.06105 | 0.06088 | 0.06071 | 0.06053 | 0.06036 | 0.06019 | 0.06002 | 0.05985 | 0.05968 |
| **26** | 0.06015 | 0.05998 | 0.05981 | 0.05964 | 0.05947 | 0.05931 | 0.05914 | 0.05897 | 0.05881 |
| **27** | 0.05927 | 0.05911 | 0.05894 | 0.05878 | 0.05861 | 0.05845 | 0.05829 | 0.05812 | 0.05796 |
| **28** | 0.05843 | 0.05827 | 0.05811 | 0.05795 | 0.05779 | 0.05763 | 0.05747 | 0.05731 | 0.05715 |
| **29** | 0.05762 | 0.05746 | 0.05730 | 0.05715 | 0.05699 | 0.05683 | 0.05667 | 0.05652 | 0.05636 |
| **30** | 0.05684 | 0.05668 | 0.05653 | 0.05637 | 0.05622 | 0.05606 | 0.05591 | 0.05576 | 0.05560 |
| **31** | 0.05608 | 0.05593 | 0.05578 | 0.05563 | 0.05548 | 0.05532 | 0.05517 | 0.05502 | 0.05487 |
| **32** | 0.05536 | 0.05521 | 0.05506 | 0.05491 | 0.05476 | 0.05461 | 0.05446 | 0.05432 | 0.05417 |
| **33** | 0.05466 | 0.05451 | 0.05436 | 0.05421 | 0.05407 | 0.05392 | 0.05378 | 0.05363 | 0.05349 |
| **34** | 0.05398 | 0.05383 | 0.05369 | 0.05355 | 0.05340 | 0.05326 | 0.05312 | 0.05297 | 0.05283 |
| **35** | 0.05333 | 0.05318 | 0.05304 | 0.05290 | 0.05276 | 0.05262 | 0.05248 | 0.05234 | 0.05220 |
| **36** | 0.05270 | 0.05256 | 0.05242 | 0.05228 | 0.05214 | 0.05200 | 0.05186 | 0.05172 | 0.05159 |
| **37** | 0.05209 | 0.05195 | 0.05181 | 0.05168 | 0.05154 | 0.05140 | 0.05127 | 0.05113 | 0.05100 |
| **38** | 0.05150 | 0.05137 | 0.05123 | 0.05110 | 0.05096 | 0.05083 | 0.05069 | 0.05056 | 0.05043 |
| **39** | 0.05094 | 0.05081 | 0.05067 | 0.05054 | 0.05041 | 0.05027 | 0.05014 | 0.05001 | 0.04988 |
| **40** | 0.05039 | 0.05026 | 0.05013 | 0.05000 | 0.04987 | 0.04974 | 0.04961 | 0.04948 | 0.04935 |

*Source*: Based on Hamme and Emerson (2004) and Eq. (1.16).

**Table 2.39** Bunsen Coefficient ($\beta'$) for Carbon Dioxide as a Function of Temperature and Salinity, 0–40 g/kg (mg/(L mmHg))

| Temperature (°C) | Salinity (g/kg) | | | | | | | | |
|---|---|---|---|---|---|---|---|---|---|
| | 0.0 | 5.0 | 10.0 | 15.0 | 20.0 | 25.0 | 30.0 | 35.0 | 40.0 |
| 0 | 4.4924 | 4.3769 | 4.2643 | 4.1546 | 4.0478 | 3.9437 | 3.8423 | 3.7435 | 3.6472 |
| 1 | 4.3188 | 4.2082 | 4.1003 | 3.9952 | 3.8928 | 3.7931 | 3.6959 | 3.6012 | 3.5089 |
| 2 | 4.1544 | 4.0483 | 3.9450 | 3.8442 | 3.7461 | 3.6504 | 3.5572 | 3.4664 | 3.3778 |
| 3 | 3.9986 | 3.8969 | 3.7977 | 3.7011 | 3.6070 | 3.5152 | 3.4258 | 3.3386 | 3.2537 |
| 4 | 3.8508 | 3.7532 | 3.6581 | 3.5654 | 3.4751 | 3.3870 | 3.3012 | 3.2176 | 3.1361 |
| 5 | 3.7105 | 3.6169 | 3.5256 | 3.4367 | 3.3500 | 3.2654 | 3.1831 | 3.1028 | 3.0245 |
| 6 | 3.5773 | 3.4874 | 3.3998 | 3.3144 | 3.2312 | 3.1500 | 3.0709 | 2.9938 | 2.9186 |
| 7 | 3.4508 | 3.3645 | 3.2804 | 3.1984 | 3.1184 | 3.0404 | 2.9644 | 2.8903 | 2.8181 |
| 8 | 3.3305 | 3.2476 | 3.1668 | 3.0881 | 3.0112 | 2.9363 | 2.8633 | 2.7920 | 2.7226 |
| 9 | 3.2161 | 3.1365 | 3.0589 | 2.9832 | 2.9094 | 2.8373 | 2.7671 | 2.6986 | 2.6318 |
| 10 | 3.1073 | 3.0308 | 2.9562 | 2.8834 | 2.8125 | 2.7432 | 2.6757 | 2.6098 | 2.5456 |
| 11 | 3.0038 | 2.9302 | 2.8585 | 2.7885 | 2.7203 | 2.6537 | 2.5887 | 2.5253 | 2.4635 |
| 12 | 2.9052 | 2.8345 | 2.7655 | 2.6982 | 2.6325 | 2.5684 | 2.5059 | 2.4449 | 2.3854 |
| 13 | 2.8112 | 2.7432 | 2.6768 | 2.6121 | 2.5489 | 2.4872 | 2.4270 | 2.3683 | 2.3110 |
| 14 | 2.7217 | 2.6562 | 2.5924 | 2.5300 | 2.4692 | 2.4098 | 2.3519 | 2.2954 | 2.2402 |
| 15 | 2.6363 | 2.5733 | 2.5118 | 2.4518 | 2.3933 | 2.3361 | 2.2803 | 2.2258 | 2.1727 |
| 16 | 2.5548 | 2.4942 | 2.4350 | 2.3772 | 2.3208 | 2.2658 | 2.2120 | 2.1596 | 2.1083 |
| 17 | 2.4770 | 2.4187 | 2.3617 | 2.3061 | 2.2517 | 2.1987 | 2.1469 | 2.0963 | 2.0469 |
| 18 | 2.4028 | 2.3466 | 2.2917 | 2.2381 | 2.1858 | 2.1347 | 2.0847 | 2.0360 | 1.9884 |
| 19 | 2.3319 | 2.2777 | 2.2249 | 2.1732 | 2.1228 | 2.0735 | 2.0254 | 1.9784 | 1.9325 |
| 20 | 2.2641 | 2.2119 | 2.1610 | 2.1112 | 2.0626 | 2.0151 | 1.9687 | 1.9234 | 1.8791 |
| 21 | 2.1993 | 2.1490 | 2.1000 | 2.0520 | 2.0051 | 1.9593 | 1.9146 | 1.8708 | 1.8281 |
| 22 | 2.1373 | 2.0889 | 2.0416 | 1.9953 | 1.9501 | 1.9060 | 1.8628 | 1.8206 | 1.7794 |
| 23 | 2.0780 | 2.0313 | 1.9857 | 1.9411 | 1.8976 | 1.8549 | 1.8133 | 1.7726 | 1.7328 |
| 24 | 2.0212 | 1.9763 | 1.9323 | 1.8893 | 1.8472 | 1.8061 | 1.7659 | 1.7266 | 1.6882 |
| 25 | 1.9669 | 1.9235 | 1.8811 | 1.8396 | 1.7991 | 1.7594 | 1.7206 | 1.6827 | 1.6456 |
| 26 | 1.9148 | 1.8730 | 1.8321 | 1.7921 | 1.7530 | 1.7147 | 1.6773 | 1.6406 | 1.6048 |
| 27 | 1.8649 | 1.8246 | 1.7852 | 1.7466 | 1.7088 | 1.6719 | 1.6357 | 1.6004 | 1.5658 |
| 28 | 1.8171 | 1.7782 | 1.7402 | 1.7030 | 1.6665 | 1.6309 | 1.5960 | 1.5618 | 1.5284 |
| 29 | 1.7713 | 1.7338 | 1.6971 | 1.6611 | 1.6260 | 1.5915 | 1.5578 | 1.5249 | 1.4926 |
| 30 | 1.7273 | 1.6911 | 1.6557 | 1.6210 | 1.5871 | 1.5538 | 1.5213 | 1.4895 | 1.4583 |
| 31 | 1.6851 | 1.6502 | 1.6160 | 1.5826 | 1.5498 | 1.5177 | 1.4863 | 1.4555 | 1.4254 |
| 32 | 1.6445 | 1.6109 | 1.5779 | 1.5456 | 1.5140 | 1.4830 | 1.4527 | 1.4230 | 1.3939 |
| 33 | 1.6056 | 1.5732 | 1.5414 | 1.5102 | 1.4797 | 1.4498 | 1.4205 | 1.3918 | 1.3636 |
| 34 | 1.5683 | 1.5370 | 1.5063 | 1.4762 | 1.4467 | 1.4179 | 1.3896 | 1.3618 | 1.3346 |
| 35 | 1.5324 | 1.5022 | 1.4726 | 1.4436 | 1.4151 | 1.3872 | 1.3599 | 1.3331 | 1.3068 |
| 36 | 1.4979 | 1.4687 | 1.4402 | 1.4122 | 1.3847 | 1.3578 | 1.3314 | 1.3055 | 1.2801 |
| 37 | 1.4647 | 1.4366 | 1.4091 | 1.3820 | 1.3555 | 1.3295 | 1.3040 | 1.2790 | 1.2545 |
| 38 | 1.4328 | 1.4057 | 1.3791 | 1.3530 | 1.3275 | 1.3024 | 1.2777 | 1.2536 | 1.2299 |
| 39 | 1.4021 | 1.3760 | 1.3503 | 1.3252 | 1.3005 | 1.2763 | 1.2525 | 1.2292 | 1.2063 |
| 40 | 1.3726 | 1.3474 | 1.3227 | 1.2984 | 1.2746 | 1.2512 | 1.2282 | 1.2057 | 1.1836 |

*Source*: Based on Weiss (1974).

**Table 2.40** Bunsen Coefficient ($\beta'$) for Carbon Dioxide as a Function of Temperature and Salinity, 33–37 g/kg (mg/(L atm))

| Temperature (°C) | Salinity (g/kg) | | | | | | | | |
|---|---|---|---|---|---|---|---|---|---|
| | **33.0** | **33.5** | **34.0** | **34.5** | **35.0** | **35.5** | **36.0** | **36.5** | **37.0** |
| **0** | 3.7827 | 3.7729 | 3.7630 | 3.7533 | 3.7435 | 3.7338 | 3.7240 | 3.7143 | 3.7047 |
| **1** | 3.6387 | 3.6293 | 3.6199 | 3.6105 | 3.6012 | 3.5918 | 3.5825 | 3.5732 | 3.5639 |
| **2** | 3.5024 | 3.4934 | 3.4843 | 3.4753 | 3.4664 | 3.4574 | 3.4485 | 3.4396 | 3.4307 |
| **3** | 3.3732 | 3.3646 | 3.3559 | 3.3473 | 3.3386 | 3.3301 | 3.3215 | 3.3129 | 3.3044 |
| **4** | 3.2508 | 3.2424 | 3.2341 | 3.2258 | 3.2176 | 3.2093 | 3.2011 | 3.1929 | 3.1847 |
| **5** | 3.1346 | 3.1266 | 3.1186 | 3.1107 | 3.1028 | 3.0948 | 3.0869 | 3.0791 | 3.0712 |
| **6** | 3.0244 | 3.0167 | 3.0091 | 3.0014 | 2.9938 | 2.9862 | 2.9786 | 2.9710 | 2.9635 |
| **7** | 2.9197 | 2.9124 | 2.9050 | 2.8976 | 2.8903 | 2.8830 | 2.8757 | 2.8684 | 2.8612 |
| **8** | 2.8203 | 2.8132 | 2.8061 | 2.7991 | 2.7920 | 2.7850 | 2.7780 | 2.7710 | 2.7640 |
| **9** | 2.7258 | 2.7190 | 2.7122 | 2.7054 | 2.6986 | 2.6919 | 2.6851 | 2.6784 | 2.6717 |
| **10** | 2.6360 | 2.6294 | 2.6228 | 2.6163 | 2.6098 | 2.6033 | 2.5968 | 2.5904 | 2.5839 |
| **11** | 2.5505 | 2.5442 | 2.5379 | 2.5316 | 2.5253 | 2.5191 | 2.5128 | 2.5066 | 2.5004 |
| **12** | 2.4691 | 2.4630 | 2.4570 | 2.4509 | 2.4449 | 2.4389 | 2.4329 | 2.4269 | 2.4209 |
| **13** | 2.3916 | 2.3858 | 2.3799 | 2.3741 | 2.3683 | 2.3625 | 2.3567 | 2.3510 | 2.3452 |
| **14** | 2.3178 | 2.3122 | 2.3066 | 2.3010 | 2.2954 | 2.2898 | 2.2842 | 2.2787 | 2.2731 |
| **15** | 2.2475 | 2.2420 | 2.2366 | 2.2312 | 2.2258 | 2.2205 | 2.2151 | 2.2098 | 2.2044 |
| **16** | 2.1804 | 2.1752 | 2.1699 | 2.1647 | 2.1596 | 2.1544 | 2.1492 | 2.1441 | 2.1389 |
| **17** | 2.1164 | 2.1114 | 2.1063 | 2.1013 | 2.0963 | 2.0913 | 2.0864 | 2.0814 | 2.0764 |
| **18** | 2.0553 | 2.0505 | 2.0456 | 2.0408 | 2.0360 | 2.0312 | 2.0264 | 2.0216 | 2.0168 |
| **19** | 1.9971 | 1.9924 | 1.9877 | 1.9830 | 1.9784 | 1.9737 | 1.9691 | 1.9645 | 1.9599 |
| **20** | 1.9414 | 1.9369 | 1.9324 | 1.9279 | 1.9234 | 1.9189 | 1.9144 | 1.9100 | 1.9055 |
| **21** | 1.8882 | 1.8838 | 1.8795 | 1.8752 | 1.8708 | 1.8665 | 1.8622 | 1.8579 | 1.8536 |
| **22** | 1.8374 | 1.8332 | 1.8290 | 1.8248 | 1.8206 | 1.8164 | 1.8123 | 1.8081 | 1.8040 |
| **23** | 1.7888 | 1.7847 | 1.7806 | 1.7766 | 1.7726 | 1.7686 | 1.7645 | 1.7605 | 1.7565 |
| **24** | 1.7423 | 1.7383 | 1.7344 | 1.7305 | 1.7266 | 1.7228 | 1.7189 | 1.7150 | 1.7112 |
| **25** | 1.6978 | 1.6940 | 1.6902 | 1.6865 | 1.6827 | 1.6790 | 1.6752 | 1.6715 | 1.6678 |
| **26** | 1.6552 | 1.6515 | 1.6479 | 1.6443 | 1.6406 | 1.6370 | 1.6334 | 1.6298 | 1.6262 |
| **27** | 1.6144 | 1.6109 | 1.6074 | 1.6039 | 1.6004 | 1.5969 | 1.5934 | 1.5899 | 1.5864 |
| **28** | 1.5754 | 1.5720 | 1.5686 | 1.5652 | 1.5618 | 1.5584 | 1.5551 | 1.5517 | 1.5484 |
| **29** | 1.5380 | 1.5347 | 1.5314 | 1.5281 | 1.5249 | 1.5216 | 1.5184 | 1.5151 | 1.5119 |
| **30** | 1.5021 | 1.4989 | 1.4958 | 1.4926 | 1.4895 | 1.4863 | 1.4832 | 1.4800 | 1.4769 |
| **31** | 1.4677 | 1.4647 | 1.4616 | 1.4586 | 1.4555 | 1.4525 | 1.4494 | 1.4464 | 1.4434 |
| **32** | 1.4348 | 1.4318 | 1.4289 | 1.4259 | 1.4230 | 1.4200 | 1.4171 | 1.4142 | 1.4113 |
| **33** | 1.4032 | 1.4003 | 1.3975 | 1.3946 | 1.3918 | 1.3889 | 1.3861 | 1.3833 | 1.3804 |
| **34** | 1.3728 | 1.3701 | 1.3673 | 1.3646 | 1.3618 | 1.3591 | 1.3563 | 1.3536 | 1.3509 |
| **35** | 1.3437 | 1.3411 | 1.3384 | 1.3357 | 1.3331 | 1.3304 | 1.3278 | 1.3251 | 1.3225 |
| **36** | 1.3158 | 1.3132 | 1.3106 | 1.3081 | 1.3055 | 1.3029 | 1.3004 | 1.2978 | 1.2953 |
| **37** | 1.2890 | 1.2865 | 1.2840 | 1.2815 | 1.2790 | 1.2765 | 1.2741 | 1.2716 | 1.2691 |
| **38** | 1.2632 | 1.2608 | 1.2584 | 1.2560 | 1.2536 | 1.2512 | 1.2488 | 1.2464 | 1.2440 |
| **39** | 1.2384 | 1.2361 | 1.2338 | 1.2315 | 1.2292 | 1.2268 | 1.2245 | 1.2222 | 1.2199 |
| **40** | 1.2147 | 1.2124 | 1.2102 | 1.2079 | 1.2057 | 1.2035 | 1.2012 | 1.1990 | 1.1968 |

*Source*: Based on Weiss (1974).

**Table 2.41** Vapor Pressure of Seawater in mmHg

| Temperature (°C) | Salinity (g/kg) | | | | | | | | |
|---|---|---|---|---|---|---|---|---|---|
| | **0.0** | **5.0** | **10.0** | **15.0** | **20.0** | **25.0** | **30.0** | **35.0** | **40.0** |
| **0** | 4.581 | 4.569 | 4.557 | 4.545 | 4.533 | 4.520 | 4.508 | 4.495 | 4.482 |
| **1** | 4.925 | 4.912 | 4.899 | 4.886 | 4.873 | 4.860 | 4.846 | 4.833 | 4.819 |
| **2** | 5.291 | 5.278 | 5.264 | 5.250 | 5.236 | 5.222 | 5.207 | 5.192 | 5.177 |
| **3** | 5.682 | 5.667 | 5.653 | 5.638 | 5.623 | 5.607 | 5.592 | 5.576 | 5.560 |
| **4** | 6.098 | 6.082 | 6.066 | 6.050 | 6.034 | 6.018 | 6.001 | 5.984 | 5.967 |
| **5** | 6.541 | 6.524 | 6.507 | 6.490 | 6.472 | 6.455 | 6.437 | 6.419 | 6.400 |
| **6** | 7.012 | 6.993 | 6.975 | 6.957 | 6.938 | 6.919 | 6.900 | 6.881 | 6.861 |
| **7** | 7.512 | 7.493 | 7.473 | 7.453 | 7.433 | 7.413 | 7.393 | 7.372 | 7.350 |
| **8** | 8.044 | 8.023 | 8.002 | 7.981 | 7.960 | 7.938 | 7.916 | 7.893 | 7.870 |
| **9** | 8.608 | 8.586 | 8.564 | 8.541 | 8.518 | 8.495 | 8.472 | 8.448 | 8.423 |
| **10** | 9.208 | 9.184 | 9.160 | 9.136 | 9.111 | 9.087 | 9.061 | 9.036 | 9.009 |
| **11** | 9.843 | 9.818 | 9.792 | 9.766 | 9.740 | 9.714 | 9.687 | 9.659 | 9.631 |
| **12** | 10.517 | 10.490 | 10.463 | 10.435 | 10.407 | 10.379 | 10.350 | 10.321 | 10.291 |
| **13** | 11.232 | 11.202 | 11.173 | 11.144 | 11.114 | 11.084 | 11.053 | 11.022 | 10.990 |
| **14** | 11.988 | 11.957 | 11.926 | 11.894 | 11.863 | 11.830 | 11.797 | 11.764 | 11.730 |
| **15** | 12.789 | 12.756 | 12.722 | 12.689 | 12.655 | 12.621 | 12.586 | 12.550 | 12.513 |
| **16** | 13.636 | 13.601 | 13.565 | 13.530 | 13.494 | 13.457 | 13.420 | 13.382 | 13.343 |
| **17** | 14.533 | 14.495 | 14.457 | 14.419 | 14.381 | 14.342 | 14.302 | 14.261 | 14.220 |
| **18** | 15.480 | 15.440 | 15.400 | 15.359 | 15.318 | 15.277 | 15.234 | 15.191 | 15.147 |
| **19** | 16.482 | 16.439 | 16.396 | 16.353 | 16.309 | 16.265 | 16.220 | 16.173 | 16.126 |
| **20** | 17.539 | 17.493 | 17.448 | 17.402 | 17.355 | 17.308 | 17.260 | 17.211 | 17.161 |
| **21** | 18.656 | 18.607 | 18.558 | 18.509 | 18.460 | 18.410 | 18.359 | 18.307 | 18.253 |
| **22** | 19.834 | 19.782 | 19.730 | 19.678 | 19.626 | 19.573 | 19.518 | 19.463 | 19.406 |
| **23** | 21.076 | 21.021 | 20.966 | 20.911 | 20.855 | 20.799 | 20.741 | 20.682 | 20.622 |
| **24** | 22.386 | 22.328 | 22.269 | 22.211 | 22.152 | 22.091 | 22.030 | 21.968 | 21.904 |
| **25** | 23.767 | 23.705 | 23.643 | 23.580 | 23.518 | 23.454 | 23.389 | 23.322 | 23.254 |
| **26** | 25.221 | 25.155 | 25.089 | 25.023 | 24.956 | 24.889 | 24.820 | 24.749 | 24.677 |
| **27** | 26.752 | 26.682 | 26.612 | 26.542 | 26.472 | 26.400 | 26.326 | 26.252 | 26.175 |
| **28** | 28.363 | 28.289 | 28.215 | 28.141 | 28.066 | 27.990 | 27.912 | 27.833 | 27.752 |
| **29** | 30.058 | 29.980 | 29.902 | 29.823 | 29.743 | 29.663 | 29.580 | 29.496 | 29.410 |
| **30** | 31.841 | 31.758 | 31.675 | 31.592 | 31.507 | 31.422 | 31.335 | 31.246 | 31.155 |
| **31** | 33.715 | 33.627 | 33.539 | 33.451 | 33.362 | 33.271 | 33.179 | 33.085 | 32.988 |
| **32** | 35.684 | 35.591 | 35.498 | 35.404 | 35.310 | 35.214 | 35.116 | 35.017 | 34.915 |
| **33** | 37.752 | 37.653 | 37.555 | 37.456 | 37.356 | 37.255 | 37.152 | 37.046 | 36.938 |
| **34** | 39.923 | 39.819 | 39.715 | 39.610 | 39.505 | 39.397 | 39.288 | 39.177 | 39.062 |
| **35** | 42.201 | 42.091 | 41.982 | 41.871 | 41.759 | 41.646 | 41.530 | 41.412 | 41.292 |
| **36** | 44.592 | 44.476 | 44.359 | 44.243 | 44.125 | 44.005 | 43.883 | 43.758 | 43.631 |
| **37** | 47.098 | 46.976 | 46.853 | 46.730 | 46.605 | 46.479 | 46.350 | 46.218 | 46.083 |
| **38** | 49.726 | 49.597 | 49.467 | 49.337 | 49.205 | 49.072 | 48.936 | 48.797 | 48.654 |
| **39** | 52.480 | 52.343 | 52.206 | 52.069 | 51.930 | 51.789 | 51.645 | 51.498 | 51.349 |
| **40** | 55.364 | 55.219 | 55.075 | 54.930 | 54.784 | 54.635 | 54.484 | 54.329 | 54.171 |

*Source*: Based on Ambrose and Lawrenson (1972).

**Table 2.42** Vapor Pressure of Seawater in kPa

| Temperature (°C) | Salinity (g/kg) | | | | | | | | |
|---|---|---|---|---|---|---|---|---|---|
| | 0.0 | 5.0 | 10.0 | 15.0 | 20.0 | 25.0 | 30.0 | 35.0 | 40.0 |
| 0 | 0.6107 | 0.6091 | 0.6075 | 0.6059 | 0.6043 | 0.6027 | 0.6010 | 0.5993 | 0.5975 |
| 1 | 0.6566 | 0.6549 | 0.6531 | 0.6514 | 0.6497 | 0.6479 | 0.6461 | 0.6443 | 0.6424 |
| 2 | 0.7055 | 0.7036 | 0.7018 | 0.6999 | 0.6981 | 0.6962 | 0.6942 | 0.6923 | 0.6903 |
| 3 | 0.7576 | 0.7556 | 0.7536 | 0.7516 | 0.7496 | 0.7476 | 0.7455 | 0.7434 | 0.7412 |
| 4 | 0.8130 | 0.8109 | 0.8088 | 0.8067 | 0.8045 | 0.8023 | 0.8001 | 0.7978 | 0.7955 |
| 5 | 0.8720 | 0.8698 | 0.8675 | 0.8652 | 0.8629 | 0.8606 | 0.8582 | 0.8557 | 0.8532 |
| 6 | 0.9348 | 0.9324 | 0.9299 | 0.9275 | 0.9250 | 0.9225 | 0.9199 | 0.9173 | 0.9147 |
| 7 | 1.0015 | 0.9989 | 0.9963 | 0.9937 | 0.9910 | 0.9884 | 0.9856 | 0.9828 | 0.9799 |
| 8 | 1.0724 | 1.0696 | 1.0668 | 1.0640 | 1.0612 | 1.0583 | 1.0554 | 1.0524 | 1.0493 |
| 9 | 1.1477 | 1.1447 | 1.1417 | 1.1387 | 1.1357 | 1.1326 | 1.1295 | 1.1262 | 1.1230 |
| 10 | 1.2276 | 1.2244 | 1.2212 | 1.2180 | 1.2147 | 1.2114 | 1.2081 | 1.2046 | 1.2011 |
| 11 | 1.3123 | 1.3089 | 1.3055 | 1.3021 | 1.2986 | 1.2951 | 1.2915 | 1.2878 | 1.2841 |
| 12 | 1.4022 | 1.3985 | 1.3949 | 1.3912 | 1.3875 | 1.3837 | 1.3799 | 1.3760 | 1.3720 |
| 13 | 1.4974 | 1.4935 | 1.4896 | 1.4857 | 1.4817 | 1.4777 | 1.4736 | 1.4694 | 1.4652 |
| 14 | 1.5983 | 1.5941 | 1.5900 | 1.5858 | 1.5815 | 1.5772 | 1.5729 | 1.5684 | 1.5638 |
| 15 | 1.7051 | 1.7006 | 1.6962 | 1.6917 | 1.6872 | 1.6826 | 1.6779 | 1.6732 | 1.6683 |
| 16 | 1.8180 | 1.8133 | 1.8086 | 1.8038 | 1.7990 | 1.7941 | 1.7891 | 1.7841 | 1.7789 |
| 17 | 1.9375 | 1.9325 | 1.9274 | 1.9224 | 1.9173 | 1.9120 | 1.9067 | 1.9013 | 1.8958 |
| 18 | 2.0639 | 2.0585 | 2.0531 | 2.0477 | 2.0423 | 2.0367 | 2.0311 | 2.0253 | 2.0194 |
| 19 | 2.1974 | 2.1916 | 2.1859 | 2.1802 | 2.1744 | 2.1684 | 2.1624 | 2.1563 | 2.1500 |
| 20 | 2.3384 | 2.3323 | 2.3262 | 2.3201 | 2.3139 | 2.3076 | 2.3012 | 2.2946 | 2.2880 |
| 21 | 2.4872 | 2.4807 | 2.4742 | 2.4677 | 2.4611 | 2.4545 | 2.4477 | 2.4407 | 2.4336 |
| 22 | 2.6443 | 2.6374 | 2.6305 | 2.6236 | 2.6166 | 2.6095 | 2.6022 | 2.5948 | 2.5873 |
| 23 | 2.8099 | 2.8026 | 2.7953 | 2.7879 | 2.7805 | 2.7729 | 2.7652 | 2.7574 | 2.7493 |
| 24 | 2.9846 | 2.9768 | 2.9690 | 2.9612 | 2.9533 | 2.9453 | 2.9371 | 2.9288 | 2.9202 |
| 25 | 3.1686 | 3.1603 | 3.1521 | 3.1438 | 3.1354 | 3.1269 | 3.1182 | 3.1094 | 3.1003 |
| 26 | 3.3625 | 3.3537 | 3.3450 | 3.3362 | 3.3273 | 3.3182 | 3.3090 | 3.2996 | 3.2900 |
| 27 | 3.5666 | 3.5573 | 3.5480 | 3.5387 | 3.5292 | 3.5197 | 3.5099 | 3.4999 | 3.4897 |
| 28 | 3.7814 | 3.7716 | 3.7617 | 3.7518 | 3.7418 | 3.7317 | 3.7213 | 3.7107 | 3.6999 |
| 29 | 4.0074 | 3.9970 | 3.9866 | 3.9761 | 3.9655 | 3.9547 | 3.9437 | 3.9325 | 3.9211 |
| 30 | 4.2451 | 4.2341 | 4.2230 | 4.2119 | 4.2007 | 4.1892 | 4.1776 | 4.1658 | 4.1536 |
| 31 | 4.4949 | 4.4832 | 4.4715 | 4.4598 | 4.4479 | 4.4358 | 4.4235 | 4.4109 | 4.3981 |
| 32 | 4.7574 | 4.7450 | 4.7327 | 4.7202 | 4.7076 | 4.6948 | 4.6818 | 4.6685 | 4.6549 |
| 33 | 5.0332 | 5.0200 | 5.0069 | 4.9938 | 4.9804 | 4.9669 | 4.9531 | 4.9391 | 4.9247 |
| 34 | 5.3226 | 5.3087 | 5.2949 | 5.2809 | 5.2669 | 5.2526 | 5.2380 | 5.2231 | 5.2079 |
| 35 | 5.6264 | 5.6117 | 5.5971 | 5.5823 | 5.5675 | 5.5523 | 5.5369 | 5.5212 | 5.5051 |
| 36 | 5.9451 | 5.9296 | 5.9141 | 5.8985 | 5.8828 | 5.8668 | 5.8506 | 5.8339 | 5.8169 |
| 37 | 6.2793 | 6.2629 | 6.2466 | 6.2301 | 6.2135 | 6.1966 | 6.1794 | 6.1619 | 6.1439 |
| 38 | 6.6296 | 6.6123 | 6.5951 | 6.5777 | 6.5602 | 6.5424 | 6.5242 | 6.5057 | 6.4867 |
| 39 | 6.9967 | 6.9785 | 6.9602 | 6.9419 | 6.9234 | 6.9046 | 6.8855 | 6.8659 | 6.8459 |
| 40 | 7.3812 | 7.3620 | 7.3428 | 7.3235 | 7.3039 | 7.2841 | 7.2639 | 7.2432 | 7.2221 |

*Source*: Based on Ambrose and Lawrenson (1972).

**Table 2.43** The Hydrostatic Pressure of Water in mmHg/m as a Function of Temperature and Salinity

| Temperature (°C) | Salinity (g/kg) | | | | | | | | |
|---|---|---|---|---|---|---|---|---|---|
| | **0.0** | **5.0** | **10.0** | **15.0** | **20.0** | **25.0** | **30.0** | **35.0** | **40.0** |
| **0** | 73.544 | 73.844 | 74.141 | 74.438 | 74.734 | 75.030 | 75.327 | 75.623 | 75.921 |
| **1** | 73.549 | 73.847 | 74.143 | 74.438 | 74.733 | 75.028 | 75.323 | 75.619 | 75.915 |
| **2** | 73.552 | 73.848 | 74.143 | 74.437 | 74.731 | 75.025 | 75.319 | 75.613 | 75.908 |
| **3** | 73.554 | 73.849 | 74.142 | 74.435 | 74.728 | 75.021 | 75.314 | 75.607 | 75.901 |
| **4** | 73.554 | 73.848 | 74.141 | 74.432 | 74.724 | 75.015 | 75.307 | 75.600 | 75.893 |
| **5** | 73.553 | 73.846 | 74.138 | 74.428 | 74.719 | 75.009 | 75.300 | 75.592 | 75.884 |
| **6** | 73.552 | 73.844 | 74.134 | 74.423 | 74.713 | 75.002 | 75.292 | 75.583 | 75.874 |
| **7** | 73.549 | 73.840 | 74.129 | 74.417 | 74.706 | 74.994 | 75.283 | 75.573 | 75.863 |
| **8** | 73.545 | 73.835 | 74.123 | 74.410 | 74.698 | 74.985 | 75.273 | 75.562 | 75.851 |
| **9** | 73.540 | 73.829 | 74.116 | 74.402 | 74.689 | 74.976 | 75.263 | 75.551 | 75.839 |
| **10** | 73.534 | 73.822 | 74.108 | 74.393 | 74.679 | 74.965 | 75.251 | 75.538 | 75.826 |
| **11** | 73.527 | 73.814 | 74.099 | 74.384 | 74.669 | 74.954 | 75.239 | 75.525 | 75.812 |
| **12** | 73.519 | 73.805 | 74.089 | 74.373 | 74.657 | 74.942 | 75.226 | 75.512 | 75.798 |
| **13** | 73.510 | 73.795 | 74.079 | 74.362 | 74.645 | 74.929 | 75.213 | 75.497 | 75.783 |
| **14** | 73.500 | 73.785 | 74.067 | 74.350 | 74.632 | 74.915 | 75.198 | 75.482 | 75.767 |
| **15** | 73.490 | 73.773 | 74.055 | 74.337 | 74.618 | 74.900 | 75.183 | 75.466 | 75.750 |
| **16** | 73.478 | 73.761 | 74.042 | 74.323 | 74.604 | 74.885 | 75.167 | 75.450 | 75.733 |
| **17** | 73.466 | 73.748 | 74.028 | 74.308 | 74.589 | 74.869 | 75.151 | 75.433 | 75.715 |
| **18** | 73.453 | 73.734 | 74.013 | 74.293 | 74.573 | 74.853 | 75.134 | 75.415 | 75.697 |
| **19** | 73.439 | 73.719 | 73.998 | 74.277 | 74.556 | 74.836 | 75.116 | 75.396 | 75.678 |
| **20** | 73.424 | 73.704 | 73.982 | 74.260 | 74.539 | 74.818 | 75.097 | 75.377 | 75.658 |
| **21** | 73.408 | 73.687 | 73.965 | 74.243 | 74.521 | 74.799 | 75.078 | 75.358 | 75.638 |
| **22** | 73.392 | 73.670 | 73.948 | 74.225 | 74.502 | 74.780 | 75.058 | 75.337 | 75.617 |
| **23** | 73.375 | 73.653 | 73.929 | 74.206 | 74.483 | 74.760 | 75.038 | 75.316 | 75.596 |
| **24** | 73.357 | 73.634 | 73.911 | 74.187 | 74.463 | 74.740 | 75.017 | 75.295 | 75.574 |
| **25** | 73.339 | 73.615 | 73.891 | 74.167 | 74.442 | 74.719 | 74.995 | 75.273 | 75.551 |
| **26** | 73.320 | 73.596 | 73.871 | 74.146 | 74.421 | 74.697 | 74.973 | 75.250 | 75.528 |
| **27** | 73.300 | 73.575 | 73.850 | 74.124 | 74.399 | 74.675 | 74.950 | 75.227 | 75.505 |
| **28** | 73.279 | 73.554 | 73.828 | 74.103 | 74.377 | 74.652 | 74.927 | 75.203 | 75.480 |
| **29** | 73.258 | 73.533 | 73.806 | 74.080 | 74.354 | 74.628 | 74.903 | 75.179 | 75.456 |
| **30** | 73.236 | 73.510 | 73.784 | 74.057 | 74.330 | 74.604 | 74.879 | 75.154 | 75.430 |
| **31** | 73.214 | 73.487 | 73.760 | 74.033 | 74.306 | 74.580 | 74.854 | 75.129 | 75.405 |
| **32** | 73.190 | 73.464 | 73.736 | 74.009 | 74.281 | 74.555 | 74.828 | 75.103 | 75.378 |
| **33** | 73.167 | 73.440 | 73.712 | 73.984 | 74.256 | 74.529 | 74.802 | 75.077 | 75.351 |
| **34** | 73.142 | 73.415 | 73.687 | 73.958 | 74.230 | 74.503 | 74.776 | 75.050 | 75.324 |
| **35** | 73.117 | 73.390 | 73.661 | 73.932 | 74.204 | 74.476 | 74.749 | 75.022 | 75.296 |
| **36** | 73.092 | 73.364 | 73.635 | 73.906 | 74.177 | 74.449 | 74.721 | 74.994 | 75.268 |
| **37** | 73.066 | 73.337 | 73.608 | 73.879 | 74.150 | 74.421 | 74.693 | 74.966 | 75.239 |
| **38** | 73.039 | 73.310 | 73.581 | 73.851 | 74.122 | 74.393 | 74.665 | 74.937 | 75.210 |
| **39** | 73.012 | 73.283 | 73.553 | 73.823 | 74.093 | 74.364 | 74.636 | 74.908 | 75.181 |
| **40** | 72.984 | 73.255 | 73.525 | 73.795 | 74.065 | 74.335 | 74.606 | 74.878 | 75.151 |

*Source*: Based on Millero and Poisson (1981).

**Table 2.44** The Hydrostatic Pressure of Water in kPa/m as a Function of Temperature and Salinity

| Temperature (°C) | Salinity (g/kg) | | | | | | | | |
|---|---|---|---|---|---|---|---|---|---|
| | **0.0** | **5.0** | **10.0** | **15.0** | **20.0** | **25.0** | **30.0** | **35.0** | **40.0** |
| **0** | 9.805 | 9.845 | 9.885 | 9.924 | 9.964 | 10.003 | 10.043 | 10.082 | 10.122 |
| **1** | 9.806 | 9.845 | 9.885 | 9.924 | 9.964 | 10.003 | 10.042 | 10.082 | 10.121 |
| **2** | 9.806 | 9.846 | 9.885 | 9.924 | 9.963 | 10.002 | 10.042 | 10.081 | 10.120 |
| **3** | 9.806 | 9.846 | 9.885 | 9.924 | 9.963 | 10.002 | 10.041 | 10.080 | 10.119 |
| **4** | 9.806 | 9.846 | 9.885 | 9.923 | 9.962 | 10.001 | 10.040 | 10.079 | 10.118 |
| **5** | 9.806 | 9.845 | 9.884 | 9.923 | 9.962 | 10.000 | 10.039 | 10.078 | 10.117 |
| **6** | 9.806 | 9.845 | 9.884 | 9.922 | 9.961 | 9.999 | 10.038 | 10.077 | 10.116 |
| **7** | 9.806 | 9.844 | 9.883 | 9.921 | 9.960 | 9.998 | 10.037 | 10.076 | 10.114 |
| **8** | 9.805 | 9.844 | 9.882 | 9.921 | 9.959 | 9.997 | 10.036 | 10.074 | 10.113 |
| **9** | 9.805 | 9.843 | 9.881 | 9.919 | 9.958 | 9.996 | 10.034 | 10.073 | 10.111 |
| **10** | 9.804 | 9.842 | 9.880 | 9.918 | 9.956 | 9.995 | 10.033 | 10.071 | 10.109 |
| **11** | 9.803 | 9.841 | 9.879 | 9.917 | 9.955 | 9.993 | 10.031 | 10.069 | 10.107 |
| **12** | 9.802 | 9.840 | 9.878 | 9.916 | 9.953 | 9.991 | 10.029 | 10.067 | 10.106 |
| **13** | 9.801 | 9.839 | 9.876 | 9.914 | 9.952 | 9.990 | 10.028 | 10.065 | 10.104 |
| **14** | 9.799 | 9.837 | 9.875 | 9.912 | 9.950 | 9.988 | 10.026 | 10.063 | 10.101 |
| **15** | 9.798 | 9.836 | 9.873 | 9.911 | 9.948 | 9.986 | 10.024 | 10.061 | 10.099 |
| **16** | 9.796 | 9.834 | 9.871 | 9.909 | 9.946 | 9.984 | 10.021 | 10.059 | 10.097 |
| **17** | 9.795 | 9.832 | 9.870 | 9.907 | 9.944 | 9.982 | 10.019 | 10.057 | 10.095 |
| **18** | 9.793 | 9.830 | 9.868 | 9.905 | 9.942 | 9.980 | 10.017 | 10.054 | 10.092 |
| **19** | 9.791 | 9.828 | 9.866 | 9.903 | 9.940 | 9.977 | 10.015 | 10.052 | 10.090 |
| **20** | 9.789 | 9.826 | 9.863 | 9.901 | 9.938 | 9.975 | 10.012 | 10.049 | 10.087 |
| **21** | 9.787 | 9.824 | 9.861 | 9.898 | 9.935 | 9.972 | 10.010 | 10.047 | 10.084 |
| **22** | 9.785 | 9.822 | 9.859 | 9.896 | 9.933 | 9.970 | 10.007 | 10.044 | 10.081 |
| **23** | 9.783 | 9.820 | 9.856 | 9.893 | 9.930 | 9.967 | 10.004 | 10.041 | 10.079 |
| **24** | 9.780 | 9.817 | 9.854 | 9.891 | 9.928 | 9.964 | 10.001 | 10.039 | 10.076 |
| **25** | 9.778 | 9.815 | 9.851 | 9.888 | 9.925 | 9.962 | 9.999 | 10.036 | 10.073 |
| **26** | 9.775 | 9.812 | 9.849 | 9.885 | 9.922 | 9.959 | 9.996 | 10.033 | 10.070 |
| **27** | 9.772 | 9.809 | 9.846 | 9.882 | 9.919 | 9.956 | 9.993 | 10.029 | 10.066 |
| **28** | 9.770 | 9.806 | 9.843 | 9.880 | 9.916 | 9.953 | 9.989 | 10.026 | 10.063 |
| **29** | 9.767 | 9.804 | 9.840 | 9.877 | 9.913 | 9.950 | 9.986 | 10.023 | 10.060 |
| **30** | 9.764 | 9.801 | 9.837 | 9.873 | 9.910 | 9.946 | 9.983 | 10.020 | 10.057 |
| **31** | 9.761 | 9.798 | 9.834 | 9.870 | 9.907 | 9.943 | 9.980 | 10.016 | 10.053 |
| **32** | 9.758 | 9.794 | 9.831 | 9.867 | 9.903 | 9.940 | 9.976 | 10.013 | 10.050 |
| **33** | 9.755 | 9.791 | 9.827 | 9.864 | 9.900 | 9.936 | 9.973 | 10.009 | 10.046 |
| **34** | 9.751 | 9.788 | 9.824 | 9.860 | 9.897 | 9.933 | 9.969 | 10.006 | 10.042 |
| **35** | 9.748 | 9.784 | 9.821 | 9.857 | 9.893 | 9.929 | 9.966 | 10.002 | 10.039 |
| **36** | 9.745 | 9.781 | 9.817 | 9.853 | 9.889 | 9.926 | 9.962 | 9.998 | 10.035 |
| **37** | 9.741 | 9.778 | 9.814 | 9.850 | 9.886 | 9.922 | 9.958 | 9.995 | 10.031 |
| **38** | 9.738 | 9.774 | 9.810 | 9.846 | 9.882 | 9.918 | 9.954 | 9.991 | 10.027 |
| **39** | 9.734 | 9.770 | 9.806 | 9.842 | 9.878 | 9.914 | 9.951 | 9.987 | 10.023 |
| **40** | 9.730 | 9.766 | 9.802 | 9.838 | 9.874 | 9.911 | 9.947 | 9.983 | 10.019 |

*Source*: Based on Millero and Poisson (1981).

**Table 2.45** Oxygen in mmHg per mg/L as a Function of Temperature and Salinity, 0–40 g/kg

| Temperature (°C) | Salinity (g/kg) | | | | | | | | |
|---|---|---|---|---|---|---|---|---|---|
| | **0.0** | **5.0** | **10.0** | **15.0** | **20.0** | **25.0** | **30.0** | **35.0** | **40.0** |
| **0** | 10.822 | 11.207 | 11.605 | 12.018 | 12.446 | 12.890 | 13.350 | 13.827 | 14.321 |
| **1** | 11.126 | 11.517 | 11.922 | 12.343 | 12.779 | 13.230 | 13.698 | 14.182 | 14.684 |
| **2** | 11.431 | 11.829 | 12.242 | 12.670 | 13.113 | 13.571 | 14.047 | 14.539 | 15.049 |
| **3** | 11.739 | 12.144 | 12.564 | 12.998 | 13.449 | 13.915 | 14.397 | 14.897 | 15.415 |
| **4** | 12.048 | 12.460 | 12.887 | 13.329 | 13.786 | 14.259 | 14.750 | 15.257 | 15.782 |
| **5** | 12.359 | 12.778 | 13.211 | 13.660 | 14.125 | 14.605 | 15.103 | 15.618 | 16.150 |
| **6** | 12.671 | 13.097 | 13.537 | 13.993 | 14.465 | 14.953 | 15.457 | 15.979 | 16.519 |
| **7** | 12.985 | 13.417 | 13.865 | 14.327 | 14.806 | 15.301 | 15.812 | 16.342 | 16.889 |
| **8** | 13.300 | 13.739 | 14.193 | 14.662 | 15.148 | 15.650 | 16.168 | 16.705 | 17.259 |
| **9** | 13.616 | 14.061 | 14.522 | 14.998 | 15.490 | 15.999 | 16.525 | 17.068 | 17.630 |
| **10** | 13.933 | 14.384 | 14.851 | 15.334 | 15.833 | 16.349 | 16.881 | 17.432 | 18.001 |
| **11** | 14.250 | 14.708 | 15.182 | 15.671 | 16.176 | 16.699 | 17.238 | 17.795 | 18.371 |
| **12** | 14.568 | 15.032 | 15.512 | 16.008 | 16.520 | 17.049 | 17.595 | 18.159 | 18.741 |
| **13** | 14.886 | 15.356 | 15.842 | 16.345 | 16.863 | 17.399 | 17.951 | 18.522 | 19.111 |
| **14** | 15.204 | 15.681 | 16.173 | 16.681 | 17.206 | 17.748 | 18.307 | 18.884 | 19.480 |
| **15** | 15.522 | 16.005 | 16.503 | 17.018 | 17.549 | 18.097 | 18.663 | 19.246 | 19.849 |
| **16** | 15.840 | 16.328 | 16.833 | 17.354 | 17.891 | 18.445 | 19.017 | 19.607 | 20.216 |
| **17** | 16.157 | 16.652 | 17.162 | 17.689 | 18.232 | 18.793 | 19.371 | 19.967 | 20.582 |
| **18** | 16.474 | 16.974 | 17.491 | 18.024 | 18.573 | 19.139 | 19.723 | 20.326 | 20.947 |
| **19** | 16.790 | 17.296 | 17.819 | 18.357 | 18.912 | 19.485 | 20.075 | 20.683 | 21.310 |
| **20** | 17.106 | 17.617 | 18.145 | 18.689 | 19.250 | 19.829 | 20.425 | 21.039 | 21.672 |
| **21** | 17.420 | 17.937 | 18.471 | 19.020 | 19.587 | 20.171 | 20.773 | 21.393 | 22.032 |
| **22** | 17.733 | 18.256 | 18.795 | 19.350 | 19.922 | 20.512 | 21.119 | 21.745 | 22.390 |
| **23** | 18.045 | 18.573 | 19.117 | 19.678 | 20.256 | 20.851 | 21.464 | 22.096 | 22.746 |
| **24** | 18.356 | 18.889 | 19.438 | 20.005 | 20.588 | 21.188 | 21.807 | 22.444 | 23.100 |
| **25** | 18.664 | 19.203 | 19.758 | 20.329 | 20.917 | 21.523 | 22.147 | 22.790 | 23.451 |
| **26** | 18.971 | 19.515 | 20.075 | 20.652 | 21.245 | 21.856 | 22.485 | 23.133 | 23.800 |
| **27** | 19.277 | 19.825 | 20.390 | 20.972 | 21.571 | 22.187 | 22.821 | 23.474 | 24.146 |
| **28** | 19.580 | 20.133 | 20.703 | 21.290 | 21.894 | 22.515 | 23.154 | 23.812 | 24.489 |
| **29** | 19.881 | 20.439 | 21.014 | 21.606 | 22.214 | 22.840 | 23.485 | 24.148 | 24.830 |
| **30** | 20.180 | 20.743 | 21.322 | 21.919 | 22.532 | 23.163 | 23.812 | 24.480 | 25.167 |
| **31** | 20.477 | 21.044 | 21.628 | 22.229 | 22.847 | 23.483 | 24.137 | 24.810 | 25.501 |
| **32** | 20.771 | 21.343 | 21.931 | 22.537 | 23.160 | 23.800 | 24.459 | 25.136 | 25.832 |
| **33** | 21.062 | 21.639 | 22.232 | 22.842 | 23.469 | 24.114 | 24.777 | 25.459 | 26.160 |
| **34** | 21.351 | 21.932 | 22.529 | 23.144 | 23.775 | 24.425 | 25.092 | 25.779 | 26.484 |
| **35** | 21.637 | 22.222 | 22.824 | 23.442 | 24.078 | 24.732 | 25.404 | 26.095 | 26.805 |
| **36** | 21.920 | 22.509 | 23.115 | 23.738 | 24.378 | 25.036 | 25.712 | 26.407 | 27.122 |
| **37** | 22.200 | 22.793 | 23.403 | 24.030 | 24.675 | 25.337 | 26.017 | 26.716 | 27.435 |
| **38** | 22.477 | 23.074 | 23.688 | 24.319 | 24.968 | 25.634 | 26.318 | 27.022 | 27.744 |
| **39** | 22.750 | 23.352 | 23.969 | 24.604 | 25.257 | 25.927 | 26.616 | 27.323 | 28.049 |
| **40** | 23.021 | 23.626 | 24.248 | 24.886 | 25.543 | 26.217 | 26.909 | 27.620 | 28.351 |

**Table 2.46** Oxygen in mmHg per mg/L as a Function of Temperature and Salinity, 33–37 g/kg

| Temperature (°C) | Salinity (g/kg) | | | | | | | | |
|---|---|---|---|---|---|---|---|---|---|
| | **33.0** | **33.5** | **34.0** | **34.5** | **35.0** | **35.5** | **36.0** | **36.5** | **37.0** |
| **0** | 13.634 | 13.682 | 13.730 | 13.779 | 13.827 | 13.876 | 13.924 | 13.973 | 14.023 |
| **1** | 13.986 | 14.035 | 14.084 | 14.133 | 14.182 | 14.231 | 14.281 | 14.331 | 14.381 |
| **2** | 14.340 | 14.389 | 14.439 | 14.489 | 14.539 | 14.589 | 14.639 | 14.690 | 14.741 |
| **3** | 14.695 | 14.745 | 14.796 | 14.846 | 14.897 | 14.948 | 14.999 | 15.051 | 15.102 |
| **4** | 15.052 | 15.103 | 15.154 | 15.205 | 15.257 | 15.309 | 15.360 | 15.412 | 15.465 |
| **5** | 15.410 | 15.461 | 15.513 | 15.565 | 15.618 | 15.670 | 15.723 | 15.775 | 15.828 |
| **6** | 15.768 | 15.821 | 15.873 | 15.926 | 15.979 | 16.032 | 16.086 | 16.139 | 16.193 |
| **7** | 16.128 | 16.181 | 16.234 | 16.288 | 16.342 | 16.396 | 16.450 | 16.504 | 16.559 |
| **8** | 16.488 | 16.542 | 16.596 | 16.650 | 16.705 | 16.759 | 16.814 | 16.869 | 16.924 |
| **9** | 16.849 | 16.903 | 16.958 | 17.013 | 17.068 | 17.124 | 17.179 | 17.235 | 17.291 |
| **10** | 17.209 | 17.265 | 17.320 | 17.376 | 17.432 | 17.488 | 17.544 | 17.600 | 17.657 |
| **11** | 17.570 | 17.626 | 17.682 | 17.739 | 17.795 | 17.852 | 17.909 | 17.966 | 18.023 |
| **12** | 17.931 | 17.988 | 18.045 | 18.102 | 18.159 | 18.216 | 18.274 | 18.332 | 18.390 |
| **13** | 18.291 | 18.349 | 18.406 | 18.464 | 18.522 | 18.580 | 18.638 | 18.697 | 18.755 |
| **14** | 18.651 | 18.709 | 18.768 | 18.826 | 18.884 | 18.943 | 19.002 | 19.061 | 19.120 |
| **15** | 19.011 | 19.069 | 19.128 | 19.187 | 19.246 | 19.306 | 19.365 | 19.425 | 19.485 |
| **16** | 19.369 | 19.428 | 19.488 | 19.547 | 19.607 | 19.667 | 19.727 | 19.788 | 19.848 |
| **17** | 19.726 | 19.786 | 19.846 | 19.907 | 19.967 | 20.028 | 20.089 | 20.150 | 20.211 |
| **18** | 20.083 | 20.143 | 20.204 | 20.265 | 20.326 | 20.387 | 20.449 | 20.510 | 20.572 |
| **19** | 20.438 | 20.499 | 20.560 | 20.621 | 20.683 | 20.745 | 20.807 | 20.869 | 20.932 |
| **20** | 20.791 | 20.853 | 20.915 | 20.977 | 21.039 | 21.101 | 21.164 | 21.227 | 21.290 |
| **21** | 21.143 | 21.205 | 21.268 | 21.330 | 21.393 | 21.456 | 21.519 | 21.583 | 21.646 |
| **22** | 21.493 | 21.556 | 21.619 | 21.682 | 21.745 | 21.809 | 21.873 | 21.937 | 22.001 |
| **23** | 21.841 | 21.904 | 21.968 | 22.032 | 22.096 | 22.160 | 22.224 | 22.289 | 22.353 |
| **24** | 22.187 | 22.251 | 22.315 | 22.379 | 22.444 | 22.508 | 22.573 | 22.638 | 22.704 |
| **25** | 22.530 | 22.595 | 22.660 | 22.724 | 22.790 | 22.855 | 22.920 | 22.986 | 23.052 |
| **26** | 22.872 | 22.937 | 23.002 | 23.067 | 23.133 | 23.199 | 23.265 | 23.331 | 23.397 |
| **27** | 23.211 | 23.276 | 23.342 | 23.408 | 23.474 | 23.540 | 23.607 | 23.673 | 23.740 |
| **28** | 23.547 | 23.613 | 23.679 | 23.746 | 23.812 | 23.879 | 23.946 | 24.013 | 24.081 |
| **29** | 23.880 | 23.947 | 24.013 | 24.080 | 24.148 | 24.215 | 24.282 | 24.350 | 24.418 |
| **30** | 24.211 | 24.278 | 24.345 | 24.412 | 24.480 | 24.548 | 24.616 | 24.684 | 24.753 |
| **31** | 24.538 | 24.606 | 24.674 | 24.741 | 24.810 | 24.878 | 24.946 | 25.015 | 25.084 |
| **32** | 24.863 | 24.931 | 24.999 | 25.067 | 25.136 | 25.205 | 25.274 | 25.343 | 25.412 |
| **33** | 25.184 | 25.252 | 25.321 | 25.390 | 25.459 | 25.528 | 25.598 | 25.667 | 25.737 |
| **34** | 25.502 | 25.571 | 25.640 | 25.709 | 25.779 | 25.848 | 25.918 | 25.988 | 26.058 |
| **35** | 25.816 | 25.886 | 25.955 | 26.025 | 26.095 | 26.165 | 26.235 | 26.306 | 26.376 |
| **36** | 26.127 | 26.197 | 26.267 | 26.337 | 26.407 | 26.478 | 26.549 | 26.620 | 26.691 |
| **37** | 26.434 | 26.505 | 26.575 | 26.646 | 26.716 | 26.787 | 26.858 | 26.930 | 27.001 |
| **38** | 26.738 | 26.809 | 26.879 | 26.950 | 27.022 | 27.093 | 27.165 | 27.236 | 27.308 |
| **39** | 27.038 | 27.109 | 27.180 | 27.251 | 27.323 | 27.395 | 27.467 | 27.539 | 27.611 |
| **40** | 27.334 | 27.405 | 27.477 | 27.548 | 27.620 | 27.693 | 27.765 | 27.838 | 27.910 |

**Table 2.47** Nitrogen in mmHg per mg/L as a Function of Temperature and Salinity, 0–40 g/kg

| Temperature (°C) | Salinity (g/kg) | | | | | | | | |
|---|---|---|---|---|---|---|---|---|---|
| | **0.0** | **5.0** | **10.0** | **15.0** | **20.0** | **25.0** | **30.0** | **35.0** | **40.0** |
| **0** | 25.359 | 26.321 | 27.321 | 28.359 | 29.438 | 30.558 | 31.721 | 32.929 | 34.183 |
| **1** | 26.022 | 27.000 | 28.016 | 29.070 | 30.165 | 31.302 | 32.482 | 33.707 | 34.979 |
| **2** | 26.688 | 27.681 | 28.713 | 29.783 | 30.895 | 32.048 | 33.245 | 34.487 | 35.775 |
| **3** | 27.355 | 28.364 | 29.411 | 30.498 | 31.625 | 32.795 | 34.008 | 35.266 | 36.571 |
| **4** | 28.024 | 29.048 | 30.111 | 31.213 | 32.356 | 33.542 | 34.771 | 36.046 | 37.367 |
| **5** | 28.695 | 29.734 | 30.811 | 31.929 | 33.087 | 34.289 | 35.534 | 36.824 | 38.162 |
| **6** | 29.366 | 30.419 | 31.512 | 32.645 | 33.818 | 35.035 | 36.296 | 37.602 | 38.956 |
| **7** | 30.037 | 31.105 | 32.212 | 33.360 | 34.549 | 35.780 | 37.056 | 38.378 | 39.748 |
| **8** | 30.708 | 31.790 | 32.912 | 34.074 | 35.278 | 36.524 | 37.815 | 39.152 | 40.537 |
| **9** | 31.378 | 32.474 | 33.610 | 34.787 | 36.005 | 37.266 | 38.572 | 39.924 | 41.324 |
| **10** | 32.047 | 33.157 | 34.307 | 35.498 | 36.730 | 38.006 | 39.326 | 40.693 | 42.107 |
| **11** | 32.714 | 33.838 | 35.002 | 36.206 | 37.453 | 38.743 | 40.077 | 41.458 | 42.887 |
| **12** | 33.379 | 34.517 | 35.694 | 36.912 | 38.173 | 39.476 | 40.825 | 42.220 | 43.663 |
| **13** | 34.042 | 35.193 | 36.383 | 37.615 | 38.889 | 40.206 | 41.569 | 42.978 | 44.434 |
| **14** | 34.702 | 35.866 | 37.069 | 38.314 | 39.601 | 40.932 | 42.308 | 43.731 | 45.201 |
| **15** | 35.359 | 36.535 | 37.751 | 39.009 | 40.309 | 41.653 | 43.043 | 44.478 | 45.962 |
| **16** | 36.011 | 37.200 | 38.429 | 39.700 | 41.013 | 42.370 | 43.772 | 45.221 | 46.718 |
| **17** | 36.660 | 37.861 | 39.102 | 40.386 | 41.711 | 43.081 | 44.496 | 45.958 | 47.468 |
| **18** | 37.303 | 38.517 | 39.771 | 41.066 | 42.404 | 43.786 | 45.214 | 46.688 | 48.211 |
| **19** | 37.942 | 39.167 | 40.433 | 41.741 | 43.091 | 44.486 | 45.926 | 47.413 | 48.948 |
| **20** | 38.575 | 39.812 | 41.090 | 42.410 | 43.772 | 45.179 | 46.631 | 48.130 | 49.677 |
| **21** | 39.203 | 40.451 | 41.740 | 43.072 | 44.446 | 45.865 | 47.329 | 48.840 | 50.400 |
| **22** | 39.824 | 41.083 | 42.384 | 43.727 | 45.113 | 46.543 | 48.019 | 49.542 | 51.114 |
| **23** | 40.438 | 41.709 | 43.021 | 44.375 | 45.773 | 47.215 | 48.702 | 50.237 | 51.820 |
| **24** | 41.045 | 42.327 | 43.650 | 45.016 | 46.425 | 47.878 | 49.377 | 50.923 | 52.518 |
| **25** | 41.645 | 42.938 | 44.272 | 45.649 | 47.068 | 48.533 | 50.043 | 51.601 | 53.207 |
| **26** | 42.237 | 43.540 | 44.885 | 46.273 | 47.704 | 49.179 | 50.700 | 52.269 | 53.887 |
| **27** | 42.820 | 44.134 | 45.490 | 46.888 | 48.330 | 49.816 | 51.349 | 52.929 | 54.558 |
| **28** | 43.395 | 44.720 | 46.086 | 47.495 | 48.947 | 50.444 | 51.988 | 53.579 | 55.218 |
| **29** | 43.961 | 45.296 | 46.673 | 48.092 | 49.555 | 51.063 | 52.617 | 54.219 | 55.869 |
| **30** | 44.518 | 45.863 | 47.250 | 48.679 | 50.153 | 51.671 | 53.236 | 54.848 | 56.510 |
| **31** | 45.065 | 46.420 | 47.817 | 49.256 | 50.740 | 52.269 | 53.844 | 55.467 | 57.140 |
| **32** | 45.601 | 46.966 | 48.373 | 49.823 | 51.317 | 52.856 | 54.442 | 56.076 | 57.759 |
| **33** | 46.128 | 47.502 | 48.919 | 50.379 | 51.883 | 53.433 | 55.029 | 56.673 | 58.367 |
| **34** | 46.643 | 48.027 | 49.454 | 50.924 | 52.438 | 53.997 | 55.604 | 57.259 | 58.963 |
| **35** | 47.147 | 48.541 | 49.977 | 51.457 | 52.981 | 54.551 | 56.167 | 57.832 | 59.547 |
| **36** | 47.639 | 49.042 | 50.488 | 51.978 | 53.512 | 55.092 | 56.719 | 58.394 | 60.119 |
| **37** | 48.119 | 49.532 | 50.987 | 52.487 | 54.030 | 55.620 | 57.257 | 58.943 | 60.679 |
| **38** | 48.587 | 50.009 | 51.474 | 52.983 | 54.536 | 56.136 | 57.783 | 59.479 | 61.225 |
| **39** | 49.042 | 50.473 | 51.947 | 53.466 | 55.029 | 56.639 | 58.296 | 60.002 | 61.759 |
| **40** | 49.483 | 50.924 | 52.407 | 53.935 | 55.508 | 57.128 | 58.795 | 60.512 | 62.279 |

**Table 2.48** Nitrogen in mmHg per mg/L as a Function of Temperature and Salinity, 33–37 g/kg

| Temperature (°C) | Salinity (g/kg) | | | | | | | | |
|---|---|---|---|---|---|---|---|---|---|
| | **33.0** | **33.5** | **34.0** | **34.5** | **35.0** | **35.5** | **36.0** | **36.5** | **37.0** |
| **0** | 32.440 | 32.562 | 32.684 | 32.806 | 32.929 | 33.052 | 33.176 | 33.300 | 33.425 |
| **1** | 33.212 | 33.335 | 33.459 | 33.583 | 33.707 | 33.832 | 33.958 | 34.084 | 34.210 |
| **2** | 33.984 | 34.109 | 34.234 | 34.360 | 34.487 | 34.613 | 34.740 | 34.868 | 34.996 |
| **3** | 34.757 | 34.884 | 35.011 | 35.138 | 35.266 | 35.395 | 35.523 | 35.653 | 35.783 |
| **4** | 35.530 | 35.658 | 35.787 | 35.916 | 36.046 | 36.176 | 36.306 | 36.437 | 36.568 |
| **5** | 36.303 | 36.432 | 36.563 | 36.693 | 36.824 | 36.956 | 37.088 | 37.221 | 37.354 |
| **6** | 37.074 | 37.205 | 37.337 | 37.469 | 37.602 | 37.735 | 37.869 | 38.003 | 38.138 |
| **7** | 37.844 | 37.977 | 38.110 | 38.244 | 38.378 | 38.513 | 38.648 | 38.784 | 38.920 |
| **8** | 38.612 | 38.746 | 38.881 | 39.017 | 39.152 | 39.289 | 39.425 | 39.563 | 39.700 |
| **9** | 39.378 | 39.514 | 39.650 | 39.787 | 39.924 | 40.062 | 40.200 | 40.339 | 40.478 |
| **10** | 40.141 | 40.278 | 40.416 | 40.554 | 40.693 | 40.832 | 40.972 | 41.112 | 41.253 |
| **11** | 40.900 | 41.039 | 41.178 | 41.318 | 41.458 | 41.599 | 41.740 | 41.882 | 42.024 |
| **12** | 41.656 | 41.797 | 41.937 | 42.078 | 42.220 | 42.362 | 42.505 | 42.648 | 42.791 |
| **13** | 42.408 | 42.550 | 42.692 | 42.835 | 42.978 | 43.121 | 43.265 | 43.410 | 43.554 |
| **14** | 43.156 | 43.299 | 43.442 | 43.586 | 43.731 | 43.875 | 44.021 | 44.167 | 44.313 |
| **15** | 43.898 | 44.043 | 44.188 | 44.333 | 44.478 | 44.625 | 44.771 | 44.919 | 45.066 |
| **16** | 44.636 | 44.781 | 44.927 | 45.074 | 45.221 | 45.369 | 45.517 | 45.665 | 45.814 |
| **17** | 45.367 | 45.514 | 45.662 | 45.810 | 45.958 | 46.107 | 46.256 | 46.406 | 46.556 |
| **18** | 46.093 | 46.241 | 46.390 | 46.539 | 46.688 | 46.839 | 46.989 | 47.140 | 47.292 |
| **19** | 46.812 | 46.962 | 47.111 | 47.262 | 47.413 | 47.564 | 47.716 | 47.868 | 48.021 |
| **20** | 47.524 | 47.675 | 47.826 | 47.978 | 48.130 | 48.282 | 48.435 | 48.589 | 48.743 |
| **21** | 48.230 | 48.382 | 48.534 | 48.687 | 48.840 | 48.994 | 49.148 | 49.303 | 49.458 |
| **22** | 48.927 | 49.080 | 49.234 | 49.388 | 49.542 | 49.697 | 49.853 | 50.009 | 50.165 |
| **23** | 49.617 | 49.771 | 49.926 | 50.081 | 50.237 | 50.393 | 50.550 | 50.707 | 50.864 |
| **24** | 50.299 | 50.454 | 50.610 | 50.766 | 50.923 | 51.080 | 51.238 | 51.396 | 51.555 |
| **25** | 50.972 | 51.128 | 51.285 | 51.443 | 51.601 | 51.759 | 51.918 | 52.077 | 52.237 |
| **26** | 51.636 | 51.794 | 51.952 | 52.110 | 52.269 | 52.429 | 52.589 | 52.749 | 52.910 |
| **27** | 52.291 | 52.450 | 52.609 | 52.769 | 52.929 | 53.089 | 53.251 | 53.412 | 53.574 |
| **28** | 52.936 | 53.096 | 53.257 | 53.417 | 53.579 | 53.740 | 53.903 | 54.065 | 54.229 |
| **29** | 53.572 | 53.733 | 53.894 | 54.056 | 54.219 | 54.381 | 54.545 | 54.709 | 54.873 |
| **30** | 54.197 | 54.359 | 54.522 | 54.685 | 54.848 | 55.012 | 55.177 | 55.342 | 55.507 |
| **31** | 54.812 | 54.975 | 55.139 | 55.303 | 55.467 | 55.632 | 55.798 | 55.964 | 56.130 |
| **32** | 55.416 | 55.581 | 55.745 | 55.910 | 56.076 | 56.242 | 56.408 | 56.575 | 56.743 |
| **33** | 56.009 | 56.175 | 56.340 | 56.506 | 56.673 | 56.840 | 57.008 | 57.176 | 57.344 |
| **34** | 56.591 | 56.757 | 56.924 | 57.091 | 57.259 | 57.427 | 57.595 | 57.765 | 57.934 |
| **35** | 57.160 | 57.328 | 57.495 | 57.664 | 57.832 | 58.002 | 58.171 | 58.341 | 58.512 |
| **36** | 57.718 | 57.886 | 58.055 | 58.224 | 58.394 | 58.564 | 58.735 | 58.906 | 59.078 |
| **37** | 58.263 | 58.432 | 58.602 | 58.772 | 58.943 | 59.114 | 59.286 | 59.458 | 59.631 |
| **38** | 58.795 | 58.965 | 59.136 | 59.307 | 59.479 | 59.652 | 59.824 | 59.998 | 60.172 |
| **39** | 59.314 | 59.485 | 59.657 | 59.829 | 60.002 | 60.176 | 60.349 | 60.524 | 60.699 |
| **40** | 59.819 | 59.991 | 60.164 | 60.338 | 60.512 | 60.686 | 60.861 | 61.036 | 61.212 |

**Table 2.49** Argon in mmHg per mg/L as a Function of Temperature and Salinity, 0−40 g/kg

| Temperature (°C) | Salinity (g/kg) | | | | | | | | |
|---|---|---|---|---|---|---|---|---|---|
| | **0.0** | **5.0** | **10.0** | **15.0** | **20.0** | **25.0** | **30.0** | **35.0** | **40.0** |
| **0** | 7.921 | 8.200 | 8.489 | 8.788 | 9.098 | 9.419 | 9.751 | 10.095 | 10.452 |
| **1** | 8.141 | 8.425 | 8.718 | 9.023 | 9.338 | 9.665 | 10.002 | 10.352 | 10.715 |
| **2** | 8.362 | 8.651 | 8.950 | 9.259 | 9.580 | 9.912 | 10.255 | 10.610 | 10.978 |
| **3** | 8.585 | 8.879 | 9.183 | 9.497 | 9.823 | 10.160 | 10.509 | 10.869 | 11.243 |
| **4** | 8.809 | 9.107 | 9.417 | 9.736 | 10.067 | 10.409 | 10.763 | 11.129 | 11.508 |
| **5** | 9.034 | 9.338 | 9.652 | 9.976 | 10.312 | 10.660 | 11.019 | 11.390 | 11.774 |
| **6** | 9.260 | 9.569 | 9.887 | 10.217 | 10.558 | 10.910 | 11.275 | 11.651 | 12.041 |
| **7** | 9.488 | 9.801 | 10.124 | 10.459 | 10.805 | 11.162 | 11.531 | 11.913 | 12.308 |
| **8** | 9.715 | 10.033 | 10.362 | 10.701 | 11.052 | 11.414 | 11.789 | 12.175 | 12.575 |
| **9** | 9.944 | 10.266 | 10.600 | 10.944 | 11.299 | 11.667 | 12.046 | 12.438 | 12.842 |
| **10** | 10.173 | 10.500 | 10.838 | 11.187 | 11.547 | 11.919 | 12.303 | 12.700 | 13.110 |
| **11** | 10.403 | 10.734 | 11.077 | 11.430 | 11.795 | 12.172 | 12.561 | 12.962 | 13.377 |
| **12** | 10.633 | 10.969 | 11.316 | 11.674 | 12.043 | 12.425 | 12.818 | 13.225 | 13.644 |
| **13** | 10.863 | 11.203 | 11.555 | 11.917 | 12.291 | 12.677 | 13.076 | 13.487 | 13.911 |
| **14** | 11.093 | 11.438 | 11.793 | 12.161 | 12.539 | 12.930 | 13.333 | 13.748 | 14.177 |
| **15** | 11.323 | 11.672 | 12.032 | 12.404 | 12.787 | 13.182 | 13.589 | 14.009 | 14.442 |
| **16** | 11.553 | 11.906 | 12.270 | 12.646 | 13.034 | 13.433 | 13.845 | 14.270 | 14.707 |
| **17** | 11.782 | 12.140 | 12.508 | 12.888 | 13.280 | 13.684 | 14.100 | 14.529 | 14.971 |
| **18** | 12.011 | 12.373 | 12.746 | 13.130 | 13.526 | 13.934 | 14.355 | 14.788 | 15.235 |
| **19** | 12.240 | 12.605 | 12.982 | 13.371 | 13.771 | 14.183 | 14.608 | 15.046 | 15.497 |
| **20** | 12.468 | 12.837 | 13.218 | 13.611 | 14.015 | 14.432 | 14.861 | 15.303 | 15.758 |
| **21** | 12.695 | 13.068 | 13.453 | 13.850 | 14.258 | 14.679 | 15.112 | 15.558 | 16.018 |
| **22** | 12.921 | 13.298 | 13.687 | 14.088 | 14.500 | 14.925 | 15.362 | 15.812 | 16.276 |
| **23** | 13.146 | 13.527 | 13.920 | 14.325 | 14.741 | 15.170 | 15.611 | 16.065 | 16.533 |
| **24** | 13.369 | 13.755 | 14.152 | 14.560 | 14.980 | 15.413 | 15.858 | 16.317 | 16.788 |
| **25** | 13.592 | 13.981 | 14.382 | 14.794 | 15.218 | 15.655 | 16.104 | 16.566 | 17.042 |
| **26** | 13.813 | 14.206 | 14.611 | 15.027 | 15.455 | 15.895 | 16.348 | 16.815 | 17.294 |
| **27** | 14.033 | 14.430 | 14.838 | 15.258 | 15.689 | 16.134 | 16.591 | 17.061 | 17.544 |
| **28** | 14.251 | 14.651 | 15.063 | 15.487 | 15.922 | 16.370 | 16.831 | 17.305 | 17.792 |
| **29** | 14.468 | 14.871 | 15.287 | 15.714 | 16.153 | 16.605 | 17.070 | 17.548 | 18.039 |
| **30** | 14.682 | 15.089 | 15.508 | 15.939 | 16.382 | 16.838 | 17.306 | 17.788 | 18.283 |
| **31** | 14.895 | 15.306 | 15.728 | 16.163 | 16.609 | 17.069 | 17.541 | 18.026 | 18.525 |
| **32** | 15.105 | 15.520 | 15.946 | 16.384 | 16.834 | 17.297 | 17.773 | 18.262 | 18.764 |
| **33** | 15.313 | 15.731 | 16.161 | 16.603 | 17.057 | 17.523 | 18.002 | 18.495 | 19.001 |
| **34** | 15.519 | 15.941 | 16.374 | 16.819 | 17.277 | 17.747 | 18.230 | 18.726 | 19.236 |
| **35** | 15.723 | 16.148 | 16.584 | 17.033 | 17.494 | 17.968 | 18.454 | 18.954 | 19.468 |
| **36** | 15.924 | 16.352 | 16.792 | 17.244 | 17.709 | 18.186 | 18.676 | 19.180 | 19.697 |
| **37** | 16.122 | 16.554 | 16.997 | 17.453 | 17.921 | 18.401 | 18.895 | 19.402 | 19.923 |
| **38** | 16.318 | 16.752 | 17.199 | 17.658 | 18.130 | 18.614 | 19.111 | 19.622 | 20.147 |
| **39** | 16.510 | 16.948 | 17.398 | 17.861 | 18.336 | 18.823 | 19.324 | 19.839 | 20.367 |
| **40** | 16.699 | 17.141 | 17.594 | 18.060 | 18.539 | 19.030 | 19.534 | 20.052 | 20.584 |

**Table 2.50** Argon in mmHg per mg/L as a Function of Temperature and Salinity, 33–37 g/kg

| Temperature (°C) | Salinity (g/kg) | | | | | | | | |
|---|---|---|---|---|---|---|---|---|---|
| | **33.0** | **33.5** | **34.0** | **34.5** | **35.0** | **35.5** | **36.0** | **36.5** | **37.0** |
| **0** | 9.956 | 9.991 | 10.026 | 10.061 | 10.095 | 10.131 | 10.166 | 10.201 | 10.237 |
| **1** | 10.211 | 10.246 | 10.281 | 10.317 | 10.352 | 10.388 | 10.424 | 10.460 | 10.496 |
| **2** | 10.467 | 10.503 | 10.538 | 10.574 | 10.610 | 10.647 | 10.683 | 10.719 | 10.756 |
| **3** | 10.724 | 10.760 | 10.796 | 10.833 | 10.869 | 10.906 | 10.943 | 10.980 | 11.017 |
| **4** | 10.981 | 11.018 | 11.055 | 11.092 | 11.129 | 11.167 | 11.204 | 11.242 | 11.279 |
| **5** | 11.240 | 11.277 | 11.315 | 11.352 | 11.390 | 11.428 | 11.466 | 11.504 | 11.542 |
| **6** | 11.499 | 11.537 | 11.575 | 11.613 | 11.651 | 11.690 | 11.728 | 11.767 | 11.806 |
| **7** | 11.759 | 11.797 | 11.836 | 11.874 | 11.913 | 11.952 | 11.991 | 12.030 | 12.069 |
| **8** | 12.019 | 12.058 | 12.097 | 12.136 | 12.175 | 12.215 | 12.254 | 12.294 | 12.334 |
| **9** | 12.279 | 12.319 | 12.358 | 12.398 | 12.438 | 12.477 | 12.517 | 12.558 | 12.598 |
| **10** | 12.540 | 12.580 | 12.620 | 12.660 | 12.700 | 12.740 | 12.781 | 12.822 | 12.862 |
| **11** | 12.800 | 12.841 | 12.881 | 12.922 | 12.962 | 13.003 | 13.044 | 13.085 | 13.127 |
| **12** | 13.061 | 13.102 | 13.142 | 13.184 | 13.225 | 13.266 | 13.307 | 13.349 | 13.391 |
| **13** | 13.321 | 13.362 | 13.403 | 13.445 | 13.487 | 13.528 | 13.570 | 13.612 | 13.655 |
| **14** | 13.580 | 13.622 | 13.664 | 13.706 | 13.748 | 13.791 | 13.833 | 13.875 | 13.918 |
| **15** | 13.840 | 13.882 | 13.924 | 13.967 | 14.009 | 14.052 | 14.095 | 14.138 | 14.181 |
| **16** | 14.098 | 14.141 | 14.184 | 14.227 | 14.270 | 14.313 | 14.356 | 14.400 | 14.443 |
| **17** | 14.356 | 14.399 | 14.443 | 14.486 | 14.529 | 14.573 | 14.617 | 14.661 | 14.705 |
| **18** | 14.613 | 14.657 | 14.700 | 14.744 | 14.788 | 14.832 | 14.876 | 14.921 | 14.965 |
| **19** | 14.869 | 14.913 | 14.957 | 15.002 | 15.046 | 15.090 | 15.135 | 15.180 | 15.225 |
| **20** | 15.124 | 15.169 | 15.213 | 15.258 | 15.303 | 15.348 | 15.393 | 15.438 | 15.483 |
| **21** | 15.378 | 15.423 | 15.468 | 15.513 | 15.558 | 15.604 | 15.649 | 15.695 | 15.740 |
| **22** | 15.631 | 15.676 | 15.721 | 15.767 | 15.812 | 15.858 | 15.904 | 15.950 | 15.996 |
| **23** | 15.882 | 15.928 | 15.973 | 16.019 | 16.065 | 16.112 | 16.158 | 16.204 | 16.251 |
| **24** | 16.132 | 16.178 | 16.224 | 16.270 | 16.317 | 16.363 | 16.410 | 16.457 | 16.504 |
| **25** | 16.380 | 16.426 | 16.473 | 16.520 | 16.566 | 16.613 | 16.661 | 16.708 | 16.755 |
| **26** | 16.626 | 16.673 | 16.720 | 16.767 | 16.815 | 16.862 | 16.909 | 16.957 | 17.005 |
| **27** | 16.871 | 16.918 | 16.966 | 17.013 | 17.061 | 17.109 | 17.156 | 17.204 | 17.253 |
| **28** | 17.114 | 17.162 | 17.209 | 17.257 | 17.305 | 17.353 | 17.402 | 17.450 | 17.498 |
| **29** | 17.355 | 17.403 | 17.451 | 17.499 | 17.548 | 17.596 | 17.645 | 17.693 | 17.742 |
| **30** | 17.594 | 17.642 | 17.690 | 17.739 | 17.788 | 17.837 | 17.886 | 17.935 | 17.984 |
| **31** | 17.830 | 17.879 | 17.928 | 17.977 | 18.026 | 18.075 | 18.125 | 18.174 | 18.224 |
| **32** | 18.064 | 18.114 | 18.163 | 18.212 | 18.262 | 18.311 | 18.361 | 18.411 | 18.461 |
| **33** | 18.296 | 18.346 | 18.395 | 18.445 | 18.495 | 18.545 | 18.595 | 18.645 | 18.696 |
| **34** | 18.526 | 18.576 | 18.626 | 18.676 | 18.726 | 18.776 | 18.827 | 18.877 | 18.928 |
| **35** | 18.753 | 18.803 | 18.853 | 18.904 | 18.954 | 19.005 | 19.056 | 19.107 | 19.158 |
| **36** | 18.977 | 19.027 | 19.078 | 19.129 | 19.180 | 19.231 | 19.282 | 19.333 | 19.385 |
| **37** | 19.198 | 19.249 | 19.300 | 19.351 | 19.402 | 19.454 | 19.505 | 19.557 | 19.609 |
| **38** | 19.416 | 19.467 | 19.519 | 19.570 | 19.622 | 19.674 | 19.726 | 19.778 | 19.830 |
| **39** | 19.631 | 19.683 | 19.735 | 19.787 | 19.839 | 19.891 | 19.943 | 19.996 | 20.048 |
| **40** | 19.843 | 19.895 | 19.948 | 20.000 | 20.052 | 20.105 | 20.157 | 20.210 | 20.263 |

**Table 2.51** Nitrogen + Argon in mmHg per mg/L as a Function of Temperature and Salinity, 0–40 g/kg

| Temperature (°C) | Salinity (g/kg) | | | | | | | | |
|---|---|---|---|---|---|---|---|---|---|
| | **0.0** | **5.0** | **10.0** | **15.0** | **20.0** | **25.0** | **30.0** | **35.0** | **40.0** |
| **0** | 24.716 | 25.651 | 26.623 | 27.632 | 28.680 | 29.768 | 30.898 | 32.071 | 33.290 |
| **1** | 25.363 | 26.314 | 27.301 | 28.326 | 29.391 | 30.495 | 31.642 | 32.832 | 34.067 |
| **2** | 26.014 | 26.979 | 27.982 | 29.023 | 30.103 | 31.224 | 32.387 | 33.593 | 34.845 |
| **3** | 26.666 | 27.647 | 28.665 | 29.721 | 30.817 | 31.953 | 33.132 | 34.355 | 35.623 |
| **4** | 27.320 | 28.316 | 29.349 | 30.420 | 31.531 | 32.683 | 33.878 | 35.116 | 36.400 |
| **5** | 27.975 | 28.985 | 30.033 | 31.119 | 32.246 | 33.413 | 34.623 | 35.877 | 37.177 |
| **6** | 28.631 | 29.655 | 30.718 | 31.819 | 32.960 | 34.143 | 35.368 | 36.638 | 37.953 |
| **7** | 29.287 | 30.326 | 31.402 | 32.518 | 33.674 | 34.871 | 36.112 | 37.396 | 38.727 |
| **8** | 29.943 | 30.996 | 32.086 | 33.216 | 34.387 | 35.599 | 36.854 | 38.153 | 39.499 |
| **9** | 30.598 | 31.665 | 32.769 | 33.913 | 35.098 | 36.324 | 37.594 | 38.908 | 40.268 |
| **10** | 31.253 | 32.332 | 33.451 | 34.609 | 35.807 | 37.048 | 38.331 | 39.660 | 41.034 |
| **11** | 31.905 | 32.999 | 34.131 | 35.302 | 36.514 | 37.768 | 39.066 | 40.408 | 41.797 |
| **12** | 32.556 | 33.663 | 34.808 | 35.993 | 37.218 | 38.486 | 39.797 | 41.154 | 42.556 |
| **13** | 33.205 | 34.324 | 35.482 | 36.680 | 37.919 | 39.200 | 40.525 | 41.895 | 43.311 |
| **14** | 33.851 | 34.982 | 36.153 | 37.364 | 38.616 | 39.910 | 41.248 | 42.632 | 44.061 |
| **15** | 34.493 | 35.637 | 36.821 | 38.045 | 39.309 | 40.617 | 41.967 | 43.364 | 44.807 |
| **16** | 35.132 | 36.289 | 37.485 | 38.721 | 39.998 | 41.318 | 42.682 | 44.091 | 45.546 |
| **17** | 35.767 | 36.936 | 38.144 | 39.392 | 40.682 | 42.014 | 43.390 | 44.812 | 46.281 |
| **18** | 36.397 | 37.578 | 38.798 | 40.059 | 41.360 | 42.705 | 44.094 | 45.528 | 47.009 |
| **19** | 37.023 | 38.215 | 39.447 | 40.720 | 42.033 | 43.390 | 44.791 | 46.237 | 47.730 |
| **20** | 37.643 | 38.847 | 40.091 | 41.375 | 42.700 | 44.069 | 45.481 | 46.940 | 48.445 |
| **21** | 38.258 | 39.473 | 40.728 | 42.024 | 43.361 | 44.741 | 46.165 | 47.635 | 49.152 |
| **22** | 38.867 | 40.093 | 41.359 | 42.666 | 44.015 | 45.406 | 46.842 | 48.324 | 49.853 |
| **23** | 39.469 | 40.706 | 41.984 | 43.302 | 44.661 | 46.064 | 47.512 | 49.005 | 50.545 |
| **24** | 40.065 | 41.313 | 42.601 | 43.930 | 45.301 | 46.715 | 48.173 | 49.678 | 51.229 |
| **25** | 40.653 | 41.912 | 43.211 | 44.550 | 45.932 | 47.357 | 48.827 | 50.342 | 51.905 |
| **26** | 41.234 | 42.503 | 43.812 | 45.163 | 46.555 | 47.991 | 49.472 | 50.998 | 52.572 |
| **27** | 41.807 | 43.086 | 44.406 | 45.767 | 47.170 | 48.617 | 50.108 | 51.645 | 53.230 |
| **28** | 42.371 | 43.661 | 44.991 | 46.362 | 47.776 | 49.233 | 50.735 | 52.283 | 53.879 |
| **29** | 42.927 | 44.227 | 45.567 | 46.949 | 48.373 | 49.840 | 51.353 | 52.912 | 54.518 |
| **30** | 43.474 | 44.783 | 46.134 | 47.525 | 48.960 | 50.438 | 51.961 | 53.530 | 55.147 |
| **31** | 44.011 | 45.330 | 46.691 | 48.093 | 49.537 | 51.025 | 52.559 | 54.138 | 55.766 |
| **32** | 44.539 | 45.868 | 47.238 | 48.649 | 50.104 | 51.602 | 53.146 | 54.736 | 56.374 |
| **33** | 45.056 | 46.395 | 47.774 | 49.196 | 50.660 | 52.169 | 53.723 | 55.323 | 56.971 |
| **34** | 45.563 | 46.911 | 48.300 | 49.732 | 51.206 | 52.724 | 54.288 | 55.899 | 57.558 |
| **35** | 46.059 | 47.416 | 48.815 | 50.256 | 51.740 | 53.269 | 54.843 | 56.463 | 58.132 |
| **36** | 46.543 | 47.910 | 49.319 | 50.769 | 52.263 | 53.801 | 55.385 | 57.016 | 58.695 |
| **37** | 47.016 | 48.392 | 49.810 | 51.270 | 52.774 | 54.322 | 55.915 | 57.557 | 59.246 |
| **38** | 47.477 | 48.862 | 50.289 | 51.759 | 53.272 | 54.830 | 56.433 | 58.085 | 59.785 |
| **39** | 47.926 | 49.320 | 50.756 | 52.235 | 53.757 | 55.325 | 56.939 | 58.600 | 60.310 |
| **40** | 48.361 | 49.764 | 51.210 | 52.698 | 54.230 | 55.807 | 57.431 | 59.102 | 60.822 |

**Table 2.52** Nitrogen + Argon in mmHg per mg/L as a Function of Temperature and Salinity, 33–37 g/kg

| Temperature (°C) | Salinity (g/kg) | | | | | | | | |
|---|---|---|---|---|---|---|---|---|---|
| | **33.0** | **33.5** | **34.0** | **34.5** | **35.0** | **35.5** | **36.0** | **36.5** | **37.0** |
| **0** | 31.597 | 31.715 | 31.833 | 31.952 | 32.071 | 32.191 | 32.311 | 32.432 | 32.553 |
| **1** | 32.350 | 32.470 | 32.590 | 32.711 | 32.832 | 32.953 | 33.075 | 33.197 | 33.320 |
| **2** | 33.105 | 33.226 | 33.348 | 33.470 | 33.593 | 33.716 | 33.840 | 33.964 | 34.088 |
| **3** | 33.860 | 33.983 | 34.107 | 34.230 | 34.355 | 34.479 | 34.605 | 34.730 | 34.856 |
| **4** | 34.615 | 34.740 | 34.865 | 34.990 | 35.116 | 35.243 | 35.369 | 35.497 | 35.624 |
| **5** | 35.370 | 35.496 | 35.623 | 35.750 | 35.877 | 36.005 | 36.134 | 36.263 | 36.392 |
| **6** | 36.124 | 36.252 | 36.380 | 36.509 | 36.638 | 36.767 | 36.897 | 37.027 | 37.158 |
| **7** | 36.877 | 37.006 | 37.136 | 37.266 | 37.396 | 37.527 | 37.659 | 37.791 | 37.923 |
| **8** | 37.628 | 37.759 | 37.890 | 38.021 | 38.153 | 38.286 | 38.419 | 38.552 | 38.686 |
| **9** | 38.377 | 38.509 | 38.641 | 38.774 | 38.908 | 39.042 | 39.176 | 39.311 | 39.446 |
| **10** | 39.123 | 39.256 | 39.390 | 39.525 | 39.660 | 39.795 | 39.931 | 40.067 | 40.204 |
| **11** | 39.866 | 40.001 | 40.136 | 40.272 | 40.408 | 40.545 | 40.682 | 40.820 | 40.958 |
| **12** | 40.606 | 40.742 | 40.879 | 41.016 | 41.154 | 41.292 | 41.430 | 41.569 | 41.709 |
| **13** | 41.341 | 41.479 | 41.617 | 41.756 | 41.895 | 42.034 | 42.174 | 42.315 | 42.456 |
| **14** | 42.073 | 42.212 | 42.351 | 42.491 | 42.632 | 42.772 | 42.914 | 43.056 | 43.198 |
| **15** | 42.800 | 42.940 | 43.081 | 43.222 | 43.364 | 43.506 | 43.648 | 43.792 | 43.935 |
| **16** | 43.522 | 43.663 | 43.805 | 43.948 | 44.091 | 44.234 | 44.378 | 44.522 | 44.667 |
| **17** | 44.238 | 44.381 | 44.524 | 44.668 | 44.812 | 44.957 | 45.102 | 45.248 | 45.394 |
| **18** | 44.948 | 45.093 | 45.237 | 45.382 | 45.528 | 45.674 | 45.820 | 45.967 | 46.114 |
| **19** | 45.653 | 45.798 | 45.944 | 46.090 | 46.237 | 46.384 | 46.532 | 46.680 | 46.829 |
| **20** | 46.351 | 46.497 | 46.644 | 46.792 | 46.940 | 47.088 | 47.237 | 47.386 | 47.536 |
| **21** | 47.042 | 47.190 | 47.338 | 47.486 | 47.635 | 47.785 | 47.935 | 48.086 | 48.237 |
| **22** | 47.726 | 47.875 | 48.024 | 48.174 | 48.324 | 48.475 | 48.626 | 48.778 | 48.930 |
| **23** | 48.402 | 48.552 | 48.702 | 48.853 | 49.005 | 49.157 | 49.309 | 49.462 | 49.615 |
| **24** | 49.070 | 49.221 | 49.373 | 49.525 | 49.678 | 49.831 | 49.984 | 50.138 | 50.293 |
| **25** | 49.730 | 49.883 | 50.035 | 50.189 | 50.342 | 50.496 | 50.651 | 50.806 | 50.962 |
| **26** | 50.382 | 50.535 | 50.689 | 50.844 | 50.998 | 51.153 | 51.309 | 51.465 | 51.622 |
| **27** | 51.025 | 51.179 | 51.334 | 51.490 | 51.645 | 51.802 | 51.958 | 52.116 | 52.274 |
| **28** | 51.658 | 51.814 | 51.970 | 52.126 | 52.283 | 52.441 | 52.598 | 52.757 | 52.916 |
| **29** | 52.282 | 52.439 | 52.596 | 52.754 | 52.912 | 53.070 | 53.229 | 53.388 | 53.548 |
| **30** | 52.897 | 53.054 | 53.212 | 53.371 | 53.530 | 53.690 | 53.850 | 54.010 | 54.171 |
| **31** | 53.501 | 53.659 | 53.819 | 53.978 | 54.138 | 54.299 | 54.460 | 54.621 | 54.783 |
| **32** | 54.094 | 54.254 | 54.414 | 54.575 | 54.736 | 54.898 | 55.060 | 55.222 | 55.385 |
| **33** | 54.677 | 54.838 | 54.999 | 55.161 | 55.323 | 55.486 | 55.649 | 55.813 | 55.977 |
| **34** | 55.249 | 55.411 | 55.573 | 55.736 | 55.899 | 56.063 | 56.227 | 56.391 | 56.557 |
| **35** | 55.809 | 55.972 | 56.135 | 56.299 | 56.463 | 56.628 | 56.793 | 56.959 | 57.125 |
| **36** | 56.358 | 56.522 | 56.686 | 56.851 | 57.016 | 57.182 | 57.348 | 57.515 | 57.682 |
| **37** | 56.894 | 57.059 | 57.225 | 57.390 | 57.557 | 57.723 | 57.891 | 58.058 | 58.227 |
| **38** | 57.418 | 57.584 | 57.751 | 57.917 | 58.085 | 58.252 | 58.421 | 58.589 | 58.759 |
| **39** | 57.930 | 58.096 | 58.264 | 58.432 | 58.600 | 58.769 | 58.938 | 59.108 | 59.278 |
| **40** | 58.428 | 58.595 | 58.764 | 58.933 | 59.102 | 59.272 | 59.442 | 59.613 | 59.784 |

**Table 2.53** Carbon Dioxide in mmHg per mg/L as a Function of Temperature and Salinity, 0–40 g/kg

| Temperature (°C) | Salinity (g/kg) | | | | | | | | |
|---|---|---|---|---|---|---|---|---|---|
| | **0.0** | **5.0** | **10.0** | **15.0** | **20.0** | **25.0** | **30.0** | **35.0** | **40.0** |
| **0** | 0.22260 | 0.22847 | 0.23451 | 0.24069 | 0.24705 | 0.25357 | 0.26026 | 0.26713 | 0.27418 |
| **1** | 0.23154 | 0.23763 | 0.24388 | 0.25030 | 0.25688 | 0.26364 | 0.27057 | 0.27769 | 0.28499 |
| **2** | 0.24071 | 0.24701 | 0.25349 | 0.26013 | 0.26695 | 0.27394 | 0.28112 | 0.28849 | 0.29605 |
| **3** | 0.25009 | 0.25662 | 0.26332 | 0.27019 | 0.27724 | 0.28448 | 0.29190 | 0.29952 | 0.30734 |
| **4** | 0.25969 | 0.26644 | 0.27337 | 0.28047 | 0.28776 | 0.29524 | 0.30292 | 0.31079 | 0.31887 |
| **5** | 0.26951 | 0.27648 | 0.28364 | 0.29098 | 0.29851 | 0.30624 | 0.31416 | 0.32229 | 0.33064 |
| **6** | 0.27954 | 0.28674 | 0.29413 | 0.30171 | 0.30948 | 0.31746 | 0.32564 | 0.33403 | 0.34263 |
| **7** | 0.28979 | 0.29722 | 0.30484 | 0.31266 | 0.32068 | 0.32890 | 0.33733 | 0.34598 | 0.35485 |
| **8** | 0.30026 | 0.30792 | 0.31577 | 0.32383 | 0.33209 | 0.34056 | 0.34925 | 0.35816 | 0.36730 |
| **9** | 0.31093 | 0.31882 | 0.32691 | 0.33521 | 0.34372 | 0.35244 | 0.36139 | 0.37056 | 0.37996 |
| **10** | 0.32182 | 0.32994 | 0.33827 | 0.34681 | 0.35556 | 0.36454 | 0.37374 | 0.38317 | 0.39284 |
| **11** | 0.33291 | 0.34127 | 0.34983 | 0.35861 | 0.36761 | 0.37684 | 0.38630 | 0.39599 | 0.40593 |
| **12** | 0.34421 | 0.35280 | 0.36160 | 0.37062 | 0.37987 | 0.38935 | 0.39906 | 0.40902 | 0.41922 |
| **13** | 0.35572 | 0.36454 | 0.37358 | 0.38284 | 0.39233 | 0.40206 | 0.41203 | 0.42224 | 0.43271 |
| **14** | 0.36742 | 0.37647 | 0.38575 | 0.39525 | 0.40499 | 0.41496 | 0.42519 | 0.43566 | 0.44639 |
| **15** | 0.37932 | 0.38861 | 0.39812 | 0.40786 | 0.41784 | 0.42806 | 0.43854 | 0.44927 | 0.46026 |
| **16** | 0.39142 | 0.40093 | 0.41068 | 0.42066 | 0.43088 | 0.44135 | 0.45207 | 0.46306 | 0.47431 |
| **17** | 0.40371 | 0.41345 | 0.42342 | 0.43364 | 0.44410 | 0.45482 | 0.46579 | 0.47703 | 0.48853 |
| **18** | 0.41618 | 0.42615 | 0.43635 | 0.44680 | 0.45750 | 0.46846 | 0.47968 | 0.49116 | 0.50292 |
| **19** | 0.42884 | 0.43903 | 0.44946 | 0.46014 | 0.47108 | 0.48227 | 0.49373 | 0.50546 | 0.51747 |
| **20** | 0.44168 | 0.45209 | 0.46275 | 0.47366 | 0.48482 | 0.49625 | 0.50795 | 0.51992 | 0.53217 |
| **21** | 0.45470 | 0.46532 | 0.47620 | 0.48733 | 0.49873 | 0.51038 | 0.52231 | 0.53452 | 0.54702 |
| **22** | 0.46788 | 0.47872 | 0.48982 | 0.50117 | 0.51279 | 0.52467 | 0.53683 | 0.54927 | 0.56200 |
| **23** | 0.48123 | 0.49229 | 0.50359 | 0.51516 | 0.52700 | 0.53910 | 0.55148 | 0.56415 | 0.57711 |
| **24** | 0.49475 | 0.50601 | 0.51752 | 0.52930 | 0.54135 | 0.55367 | 0.56627 | 0.57916 | 0.59234 |
| **25** | 0.50842 | 0.51988 | 0.53160 | 0.54358 | 0.55584 | 0.56837 | 0.58118 | 0.59428 | 0.60768 |
| **26** | 0.52224 | 0.53390 | 0.54582 | 0.55800 | 0.57046 | 0.58319 | 0.59621 | 0.60952 | 0.62312 |
| **27** | 0.53621 | 0.54806 | 0.56017 | 0.57255 | 0.58520 | 0.59813 | 0.61134 | 0.62485 | 0.63866 |
| **28** | 0.55032 | 0.56236 | 0.57465 | 0.58722 | 0.60006 | 0.61318 | 0.62658 | 0.64028 | 0.65428 |
| **29** | 0.56457 | 0.57678 | 0.58926 | 0.60200 | 0.61502 | 0.62832 | 0.64191 | 0.65580 | 0.66998 |
| **30** | 0.57895 | 0.59133 | 0.60398 | 0.61689 | 0.63009 | 0.64356 | 0.65733 | 0.67139 | 0.68574 |
| **31** | 0.59345 | 0.60600 | 0.61881 | 0.63189 | 0.64525 | 0.65889 | 0.67282 | 0.68704 | 0.70157 |
| **32** | 0.60807 | 0.62077 | 0.63374 | 0.64698 | 0.66049 | 0.67429 | 0.68837 | 0.70275 | 0.71743 |
| **33** | 0.62280 | 0.63565 | 0.64877 | 0.66215 | 0.67581 | 0.68976 | 0.70399 | 0.71851 | 0.73334 |
| **34** | 0.63764 | 0.65063 | 0.66388 | 0.67741 | 0.69121 | 0.70529 | 0.71965 | 0.73431 | 0.74927 |
| **35** | 0.65258 | 0.66570 | 0.67908 | 0.69274 | 0.70666 | 0.72087 | 0.73536 | 0.75014 | 0.76522 |
| **36** | 0.66761 | 0.68085 | 0.69436 | 0.70813 | 0.72217 | 0.73649 | 0.75110 | 0.76599 | 0.78118 |
| **37** | 0.68274 | 0.69609 | 0.70970 | 0.72357 | 0.73772 | 0.75215 | 0.76686 | 0.78185 | 0.79714 |
| **38** | 0.69794 | 0.71139 | 0.72510 | 0.73907 | 0.75331 | 0.76783 | 0.78263 | 0.79771 | 0.81309 |
| **39** | 0.71321 | 0.72675 | 0.74055 | 0.75461 | 0.76893 | 0.78353 | 0.79841 | 0.81357 | 0.82901 |
| **40** | 0.72855 | 0.74217 | 0.75605 | 0.77018 | 0.78457 | 0.79924 | 0.81418 | 0.82940 | 0.84490 |

**Table 2.54** Carbon Dioxide in mmHg per mg/L as a Function of Temperature and Salinity, 33–37 g/kg

| Temperature (°C) | Salinity (g/kg) | | | | | | | | |
|---|---|---|---|---|---|---|---|---|---|
| | **33.0** | **33.5** | **34.0** | **34.5** | **35.0** | **35.5** | **36.0** | **36.5** | **37.0** |
| **0** | 0.26436 | 0.26505 | 0.26574 | 0.26644 | 0.26713 | 0.26783 | 0.26853 | 0.26923 | 0.26993 |
| **1** | 0.27482 | 0.27553 | 0.27625 | 0.27697 | 0.27769 | 0.27841 | 0.27913 | 0.27986 | 0.28059 |
| **2** | 0.28552 | 0.28626 | 0.28700 | 0.28774 | 0.28849 | 0.28923 | 0.28998 | 0.29073 | 0.29149 |
| **3** | 0.29645 | 0.29722 | 0.29798 | 0.29875 | 0.29952 | 0.30030 | 0.30107 | 0.30185 | 0.30263 |
| **4** | 0.30762 | 0.30841 | 0.30920 | 0.31000 | 0.31079 | 0.31159 | 0.31239 | 0.31319 | 0.31400 |
| **5** | 0.31902 | 0.31983 | 0.32065 | 0.32147 | 0.32229 | 0.32312 | 0.32395 | 0.32477 | 0.32561 |
| **6** | 0.33064 | 0.33149 | 0.33233 | 0.33318 | 0.33403 | 0.33488 | 0.33573 | 0.33658 | 0.33744 |
| **7** | 0.34250 | 0.34336 | 0.34424 | 0.34511 | 0.34598 | 0.34686 | 0.34774 | 0.34862 | 0.34950 |
| **8** | 0.35457 | 0.35547 | 0.35636 | 0.35726 | 0.35816 | 0.35907 | 0.35997 | 0.36088 | 0.36179 |
| **9** | 0.36686 | 0.36778 | 0.36871 | 0.36963 | 0.37056 | 0.37149 | 0.37242 | 0.37336 | 0.37429 |
| **10** | 0.37937 | 0.38032 | 0.38127 | 0.38222 | 0.38317 | 0.38413 | 0.38509 | 0.38605 | 0.38701 |
| **11** | 0.39208 | 0.39306 | 0.39403 | 0.39501 | 0.39599 | 0.39697 | 0.39796 | 0.39895 | 0.39994 |
| **12** | 0.40501 | 0.40600 | 0.40701 | 0.40801 | 0.40902 | 0.41003 | 0.41104 | 0.41205 | 0.41307 |
| **13** | 0.41813 | 0.41915 | 0.42018 | 0.42121 | 0.42224 | 0.42328 | 0.42432 | 0.42536 | 0.42640 |
| **14** | 0.43144 | 0.43249 | 0.43355 | 0.43460 | 0.43566 | 0.43672 | 0.43779 | 0.43885 | 0.43992 |
| **15** | 0.44495 | 0.44602 | 0.44710 | 0.44818 | 0.44927 | 0.45036 | 0.45145 | 0.45254 | 0.45363 |
| **16** | 0.45863 | 0.45974 | 0.46084 | 0.46195 | 0.46306 | 0.46417 | 0.46529 | 0.46641 | 0.46753 |
| **17** | 0.47250 | 0.47363 | 0.47476 | 0.47589 | 0.47703 | 0.47816 | 0.47931 | 0.48045 | 0.48160 |
| **18** | 0.48654 | 0.48769 | 0.48884 | 0.49000 | 0.49116 | 0.49233 | 0.49349 | 0.49466 | 0.49583 |
| **19** | 0.50074 | 0.50191 | 0.50309 | 0.50428 | 0.50546 | 0.50665 | 0.50784 | 0.50904 | 0.51023 |
| **20** | 0.51510 | 0.51630 | 0.51750 | 0.51871 | 0.51992 | 0.52113 | 0.52235 | 0.52356 | 0.52479 |
| **21** | 0.52961 | 0.53083 | 0.53206 | 0.53329 | 0.53452 | 0.53576 | 0.53700 | 0.53824 | 0.53949 |
| **22** | 0.54426 | 0.54551 | 0.54676 | 0.54801 | 0.54927 | 0.55053 | 0.55179 | 0.55306 | 0.55433 |
| **23** | 0.55905 | 0.56032 | 0.56159 | 0.56287 | 0.56415 | 0.56543 | 0.56672 | 0.56801 | 0.56930 |
| **24** | 0.57397 | 0.57526 | 0.57656 | 0.57786 | 0.57916 | 0.58046 | 0.58177 | 0.58308 | 0.58439 |
| **25** | 0.58901 | 0.59032 | 0.59164 | 0.59296 | 0.59428 | 0.59561 | 0.59694 | 0.59827 | 0.59960 |
| **26** | 0.60416 | 0.60549 | 0.60683 | 0.60817 | 0.60952 | 0.61086 | 0.61221 | 0.61357 | 0.61492 |
| **27** | 0.61941 | 0.62077 | 0.62213 | 0.62349 | 0.62485 | 0.62622 | 0.62759 | 0.62896 | 0.63034 |
| **28** | 0.63477 | 0.63614 | 0.63752 | 0.63890 | 0.64028 | 0.64167 | 0.64306 | 0.64445 | 0.64585 |
| **29** | 0.65021 | 0.65160 | 0.65300 | 0.65439 | 0.65580 | 0.65720 | 0.65861 | 0.66002 | 0.66143 |
| **30** | 0.66573 | 0.66714 | 0.66855 | 0.66997 | 0.67139 | 0.67281 | 0.67423 | 0.67566 | 0.67709 |
| **31** | 0.68132 | 0.68274 | 0.68417 | 0.68560 | 0.68704 | 0.68848 | 0.68992 | 0.69137 | 0.69281 |
| **32** | 0.69697 | 0.69841 | 0.69985 | 0.70130 | 0.70275 | 0.70421 | 0.70567 | 0.70713 | 0.70859 |
| **33** | 0.71267 | 0.71413 | 0.71559 | 0.71705 | 0.71851 | 0.71998 | 0.72146 | 0.72293 | 0.72441 |
| **34** | 0.72841 | 0.72989 | 0.73136 | 0.73283 | 0.73431 | 0.73580 | 0.73728 | 0.73877 | 0.74026 |
| **35** | 0.74419 | 0.74568 | 0.74716 | 0.74865 | 0.75014 | 0.75164 | 0.75314 | 0.75464 | 0.75614 |
| **36** | 0.76000 | 0.76149 | 0.76299 | 0.76449 | 0.76599 | 0.76750 | 0.76901 | 0.77052 | 0.77203 |
| **37** | 0.77582 | 0.77732 | 0.77883 | 0.78034 | 0.78185 | 0.78337 | 0.78489 | 0.78641 | 0.78793 |
| **38** | 0.79165 | 0.79316 | 0.79467 | 0.79619 | 0.79771 | 0.79924 | 0.80076 | 0.80229 | 0.80383 |
| **39** | 0.80747 | 0.80899 | 0.81051 | 0.81204 | 0.81357 | 0.81510 | 0.81663 | 0.81817 | 0.81971 |
| **40** | 0.82328 | 0.82480 | 0.82633 | 0.82786 | 0.82940 | 0.83094 | 0.83248 | 0.83402 | 0.83557 |

# 3 Supersaturation of Gases

The previous two chapters have focused on computation of saturation concentrations between the gas and the liquid phases. While saturation concentrations are important, in the real world waters are rarely in equilibrium with the gas phase. Measured dissolved gas levels may be greater than the equilibrium concentration, supersaturated, less than the equilibrium concentration, or undersaturated. Supersaturated water will tend to lose gas to the atmosphere, but the rate may be very slow and gas transfer only occurs across the air–water interface. Gas supersaturation may result in a disease called gas bubble disease (GBD). This disease results from the formation of gas bubbles in the blood and tissues of aquatic animals.

Dissolved gases may become supersaturated due to a number of natural and human causes (Weitkamp and Katz, 1980; Colt, 1986). Well water or spring may contain high concentrations of nitrogen, argon, and carbon dioxide and little oxygen. Water falling over dams or waterfalls can result in lethal concentrations of dissolved gases by entrainment of air bubbles. In culture systems, air leaks on the suction side of a pump or the use of some types of aeration can produce lethal levels of gas supersaturation. The heating of water or mixing of waters of different temperatures can also produce gas supersaturation. Photosynthesis can produce high concentrations of oxygen that are lethal in some conditions.

In hydrologic and oceanographic research, dissolved gases may be used as tracers. Because of different solubilities and temperature dependencies, measurement of gases can also be used to study important physical and biological processes (Hamme and Emerson, 2004).

The production, reduction, and reporting of gas supersaturation values are discussed in this section. In contrast to the previous sections, in which gas solubilities and gas concentrations were discussed in concentration units (mol/kg or mg/L), the impacts of gas supersaturation depends on pressure.

## Computation of Supersaturation for Individual Gases in Oceanography

As previously discussed, supersaturation of individual gases (or ratios of the supersaturation of pairs of gases) can be used to study important oceanographic processes. Because of the large range of depths encountered, chemical

**Dissolved Gas Concentration in Water. DOI: 10.1016/B978-0-12-415916-7.00003-6**

oceanographers compute individual gas supersaturation from the following equation (Millero, 1996):

$$\Delta_i = \left[\frac{C^\dagger}{C^\dagger_{o,s,\theta}}\right]100 \tag{3.1}$$

where

$\Delta_i$ = percent supersaturation of gas $i$
$C^\dagger$ = measured concentration of gas $i$
$C^\dagger_{o,s,\theta}$ = standard air solubility concentration of gas $i$ at the measured salinity and potential temperature $\theta$.

The potential temperature is the temperature of water allowed to adiabatically expand (no heat transfer to the surroundings) to the water surface. The potential temperature is less than the *in situ* water temperature and can be determined from tables or regression equations.

## Computation and Reporting of Gas Supersaturation

Gas supersaturation levels are reported in terms of several unique parameters. Concentrations of gases in terms of μmol/kg, mg/L, or percent saturation are not too significant in gas supersaturation work because formation of gas bubbles depends on pressure, rather than on concentrations. The physics of dissolved gases, the physics and physiological basis of GBD, gas analysis, and the computation of supersaturation levels will be discussed in this section.

### *Physics of Dissolved Gases*

The sum of the partial pressures of all dissolved gases in the liquid and gas phases is equal to:

*Liquid Phase*

$$\text{Total gas pressure} = \sum_{i=1}^{n} P_i^l + P_{wv}, \tag{3.2}$$

*Gas Phase*

$$\text{Barometric pressure} = \sum_{i=1}^{n} P_i^g + P_{wv} \tag{3.3}$$

where

$P_i^l$ = partial pressure (or gas tension) of the $i$th gas in the liquid phase (kPa, atm, or mmHg)
$P_i^g$ = partial pressure of the $i$th gas in the gas phase (kPa, atm, or mmHg)
$P_{wv}$ = vapor pressure of water (kPa, atm, or mmHg).

In gas supersaturation work as it relates to bubble formation or GBD, only the partial pressures of oxygen, nitrogen, argon, carbon dioxide, and water vapor are considered. Under many conditions, the contribution of carbon dioxide can be ignored. As previously discussed, the supersaturation of other "minor" gases may be important to understanding basic chemical and biological processes that occur, but they do not make significant contributions to total gas pressure (TGP) or barometric pressure.

For the *i*th gas, the values of $P_i^l$ and $P_i^g$ are equal to

$$P_i^l = \left[\frac{C_i}{\beta_i}\right] A_i \tag{3.4}$$

$$P_i^g = \chi_i(\mathrm{BP} - P_{wv}) \tag{3.5}$$

At equilibrium, the partial pressure (or gas tension) of a gas in the liquid phase is equal to its partial pressure in the gas phase. Depending on the magnitude of the two parameters, three conditions may occur:

$$P_i^l > P_i^g: \text{ supersaturated} \tag{3.6}$$

$$P_i^l = P_i^g: \text{ equilibrium} \tag{3.7}$$

$$P_i^l < P_i^g: \text{ undersaturated} \tag{3.8}$$

---

**Example 3-1**

Compute partial pressures and gas tensions of the major atmospheric gases at 12.9°C. Assume that the barometric pressure is 760 mmHg and freshwater conditions. Use Eqs (3.2) and (3.3). Compute the TGP and barometric pressure from this information.

| Value | Table | Value | Table |
|---|---|---|---|
| $C^*_{O_2} = 10.560$ | 1.9 | $\beta_{O_2} = 0.03581$ | 1.32 |
| $C^*_{N_2} = 17.210$ | 1.10 | $\beta_{N_2} = 0.01789$ | 1.33 |
| $C^*_{Ar} = 0.6452$ | 1.11 | $\beta_{Ar} = 0.03930$ | 1.34 |
| $C^*_{CO_2} = 0.8205$ | 1.12 | $\beta_{CO_2} = 1.0843$ | 1.35 |
| $A_{O_2} = 0.5318$ | D-1 | $\chi_{O_2} = 0.20946$ | D-1 |
| $A_{N_2} = 0.6078$ | D-1 | $\chi_{N_2} = 0.78084$ | D-1 |
| $A_{Ar} = 0.4260$ | D-1 | $\chi_{Ar} = 0.00934$ | D-1 |
| $A_{CO_2} = 0.3845$ | D-1 | $\chi_{CO_2} = 0.000390$ | D-1 |
| $P_{wv} = 11.158$ | 1.21 | | |

Equation (3.4):

$$P_i^l = \left[\frac{C_i}{\beta_i}\right] A_i$$

$$P_{O_2}^l = \left[\frac{10.560}{0.03581}\right] 0.5318 = 156.82 \text{ mmHg}$$

$$P_{N_2}^l = \left[\frac{17.210}{0.01789}\right] 0.6078 = 584.70 \text{ mmHg}$$

$$P_{Ar}^l = \left[\frac{0.6452}{0.03930}\right] 0.4260 = 6.99 \text{ mmHg}$$

$$P_{CO_2}^l = \left[\frac{0.8205}{1.0843}\right] 0.3845 = 0.29 \text{ mmHg}$$

Equation (3.5):

$$P_i^g = \chi_i(\text{BP} - P_{wv})$$

$$P_{O_2}^g = 0.20946(760 - 11.158) = 156.85 \text{ mmHg}$$

$$P_{N_2}^g = 0.78084(760 - 11.158) = 584.73 \text{ mmHg}$$

$$P_{Ar}^g = 0.00934(760 - 11.158) = 6.99 \text{ mmHg}$$

$$P_{CO_2}^g = 0.000390(760 - 11.158) = 0.29 \text{ mmHg}$$

**Compute TGP and Barometric Pressure from Partial Pressures**

| Gas | TGP (Eq. (3.2) (mmHg)) | Barometric Pressure (Eq. (3.3) (mmHg)) |
|---|---|---|
| $O_2$ | 156.82 | 156.85 |
| $N_2$ | 584.70 | 584.73 |
| Ar | 6.99 | 6.99 |
| $CO_2$ | 0.29 | 0.29 |
| $H_2O$ | 11.158 | 11.158 |
| Total (3 place) | **759.958** | **760.018** |
| Total (1 place) | **760.0** | **760.0** |

The differences between the two parameters are due to uncertainty in the air solubility and Bunsen coefficients, and to the assumption that atmospheric gases can be considered ideal gases. In gas supersaturation work, TGPs are typically reported to the nearest 0.1 mmHg, so these two parameters can be considered equal.

The difference between the total gas pressure (TGP) and the local barometric pressure is called the $\Delta P$:

$$\Delta P = \text{Total gas pressure} - \text{BP} \tag{3.9}$$

or

$$\text{Total gas pressure} = \text{BP} + \Delta P \tag{3.10}$$

TGP is the absolute pressure of the sum of the partial pressures + water vapor; $\Delta P$ is the gauge pressure. TGP may also be expressed as a percent of the local barometric pressure:

$$\text{Total gas pressure (\%)} = \left[\frac{\text{BP} + \Delta P}{\text{BP}}\right] 100 \tag{3.11}$$

$\Delta P$ can be directly measured by several types of instruments. Similar to partial pressures, three conditions can occur for $\Delta P$ and TPG(%):

$$\Delta P > 0 \text{ or TGP(\%)} > 100 \text{ supersaturated} \tag{3.12}$$

$$\Delta P = 0 \text{ or TGP(\%)} = 100 \text{ equilibrium} \tag{3.13}$$

$$\Delta P < 0 \text{ or TGP(\%)} < 100 \text{ undersaturated} \tag{3.14}$$

**Example 3-2**

If the measured $\Delta P = 156$ mmHg, compute the TGP (mmHg) and TGP (%) at sea level and at 9340 ft. Use Eqs (1.9), (3.10), and (3.11).

*Barometric Pressure at Sea Level*

Assume that barometric pressure = 760 mmHg.

*Barometric Pressure at 9340 ft*

Equation (1.9):

$$\log_{10} \text{BP} = 2.880814 - \frac{h'}{63{,}718.2}$$

$$\log_{10} \text{BP} = 2.880814 - \frac{9340}{63{,}718.2} = 542.29 \text{ mmHg}$$

*TGP at Sea Level*

Equation (3.10):

Total gas pressure = BP + $\Delta P$

Total gas pressure = 760 + 156 = 916.0 mmHg

Equation (3.11):

$$\text{Total gas pressure (\%)} = \left[\frac{\text{BP} + \Delta P}{\text{BP}}\right]100$$

$$\text{Total gas pressure (\%)} = \left[\frac{760 + 156}{760}\right]100 = 120.5\%$$

*TGP at 9340 ft*

Total gas pressure = 542.29 + 156 = 698.3 mmHg

$$\text{Total gas pressure (\%)} = \left[\frac{542.29 + 156}{542.29}\right]100 = 128.8\%$$

In a similar manner, a differential pressure for a single gas can be defined as:

$$\Delta P_i = P_i^l - P_i^g \tag{3.15}$$

or

$$\Delta P_i = \frac{C_i}{\beta_i}(A_i) - (\chi_i(\text{BP} - P_{wv})) \tag{3.16}$$

The sum of all the $\Delta P_i$ s is equal to $\Delta P$.
The percent of saturation for a single gas is equal to:

$$\text{Percent saturation} = \left[\frac{P_i^l}{P_i^g}\right]100 \tag{3.17}$$

The percent saturation of the four major atmospheric gases will be abbreviated $O_2$ (%), $N_2$ (%), Ar (%), and $CO_2$ (%). In some types of gas analysis, nitrogen and argon are determined together. The symbols $\Delta P_{N_2+Ar}$ and $N_2$+Ar (%) represent the $\Delta P$ and percent saturation of this composite gas. TGP as percent of barometric pressure will be abbreviated as TGP (%).

## The Physics and Physiological Basis of GBD

Studies in hyperbaric physiology have shown that initial bubble formation depends on $\Delta P$ (D'Aoust and Clark, 1980). The $\Delta P$ value is the pressure that inflates bubbles. If $\Delta P \leq 0$, then bubbles cannot form regardless of the degree of supersaturation of a single gas. [There may be some very special cases in mixed gas diving where this may not be true, but this section will be restricted to aquatic animals exposed to pressures near 1 atm.] The impact of a given $\Delta P$ value may depend on the composition of the dissolved gases. Therefore, it is necessary to include information on partial pressures of $\Delta P$s of individual gases.

While the risk to aquatic animals depends on $\Delta P$, water quality standards in the US are written in terms of TGP (%). While it is possible to readily convert between the two units, it has never been clearly determined how a water quality standard of 110% TGP should be adjusted to elevation. At 600 mmHg, 110% is equal to 60 mmHg compared to 76 mmHg at sea level.

While typically bubbles cannot form when $\Delta P \leq 0$, it may be possible to form bubbles inside an animal when the $\Delta P$ in the water is less than zero. The condition for bubble formation ($\Delta P \geq 0$) should be applied inside the animal where bubbles form and not to the ambient water. Some fish have the ability to generate very high partial pressures in the swim bladder and eyes, even when the dissolved gas concentrations in the surrounding water are close to equilibrium.

## Dissolved Gas Analysis

Direct measurement of $\Delta P$ is the preferred method of analysis (Colt, 1983). The instruments used for this type of analysis are commonly referred to as "Weiss saturometers." This is a misnomer as these instruments measure $\Delta P$, not saturation. These instruments consist of a gas permeable silicone rubber tubing connected to a pressure measuring device. The tubing is permeable to all the dissolved gases, including water vapor, and therefore can be used to measure $\Delta P$ or TGP directly. This type of analysis will be referred to as the membrane diffusion method (MDM). Additional information on this type of instrument is presented by Bouck (1982), D'Aoust and Clark (1980), and Fickeisen et al. (1975).

Membrane diffusion instruments are subject to at least four types of errors or operational problems. When a unit is moved from one water body to another (or from air to water), a finite amount of gas must diffuse from the water into the tubing (or the reverse). Assuming that air is an ideal gas, the moles of gas that must be transferred are equal to

$$\Delta n = \frac{V}{R}\left[\frac{\mathrm{TGP}_1}{T_2} - \frac{\mathrm{TGP}_2}{T_2}\right] \quad (3.18)$$

where

$\Delta n$ = gas transferred into tubing (mol)
$V$ = volume inside tubing (L)

$R$ = gas constant (0.082057463 L atm/mol K)
$TGP_1$ = initial total gas pressure (atm)
$TGP_2$ = final total gas pressure (atm)
$T_1$ = initial temperature (K)
$T_2$ = final temperature (K).

The transfer of gas through the membrane is proportional to the surface area and pressure difference. Therefore, equilibrium between the two TGPs occurs only as time $\to \infty$. The time required for the pressure inside the tubing to approach the final reading ($TGP_2$) within measurement error ranges from 5 to 30 min. The design of TGP sensors involves trade-offs between internal "dead space" volume, wall thickness, and membrane surface area. If the reading is taken too soon, it will be in error. In high TGPs, bubbles may form on the outside of the tubing; this will cause a significantly low reading error. To reduce this error, the sensing unit should be submerged as far as possible and/or bubbles should be dislodged by shaking or moving the special pumping systems on some units. Errors in the pressure reading can occur due to either errors in the pressure sensing unit or due to leaks or tears in the tubing. The accuracy of the pressure sensor can be checked by calibration against a pressure standard. Leaks or tears in the tubing are much more difficult to detect, as they may only occur for certain TGP values (typically for $\Delta P$s $>$ than a given value). Calibration of the pressure sensor will typically not detect problems with the integrity of the tubing. Some manufacturers have suggested techniques for detection of leaks or tears. Bouck (1982) developed a system to test the overall accuracy and time response of membrane diffusion instruments, but nothing is commercially available at this time with this capacity.

Once the $\Delta P$ or TGP has been measured, information on component gas levels is typically needed. Dissolved oxygen and carbon dioxide concentrations can be determined by standard analytical methods (D'Aoust and Clark, 1980). Nitrogen and argon gases are inert and, therefore, are difficult to measure directly, except by use of a gas chromatograph or mass spectrometer. In the MDM, the sum of the gas tensions of nitrogen and argon are determined by difference. Since only the sum is computed, it is convenient to consider nitrogen and argon together in gas supersaturation work. This gas should be referred to as nitrogen + argon ($N_2$ + Ar), but some workers use just $N_2$. At equilibrium, dissolved nitrogen + argon gas is 99% nitrogen gas on a pressure basis.

Assuming that only the major atmospheric gases are present, Eqs (3.2) and (3.10) can be combined to give:

$$\mathrm{BP} + \Delta P = P^l_{\mathrm{O_2}} + P^l_{\mathrm{N_2}} + P^l_{\mathrm{Ar}} + P^l_{\mathrm{CO_2}} + P_{wv} \tag{3.19}$$

or

$$P^l_{\mathrm{N_2}} + P^l_{\mathrm{Ar}} + P^l_{\mathrm{CO_2}} = \mathrm{BP} + \Delta P - P^l_{\mathrm{O_2}} - P_{wv} \tag{3.20}$$

Substitution of Eq. (3.4) for the partial pressure of oxygen into Eq. (3.20) gives:

$$P^l_{N_2+Ar+CO_2} = BP + \Delta P - \frac{C_{O_2}}{\beta_{O_2}}(0.5318) - P_{wv} \quad (3.21)$$

The partial pressure of nitrogen + argon + carbon dioxide in the gas phase is equal to:

$$P^g_{N_2+Ar+CO_2} = \chi_{N_2+Ar}(BP - P_{wv}) \quad (3.22)$$

or

$$P^g_{N_2+Ar+CO_2} = 0.7905(BP - P_{wv}) \quad (3.23)$$

With the same assumptions, Eq. (3.15) can be rewritten as

$$\Delta P_{N_2+Ar+CO_2} = P^l_{N_2+Ar+CO_2} - P^g_{N_2+Ar+CO_2} \quad (3.24)$$

and

$$\Delta P = \Delta P_{O_2} + \Delta P_{N_2+Ar+CO_2} \quad (3.25)$$

In natural waters, the partial pressure of carbon dioxide is typically very small in comparison to other major atmospheric gases. In this book, $P^l_{N_2+Ar+CO_2}$ and $P^g_{N_2+Ar+CO_2}$ will be referred to as $P^l_{N_2+Ar}$ and $P^g_{N_2+Ar}$ because of the small contribution of carbon dioxide. Many authors further shorten this to only $P^l_{N_2}$ and $P^g_{N_2}$. It is important to understand that the measured TGP will include the contribution of all dissolved gases whether or not they are included in Eq. (3.19). Some waters contain measurable amounts of methane or hydrogen. If $P^l_{N_2+Ar+CO_2}$ is computed from Eq. (3.21), their partial pressure contribution will be included in this term. If total gas pressure is computed from concentrations, the omission of these additional gases can result in significant errors.

## Example 3-3

Compute the total gas pressure and partial pressures of oxygen, nitrogen + argon, and carbon dioxide when the barometric pressure is 732.0 mmHg, $\Delta P$ is 121 mmHg, water temperature is 13.4°C, dissolved oxygen concentration is 7.39 mg/L, and the dissolved carbon dioxide concentration is 6.32 mg/L. Use the equations listed in Table 3.1, and compare the partial pressure of nitrogen + argon with that computed from Eq. (3.20). Compare the TGP with the sum of the partial pressures of the gases in liquid phase plus the vapor pressure of water.

**Table 3.1** Recommended Formulae for the Computation of Gas Supersaturation Levels[a]

| Gas | Pressure (mmHg) | $\Delta P$ (mmHg) | Percent Saturation (%) |
|---|---|---|---|
| Total | $BP + \Delta P$ | $\Delta P$ | $\left[\frac{BP + \Delta P}{BP}\right]100$ |
| $N_2 + Ar$ | $BP + \Delta P - \frac{C_{O_2}}{\beta_{O_2}}(0.5318) - P_{wv}$ | $BP + \Delta P - \frac{C_{O_2}}{\beta_{O_2}}(0.5318) - P_{wv} - 0.7902(BP - P_{wv})$ | $\left[\frac{BP + \Delta P - \frac{C_{O_2}}{\beta_{O_2}}(0.5318) - P_{wv}}{0.7905(BP - P_{wv})}\right]100$ |
| $O_2$ | $\frac{C_{O_2}}{\beta_{O_2}}(0.5318)$ | $\frac{C_{O_2}}{\beta_{O_2}}(0.5318) - 0.20946(BP - P_{wv})$ | $\left[\frac{\frac{C_{O_2}}{\beta_{O_2}}(0.5318)}{0.20946(BP - P_{wv})}\right]100$ |
| $CO_2$ | $\frac{C_{CO_2}}{\beta_{O_2}}(0.3845)$ | $\frac{C_{CO_2}}{\beta_{CO_2}}(0.3845) - \chi_{CO_2}(BP - P_{wv})^{b}$ | $\left[\frac{\frac{C_{CO_2}}{\beta_{CO_2}}(0.3845)}{\chi_{CO_2}(BP - P_{wv})}\right]100^{b}$ |

[a]BP = barometric pressure in mmHg; $\Delta P$ = differential pressure in mmHg measured by the MDM; $C$ = concentration of gas in mg/L; β = Bunsen coefficient of gas at ambient temperature and pressure in L/(L atm); $P_{wv}$ = water vapor pressure in mmHg.
[b]Mole fraction of carbon dioxide is changing; see Figure 1.3{23} for projected future values.
*Source*: Colt, 1983.

### Data Needed

| Parameter | Value | Source |
|---|---|---|
| Temperature | 13.4°C | Given |
| BP | 732 mmHg | Given |
| $\Delta P$ | 121 mmHg | Given |
| $C_{O_2}$ | 7.39 mg/L | Given |
| $C_{CO_2}$ | 0.632 mg/L | Given |
| $\beta_{O_2}$ | 0.03543 L/(L atm) | Table 1.32 |
| $\beta_{CO_2}$ | 1.0668 L/(L atm) | Table 1.35 |
| $P_{wv}$ | 11.529 mmHg | Table 1.21 |

*TGP (mmHg)*

$$\text{Total gas pressure (mmHg)} = \text{BP} + \Delta P$$

$$\text{Total gas pressure (mmHg)} = 732 + 121 = 853 \text{ mmHg}$$

*Partial Pressure of Oxygen (Table 3.1; liquid phase)*

$$P^l_{O_2} = \frac{C_{O_2}}{\beta_{O_2}}(0.5318)$$

$$P^l_{O_2} = \frac{7.39}{0.03543}(0.5318) = 110.9 \text{ mmHg}$$

*Partial Pressure of Carbon Dioxide (Table 3.1; liquid phase)*

$$P^l_{CO_2} = \frac{C_{CO_2}}{\beta_{CO_2}}(0.3845)$$

$$P^l_{CO_2} = \frac{0.632}{1.0668}(0.3845) = 0.23 \text{ mmHg}$$

*Partial Pressure of Nitrogen + Argon (Table 3.1; liquid phase)*

$$P^l_{N_2+Ar} = \text{BP} + \Delta P - \frac{C_{O_2}}{\beta_{O_2}}(0.5318) - P_{wv}$$

$$P^l_{N_2+Ar} = 732 + 121 - 110.9 - 11.529 = 730.6 \text{ mmHg}$$

*Partial Pressure of Nitrogen + Argon (Eq. (3.19); liquid phase)*

$$P^{l}_{N_2+Ar} = BP + \Delta P - \frac{C_{O_2}}{\beta_{O_2}}(0.5318) - \frac{C_{CO_2}}{\beta_{CO_2}}(0.3845) - P_{wv}$$

$$P^{l}_{N_2+Ar} = 732 + 121 - 110.9 - 0.23 - 11.529 = 730.4 \text{ mmHg}$$

**Sample Reporting Format**

| Temperature (°C) | BP (mmHg) | TGP (mmHg) | DO (mg/L) | Salinity (g/kg) | Partial Pressure (mmHg) | | |
|---|---|---|---|---|---|---|---|
| | | | | | $N_2 + Ar$ | $O_2$ | $CO_2$ |
| 13.4 | 732.0 | 853 | 7.39 | 0.0 | 730.6 | 110.9 | 2.3 |

**Compare the TGP with the Sum of Partial Pressures**

| Gas | Sum of Partial Pressures | Gas | Equation (3.21) |
|---|---|---|---|
| $P_{N_2+Ar}$ | 730.6 | $P_{N_2+Ar+CO_2}$ | 730.6 |
| $P_{O_2}$ | 110.9 | $P_{O_2}$ | 110.9 |
| $P_{CO_2}$ | 0.23 | | |
| $P_{wv}$ | 11.5 | $P_{wv}$ | 11.5 |
| **Total** | **853.2** | **Total** | **853.0** |

TGP = 853.0 mmHg (computed above)

The equation for partial pressure of $N_2$ + Ar given in Table 3.1 includes the partial pressure of carbon dioxide. Therefore, in the sum of partial pressures (column 2 above), the contribution of carbon dioxide has been counted twice. For a carbon dioxide concentration of 0.632 mg/L, the partial pressure of carbon dioxide is only equal to 0.23 mmHg. For typical waters, this error is generally insignificant.

## Example 3-4

Compute the $\Delta P$, $\Delta P_{O_2}$, $\Delta P_{N_2+Ar}$, and $\Delta P_{CO_2}$ for the conditions listed in Example 3-3. Use the equations listed in Table 3.1. Compare the sum of the individual $\Delta P$s, $\Delta P$ computed from Eq. (3.25), to the given $\Delta P$.

$\Delta P$ (mmHg)

$\Delta P$ = 121 mmHg (given)

$\Delta P_{O_2}$ (mmHg)

$$\Delta P_{O_2} = \frac{C_{O_2}}{\beta_{O_2}}(0.5318) - 0.20946(\text{BP} - P_{wv})$$

$$\Delta P_{O_2} = \frac{7.39}{0.03543}(0.5318) - 0.20946(732 - 11.529) = -39.99 \text{ mmHg}$$

$\Delta P_{N_2+Ar}$ (mmHg)

$$\Delta P_{N_2+Ar} = \text{BP} + \Delta P - \frac{C_{O_2}}{\beta_{O_2}}(0.5318) - P_{wv} - 0.7905(\text{BP} - P_{wv})$$

$$\Delta P_{N_2+Ar} = 732 + 121 - \frac{7.39}{0.03543}(0.5318) - 11.529 - 0.7905(732 - 11.529) = 161.02 \text{ mmHg}$$

$\Delta P_{CO_2}$ (mmHg)

$$\Delta P_{CO_2} = \frac{C_{CO_2}}{\beta_{CO_2}}(0.3845) - \chi_{CO_2}(\text{BP} - P_{wv})$$

$$\Delta P_{CO_2} = \frac{0.632}{1.0668}(0.3845) - 0.000390(732 - 11.529) = -0.05 \text{ mmHg}$$

**Sample Reporting Format**

| Temperature (°C) | BP (mmHg) | $\Delta P$ (mmHg) | DO (mg/L) | Salinity (g/kg) | $\Delta P$ (mmHg) | | |
|---|---|---|---|---|---|---|---|
| | | | | | $N_2 + Ar$ | $O_2$ | $CO_2$ |
| 13.4 | 732.0 | 853 | 7.39 | 0.0 | +161.02 | −39.999 | −0.05 |

$\Delta P = 121$ mmHg (given)

| Gas | Sum of $\Delta P$s | Equation (3.24) |
|---|---|---|
| $\Delta P_{N_2+Ar}$ | 161.02 | 161.02 |
| $\Delta P_{O_2}$ | −39.99 | −39.99 |
| $\Delta P_{CO_2}$ | −0.05 | |
| **Total** | **120.98** | **121.03** |

The small difference between the sum of the individual $\Delta P$s, Eq. (3.25), and the given $\Delta P$ is due to the fact that the $\Delta P_{CO_2}$ has been included twice.

### Computation and Reporting of Gas Supersaturation Levels

Gas supersaturation can be reported in terms of TGP, $\Delta P$, or percent of local barometric pressure. Recommended formulae for the computation of gas supersaturation are presented in Table 3.1{142}. The $\Delta P$ method is the preferred method (Colt, 1983). Barometric pressure, water temperature, and salinity must also be reported.

---

#### Example 3-5

Compute the TGP in percent and percent saturation for oxygen, nitrogen + argon, and carbon dioxide for the conditions listed in Example 3-3. Use the equations listed in Table 3.1.

*TGP (%)*

$$\text{TGP (\%)} = \left[\frac{\text{BP} + \Delta P}{\text{BP}}\right] 100$$

$$\text{TGP (\%)} = \left[\frac{732 + 121}{732}\right] 100 = 116.5\%$$

*Oxygen (%)*

$$\text{O}_2\ (\%) = \left[\frac{\frac{C_{O_2}}{\beta_{O_2}}(0.5318)}{0.20946(\text{BP} - P_{wv})}\right] 100$$

$$\text{O}_2\ (\%) = \left[\frac{\frac{7.39}{0.03543}(0.5318)}{0.20946(732 - 11.529)}\right] 100 = 73.5\%$$

*Nitrogen + Argon (%)*

$$\text{N}_2 + \text{Ar (\%)} = \left[\frac{\text{BP} + \Delta P - \frac{C_{O_2}}{\beta_{O_2}}(0.5318) - P_{wv}}{0.7905(\text{BP} - P_{wv})}\right] 100$$

$$\text{N}_2 + \text{Ar (\%)} = \left[\frac{732 + 121 - \frac{7.39}{0.03543}(0.5318) - 11.529}{0.7905(732 - 11.529)}\right] 100 = 128.3\%$$

*Carbon Dioxide (%)*

$$CO_2\ (\%) = \left[\frac{\frac{C_{CO_2}}{\beta_{CO_2}}(0.3845)}{\chi_{CO_2}(BP - P_{wv})}\right]100$$

$$CO_2\ (\%) = \left[\frac{\frac{0.632}{1.0668}(0.3845)}{0.000390(7.32 - 11.529)}\right]100 = 81.1\%$$

**Sample Reporting Format**

| Temperature (°C) | BP (mmHg) | TGP (%) | DO (mg/L) | Salinity (g/kg) | Percent of Saturation | | |
|---|---|---|---|---|---|---|---|
| | | | | | $N_2 + Ar$ | $O_2$ | $CO_2$ |
| 13.4 | 732.0 | 116.6 | 7.39 | 0.0 | 128.3 | 73.5 | 81.1 |

Computation of the parameters in Table 3.1{142} will require the Bunsen coefficient for oxygen (Tables 1.32{55}, 2.25{102}, 2.26{103}); the Bunsen coefficients for carbon dioxide (Tables 1.35{58}, 2.31{108}, 2.32{109}); and the vapor pressure of water (Tables 1.21{44} and 2.41{118}). It is common in some fields to use the abbreviation DO for the dissolved oxygen concentration in mg/L and the abbreviation DC for dissolved carbon dioxide in mg/L.

## *Computation of Standard Gas Supersaturation Parameters from Concentration Units*

In several types of analysis, the concentration of the gases is measured in concentration units, and $\Delta P$ or TGP has to be computed. It is also necessary to make these conversions when modeling gas transfer. In general, it is necessary to compute the final concentration of oxygen, nitrogen + argon, and carbon dioxide from separate mass transfer equations, and then to compute the resulting $\Delta P$ or TGP. The equations for the computation of standard gas supersaturation parameters in terms of gas concentration measured in mg/L are listed below (Colt, 1983):

$$\Delta P\ (\text{mmHg}) = \frac{C_{O_2}}{\beta_{O_2}}(0.5318) + \frac{C_{N_2}}{\beta_{N_2}}(0.6078) + \frac{C_{Ar}}{\beta_{Ar}}(0.4260) + \frac{C_{CO_2}}{\beta_{CO_2}}(0.3845) + P_{wv} - BP \tag{3.26}$$

$$\mathrm{TGP(mmHg)} = \left[\frac{C_{O_2}}{\beta_{O_2}}(0.5318) + \frac{C_{N_2}}{\beta_{N_2}}(0.6078) + \frac{C_{Ar}}{\beta_{Ar}}(0.4260) + \frac{C_{CO_2}}{\beta_{CO_2}}(0.3845) + P_{wv}\right] \tag{3.27}$$

$$\mathrm{TGP(\%)} = \left[\frac{\frac{C_{O_2}}{\beta_{O_2}}(0.5318) + \frac{C_{N_2}}{\beta_{N_2}}(0.6078) + \frac{C_{Ar}}{\beta_{Ar}}(0.4260) + \frac{C_{CO_2}}{\beta_{CO_2}}(0.3845) + P_{wv}}{\mathrm{BP}}\right]100 \tag{3.28}$$

$$N_2 + \mathrm{Ar(mmHg)} = \left[\frac{\frac{C_{N_2}}{\beta_{N_2}}(0.6078) + \frac{C_{Ar}}{\beta_{Ar}}(0.4260)}{0.7902(\mathrm{BP} - P_{wv})}\right]100 \tag{3.29}$$

These equations are derived from substituting Eq. (3.4) into Eq. (3.2).

## Example 3-6

Compute $\Delta P$, total gas pressure, $N_2$ + Ar(mmHg), and $N_2$ + Ar(%) when BP = 745 mmHg, water temperature = 7.3°C, $C_{O_2}$ = 9.41 mg/L, $C_{N_2}$ = 23.11 mg/L, $C_{Ar}$ = 0.8816, and $C_{CO_2}$ = 0.96 mg/L. Use Eqs (3.26)–(3.29).

### Data Needed

| Value | Table | Value | Table |
|---|---|---|---|
| $C^*_{O_2} = 9.41$ | Given | $\beta_{O_2} = 0.04066$ | 1.32 |
| $C^*_{N_2} = 23.11$ | Given | $\beta_{N_2} = 0.02010$ | 1.33 |
| $C^*_{Ar} = 0.8816$ | Given | $\beta_{Ar} = 0.04458$ | 1.34 |
| $C^*_{CO_2} = 0.9600$ | Given | $\beta_{CO_2} = 1.3126$ | 1.35 |
| $P_{wv} = 7.668$ | 1.21 | BP = 745 | Given |

$\Delta P$

$$\Delta P\,(\mathrm{mmHg}) = \frac{C_{O_2}}{\beta_{O_2}}(0.5318) + \frac{C_{N_2}}{\beta_{N_2}}(0.6078) + \frac{C_{Ar}}{\beta_{Ar}}(0.4260) + \frac{C_{CO_2}}{\beta_{CO_2}}(0.3845) + P_{wv} - \mathrm{BP}$$

$$\Delta P\,(\mathrm{mmHg}) = \frac{9.41}{0.04066}(0.5318) + \frac{23.11}{0.02010}(0.6078) + \frac{0.8816}{0.04458}(0.4260) + \frac{0.9600}{1.3126}(0.3845) + 7.668 - 745$$

$$\Delta P\,(\mathrm{mmHg}) = +93.3\ \mathrm{mmHg}$$

*TGP (mmHg)*

$$\text{TGP (mmHg)} = \frac{9.41}{0.04066}(0.5318) + \frac{23.11}{0.02010}(0.6078) + \frac{0.8816}{0.04458}(0.4260) + \frac{0.9600}{1.3126}(0.3845) + 7.668$$

$$\text{TGP (mmHg)} = 838.3 \text{ mmHg}$$

*$N_2 + Ar$ (mmHg)*

$$N_2 + \text{Ar (mmHg)} = \frac{9.41}{0.04066}(0.5318) + \frac{23.11}{0.02010}(0.6078)$$

$$N_2 + \text{Ar (mmHg)} = 707.2 \text{ mmHg}$$

*$N_2 + Ar$ (%)*

$$N_2 + \text{Ar (\%)} = \left[\frac{\frac{C_{N_2}}{\beta_{N_2}}(0.6078) + \frac{C_{Ar}}{\beta_{Ar}}(0.4260)}{0.7905(\text{BP} - P_{wv})}\right]100$$

$$N_2 + \text{Ar (\%)} = \left[\frac{\frac{9.41}{0.04066}(0.5318) + \frac{23.11}{0.02010}(0.6078)}{0.7905(745 - 7.668)}\right]100$$

$$N_2 + \text{Ar (\%)} = 121.3\%$$

In the van Slyke method (Beiningen, 1973), the volume of nitrogen and argon is measured together. In the gas chromatography method with molecular sieve columns, oxygen and argon are measured together (D'Aoust and Clark, 1980). For these analytical procedures, the values of $A$ and $\beta$ used in Eq. (3.4) depend on the physical properties of both gases.

For the case where nitrogen + argon are determined together in mg/L, the apparent values of $A_{N_2+Ar}$ and $\beta_{N_2+Ar}$ are

$$\beta_{N_2+Ar} = \frac{\beta_{N_2}\chi_{N_2} + \beta_{Ar}\chi_{Ar}}{\chi_{N_2} + \chi_{Ar}} \tag{3.30}$$

$$A_{N_2+Ar} = \frac{760}{1000}\left[\frac{\beta_{N_2}\chi_{N_2} + \beta_{Ar}\chi_{Ar}}{1.25043\beta_{N_2}\chi_{N_2} + 1.784198\beta_{Ar}\chi_{Ar}}\right] \tag{3.31}$$

where

$N_2 + Ar$ = refers to the composite of nitrogen + argon
$\beta$ = Bunsen coefficients of the individual gases (L/(L atm))
$\chi$ = mole fraction of the individual gases (dimensionless).

The values of the mole fractions are commonly assumed to be equal to the values in air.

## Example 3-7

Compute the values of $\beta_{N_2+Ar}$ and $A_{N_2+Ar}$ from Eqs (3.30) and (3.31) for a water temperature of 7.3°C and salinity = 0.0 g/kg.

**Data Needed**

| Value | Table | Value | Table |
|---|---|---|---|
| $\chi_{N_2} = 0.78084$ | D-1 | $\beta_{N_2} = 0.02010$ | 1.33 |
| $\chi_{Ar} = 0.00934$ | D-1 | $\beta_{Ar} = 0.04458$ | 1.34 |

$\beta_{N_2+Ar}$

$$\beta_{N_2+Ar} = \frac{\beta_{N_2}\chi_{N_2} + \beta_{Ar}\chi_{Ar}}{\chi_{N_2} + \chi_{Ar}}$$

$$\beta_{N_2+Ar} = \frac{0.02010 \times 0.78084 + 0.04458 \times 0.00934}{0.78084 + 0.00934} = 0.02039 \text{ L/(L atm)}$$

$A_{N_2+Ar}$

$$A_{N_2+Ar} = \frac{760}{1000}\left[\frac{\beta_{N_2}\chi_{N_2} + \beta_{Ar}\chi_{Ar}}{1.25040\beta_{N_2}\chi_{N_2} + 1.784195\beta_{Ar}\chi_{Ar}}\right]$$

$$A_{N_2+Ar} = \frac{760}{1000}\left[\frac{0.02010 \times 0.78084 + 0.04458 \times 0.00934}{1.25040 \times 0.02010 \times 0.78084 + 1.784195 \times 0.04458 \times 0.00934}\right]$$

$$A_{N_2+Ar} = 0.6012$$

If nitrogen and argon are measured together, then Eqs (3.26)–(3.29) can be written as:

$$\Delta P\ (\text{mmHg}) = \frac{C_{O_2}}{\beta_{O_2}}(0.5318) + \frac{C_{N_2+Ar}}{\beta_{N_2+Ar}}(A_{N_2+Ar}) + \frac{C_{CO_2}}{\beta_{CO_2}}(0.3845) + P_{wv} - \text{BP} \tag{3.32}$$

$$\text{TGP}\ (\text{mmHg}) = \left[\frac{C_{O_2}}{\beta_{O_2}}(0.5318) + \frac{C_{N_2+Ar}}{\beta_{N_2+Ar}}(A_{N_2+Ar}) + \frac{C_{CO_2}}{\beta_{CO_2}}(0.3845) + P_{wv}\right] \tag{3.33}$$

$$\text{TGP}\ (\%) = \left[\frac{\frac{C_{O_2}}{\beta_{O_2}}(0.5318) + \frac{C_{N_2+Ar}}{\beta_{N_2+Ar}}(A_{N_2+Ar}) + \frac{C_{CO_2}}{\beta_{CO_2}}(0.3845) + P_{wv}}{\text{BP}}\right]100 \tag{3.34}$$

$$N_2 + \text{Ar}\ (\text{mmHg}) = \frac{C_{N_2+Ar}}{\beta_{N_2+Ar}}(A_{N_2+Ar}) \tag{3.35}$$

$$N_2 + \text{Ar}\ (\%) = \left[\frac{\frac{C_{N_2+Ar}}{\beta_{N_2+Ar}}(A_{N_2+Ar})}{0.7905(\text{BP} - P_{wv})}\right]100 \tag{3.36}$$

Values of $A_{N_2+Ar}$ computed from Eq. (3.30) are presented in Table 3.2{152} as functions of temperature and salinity. Values of $\beta_{N_2+Ar}$ computed from Eq. (3.31) are presented in Tables 3.3–3.5{153, 154, 155} for freshwater and seawater conditions.

The computation of $\Delta P$ and $N_2$ + Ar from concentration units are not as accurate as the computation from the MDM due to the greater number of measurements required, implicit sampling and storage problems, and uncertainty in the $A_{N_2+Ar}$ and $\beta_{N_2+Ar}$ values (Colt, 1983).

**Table 3.2** $A_{N_2+Ar}$ as a Function of Temperature and Salinity

| Temperature (°C) | Salinity (g/kg) | | | | | | | | |
|---|---|---|---|---|---|---|---|---|---|
| | **0.0** | **5.0** | **10.0** | **15.0** | **20.0** | **25.0** | **30.0** | **35.0** | **40.0** |
| **0** | 0.6011 | 0.6011 | 0.6011 | 0.6010 | 0.6010 | 0.6010 | 0.6010 | 0.6010 | 0.6010 |
| **1** | 0.6011 | 0.6011 | 0.6011 | 0.6011 | 0.6010 | 0.6010 | 0.6010 | 0.6010 | 0.6010 |
| **2** | 0.6011 | 0.6011 | 0.6011 | 0.6011 | 0.6011 | 0.6010 | 0.6010 | 0.6010 | 0.6010 |
| **3** | 0.6011 | 0.6011 | 0.6011 | 0.6011 | 0.6011 | 0.6010 | 0.6010 | 0.6010 | 0.6010 |
| **4** | 0.6011 | 0.6011 | 0.6011 | 0.6011 | 0.6011 | 0.6011 | 0.6010 | 0.6010 | 0.6010 |
| **5** | 0.6012 | 0.6011 | 0.6011 | 0.6011 | 0.6011 | 0.6011 | 0.6011 | 0.6010 | 0.6010 |
| **6** | 0.6012 | 0.6011 | 0.6011 | 0.6011 | 0.6011 | 0.6011 | 0.6011 | 0.6010 | 0.6010 |
| **7** | 0.6012 | 0.6012 | 0.6011 | 0.6011 | 0.6011 | 0.6011 | 0.6011 | 0.6011 | 0.6010 |
| **8** | 0.6012 | 0.6012 | 0.6012 | 0.6011 | 0.6011 | 0.6011 | 0.6011 | 0.6011 | 0.6011 |
| **9** | 0.6012 | 0.6012 | 0.6012 | 0.6011 | 0.6011 | 0.6011 | 0.6011 | 0.6011 | 0.6011 |
| **10** | 0.6012 | 0.6012 | 0.6012 | 0.6012 | 0.6011 | 0.6011 | 0.6011 | 0.6011 | 0.6011 |
| **11** | 0.6012 | 0.6012 | 0.6012 | 0.6012 | 0.6012 | 0.6011 | 0.6011 | 0.6011 | 0.6011 |
| **12** | 0.6012 | 0.6012 | 0.6012 | 0.6012 | 0.6012 | 0.6011 | 0.6011 | 0.6011 | 0.6011 |
| **13** | 0.6012 | 0.6012 | 0.6012 | 0.6012 | 0.6012 | 0.6012 | 0.6011 | 0.6011 | 0.6011 |
| **14** | 0.6012 | 0.6012 | 0.6012 | 0.6012 | 0.6012 | 0.6012 | 0.6012 | 0.6011 | 0.6011 |
| **15** | 0.6013 | 0.6012 | 0.6012 | 0.6012 | 0.6012 | 0.6012 | 0.6012 | 0.6012 | 0.6011 |
| **16** | 0.6013 | 0.6013 | 0.6012 | 0.6012 | 0.6012 | 0.6012 | 0.6012 | 0.6012 | 0.6012 |
| **17** | 0.6013 | 0.6013 | 0.6013 | 0.6012 | 0.6012 | 0.6012 | 0.6012 | 0.6012 | 0.6012 |
| **18** | 0.6013 | 0.6013 | 0.6013 | 0.6012 | 0.6012 | 0.6012 | 0.6012 | 0.6012 | 0.6012 |
| **19** | 0.6013 | 0.6013 | 0.6013 | 0.6013 | 0.6012 | 0.6012 | 0.6012 | 0.6012 | 0.6012 |
| **20** | 0.6013 | 0.6013 | 0.6013 | 0.6013 | 0.6013 | 0.6012 | 0.6012 | 0.6012 | 0.6012 |
| **21** | 0.6013 | 0.6013 | 0.6013 | 0.6013 | 0.6013 | 0.6013 | 0.6012 | 0.6012 | 0.6012 |
| **22** | 0.6013 | 0.6013 | 0.6013 | 0.6013 | 0.6013 | 0.6013 | 0.6013 | 0.6012 | 0.6012 |
| **23** | 0.6014 | 0.6013 | 0.6013 | 0.6013 | 0.6013 | 0.6013 | 0.6013 | 0.6013 | 0.6012 |
| **24** | 0.6014 | 0.6014 | 0.6013 | 0.6013 | 0.6013 | 0.6013 | 0.6013 | 0.6013 | 0.6012 |
| **25** | 0.6014 | 0.6014 | 0.6013 | 0.6013 | 0.6013 | 0.6013 | 0.6013 | 0.6013 | 0.6013 |
| **26** | 0.6014 | 0.6014 | 0.6014 | 0.6013 | 0.6013 | 0.6013 | 0.6013 | 0.6013 | 0.6013 |
| **27** | 0.6014 | 0.6014 | 0.6014 | 0.6014 | 0.6013 | 0.6013 | 0.6013 | 0.6013 | 0.6013 |
| **28** | 0.6014 | 0.6014 | 0.6014 | 0.6014 | 0.6014 | 0.6013 | 0.6013 | 0.6013 | 0.6013 |
| **29** | 0.6014 | 0.6014 | 0.6014 | 0.6014 | 0.6014 | 0.6014 | 0.6013 | 0.6013 | 0.6013 |
| **30** | 0.6014 | 0.6014 | 0.6014 | 0.6014 | 0.6014 | 0.6014 | 0.6014 | 0.6013 | 0.6013 |
| **31** | 0.6015 | 0.6014 | 0.6014 | 0.6014 | 0.6014 | 0.6014 | 0.6014 | 0.6014 | 0.6013 |
| **32** | 0.6015 | 0.6015 | 0.6014 | 0.6014 | 0.6014 | 0.6014 | 0.6014 | 0.6014 | 0.6014 |
| **33** | 0.6015 | 0.6015 | 0.6015 | 0.6014 | 0.6014 | 0.6014 | 0.6014 | 0.6014 | 0.6014 |
| **34** | 0.6015 | 0.6015 | 0.6015 | 0.6015 | 0.6014 | 0.6014 | 0.6014 | 0.6014 | 0.6014 |
| **35** | 0.6015 | 0.6015 | 0.6015 | 0.6015 | 0.6015 | 0.6014 | 0.6014 | 0.6014 | 0.6014 |
| **36** | 0.6015 | 0.6015 | 0.6015 | 0.6015 | 0.6015 | 0.6014 | 0.6014 | 0.6014 | 0.6014 |
| **37** | 0.6015 | 0.6015 | 0.6015 | 0.6015 | 0.6015 | 0.6015 | 0.6014 | 0.6014 | 0.6014 |
| **38** | 0.6016 | 0.6015 | 0.6015 | 0.6015 | 0.6015 | 0.6015 | 0.6015 | 0.6014 | 0.6014 |
| **39** | 0.6016 | 0.6016 | 0.6015 | 0.6015 | 0.6015 | 0.6015 | 0.6015 | 0.6015 | 0.6014 |
| **40** | 0.6016 | 0.6016 | 0.6016 | 0.6015 | 0.6015 | 0.6015 | 0.6015 | 0.6015 | 0.6015 |

**Table 3.3** Bunsen Coefficient ($\beta$) for Nitrogen + Argon as a Function of Temperature (L real gas/(L atm))

| Temperature (°C) | Δt (°C) | | | | | | | | | |
|---|---|---|---|---|---|---|---|---|---|---|
| | **0.0** | **0.1** | **0.2** | **0.3** | **0.4** | **0.5** | **0.6** | **0.7** | **0.8** | **0.9** |
| **0** | .02432 | .02426 | .02419 | .02413 | .02407 | .02401 | .02394 | .02388 | .02382 | .02376 |
| **1** | .02370 | .02364 | .02358 | .02352 | .02346 | .02340 | .02334 | .02328 | .02322 | .02317 |
| **2** | .02311 | .02305 | .02299 | .02294 | .02288 | .02282 | .02277 | .02271 | .02265 | .02260 |
| **3** | .02254 | .02249 | .02243 | .02238 | .02232 | .02227 | .02222 | .02216 | .02211 | .02206 |
| **4** | .02200 | .02195 | .02190 | .02185 | .02179 | .02174 | .02169 | .02164 | .02159 | .02154 |
| **5** | .02149 | .02144 | .02139 | .02134 | .02129 | .02124 | .02119 | .02114 | .02109 | .02105 |
| **6** | .02100 | .02095 | .02090 | .02085 | .02081 | .02076 | .02071 | .02067 | .02062 | .02057 |
| **7** | .02053 | .02048 | .02044 | .02039 | .02034 | .02030 | .02025 | .02021 | .02017 | .02012 |
| **8** | .02008 | .02003 | .01999 | .01995 | .01990 | .01986 | .01982 | .01977 | .01973 | .01969 |
| **9** | .01965 | .01961 | .01956 | .01952 | .01948 | .01944 | .01940 | .01936 | .01932 | .01928 |
| **10** | .01924 | .01920 | .01916 | .01912 | .01908 | .01904 | .01900 | .01896 | .01892 | .01888 |
| **11** | .01884 | .01881 | .01877 | .01873 | .01869 | .01865 | .01862 | .01858 | .01854 | .01850 |
| **12** | .01847 | .01843 | .01839 | .01836 | .01832 | .01829 | .01825 | .01821 | .01818 | .01814 |
| **13** | .01811 | .01807 | .01804 | .01800 | .01797 | .01793 | .01790 | .01786 | .01783 | .01780 |
| **14** | .01776 | .01773 | .01769 | .01766 | .01763 | .01759 | .01756 | .01753 | .01750 | .01746 |
| **15** | .01743 | .01740 | .01737 | .01733 | .01730 | .01727 | .01724 | .01721 | .01718 | .01715 |
| **16** | .01711 | .01708 | .01705 | .01702 | .01699 | .01696 | .01693 | .01690 | .01687 | .01684 |
| **17** | .01681 | .01678 | .01675 | .01672 | .01669 | .01666 | .01664 | .01661 | .01658 | .01655 |
| **18** | .01652 | .01649 | .01646 | .01644 | .01641 | .01638 | .01635 | .01632 | .01630 | .01627 |
| **19** | .01624 | .01621 | .01619 | .01616 | .01613 | .01611 | .01608 | .01605 | .01603 | .01600 |
| **20** | .01597 | .01595 | .01592 | .01590 | .01587 | .01584 | .01582 | .01579 | .01577 | .01574 |
| **21** | .01572 | .01569 | .01567 | .01564 | .01562 | .01559 | .01557 | .01554 | .01552 | .01550 |
| **22** | .01547 | .01545 | .01542 | .01540 | .01538 | .01535 | .01533 | .01531 | .01528 | .01526 |
| **23** | .01524 | .01521 | .01519 | .01517 | .01514 | .01512 | .01510 | .01508 | .01505 | .01503 |
| **24** | .01501 | .01499 | .01497 | .01494 | .01492 | .01490 | .01488 | .01486 | .01484 | .01481 |
| **25** | .01479 | .01477 | .01475 | .01473 | .01471 | .01469 | .01467 | .01465 | .01463 | .01461 |
| **26** | .01458 | .01456 | .01454 | .01452 | .01450 | .01448 | .01446 | .01444 | .01442 | .01440 |
| **27** | .01439 | .01437 | .01435 | .01433 | .01431 | .01429 | .01427 | .01425 | .01423 | .01421 |
| **28** | .01419 | .01418 | .01416 | .01414 | .01412 | .01410 | .01408 | .01406 | .01405 | .01403 |
| **29** | .01401 | .01399 | .01397 | .01396 | .01394 | .01392 | .01390 | .01389 | .01387 | .01385 |
| **30** | .01383 | .01382 | .01380 | .01378 | .01377 | .01375 | .01373 | .01372 | .01370 | .01368 |
| **31** | .01367 | .01365 | .01363 | .01362 | .01360 | .01358 | .01357 | .01355 | .01354 | .01352 |
| **32** | .01350 | .01349 | .01347 | .01346 | .01344 | .01343 | .01341 | .01340 | .01338 | .01336 |
| **33** | .01335 | .01333 | .01332 | .01330 | .01329 | .01327 | .01326 | .01325 | .01323 | .01322 |
| **34** | .01320 | .01319 | .01317 | .01316 | .01314 | .01313 | .01312 | .01310 | .01309 | .01307 |
| **35** | .01306 | .01305 | .01303 | .01302 | .01300 | .01299 | .01298 | .01296 | .01295 | .01294 |
| **36** | .01292 | .01291 | .01290 | .01288 | .01287 | .01286 | .01285 | .01283 | .01282 | .01281 |
| **37** | .01279 | .01278 | .01277 | .01276 | .01274 | .01273 | .01272 | .01271 | .01269 | .01268 |
| **38** | .01267 | .01266 | .01265 | .01263 | .01262 | .01261 | .01260 | .01259 | .01258 | .01256 |
| **39** | .01255 | .01254 | .01253 | .01252 | .01251 | .01250 | .01248 | .01247 | .01246 | .01245 |
| **40** | .01244 | .01243 | .01242 | .01241 | .01240 | .01239 | .01237 | .01236 | .01235 | .01234 |

*Source*: Freshwater, based on Hamme and Emerson (2004).

**Table 3.4** Bunsen Coefficient ($\beta$) for Nitrogen + Argon as a Function of Temperature and Salinity, 0–40 g/kg (L real gas/(L atm))

| Temperature (°C) | Salinity (g/kg) | | | | | | | | |
|---|---|---|---|---|---|---|---|---|---|
| | **0.0** | **5.0** | **10.0** | **15.0** | **20.0** | **25.0** | **30.0** | **35.0** | **40.0** |
| **0** | 0.02432 | 0.02343 | 0.02258 | 0.02175 | 0.02096 | 0.02019 | 0.01945 | 0.01874 | 0.01805 |
| **1** | 0.02370 | 0.02284 | 0.02202 | 0.02122 | 0.02045 | 0.01971 | 0.01899 | 0.01831 | 0.01764 |
| **2** | 0.02311 | 0.02228 | 0.02148 | 0.02071 | 0.01997 | 0.01925 | 0.01856 | 0.01789 | 0.01725 |
| **3** | 0.02254 | 0.02174 | 0.02097 | 0.02022 | 0.01950 | 0.01881 | 0.01814 | 0.01749 | 0.01687 |
| **4** | 0.02200 | 0.02123 | 0.02048 | 0.01976 | 0.01906 | 0.01839 | 0.01774 | 0.01712 | 0.01651 |
| **5** | 0.02149 | 0.02074 | 0.02002 | 0.01932 | 0.01864 | 0.01799 | 0.01736 | 0.01675 | 0.01617 |
| **6** | 0.02100 | 0.02027 | 0.01957 | 0.01889 | 0.01824 | 0.01761 | 0.01699 | 0.01641 | 0.01584 |
| **7** | 0.02053 | 0.01982 | 0.01914 | 0.01849 | 0.01785 | 0.01724 | 0.01664 | 0.01607 | 0.01552 |
| **8** | 0.02008 | 0.01940 | 0.01874 | 0.01810 | 0.01748 | 0.01689 | 0.01631 | 0.01575 | 0.01522 |
| **9** | 0.01965 | 0.01899 | 0.01835 | 0.01773 | 0.01713 | 0.01655 | 0.01599 | 0.01545 | 0.01493 |
| **10** | 0.01924 | 0.01859 | 0.01797 | 0.01737 | 0.01679 | 0.01623 | 0.01568 | 0.01516 | 0.01465 |
| **11** | 0.01884 | 0.01822 | 0.01761 | 0.01703 | 0.01646 | 0.01592 | 0.01539 | 0.01488 | 0.01438 |
| **12** | 0.01847 | 0.01786 | 0.01727 | 0.01670 | 0.01615 | 0.01562 | 0.01510 | 0.01461 | 0.01412 |
| **13** | 0.01811 | 0.01752 | 0.01694 | 0.01639 | 0.01585 | 0.01534 | 0.01483 | 0.01435 | 0.01388 |
| **14** | 0.01776 | 0.01719 | 0.01663 | 0.01609 | 0.01557 | 0.01506 | 0.01457 | 0.01410 | 0.01364 |
| **15** | 0.01743 | 0.01687 | 0.01633 | 0.01580 | 0.01529 | 0.01480 | 0.01432 | 0.01386 | 0.01342 |
| **16** | 0.01711 | 0.01657 | 0.01604 | 0.01553 | 0.01503 | 0.01455 | 0.01409 | 0.01363 | 0.01320 |
| **17** | 0.01681 | 0.01628 | 0.01576 | 0.01526 | 0.01478 | 0.01431 | 0.01386 | 0.01342 | 0.01299 |
| **18** | 0.01652 | 0.01600 | 0.01550 | 0.01501 | 0.01454 | 0.01408 | 0.01363 | 0.01320 | 0.01279 |
| **19** | 0.01624 | 0.01573 | 0.01524 | 0.01477 | 0.01430 | 0.01386 | 0.01342 | 0.01300 | 0.01260 |
| **20** | 0.01597 | 0.01548 | 0.01500 | 0.01453 | 0.01408 | 0.01364 | 0.01322 | 0.01281 | 0.01241 |
| **21** | 0.01572 | 0.01523 | 0.01476 | 0.01431 | 0.01387 | 0.01344 | 0.01302 | 0.01262 | 0.01223 |
| **22** | 0.01547 | 0.01500 | 0.01454 | 0.01409 | 0.01366 | 0.01324 | 0.01284 | 0.01244 | 0.01206 |
| **23** | 0.01524 | 0.01477 | 0.01432 | 0.01389 | 0.01346 | 0.01305 | 0.01266 | 0.01227 | 0.01190 |
| **24** | 0.01501 | 0.01456 | 0.01412 | 0.01369 | 0.01327 | 0.01287 | 0.01248 | 0.01210 | 0.01174 |
| **25** | 0.01479 | 0.01435 | 0.01392 | 0.01350 | 0.01309 | 0.01270 | 0.01231 | 0.01194 | 0.01158 |
| **26** | 0.01458 | 0.01415 | 0.01373 | 0.01332 | 0.01292 | 0.01253 | 0.01215 | 0.01179 | 0.01144 |
| **27** | 0.01439 | 0.01396 | 0.01354 | 0.01314 | 0.01275 | 0.01237 | 0.01200 | 0.01164 | 0.01130 |
| **28** | 0.01419 | 0.01377 | 0.01337 | 0.01297 | 0.01259 | 0.01221 | 0.01185 | 0.01150 | 0.01116 |
| **29** | 0.01401 | 0.01360 | 0.01320 | 0.01281 | 0.01243 | 0.01207 | 0.01171 | 0.01136 | 0.01103 |
| **30** | 0.01383 | 0.01343 | 0.01304 | 0.01265 | 0.01228 | 0.01192 | 0.01157 | 0.01123 | 0.01090 |
| **31** | 0.01367 | 0.01327 | 0.01288 | 0.01251 | 0.01214 | 0.01179 | 0.01144 | 0.01111 | 0.01078 |
| **32** | 0.01350 | 0.01311 | 0.01273 | 0.01236 | 0.01200 | 0.01165 | 0.01132 | 0.01099 | 0.01067 |
| **33** | 0.01335 | 0.01296 | 0.01259 | 0.01223 | 0.01187 | 0.01153 | 0.01119 | 0.01087 | 0.01056 |
| **34** | 0.01320 | 0.01282 | 0.01245 | 0.01209 | 0.01175 | 0.01141 | 0.01108 | 0.01076 | 0.01045 |
| **35** | 0.01306 | 0.01269 | 0.01232 | 0.01197 | 0.01162 | 0.01129 | 0.01097 | 0.01065 | 0.01035 |
| **36** | 0.01292 | 0.01255 | 0.01220 | 0.01185 | 0.01151 | 0.01118 | 0.01086 | 0.01055 | 0.01025 |
| **37** | 0.01279 | 0.01243 | 0.01208 | 0.01173 | 0.01140 | 0.01107 | 0.01076 | 0.01045 | 0.01015 |
| **38** | 0.01267 | 0.01231 | 0.01196 | 0.01162 | 0.01129 | 0.01097 | 0.01066 | 0.01035 | 0.01006 |
| **39** | 0.01255 | 0.01220 | 0.01185 | 0.01152 | 0.01119 | 0.01087 | 0.01056 | 0.01026 | 0.00997 |
| **40** | 0.01244 | 0.01209 | 0.01175 | 0.01141 | 0.01109 | 0.01078 | 0.01047 | 0.01018 | 0.00989 |

*Source*: Based on Hamme and Emerson (2004).

**Table 3.5** Bunsen Coefficient ($\beta$) for Nitrogen + Argon as a Function of Temperature and Salinity, 33–37 g/kg (L real gas/(L atm))

| Temperature (°C) | Salinity (g/kg) | | | | | | | | |
|---|---|---|---|---|---|---|---|---|---|
| | **33.0** | **33.5** | **34.0** | **34.5** | **35.0** | **35.5** | **36.0** | **36.5** | **37.0** |
| **0** | 0.01902 | 0.01895 | 0.01888 | 0.01881 | 0.01874 | 0.01867 | 0.01860 | 0.01853 | 0.01846 |
| **1** | 0.01858 | 0.01851 | 0.01844 | 0.01837 | 0.01831 | 0.01824 | 0.01817 | 0.01810 | 0.01804 |
| **2** | 0.01815 | 0.01809 | 0.01802 | 0.01796 | 0.01789 | 0.01783 | 0.01776 | 0.01770 | 0.01763 |
| **3** | 0.01775 | 0.01769 | 0.01762 | 0.01756 | 0.01749 | 0.01743 | 0.01737 | 0.01731 | 0.01724 |
| **4** | 0.01736 | 0.01730 | 0.01724 | 0.01718 | 0.01712 | 0.01705 | 0.01699 | 0.01693 | 0.01687 |
| **5** | 0.01699 | 0.01693 | 0.01687 | 0.01681 | 0.01675 | 0.01669 | 0.01663 | 0.01657 | 0.01652 |
| **6** | 0.01664 | 0.01658 | 0.01652 | 0.01646 | 0.01641 | 0.01635 | 0.01629 | 0.01623 | 0.01618 |
| **7** | 0.01630 | 0.01624 | 0.01619 | 0.01613 | 0.01607 | 0.01602 | 0.01596 | 0.01590 | 0.01585 |
| **8** | 0.01597 | 0.01592 | 0.01586 | 0.01581 | 0.01575 | 0.01570 | 0.01565 | 0.01559 | 0.01554 |
| **9** | 0.01566 | 0.01561 | 0.01556 | 0.01550 | 0.01545 | 0.01540 | 0.01534 | 0.01529 | 0.01524 |
| **10** | 0.01536 | 0.01531 | 0.01526 | 0.01521 | 0.01516 | 0.01510 | 0.01505 | 0.01500 | 0.01495 |
| **11** | 0.01508 | 0.01503 | 0.01498 | 0.01493 | 0.01488 | 0.01483 | 0.01478 | 0.01473 | 0.01468 |
| **12** | 0.01480 | 0.01475 | 0.01471 | 0.01466 | 0.01461 | 0.01456 | 0.01451 | 0.01446 | 0.01441 |
| **13** | 0.01454 | 0.01449 | 0.01444 | 0.01440 | 0.01435 | 0.01430 | 0.01425 | 0.01421 | 0.01416 |
| **14** | 0.01429 | 0.01424 | 0.01419 | 0.01415 | 0.01410 | 0.01405 | 0.01401 | 0.01396 | 0.01392 |
| **15** | 0.01405 | 0.01400 | 0.01395 | 0.01391 | 0.01386 | 0.01382 | 0.01377 | 0.01373 | 0.01368 |
| **16** | 0.01381 | 0.01377 | 0.01372 | 0.01368 | 0.01363 | 0.01359 | 0.01355 | 0.01350 | 0.01346 |
| **17** | 0.01359 | 0.01355 | 0.01350 | 0.01346 | 0.01342 | 0.01337 | 0.01333 | 0.01329 | 0.01324 |
| **18** | 0.01338 | 0.01333 | 0.01329 | 0.01325 | 0.01320 | 0.01316 | 0.01312 | 0.01308 | 0.01304 |
| **19** | 0.01317 | 0.01313 | 0.01309 | 0.01304 | 0.01300 | 0.01296 | 0.01292 | 0.01288 | 0.01284 |
| **20** | 0.01297 | 0.01293 | 0.01289 | 0.01285 | 0.01281 | 0.01277 | 0.01273 | 0.01269 | 0.01265 |
| **21** | 0.01278 | 0.01274 | 0.01270 | 0.01266 | 0.01262 | 0.01258 | 0.01254 | 0.01250 | 0.01246 |
| **22** | 0.01260 | 0.01256 | 0.01252 | 0.01248 | 0.01244 | 0.01240 | 0.01236 | 0.01233 | 0.01229 |
| **23** | 0.01242 | 0.01238 | 0.01235 | 0.01231 | 0.01227 | 0.01223 | 0.01219 | 0.01216 | 0.01212 |
| **24** | 0.01225 | 0.01222 | 0.01218 | 0.01214 | 0.01210 | 0.01207 | 0.01203 | 0.01199 | 0.01196 |
| **25** | 0.01209 | 0.01205 | 0.01202 | 0.01198 | 0.01194 | 0.01191 | 0.01187 | 0.01183 | 0.01180 |
| **26** | 0.01193 | 0.01190 | 0.01186 | 0.01183 | 0.01179 | 0.01175 | 0.01172 | 0.01168 | 0.01165 |
| **27** | 0.01178 | 0.01175 | 0.01171 | 0.01168 | 0.01164 | 0.01161 | 0.01157 | 0.01154 | 0.01150 |
| **28** | 0.01164 | 0.01161 | 0.01157 | 0.01154 | 0.01150 | 0.01147 | 0.01143 | 0.01140 | 0.01136 |
| **29** | 0.01150 | 0.01147 | 0.01143 | 0.01140 | 0.01136 | 0.01133 | 0.01130 | 0.01126 | 0.01123 |
| **30** | 0.01137 | 0.01133 | 0.01130 | 0.01127 | 0.01123 | 0.01120 | 0.01117 | 0.01113 | 0.01110 |
| **31** | 0.01124 | 0.01121 | 0.01117 | 0.01114 | 0.01111 | 0.01107 | 0.01104 | 0.01101 | 0.01098 |
| **32** | 0.01112 | 0.01108 | 0.01105 | 0.01102 | 0.01099 | 0.01095 | 0.01092 | 0.01089 | 0.01086 |
| **33** | 0.01100 | 0.01097 | 0.01093 | 0.01090 | 0.01087 | 0.01084 | 0.01081 | 0.01077 | 0.01074 |
| **34** | 0.01089 | 0.01085 | 0.01082 | 0.01079 | 0.01076 | 0.01073 | 0.01070 | 0.01066 | 0.01063 |
| **35** | 0.01078 | 0.01074 | 0.01071 | 0.01068 | 0.01065 | 0.01062 | 0.01059 | 0.01056 | 0.01053 |
| **36** | 0.01067 | 0.01064 | 0.01061 | 0.01058 | 0.01055 | 0.01052 | 0.01049 | 0.01046 | 0.01043 |
| **37** | 0.01057 | 0.01054 | 0.01051 | 0.01048 | 0.01045 | 0.01042 | 0.01039 | 0.01036 | 0.01033 |
| **38** | 0.01047 | 0.01044 | 0.01041 | 0.01038 | 0.01035 | 0.01032 | 0.01030 | 0.01027 | 0.01024 |
| **39** | 0.01038 | 0.01035 | 0.01032 | 0.01029 | 0.01026 | 0.01023 | 0.01020 | 0.01018 | 0.01015 |
| **40** | 0.01029 | 0.01026 | 0.01024 | 0.01021 | 0.01018 | 0.01015 | 0.01012 | 0.01009 | 0.01006 |

*Source*: Based on Hamme and Emerson (2004).

## Example 3-8

Compute $\Delta P$ and $N_2 + Ar$ (mmHg), when BP = 745 mmHg, water temperature = 7.3°C, $C_{O_2} = 9.41$ mg/L, $C_{N_2+Ar} = 23.99$ mg/L, and $C_{CO_2} = 0.96$ mg/L. Use Eqs (3.32) and (3.35), the $\beta_{N_2+Ar}$ and $A_{N_2+Ar}$ from Example 3-7, and the Bunsen coefficient and $P_{wv}$ information presented in Example 3-6.

*ΔP*

$$\Delta P\,(\text{mmHg}) = \frac{C_{O_2}}{\beta_{O_2}}(0.5318) + \frac{C_{N_2+Ar}}{\beta_{N_2+Ar}}(A_{N_2+Ar}) + \frac{C_{CO_2}}{\beta_{CO_2}}(0.3845) + P_{wv} - \text{BP}$$

$$\Delta P\,(\text{mmHg}) = \frac{9.41}{0.04066}(0.5318) + \frac{23.99}{0.02039}(0.6012) + \frac{0.96}{1.3126}(0.3845) + 7.668 - 745$$

$$\Delta P\,(\text{mmHg}) = +93.4\ \text{mmHg}$$

*$N_2 + Ar$ (mmHg)*

$$N_2 + Ar\,(\text{mmHg}) = \frac{C_{N_2+Ar}}{\beta_{N_2+Ar}}(A_{N_2+Ar})$$

$$N_2 + Ar\,(\text{mmHg}) = \frac{23.99}{0.02039}(0.6012) = 707.4\ \text{mmHg}$$

## Example 3-9

Compute $\Delta P$ and $N_2 + Ar$ (mmHg) for Example 3-8 assuming that $\beta_{N_2}$ and $A_{N_2}$ can be used. Compare with the results from Examples 3-6 and 3-8. Use the Bunsen coefficient and $P_{wv}$ information presented in Example 3-6.

*ΔP*

$$\Delta P\,(\text{mmHg}) = \frac{C_{O_2}}{\beta_{O_2}}(0.5318) + \frac{C_{N_2+Ar}}{\beta_{N_2}}(A_{N_2}) + \frac{C_{CO_2}}{\beta_{CO_2}}(0.3845) + P_{wv} - \text{BP}$$

$$\Delta P\,(\text{mmHg}) = \frac{9.41}{0.04066}(0.5318) + \frac{23.99}{0.02010}(0.6078) + \frac{0.96}{1.3126}(0.3845) + 7.668 - 745$$

$$\Delta P\,(\text{mmHg}) = +111.5\ \text{mmHg}$$

*$N_2 + Ar$ (mmHg)*

$$N_2 + Ar\ (\text{mmHg}) = \frac{C_{N_2+Ar}}{\beta_{N_2}}(A_{N_2})$$

$$N_2 + Ar\ (\text{mmHg}) = \frac{23.99}{0.02010}(0.6078) = 725.4\ \text{mmHg}$$

*Comparison of Examples 3-6, 3-8, and 3-9*

| Example | Approach | $\Delta P$ (mmHg) | $N_2 + Ar$ (mmHg) |
|---|---|---|---|
| 3-6 | $\beta$ and $A$ values for all gases | 93.3 | 707.2 |
| 3-8 | $C_{N_2+Ar}$, $\beta_{N_2+Ar}$, and $A_{N_2+Ar}$ | 93.4 | 707.4 |
| 3-9 | $C_{N_2+Ar}$, $\beta_{N_2}$, and $A_{N_2}$ | 111.5 | 725.4 |

The $\Delta P$ and $N_2 + Ar$ can be computed from the individual gases or from $C_{N_2+Ar}$ as long as $\beta_{N_2+Ar}$ and $A_{N_2+Ar}$ are used. The computation of $N_2 + Ar$ using the Bunsen coefficient and $A$ values for nitrogen alone results in a positive error of 18.2 mmHg. This also results in a positive error of 18.2 mmHg in the $\Delta P$. In chronic exposure to gas supersaturation, $\Delta P$s in the range of 20–40 mmHg may be lethal (Bouck, 1976; Cornacchia and Colt, 1984). Under these conditions, the use of $\beta_{N_2+Ar}$ and $A_{N_2+Ar}$ is required. For $\Delta P$s greater than 150 mmHg, these corrections may be ignored.

## *Conversion of Older Reported Gas Supersaturation Data*

In the early days of gas supersaturation work, gas supersaturation data was reported in a different manner than is presently recommended. TGP as a percent of barometric pressure (TGP (%)) was commonly computed from

$$\text{TGP}'(\%) = \left[\frac{\text{BP} + \Delta P - P_{wv}}{\text{BP}}\right]100 \tag{3.37}$$

or

$$\Delta P' = \Delta P - P_{wv} \tag{3.38}$$

The conversions to the recommended forms are equal to

$$\text{TGP}\ (\%) = \text{TGP}'(\%) + \left[\frac{100P_{wv}}{\text{BP}}\right] \tag{3.39}$$

$$\Delta P = \Delta P' + P_{wv} \tag{3.40}$$

To make the conversion between Eqs (3.37) and (3.38) and Eqs (3.39) and (3.40), both barometric pressure and water temperature must be known. The difference between these two forms depends largely on water temperature and may not be critical for temperatures less than 10°C, especially if the $\Delta P$s are in the lethal range (>150 mmHg). In some work, information on $N_2 + Ar$ (%) and $O_2$ (%) is reported, but information on $\Delta P$ or TGP is omitted. $\Delta P$ can be computed for these data using:

$$\Delta P = \left[\frac{N_2 + Ar\,(\%)}{100} - 1\right] 0.7905(BP - P_{wv}) + \left[\frac{O_2\,(\%)}{100} - 1\right] 0.20496(BP - P_{wv}) \tag{3.41}$$

If the $O_2$ (%) is not known, the $\Delta P$ cannot be computed.

### Example 3-10

The TGP (%) computed from Eq. (3.37) is 105.7%. If the water temperature is 28.5°C, the barometric pressure is 650 mmHg, and the salinity is 0.0 g/kg, compute the TGP (%) and $\Delta P$ values from Eqs (3.39) and (3.40).

**Data Needed**

| Parameter | Value | Source |
|---|---|---|
| Temperature | 28.5°C | Given |
| BP | 650 mmHg | Given |
| $P_{wv}$ | 29.200 mmHg | Table 1.21 |

*TGP (%)*

$$\text{TGP}\,(\%) = \text{TGP}'(\%) + \left[\frac{100 P_{wv}}{\text{BP}}\right]$$

$$\text{TGP}\,(\%) = 105.7 + \left[\frac{100 \times 29.200}{650}\right] = 110.2\%$$

*$\Delta P$*

From Eq. (3.10), the $\Delta P$ can be computed from the TGP (%) value:

$$\Delta P = 66.3 \text{ mmHg}$$

*Comparison of $\Delta P$ and $\Delta P'$*

$$\Delta P = \Delta P' + P_{wv} \text{ or } \Delta P' = \Delta P - P_{wv}$$

$$\Delta P' = 66.3 - 29.200 = 37.1 \text{ mmHg}$$

The $\Delta P'$ value is only 44% of the $\Delta P$ value. The error between the two forms of the TPG (%) and $\Delta P$ equations is larger at higher temperatures. This example also shows why the $\Delta P$ method is preferred, as it is much easier to evaluate the importance of this change. The magnitude of a change from 105.7% to 110.2% is not as easy to understand as a change from 37.1 to 66.3 mmHg.

**Example 3-11**

Compute the $\Delta P$ from the $N_2 + Ar$ (%) and $O_2$ (%) values in Example 3-5. Compare with given $\Delta P$ in Example 3-4. Assume the barometric pressure is 732 mmHg and the water temperature is 13.4°C ($P_{wv} = 11.529$ mmHg).

$$\Delta P = \left[\frac{N_2 + Ar\,(\%)}{100} - 1\right] 0.7905(BP - P_{wv}) + \left[\frac{O_2\,(\%)}{100} - 1\right] 0.20496(BP - P_{wv})$$

$$\Delta P = \left[\frac{128.3}{100} - 1\right] 0.7905(732 - 11.529) + \left[\frac{73.5}{100} - 1\right] 0.20496(732. - 11.529)$$

$\Delta P = 121.9$ mmHg versus 121.0 as given in Example 3-4.

In the early days of gas supersaturation research, it was also thought that GBD resulted from nitrogen gas and, therefore, only $N_2$ (%) was reported. A very rough estimate of the $\Delta P$ can be computed from Eq. (3.41), if an educated guess can be made about the $O_2$ (%) value. In many cases, the old data cannot be converted because the equations used to compute the parameters are not clearly stated.

Dissolved oxygen instruments are much more common than gas supersaturation measuring equipment, so there are published reports on GBD in which only the DO is reported. When gas supersaturation results from air entrainment or heating, it may be possible to roughly estimate $N_2 + Ar$ (%) from $O_2$ (%). When the elevated oxygen concentration results from heating and photosynthesis, $O_2$ (%) may have little relationship to the $N_2 + Ar$ (%), and it is impossible to estimate $\Delta P$ or TGP.

## *Impact of Depth on Gas Supersaturation Levels*

The $\Delta P$ or TGP (%) are measured with respect to local barometric pressure. The actual $\Delta P$ or TGP (%) that an aquatic animal experiences at depth is called the uncompensated $\Delta P$ or TGP (%) (Colt, 1983) and is equal to

$$\Delta P_{\mathrm{uncomp}} = \Delta P - \rho g Z \tag{3.42}$$

$$\mathrm{TGP}_{\mathrm{uncomp}} = \left[\frac{\mathrm{BP} + \Delta P}{\mathrm{BP} + \rho g Z}\right] \tag{3.43}$$

Values of $\rho g$ are presented in Table 1.29{52} for freshwater conditions and in Table 2.43{118} for marine conditions. Blood pressure is another compensating pressure that tends to hold dissolved gas in solution. Equation (3.36) can be rewritten to include this term:

$$\Delta P_{\mathrm{uncomp}} = \Delta P - \rho g Z - P_{\mathrm{blood}} \tag{3.44}$$

where

$P_{\mathrm{blood}}$ = minimum blood pressure in animal (mmHg).

If it is assumed that the $\Delta P$ inside the animal is equal to the $\Delta P$ of the surrounding water, then GBD cannot be produced unless $\Delta P_{\mathrm{uncomp}} > 0$ or $\mathrm{TGP}_{\mathrm{uncomp}} > 100\%$. For values of $\Delta P_{\mathrm{uncomp}}$ or $\mathrm{TGP}_{\mathrm{uncomp}}$ less than these values, GBD cannot occur because bubble formation is impossible. The depth where $\Delta P_{\mathrm{uncomp}} = 0$ and $\mathrm{TGP}_{\mathrm{uncomp}} = 100\%$ is called the hydrostatic compensation depth:

$$Z_{\mathrm{uncomp}} = \frac{\Delta P}{\rho g} \tag{3.45}$$

Aquatic animals can tolerate large $\Delta P$s by remaining at depth, but rapid mortality may result if they are forced to remain at the surface. This has occurred in fish ladders in the Columbia and Snake River (Weitkamp and Katz, 1980). Very high levels of gas supersaturation are produced by air entrainment when water passes over the spillways.

The effects of depth on $\Delta P_{\mathrm{uncomp}}$ or $\mathrm{TGP}_{\mathrm{uncomp}}$ are presented in Tables 3.6{161} and 3.7{162}. Equations (3.42) and (3.43) and Tables 3.6{161} and 3.7{162} are based on the assumption that both $\Delta P$ and temperature are uniform with depth. These assumptions may not be valid in lakes and reservoirs.

**Table 3.6** The Effect of Depth on Uncompensated $\Delta P$ (Temperature = 20°C, Barometric Pressure = 760 mmHg, Salinity = 0.0 g/kg)

| **Depth (m)** | **$\Delta P$ (mmHg)** | | | | | | | | | |
|---|---|---|---|---|---|---|---|---|---|---|
| | **0** | **25** | **50** | **75** | **100** | **125** | **150** | **175** | **200** | **225** |
| **0.0** | 0.0 | 25.0 | 50.0 | 75.0 | 100.0 | 125.0 | 150.0 | 175.0 | 200.0 | 225.0 |
| **0.1** | −7.3 | 17.7 | 42.7 | 67.7 | 92.7 | 117.7 | 142.7 | 167.7 | 192.7 | 217.7 |
| **0.2** | −14.7 | 10.3 | 35.3 | 60.3 | 85.3 | 110.3 | 135.3 | 160.3 | 185.3 | 210.3 |
| **0.3** | −22.0 | 3.0 | 28.0 | 53.0 | 78.0 | 103.0 | 128.0 | 153.0 | 178.0 | 203.0 |
| **0.4** | −29.4 | −4.4 | 20.6 | 45.6 | 70.6 | 95.6 | 120.6 | 145.6 | 170.6 | 195.6 |
| **0.5** | −36.7 | −11.7 | 13.3 | 38.3 | 63.3 | 88.3 | 113.3 | 138.3 | 163.3 | 188.3 |
| **0.6** | −44.1 | −19.1 | 5.9 | 30.9 | 55.9 | 80.9 | 105.9 | 130.9 | 155.9 | 180.9 |
| **0.7** | −51.4 | −26.4 | −1.4 | 23.6 | 48.6 | 73.6 | 98.6 | 123.6 | 148.6 | 173.6 |
| **0.8** | −58.7 | −33.7 | −8.7 | 16.3 | 41.3 | 66.3 | 91.3 | 116.3 | 141.3 | 166.3 |
| **0.9** | −66.1 | −41.1 | −16.1 | 8.9 | 33.9 | 58.9 | 83.9 | 108.9 | 133.9 | 158.9 |
| **1.0** | −73.4 | −48.4 | −23.4 | 1.6 | 26.6 | 51.6 | 76.6 | 101.6 | 126.6 | 151.6 |
| **1.1** | −80.8 | −55.8 | −30.8 | −5.8 | 19.2 | 44.2 | 69.2 | 94.2 | 119.2 | 144.2 |
| **1.2** | −88.1 | −63.1 | −38.1 | −13.1 | 11.9 | 36.9 | 61.9 | 86.9 | 111.9 | 136.9 |
| **1.3** | −95.5 | −70.5 | −45.5 | −20.5 | 4.5 | 29.5 | 54.5 | 79.5 | 104.5 | 129.5 |
| **1.4** | −102.8 | −77.8 | −52.8 | −27.8 | −2.8 | 22.2 | 47.2 | 72.2 | 97.2 | 122.2 |
| **1.5** | −110.1 | −85.1 | −60.1 | −35.1 | −10.1 | 14.9 | 39.9 | 64.9 | 89.9 | 114.9 |
| **1.6** | −117.5 | −92.5 | −67.5 | −42.5 | −17.5 | 7.5 | 32.5 | 57.5 | 82.5 | 107.5 |
| **1.7** | −124.8 | −99.8 | −74.8 | −49.8 | −24.8 | 0.2 | 25.2 | 50.2 | 75.2 | 100.2 |
| **1.8** | −132.2 | −107.2 | −82.2 | −57.2 | −32.2 | −7.2 | 17.8 | 42.8 | 67.8 | 92.8 |
| **1.9** | −139.5 | −114.5 | −89.5 | −64.5 | −39.5 | −14.5 | 10.5 | 35.5 | 60.5 | 85.5 |
| **2.0** | −146.8 | −121.8 | −96.8 | −71.8 | −46.8 | −21.8 | 3.2 | 28.2 | 53.2 | 78.2 |
| **2.1** | −154.2 | −129.2 | −104.2 | −79.2 | −54.2 | −29.2 | −4.2 | 20.8 | 45.8 | 70.8 |
| **2.2** | −161.5 | −136.5 | −111.5 | −86.5 | −61.5 | −36.5 | −11.5 | 13.5 | 38.5 | 63.5 |
| **2.3** | −168.9 | −143.9 | −118.9 | −93.9 | −68.9 | −43.9 | −18.9 | 6.1 | 31.1 | 56.1 |
| **2.4** | −176.2 | −151.2 | −126.2 | −101.2 | −76.2 | −51.2 | −26.2 | −1.2 | 23.8 | 48.8 |
| **2.5** | −183.6 | −158.6 | −133.6 | −108.6 | −83.6 | −58.6 | −33.6 | −8.6 | 16.4 | 41.4 |
| **2.6** | −190.9 | −165.9 | −140.9 | −115.9 | −90.9 | −65.9 | −40.9 | −15.9 | 9.1 | 34.1 |
| **2.7** | −198.2 | −173.2 | −148.2 | −123.2 | −98.2 | −73.2 | −48.2 | −23.2 | 1.8 | 26.8 |
| **2.8** | −205.6 | −180.6 | −155.6 | −130.6 | −105.6 | −80.6 | −55.6 | −30.6 | −5.6 | 19.4 |
| **2.9** | −212.9 | −187.9 | −162.9 | −137.9 | −112.9 | −87.9 | −62.9 | −37.9 | −12.9 | 12.1 |
| **3.0** | −220.3 | −195.3 | −170.3 | −145.3 | −120.3 | −95.3 | −70.3 | −45.3 | −20.3 | 4.7 |
| **3.1** | −227.6 | −202.6 | −177.6 | −152.6 | −127.6 | −102.6 | −77.6 | −52.6 | −27.6 | −2.6 |
| **3.2** | −235.0 | −210.0 | −185.0 | −160.0 | −135.0 | −110.0 | −85.0 | −60.0 | −35.0 | −10.0 |
| **3.3** | −242.3 | −217.3 | −192.3 | −167.3 | −142.3 | −117.3 | −92.3 | −67.3 | −42.3 | −17.3 |
| **3.4** | −249.6 | −224.6 | −199.6 | −174.6 | −149.6 | −124.6 | −99.6 | −74.6 | −49.6 | −24.6 |
| **3.5** | −257.0 | −232.0 | −207.0 | −182.0 | −157.0 | −132.0 | −107.0 | −82.0 | −57.0 | −32.0 |
| **3.6** | −264.3 | −239.3 | −214.3 | −189.3 | −164.3 | −139.3 | −114.3 | −89.3 | −64.3 | −39.3 |
| **3.7** | −271.7 | −246.7 | −221.7 | −196.7 | −171.7 | −146.7 | −121.7 | −96.7 | −71.7 | −46.7 |
| **3.8** | −279.0 | −254.0 | −229.0 | −204.0 | −179.0 | −154.0 | −129.0 | −104.0 | −79.0 | −54.0 |
| **3.9** | −286.4 | −261.4 | −236.4 | −211.4 | −186.4 | −161.4 | −136.4 | −111.4 | −86.4 | −61.4 |
| **4.0** | −293.7 | −268.7 | −243.7 | −218.7 | −193.7 | −168.7 | −143.7 | −118.7 | −93.7 | −68.7 |

**Table 3.7** The Effect of Depth on Uncompensated TGP Expressed as a Percent of Local Barometric Pressure (Temperature = 20°C, Barometric Pressure = 760 mmHg, Salinity = 0.0 g/kg)

| **Depth (m)** | **TGP (%)** | | | | | | | | | |
|---|---|---|---|---|---|---|---|---|---|---|
| | **100** | **105** | **110** | **115** | **120** | **125** | **130** | **135** | **140** | **145** |
| **0.0** | 100.0 | 105.0 | 110.0 | 115.0 | 120.0 | 125.0 | 130.0 | 135.0 | 140.0 | 145.0 |
| **0.1** | 99.0 | 104.0 | 108.9 | 113.9 | 118.9 | 123.8 | 128.8 | 133.7 | 138.7 | 143.6 |
| **0.2** | 98.1 | 103.0 | 107.9 | 112.8 | 117.7 | 122.6 | 127.5 | 132.4 | 137.3 | 142.3 |
| **0.3** | 97.2 | 102.0 | 106.9 | 111.8 | 116.6 | 121.5 | 126.3 | 131.2 | 136.1 | 140.9 |
| **0.4** | 96.3 | 101.1 | 105.9 | 110.7 | 115.5 | 120.3 | 125.2 | 130.0 | 134.8 | 139.6 |
| **0.5** | 95.4 | 100.2 | 104.9 | 109.7 | 114.5 | 119.2 | 124.0 | 128.8 | 133.5 | 138.3 |
| **0.6** | 94.5 | 99.2 | 104.0 | 108.7 | 113.4 | 118.2 | 122.9 | 127.6 | 132.3 | 137.1 |
| **0.7** | 93.7 | 98.3 | 103.0 | 107.7 | 112.4 | 117.1 | 121.8 | 126.4 | 131.1 | 135.8 |
| **0.8** | 92.8 | 97.5 | 102.1 | 106.7 | 111.4 | 116.0 | 120.7 | 125.3 | 130.0 | 134.6 |
| **0.9** | 92.0 | 96.6 | 101.2 | 105.8 | 110.4 | 115.0 | 119.6 | 124.2 | 128.8 | 133.4 |
| **1.0** | 91.2 | 95.7 | 100.3 | 104.9 | 109.4 | 114.0 | 118.5 | 123.1 | 127.7 | 132.2 |
| **1.1** | 90.4 | 94.9 | 99.4 | 104.0 | 108.5 | 113.0 | 117.5 | 122.0 | 126.6 | 131.1 |
| **1.2** | 89.6 | 94.1 | 98.6 | 103.1 | 107.5 | 112.0 | 116.5 | 121.0 | 125.5 | 129.9 |
| **1.3** | 88.8 | 93.3 | 97.7 | 102.2 | 106.6 | 111.1 | 115.5 | 119.9 | 124.4 | 128.8 |
| **1.4** | 88.1 | 92.5 | 96.9 | 101.3 | 105.7 | 110.1 | 114.5 | 118.9 | 123.3 | 127.7 |
| **1.5** | 87.3 | 91.7 | 96.1 | 100.4 | 104.8 | 109.2 | 113.5 | 117.9 | 122.3 | 126.6 |
| **1.6** | 86.6 | 90.9 | 95.3 | 99.6 | 103.9 | 108.3 | 112.6 | 116.9 | 121.3 | 125.6 |
| **1.7** | 85.9 | 90.2 | 94.5 | 98.8 | 103.1 | 107.4 | 111.7 | 116.0 | 120.3 | 124.5 |
| **1.8** | 85.2 | 89.4 | 93.7 | 98.0 | 102.2 | 106.5 | 110.7 | 115.0 | 119.3 | 123.5 |
| **1.9** | 84.5 | 88.7 | 92.9 | 97.2 | 101.4 | 105.6 | 109.8 | 114.1 | 118.3 | 122.5 |
| **2.0** | 83.8 | 88.0 | 92.2 | 96.4 | 100.6 | 104.8 | 108.9 | 113.1 | 117.3 | 121.5 |
| **2.1** | 83.1 | 87.3 | 91.4 | 95.6 | 99.8 | 103.9 | 108.1 | 112.2 | 116.4 | 120.5 |
| **2.2** | 82.5 | 86.6 | 90.7 | 94.8 | 99.0 | 103.1 | 107.2 | 111.3 | 115.5 | 119.6 |
| **2.3** | 81.8 | 85.9 | 90.0 | 94.1 | 98.2 | 102.3 | 106.4 | 110.5 | 114.5 | 118.6 |
| **2.4** | 81.2 | 85.2 | 89.3 | 93.4 | 97.4 | 101.5 | 105.5 | 109.6 | 113.6 | 117.7 |
| **2.5** | 80.5 | 84.6 | 88.6 | 92.6 | 96.7 | 100.7 | 104.7 | 108.7 | 112.8 | 116.8 |
| **2.6** | 79.9 | 83.9 | 87.9 | 91.9 | 95.9 | 99.9 | 103.9 | 107.9 | 111.9 | 115.9 |
| **2.7** | 79.3 | 83.3 | 87.2 | 91.2 | 95.2 | 99.1 | 103.1 | 107.1 | 111.0 | 115.0 |
| **2.8** | 78.7 | 82.6 | 86.6 | 90.5 | 94.5 | 98.4 | 102.3 | 106.3 | 110.2 | 114.1 |
| **2.9** | 78.1 | 82.0 | 85.9 | 89.8 | 93.7 | 97.6 | 101.5 | 105.5 | 109.4 | 113.3 |
| **3.0** | 77.5 | 81.4 | 85.3 | 89.2 | 93.0 | 96.9 | 100.8 | 104.7 | 108.5 | 112.4 |
| **3.1** | 77.0 | 80.8 | 84.6 | 88.5 | 92.3 | 96.2 | 100.0 | 103.9 | 107.7 | 111.6 |
| **3.2** | 76.4 | 80.2 | 84.0 | 87.8 | 91.7 | 95.5 | 99.3 | 103.1 | 106.9 | 110.8 |
| **3.3** | 75.8 | 79.6 | 83.4 | 87.2 | 91.0 | 94.8 | 98.6 | 102.4 | 106.2 | 109.9 |
| **3.4** | 75.3 | 79.0 | 82.8 | 86.6 | 90.3 | 94.1 | 97.9 | 101.6 | 105.4 | 109.1 |
| **3.5** | 74.7 | 78.5 | 82.2 | 85.9 | 89.7 | 93.4 | 97.2 | 100.9 | 104.6 | 108.4 |
| **3.6** | 74.2 | 77.9 | 81.6 | 85.3 | 89.0 | 92.7 | 96.5 | 100.2 | 103.9 | 107.6 |
| **3.7** | 73.7 | 77.4 | 81.0 | 84.7 | 88.4 | 92.1 | 95.8 | 99.5 | 103.1 | 106.8 |
| **3.8** | 73.1 | 76.8 | 80.5 | 84.1 | 87.8 | 91.4 | 95.1 | 98.7 | 102.4 | 106.1 |
| **3.9** | 72.6 | 76.3 | 79.9 | 83.5 | 87.2 | 90.8 | 94.4 | 98.1 | 101.7 | 105.3 |
| **4.0** | 72.1 | 75.7 | 79.3 | 82.9 | 86.6 | 90.2 | 93.8 | 97.4 | 101.0 | 104.6 |

## Example 3-12

If the barometric pressure is 770.0 mmHg, the $\Delta P$ is 150.0 mmHg, and the water temperature is 19°C, compute the (1) TGP (%) and $TGP_{uncomp}$; (2) $\Delta P_{uncomp}$; and (3) compensation depth.

### Data Needed

| Parameter | Value | Source |
|---|---|---|
| BP | 770.0 mmHg | Given |
| $\Delta P$ | 150.0 mmHg | Given |
| Temperature | 19.0°C | Given |
| Z | 2 m | Given |
| $\rho g$ | 73.439 mmHg/m | Table 1.29 |

*TGP (%)*

$$TGP\ (\%) = \left[\frac{BP + \Delta P}{BP}\right] 100$$

$$TGP\ (\%) = \left[\frac{770.0 + 150.0}{770}\right] 100 = 119.5\%$$

*$TGP_{uncomp}$*

$$TGP_{uncomp} = \left[\frac{BP + \Delta P}{BP + \rho g Z}\right]$$

$$TGP_{uncomp} = \left[\frac{770 + 150}{770.0 + 73.439 \times 2}\right] = 100.3\%$$

*$\Delta P_{uncomp}$*

$$\Delta P_{uncomp} = \Delta P - \rho g Z$$

$$\Delta P_{uncomp} = 150 - 73.439 \times 2.0 = 3.1\ mmHg$$

*Compensation Depth*

$$Z_{uncomp} = \frac{\Delta P}{\rho g}$$

$$Z_{uncomp} = \frac{150.0}{73.439} = 2.04\ m$$

### Example 3-13

River water at a temperature of 9°C is heated by 10°C as it passes through a nuclear power plant. Initially, the concentration of oxygen, nitrogen, and argon are 10.8, 19.85, and 0.76 mg/L, respectively. Compute the $\Delta P$ in the effluent from the plant if the barometric pressure is 752.5 mmHg and find the depth at which the effluent must be discharged to prevent gas supersaturation problems. Ignore carbon dioxide and assume the effluent does not rise to the surface. Use Eqs (3.26) and (3.45).

**Data Needed**

| Value | Table | Value | Table |
|---|---|---|---|
| $C_{O_2} = 10.8$ mg/L | Given | $\beta_{O_2} = 0.03168$ | 1.32 |
| $C_{N_2} = 19.85$ mg/L | Given | $\beta_{N_2} = 0.01602$ | 1.33 |
| $C_{Ar} = 0.76$ mg/L | Given | $\beta_{Ar} = 0.03481$ | 1.34 |
| $P_{wv} = 16.482$ mmHg | 1.21 | BP = 752.5 mmHg | Given |
| $T_o = 9.0$°C | Given | $T_o + \Delta T = 19.0$°C | Given |
| $\rho g = 73.439$ mmHg/m | 1.29 | | |

$\Delta P$

$$\Delta P\ (\text{mmHg}) = \frac{C_{O_2}}{\beta_{O_2}}(0.5318) + \frac{C_{N_2}}{\beta_{N_2}}(0.6078) + \frac{C_{Ar}}{\beta_{Ar}}(0.4260) + P_{wv} - \text{BP}$$

$$\Delta P\ (\text{mmHg}) = \frac{10.8}{0.03168}(0.5318) + \frac{19.85}{0.01602}(0.6078) + \frac{0.76}{0.03481}(0.4260) + 16.482 - 752.5$$

$$\Delta P\ (\text{mmHg}) = 207.7\ \text{mmHg}$$

*Compensation Depth*

$$Z_{\text{uncomp}} = \frac{\Delta P}{\rho g}$$

$$Z_{\text{uncomp}} = \frac{207.7}{73.439} = 2.8\ \text{m}$$

## Impact of Physical, Chemical, and Biological Processes on Gas Supersaturation

This section discusses the impact of physical, chemical, and biological processes on gas supersaturation. Emphasis is placed on processes that commonly occur in the operation of aquatic culture systems.

## Heating and Cooling Waters

The solubility of dissolved gases decreases as the temperature is increased. If water is heated more than a few degrees C, the excess dissolved gases must be removed before the water is used in a culture system. Water can be cooled without supersaturation problems.

The percent saturation that will result when water is subjected to any temperature change can be computed from the following equation:

$$\text{Saturation (\%)} = \left[\frac{C_i}{C_f^*}\right] 100 \qquad (3.46)$$

where

$C_i$ = concentration of gas at initial temperature (mass or volume basis)
$C_f^*$ = saturation concentration of gas at final temperature (mass or volume basis).

If it is assumed that the water is initially saturated, the percent saturation can be computed from the following equation:

$$\text{Saturation (\%)} = \left[\frac{C_i^*}{C_f^*}\right] 100 \qquad (3.47)$$

Supersaturation of a single gas may not produce GBD, so the impact of heating and cooling is best described by $\Delta P$ (Eq. (3.20)). For the assumption that the water is initially at saturation, the concentrations ($C$ values) should be evaluated at the initial temperature, while the Bunsen coefficient and vapor pressure must be based on the final temperature.

$$\Delta P = \frac{C^*_{O_2,T_1}}{\beta_{O_2,T_f}}(0.5318) + \frac{C^*_{N_2+Ar,T_1}}{\beta_{N_2+Ar,T_f}}(A_{N_2\,+\,Ar}) + \frac{C^*_{CO_2,T_1}}{\beta_{CO_2,T_f}}(0.3845) + P_{ww,T_f} - BP \qquad (3.48)$$

For a general case, the concentration of the individual gases would have to be measured at the initial temperature. The $\Delta P$s resulting from heating water, based on the assumption that the water is saturated at the initial temperature, are presented in Table 3.8{166} for freshwater and Table 3.9{167} for marine conditions.

The cooling of water will produce negative $\Delta P$s. As long as the dissolved oxygen is not reduced to a critical concentration, negative $\Delta P$s do not appear to have any reported impact on aquatic animals. In pressurized, heated water systems, there is no tendency for supersaturated water to degas because of the additional pressure. Once the heated water is discharged to the atmospheric pressure, degassing will occur. The primary loss of supersaturated gases occurs across the air–water interface, and relatively high levels of gas supersaturation are amazingly stable.

**Table 3.8** $\Delta P$ in mmHg When Water at $T_o$ is Heated to $T_o + \Delta T$ (Barometric Pressure = 760 mmHg, Salinity = 0.0 g/kg)

| Temperature (°C) | Δt (°C) | | | | | | | | | |
|---|---|---|---|---|---|---|---|---|---|---|
| | 1 | 2 | 3 | 4 | 5 | 6 | 7 | 8 | 9 | 10 |
| 0 | 19.1 | 39.6 | 60.2 | 80.9 | 101.7 | 122.5 | 143.4 | 164.3 | 185.3 | 206.2 |
| 1 | 18.7 | 38.7 | 58.9 | 79.1 | 99.4 | 119.8 | 140.1 | 160.6 | 181.0 | 201.4 |
| 2 | 18.2 | 37.9 | 57.6 | 77.4 | 97.2 | 117.1 | 137.0 | 156.9 | 176.8 | 196.8 |
| 3 | 17.8 | 37.1 | 56.3 | 75.7 | 95.1 | 114.5 | 133.9 | 153.3 | 172.8 | 192.2 |
| 4 | 17.4 | 36.2 | 55.1 | 74.0 | 93.0 | 111.9 | 130.9 | 149.9 | 168.9 | 187.8 |
| 5 | 17.1 | 35.5 | 53.9 | 72.4 | 91.0 | 109.5 | 128.0 | 146.5 | 165.0 | 183.5 |
| 6 | 16.7 | 34.7 | 52.8 | 70.9 | 89.0 | 107.1 | 125.2 | 143.3 | 161.3 | 179.4 |
| 7 | 16.4 | 34.0 | 51.7 | 69.4 | 87.1 | 104.8 | 122.5 | 140.1 | 157.7 | 175.3 |
| 8 | 16.0 | 33.3 | 50.6 | 67.9 | 85.2 | 102.5 | 119.8 | 137.0 | 154.2 | 171.4 |
| 9 | 15.7 | 32.7 | 49.6 | 66.5 | 83.4 | 100.3 | 117.2 | 134.1 | 150.9 | 167.6 |
| 10 | 15.5 | 32.0 | 48.6 | 65.2 | 81.7 | 98.2 | 114.7 | 131.2 | 147.6 | 163.9 |
| 11 | 15.2 | 31.4 | 47.6 | 63.8 | 80.0 | 96.2 | 112.3 | 128.4 | 144.4 | 160.3 |
| 12 | 14.9 | 30.8 | 46.7 | 62.6 | 78.4 | 94.2 | 109.9 | 125.6 | 141.3 | 156.9 |
| 13 | 14.7 | 30.3 | 45.8 | 61.3 | 76.8 | 92.3 | 107.7 | 123.0 | 138.3 | 153.5 |
| 14 | 14.5 | 29.7 | 45.0 | 60.2 | 75.3 | 90.4 | 105.5 | 120.5 | 135.4 | 150.3 |
| 15 | 14.3 | 29.2 | 44.1 | 59.0 | 73.8 | 88.6 | 103.3 | 118.0 | 132.6 | 147.1 |
| 16 | 14.1 | 28.7 | 43.3 | 57.9 | 72.4 | 86.9 | 101.3 | 115.6 | 129.9 | 144.1 |
| 17 | 14.0 | 28.3 | 42.6 | 56.8 | 71.0 | 85.2 | 99.3 | 113.3 | 127.2 | 141.1 |
| 18 | 13.8 | 27.8 | 41.9 | 55.8 | 69.7 | 83.6 | 97.3 | 111.1 | 124.7 | 138.3 |
| 19 | 13.7 | 27.4 | 41.2 | 54.8 | 68.4 | 82.0 | 95.5 | 108.9 | 122.3 | 135.5 |
| 20 | 13.5 | 27.0 | 40.5 | 53.9 | 67.2 | 80.5 | 93.7 | 106.8 | 119.9 | 132.9 |
| 21 | 13.4 | 26.7 | 39.8 | 53.0 | 66.0 | 79.0 | 92.0 | 104.8 | 117.6 | 130.3 |
| 22 | 13.4 | 26.3 | 39.2 | 52.1 | 64.9 | 77.6 | 90.3 | 102.9 | 115.4 | 127.9 |
| 23 | 13.3 | 26.0 | 38.7 | 51.3 | 63.8 | 76.3 | 88.7 | 101.1 | 113.3 | 125.5 |
| 24 | 13.2 | 25.7 | 38.1 | 50.5 | 62.8 | 75.0 | 87.2 | 99.3 | 111.3 | 123.3 |
| 25 | 13.2 | 25.4 | 37.6 | 49.7 | 61.8 | 73.8 | 85.7 | 97.6 | 109.4 | 121.1 |
| 26 | 13.1 | 25.1 | 37.1 | 49.0 | 60.8 | 72.6 | 84.3 | 95.9 | 107.5 | 119.0 |
| 27 | 13.1 | 24.9 | 36.6 | 48.3 | 59.9 | 71.5 | 83.0 | 94.4 | 105.7 | 117.0 |
| 28 | 13.1 | 24.7 | 36.2 | 47.7 | 59.1 | 70.4 | 81.7 | 92.9 | 104.0 | 115.1 |
| 29 | 13.1 | 24.5 | 35.8 | 47.0 | 58.2 | 69.4 | 80.4 | 91.5 | 102.4 | 113.3 |
| 30 | 13.1 | 24.3 | 35.4 | 46.5 | 57.5 | 68.4 | 79.3 | 90.1 | 100.8 | 111.5 |
| 31 | 13.1 | 24.1 | 35.0 | 45.9 | 56.7 | 67.5 | 78.2 | 88.8 | 99.4 | 109.9 |
| 32 | 13.2 | 24.0 | 34.7 | 45.4 | 56.0 | 66.6 | 77.1 | 87.6 | 98.0 | 108.3 |
| 33 | 13.2 | 23.8 | 34.4 | 44.9 | 55.4 | 65.8 | 76.1 | 86.4 | 96.6 | 106.8 |
| 34 | 13.3 | 23.7 | 34.1 | 44.5 | 54.8 | 65.0 | 75.2 | 85.3 | 95.4 | 105.4 |
| 35 | 13.4 | 23.6 | 33.9 | 44.1 | 54.2 | 64.3 | 74.3 | 84.3 | 94.2 | 104.0 |
| 36 | 13.4 | 23.6 | 33.7 | 43.7 | 53.7 | 63.6 | 73.5 | 83.3 | 93.1 | 102.8 |
| 37 | 13.5 | 23.5 | 33.4 | 43.3 | 53.2 | 62.9 | 72.7 | 82.4 | 92.0 | 101.6 |
| 38 | 13.6 | 23.5 | 33.3 | 43.0 | 52.7 | 62.4 | 72.0 | 81.5 | 91.0 | 100.5 |
| 39 | 13.8 | 23.5 | 33.1 | 42.7 | 52.3 | 61.8 | 71.3 | 80.7 | 90.1 | 99.4 |
| 40 | 13.9 | 23.4 | 33.0 | 42.5 | 51.9 | 61.3 | 70.6 | 80.0 | 89.2 | 98.5 |

**Table 3.9** $\Delta P$ in mmHg When Water at $T_o$ is heated to $T_o + \Delta T$ (Barometric Pressure = 760 mmHg, Salinity = 35.0 g/kg)

| Temperature (°C) | Δt (°C) | | | | | | | | | |
|---|---|---|---|---|---|---|---|---|---|---|
| | 1 | 2 | 3 | 4 | 5 | 6 | 7 | 8 | 9 | 10 |
| 0 | 16.0 | 34.6 | 53.1 | 71.8 | 90.4 | 109.1 | 127.8 | 146.5 | 165.1 | 183.8 |
| 1 | 15.7 | 33.8 | 52.0 | 70.2 | 88.5 | 106.7 | 125.0 | 143.2 | 161.4 | 179.6 |
| 2 | 15.4 | 33.2 | 50.9 | 68.8 | 86.6 | 104.4 | 122.3 | 140.1 | 157.9 | 175.6 |
| 3 | 15.1 | 32.5 | 49.9 | 67.3 | 84.8 | 102.2 | 119.6 | 137.0 | 154.4 | 171.7 |
| 4 | 14.8 | 31.8 | 48.9 | 65.9 | 83.0 | 100.0 | 117.1 | 134.0 | 151.0 | 167.9 |
| 5 | 14.5 | 31.2 | 47.9 | 64.6 | 81.3 | 97.9 | 114.6 | 131.2 | 147.7 | 164.3 |
| 6 | 14.3 | 30.6 | 47.0 | 63.3 | 79.6 | 95.9 | 112.2 | 128.4 | 144.6 | 160.7 |
| 7 | 14.1 | 30.1 | 46.1 | 62.0 | 78.0 | 93.9 | 109.8 | 125.7 | 141.5 | 157.2 |
| 8 | 13.8 | 29.5 | 45.2 | 60.8 | 76.4 | 92.0 | 107.6 | 123.1 | 138.5 | 153.9 |
| 9 | 13.6 | 29.0 | 44.3 | 59.6 | 74.9 | 90.2 | 105.4 | 120.5 | 135.6 | 150.7 |
| 10 | 13.5 | 28.5 | 43.5 | 58.5 | 73.5 | 88.4 | 103.2 | 118.1 | 132.8 | 147.5 |
| 11 | 13.3 | 28.0 | 42.7 | 57.4 | 72.1 | 86.6 | 101.2 | 115.7 | 130.1 | 144.5 |
| 12 | 13.1 | 27.6 | 42.0 | 56.4 | 70.7 | 85.0 | 99.2 | 113.4 | 127.5 | 141.6 |
| 13 | 13.0 | 27.2 | 41.3 | 55.3 | 69.4 | 83.4 | 97.3 | 111.2 | 125.0 | 138.8 |
| 14 | 12.9 | 26.7 | 40.6 | 54.4 | 68.1 | 81.8 | 95.5 | 109.0 | 122.6 | 136.0 |
| 15 | 12.8 | 26.4 | 39.9 | 53.4 | 66.9 | 80.3 | 93.7 | 107.0 | 120.2 | 133.4 |
| 16 | 12.7 | 26.0 | 39.3 | 52.5 | 65.7 | 78.9 | 92.0 | 105.0 | 118.0 | 130.9 |
| 17 | 12.6 | 25.7 | 38.7 | 51.7 | 64.6 | 77.5 | 90.3 | 103.1 | 115.8 | 128.4 |
| 18 | 12.5 | 25.3 | 38.1 | 50.8 | 63.5 | 76.2 | 88.7 | 101.2 | 113.7 | 126.1 |
| 19 | 12.5 | 25.0 | 37.6 | 50.1 | 62.5 | 74.9 | 87.2 | 99.5 | 111.7 | 123.8 |
| 20 | 12.4 | 24.8 | 37.1 | 49.3 | 61.5 | 73.7 | 85.7 | 97.8 | 109.7 | 121.7 |
| 21 | 12.4 | 24.5 | 36.6 | 48.6 | 60.6 | 72.5 | 84.3 | 96.1 | 107.9 | 119.6 |
| 22 | 12.4 | 24.3 | 36.1 | 47.9 | 59.7 | 71.4 | 83.0 | 94.6 | 106.1 | 117.6 |
| 23 | 12.3 | 24.0 | 35.7 | 47.3 | 58.8 | 70.3 | 81.7 | 93.1 | 104.4 | 115.7 |
| 24 | 12.3 | 23.8 | 35.3 | 46.7 | 58.0 | 69.3 | 80.5 | 91.7 | 102.8 | 113.9 |
| 25 | 12.4 | 23.7 | 34.9 | 46.1 | 57.2 | 68.3 | 79.4 | 90.3 | 101.3 | 112.2 |
| 26 | 12.4 | 23.5 | 34.5 | 45.5 | 56.5 | 67.4 | 78.2 | 89.1 | 99.8 | 110.5 |
| 27 | 12.4 | 23.3 | 34.2 | 45.0 | 55.8 | 66.5 | 77.2 | 87.8 | 98.4 | 109.0 |
| 28 | 12.5 | 23.2 | 33.9 | 44.6 | 55.2 | 65.7 | 76.2 | 86.7 | 97.1 | 107.5 |
| 29 | 12.5 | 23.1 | 33.6 | 44.1 | 54.5 | 64.9 | 75.3 | 85.6 | 95.9 | 106.1 |
| 30 | 12.6 | 23.0 | 33.4 | 43.7 | 54.0 | 64.2 | 74.4 | 84.6 | 94.7 | 104.8 |
| 31 | 12.7 | 22.9 | 33.2 | 43.3 | 53.4 | 63.5 | 73.6 | 83.6 | 93.6 | 103.5 |
| 32 | 12.8 | 22.9 | 33.0 | 43.0 | 53.0 | 62.9 | 72.8 | 82.7 | 92.5 | 102.4 |
| 33 | 12.9 | 22.9 | 32.8 | 42.7 | 52.5 | 62.3 | 72.1 | 81.9 | 91.6 | 101.3 |
| 34 | 13.0 | 22.8 | 32.6 | 42.4 | 52.1 | 61.8 | 71.4 | 81.1 | 90.7 | 100.3 |
| 35 | 13.1 | 22.8 | 32.5 | 42.1 | 51.7 | 61.3 | 70.8 | 80.4 | 89.8 | 99.3 |
| 36 | 13.3 | 22.8 | 32.4 | 41.9 | 51.4 | 60.8 | 70.3 | 79.7 | 89.1 | 98.4 |
| 37 | 13.4 | 22.9 | 32.3 | 41.7 | 51.1 | 60.4 | 69.8 | 79.1 | 88.4 | 97.6 |
| 38 | 13.6 | 22.9 | 32.2 | 41.5 | 50.8 | 60.1 | 69.3 | 78.5 | 87.7 | 96.9 |
| 39 | 13.7 | 23.0 | 32.2 | 41.4 | 50.6 | 59.7 | 68.9 | 78.0 | 87.1 | 96.2 |
| 40 | 13.9 | 23.0 | 32.2 | 41.3 | 50.4 | 59.4 | 68.5 | 77.6 | 86.6 | 95.6 |

## Example 3-14

Compute the percent saturation of oxygen, nitrogen, argon, carbon dioxide, and total gas concentration in mg/L when water is heated from 9.3°C to 17.8°C. Assume that water was air saturated at 9°C and no gas is lost upon heating. Correct gas concentrations for changes in volume due to heating.

**Data Needed**

| Gas | 9.3°C | 17.8°C | Source |
|---|---|---|---|
| $O_2$ | 11.476 | 9.506 | Table 1.9 |
| $N_2$ | 18.575 | 15.642 | Table 1.10 |
| Ar | 0.7007 | 0.5813 | Table 1.11 |
| $CO_2$ | 0.9288 | 0.6995 | Table 1.12 |
| **Total** | **31.6805** | **26.4288** | |

***Individual Gases***

$$O_2\ (\%) = \left[\frac{11.476}{9.506}\right]100 = 120.7\%$$

$$N_2\ (\%) = \left[\frac{18.575}{15.642}\right]100 = 118.8\%$$

$$\text{Ar}\ (\%) = \left[\frac{0.7007}{0.5813}\right]100 = 120.7\%$$

$$CO_2\ (\%) = \left[\frac{0.9288}{0.6995}\right]100 = 132.8\%$$

***Total Dissolved Gas***

$$\text{Total gas}\ (\%) = \left[\frac{31.6805}{26.4288}\right]100 = 119.9\%$$

***Correct Final Gas Concentrations for Change in Volume***

**Data Needed**

| | 9.3°C | 17.8°C | Source |
|---|---|---|---|
| $\rho$(kg/L) | 0.999760 | 0.998634 | Table 7.1 |

If gas concentrations were expressed in mg/kg, the final concentration would be equal to the initial concentration as the mass of water is conservative:

$$C^{\dagger}_{\text{final}} = C^{\dagger}_{\text{initial}}$$

or in terms of concentration expressed in volume units:

$$\rho_{\text{final}} C^{*}_{\text{final}} = \rho_{\text{initial}} C^{*}_{\text{initial}}$$

and

$$C^{*}_{\text{final}} = \left[\frac{\rho_{\text{initial}}}{\rho_{\text{final}}}\right] C^{*}_{\text{initial}} = \left[\frac{0.999760}{0.998634}\right] C^{*}_{\text{initial}} = 1.001128\ C^{*}_{\text{initial}}$$

Applying this factor to the table values resulting in the following final concentrations:

**Data Needed**

| Gas | 9.3°C | 17.8°C | Source |
|---|---|---|---|
| $O_2$ | 11.476 | 9.517 | Table 1.9 + above |
| $N_2$ | 18.575 | 15.660 | Table 1.10 + above |
| Ar | 0.7007 | 0.5820 | Table 1.11 + above |
| $CO_2$ | 0.9288 | 0.7003 | Table 1.12 + above |
| **Total** | **31.6805** | **26.4593** | |

and

$$O_2\ (\%) = \left[\frac{11.476}{9.517}\right] 100 = 120.6\%$$

$$N_2\ (\%) = \left[\frac{18.575}{15.660}\right] 100 = 118.6\%$$

$$\text{Ar}\ (\%) = \left[\frac{0.7007}{0.5820}\right] 100 = 120.4\%$$

$$CO_2\ (\%) = \left[\frac{0.9288}{0.7003}\right] 100 = 132.6\%$$

In this case, the impact of change in density has only a small impact on the resulting gas supersaturation.

## *Mixing of Water of Different Temperatures*

The mixing of waters of different temperatures may result in supersaturation of dissolved gases. This is not as serious as the impact of heating discussed in the previous section. The mixing of waters of different temperatures results in supersaturation because the solubility of gases is not linear with temperature. Ignoring the change in heat capacity within a narrow range of temperature, the final temperature of a mixture of waters of different temperatures is equal to:

$$T_f = \frac{T_1 Q_1 + T_2 Q_2}{Q_1 + Q_2} \tag{3.49}$$

where

$T_f$ = final temperature of mixture (°C)
$T_1$ = temperature of stream #1 (°C)
$T_2$ = temperature of stream #2 (°C)
$Q_1$ = flow of stream #1 ($m^3/s$)
$Q_2$ = flow of stream #2 ($m^3/s$).

The percent saturation for a single gas of the resulting water is

$$\text{Percent saturation} = \left[\frac{C_1 \frac{Q_1}{Q_T} + C_2 \frac{Q_2}{Q_T}}{C_f^*}\right] 100 \tag{3.50}$$

where

$C_1$ = measured concentration (mg/L) of dissolved gas in stream #1
$C_2$ = measured concentration (mg/L) of dissolved gas in stream #2
$C_i^*$ = saturation concentration (mg/L) at final temperature
$Q_T$ = total flow ($m^3/s$).

If it is assumed that $C_1 = C_1^*$ and $C_2 = C_2^*$, Eq. (3.50) can be written as

$$\text{Percent saturation} = \left[\frac{C_1^* \frac{Q_1}{Q_T} + C_2^* \frac{Q_2}{Q_T}}{C_f^*}\right] 100 \tag{3.51}$$

where

$C_1^*$ = saturation concentration (mg/L) of dissolved gas at $T_1$
$C_2^*$ = saturation concentration (mg/L) of dissolved gas at $T_2$.

**Table 3.10** $\Delta P$ in mmHg Produced by Mixing Equal Flows of Water at Temperature $T_o$ and $T_o + \Delta T$ (Barometric Pressure = 760 mmHg, Salinity = 0.0 g/kg)

| Temperature (°C) | Δt (°C) | | | | | | | | | |
|---|---|---|---|---|---|---|---|---|---|---|
| | **2** | **4** | **6** | **8** | **10** | **12** | **14** | **16** | **18** | **20** |
| **0** | −0.8 | 0.5 | 2.6 | 5.5 | 9.0 | 13.2 | 17.9 | 23.1 | 28.7 | 34.8 |
| **1** | −0.9 | 0.4 | 2.5 | 5.3 | 8.7 | 12.8 | 17.3 | 22.3 | 27.8 | 33.7 |
| **2** | −0.9 | 0.4 | 2.4 | 5.1 | 8.4 | 12.3 | 16.8 | 21.6 | 26.9 | 32.6 |
| **3** | −0.9 | 0.3 | 2.3 | 4.9 | 8.2 | 12.0 | 16.2 | 21.0 | 26.1 | 31.6 |
| **4** | −0.9 | 0.3 | 2.2 | 4.8 | 7.9 | 11.6 | 15.7 | 20.3 | 25.3 | 30.6 |
| **5** | −0.9 | 0.3 | 2.1 | 4.6 | 7.7 | 11.2 | 15.3 | 19.7 | 24.5 | 29.7 |
| **6** | −0.9 | 0.3 | 2.1 | 4.5 | 7.5 | 10.9 | 14.8 | 19.1 | 23.8 | 28.8 |
| **7** | −0.9 | 0.3 | 2.0 | 4.4 | 7.3 | 10.6 | 14.4 | 18.6 | 23.1 | 27.9 |
| **8** | −0.8 | 0.3 | 2.0 | 4.3 | 7.1 | 10.3 | 14.0 | 18.0 | 22.4 | 27.1 |
| **9** | −0.8 | 0.3 | 2.0 | 4.2 | 6.9 | 10.1 | 13.6 | 17.5 | 21.8 | 26.3 |
| **10** | −0.7 | 0.3 | 2.0 | 4.1 | 6.8 | 9.8 | 13.3 | 17.1 | 21.2 | 25.5 |
| **11** | −0.6 | 0.4 | 2.0 | 4.1 | 6.6 | 9.6 | 13.0 | 16.6 | 20.6 | 24.8 |
| **12** | −0.6 | 0.5 | 2.0 | 4.0 | 6.5 | 9.4 | 12.6 | 16.2 | 20.0 | 24.1 |
| **13** | −0.5 | 0.5 | 2.0 | 4.0 | 6.4 | 9.2 | 12.4 | 15.8 | 19.5 | 23.4 |
| **14** | −0.4 | 0.6 | 2.1 | 4.0 | 6.3 | 9.1 | 12.1 | 15.4 | 19.0 | 22.8 |
| **15** | −0.3 | 0.7 | 2.1 | 4.0 | 6.3 | 8.9 | 11.8 | 15.0 | 18.5 | 22.2 |
| **16** | −0.1 | 0.8 | 2.2 | 4.0 | 6.2 | 8.8 | 11.6 | 14.7 | 18.0 | 21.6 |
| **17** | 0.0 | 0.9 | 2.3 | 4.0 | 6.2 | 8.6 | 11.4 | 14.4 | 17.6 | 21.0 |
| **18** | 0.1 | 1.0 | 2.3 | 4.1 | 6.1 | 8.5 | 11.2 | 14.1 | 17.2 | 20.5 |
| **19** | 0.3 | 1.1 | 2.4 | 4.1 | 6.1 | 8.4 | 11.0 | 13.8 | 16.8 | 20.0 |
| **20** | 0.4 | 1.3 | 2.5 | 4.2 | 6.1 | 8.3 | 10.8 | 13.5 | 16.4 | 19.5 |
| **21** | 0.6 | 1.4 | 2.6 | 4.2 | 6.1 | 8.3 | 10.7 | 13.3 | 16.1 | 19.0 |
| **22** | 0.7 | 1.5 | 2.7 | 4.3 | 6.1 | 8.2 | 10.5 | 13.1 | 15.7 | 18.5 |
| **23** | 0.9 | 1.7 | 2.9 | 4.4 | 6.1 | 8.2 | 10.4 | 12.8 | 15.4 | 18.1 |
| **24** | 1.1 | 1.9 | 3.0 | 4.4 | 6.2 | 8.1 | 10.3 | 12.6 | 15.1 | 17.7 |
| **25** | 1.3 | 2.0 | 3.1 | 4.5 | 6.2 | 8.1 | 10.2 | 12.4 | 14.8 | 17.3 |
| **26** | 1.5 | 2.2 | 3.3 | 4.6 | 6.3 | 8.1 | 10.1 | 12.3 | 14.6 | 16.9 |
| **27** | 1.6 | 2.4 | 3.4 | 4.8 | 6.3 | 8.1 | 10.0 | 12.1 | 14.3 | 16.6 |
| **28** | 1.8 | 2.6 | 3.6 | 4.9 | 6.4 | 8.1 | 10.0 | 12.0 | 14.1 | 16.2 |
| **29** | 2.1 | 2.8 | 3.8 | 5.0 | 6.5 | 8.1 | 9.9 | 11.8 | 13.8 | 15.9 |
| **30** | 2.3 | 3.0 | 3.9 | 5.1 | 6.5 | 8.1 | 9.8 | 11.7 | 13.6 | 15.6 |
| **31** | 2.5 | 3.2 | 4.1 | 5.3 | 6.6 | 8.2 | 9.8 | 11.6 | 13.4 | 15.3 |
| **32** | 2.7 | 3.4 | 4.3 | 5.4 | 6.7 | 8.2 | 9.8 | 11.5 | 13.2 | 15.0 |
| **33** | 2.9 | 3.6 | 4.5 | 5.6 | 6.8 | 8.2 | 9.8 | 11.4 | 13.0 | 14.7 |
| **34** | 3.2 | 3.8 | 4.7 | 5.7 | 6.9 | 8.3 | 9.7 | 11.3 | 12.8 | 14.4 |
| **35** | 3.4 | 4.0 | 4.9 | 5.9 | 7.1 | 8.3 | 9.7 | 11.2 | 12.7 | 14.2 |
| **36** | 3.6 | 4.3 | 5.1 | 6.1 | 7.2 | 8.4 | 9.7 | 11.1 | 12.5 | 13.9 |
| **37** | 3.9 | 4.5 | 5.3 | 6.2 | 7.3 | 8.5 | 9.7 | 11.0 | 12.4 | 13.7 |
| **38** | 4.1 | 4.7 | 5.5 | 6.4 | 7.4 | 8.6 | 9.7 | 11.0 | 12.2 | 13.4 |
| **39** | 4.4 | 5.0 | 5.7 | 6.6 | 7.6 | 8.6 | 9.8 | 10.9 | 12.1 | 13.2 |
| **40** | 4.6 | 5.2 | 5.9 | 6.7 | 7.7 | 8.7 | 9.8 | 10.9 | 11.9 | 13.0 |

**Table 3.11** $\Delta P$ in mmHg Produced by Mixing Equal Flows of Water at Temperature $T_o$ and $T_o + \Delta T$ (Barometric Pressure = 760 mmHg, Salinity = 35.0 g/kg)

| Temperature (°C) | Δt (°C) | | | | | | | | | |
|---|---|---|---|---|---|---|---|---|---|---|
| | 2 | 4 | 6 | 8 | 10 | 12 | 14 | 16 | 18 | 20 |
| 0 | −2.0 | −0.8 | 1.1 | 3.6 | 6.7 | 10.4 | 14.4 | 18.9 | 23.8 | 29.0 |
| 1 | −2.0 | −0.8 | 1.1 | 3.5 | 6.5 | 10.1 | 14.0 | 18.4 | 23.1 | 28.1 |
| 2 | −1.9 | −0.8 | 1.0 | 3.4 | 6.4 | 9.8 | 13.6 | 17.8 | 22.4 | 27.3 |
| 3 | −1.8 | −0.7 | 1.0 | 3.4 | 6.2 | 9.5 | 13.2 | 17.3 | 21.7 | 26.5 |
| 4 | −1.8 | −0.7 | 1.0 | 3.3 | 6.1 | 9.3 | 12.9 | 16.8 | 21.1 | 25.7 |
| 5 | −1.7 | −0.6 | 1.0 | 3.2 | 5.9 | 9.0 | 12.5 | 16.4 | 20.5 | 24.9 |
| 6 | −1.6 | −0.6 | 1.1 | 3.2 | 5.8 | 8.8 | 12.2 | 15.9 | 19.9 | 24.2 |
| 7 | −1.5 | −0.5 | 1.1 | 3.2 | 5.7 | 8.6 | 11.9 | 15.5 | 19.4 | 23.5 |
| 8 | −1.4 | −0.4 | 1.1 | 3.2 | 5.6 | 8.4 | 11.6 | 15.1 | 18.9 | 22.9 |
| 9 | −1.3 | −0.3 | 1.2 | 3.2 | 5.5 | 8.3 | 11.4 | 14.7 | 18.4 | 22.2 |
| 10 | −1.2 | −0.2 | 1.3 | 3.2 | 5.5 | 8.1 | 11.1 | 14.4 | 17.9 | 21.6 |
| 11 | −1.0 | −0.1 | 1.3 | 3.2 | 5.4 | 8.0 | 10.9 | 14.0 | 17.4 | 21.0 |
| 12 | −0.9 | 0.0 | 1.4 | 3.2 | 5.4 | 7.9 | 10.7 | 13.7 | 17.0 | 20.5 |
| 13 | −0.8 | 0.1 | 1.5 | 3.2 | 5.3 | 7.8 | 10.5 | 13.4 | 16.6 | 20.0 |
| 14 | −0.6 | 0.3 | 1.6 | 3.3 | 5.3 | 7.7 | 10.3 | 13.2 | 16.2 | 19.4 |
| 15 | −0.5 | 0.4 | 1.7 | 3.3 | 5.3 | 7.6 | 10.1 | 12.9 | 15.8 | 19.0 |
| 16 | −0.3 | 0.5 | 1.8 | 3.4 | 5.3 | 7.5 | 10.0 | 12.7 | 15.5 | 18.5 |
| 17 | −0.1 | 0.7 | 1.9 | 3.5 | 5.3 | 7.5 | 9.8 | 12.4 | 15.2 | 18.1 |
| 18 | 0.0 | 0.9 | 2.0 | 3.6 | 5.4 | 7.4 | 9.7 | 12.2 | 14.9 | 17.6 |
| 19 | 0.2 | 1.0 | 2.2 | 3.7 | 5.4 | 7.4 | 9.6 | 12.0 | 14.6 | 17.2 |
| 20 | 0.4 | 1.2 | 2.3 | 3.7 | 5.4 | 7.4 | 9.5 | 11.8 | 14.3 | 16.8 |
| 21 | 0.6 | 1.4 | 2.5 | 3.9 | 5.5 | 7.4 | 9.4 | 11.7 | 14.0 | 16.5 |
| 22 | 0.8 | 1.6 | 2.6 | 4.0 | 5.6 | 7.4 | 9.4 | 11.5 | 13.8 | 16.1 |
| 23 | 1.0 | 1.7 | 2.8 | 4.1 | 5.6 | 7.4 | 9.3 | 11.3 | 13.5 | 15.8 |
| 24 | 1.2 | 1.9 | 2.9 | 4.2 | 5.7 | 7.4 | 9.2 | 11.2 | 13.3 | 15.5 |
| 25 | 1.4 | 2.1 | 3.1 | 4.3 | 5.8 | 7.4 | 9.2 | 11.1 | 13.1 | 15.1 |
| 26 | 1.6 | 2.3 | 3.3 | 4.5 | 5.9 | 7.4 | 9.2 | 11.0 | 12.9 | 14.9 |
| 27 | 1.9 | 2.5 | 3.5 | 4.6 | 6.0 | 7.5 | 9.1 | 10.9 | 12.7 | 14.6 |
| 28 | 2.1 | 2.8 | 3.7 | 4.8 | 6.1 | 7.5 | 9.1 | 10.8 | 12.5 | 14.3 |
| 29 | 2.3 | 3.0 | 3.9 | 4.9 | 6.2 | 7.6 | 9.1 | 10.7 | 12.4 | 14.0 |
| 30 | 2.6 | 3.2 | 4.1 | 5.1 | 6.3 | 7.7 | 9.1 | 10.6 | 12.2 | 13.8 |
| 31 | 2.8 | 3.4 | 4.3 | 5.3 | 6.4 | 7.7 | 9.1 | 10.6 | 12.0 | 13.6 |
| 32 | 3.0 | 3.6 | 4.5 | 5.4 | 6.6 | 7.8 | 9.1 | 10.5 | 11.9 | 13.3 |
| 33 | 3.3 | 3.9 | 4.7 | 5.6 | 6.7 | 7.9 | 9.1 | 10.4 | 11.8 | 13.1 |
| 34 | 3.5 | 4.1 | 4.9 | 5.8 | 6.8 | 8.0 | 9.1 | 10.4 | 11.6 | 12.9 |
| 35 | 3.8 | 4.4 | 5.1 | 6.0 | 7.0 | 8.0 | 9.2 | 10.3 | 11.5 | 12.7 |
| 36 | 4.0 | 4.6 | 5.3 | 6.2 | 7.1 | 8.1 | 9.2 | 10.3 | 11.4 | 12.5 |
| 37 | 4.3 | 4.8 | 5.5 | 6.3 | 7.2 | 8.2 | 9.2 | 10.3 | 11.3 | 12.3 |
| 38 | 4.6 | 5.1 | 5.8 | 6.5 | 7.4 | 8.3 | 9.3 | 10.2 | 11.2 | 12.1 |
| 39 | 4.8 | 5.3 | 6.0 | 6.7 | 7.5 | 8.4 | 9.3 | 10.2 | 11.1 | 11.9 |
| 40 | 5.1 | 5.6 | 6.2 | 6.9 | 7.7 | 8.5 | 9.4 | 10.2 | 11.0 | 11.8 |

Equation (3.51) can be used without measurement of dissolved gas concentrations in streams 1 and 2. The $\Delta P$ produced by mixing two equal flows of water at $T_o$ and $T_o + \Delta T$ are presented in Table 3.10{171} for freshwater and in Table 3.11{172} for marine conditions. The final concentration of each gas is equal to $(C_1^* + C_2^*)/2$ and the Bunsen coefficient is evaluated at $T_f$:

$$\Delta P = \frac{(C^*_{O_2,T_1} + C^*_{O_2,T_2})}{2\beta_{T_f}}(0.5318) + \frac{(C^*_{N_2,T_1} + C^*_{N_2,T_2})}{2\beta_{T_f}}(0.6078) + \frac{(C^*_{Ar,T_1} + C^*_{Ar,T_2})}{2\beta_{T_f}}(0.4260) + \frac{(C^*_{CO_2,T_1} + C^*_{CO_2,T_2})}{\beta_{T_f}}(0.3845) + P_{ww,T_f} - BP \tag{3.52}$$

## Example 3-15

In a raceway system in Southern Idaho, 4°C surface water is mixed with 40°C geothermal water. The final temperature of the mixed water should be 28°C and 4000 lpm is needed. Compute the flow needed and the percent nitrogen + argon saturation in the mixed flow assuming that both waters are initially saturated. Correct gas concentrations for changes in volume due to mixing.

### Data Needed

| Temperature (°C) | $C^*_{N_2}$ (mg/L) | Source | Density (kg/L) | Source |
|---|---|---|---|---|
| 4 | 21.006 | Table 1.10 | 0.999975 | Table 7.1 |
| 28 | 13.165 | Table 1.10 | 0.996237 | Table 7.1 |
| 40 | 11.119 | Table 1.10 | 0.992220 | Table 7.1 |

*Compute Required Flows*

$$T_f = \frac{T_1Q_1 + T_2Q_2}{Q_1 + Q_2}$$

$$28.0 = \frac{4_1Q_1 + 40(4000 - Q_1)}{4000}$$

$Q_1 = 1333$ lpm

$Q_1 = 2667$ lpm

*Percent Saturation*

$$\text{Percent saturation} = \left[\frac{C_1^*\frac{Q_1}{Q_T} + C_2^*\frac{Q_2}{Q_T}}{C_f^*}\right]100$$

$$\text{Percent saturation} = \left[\frac{21.006\left[\frac{1333}{4000}\right] + 11.119\left[\frac{2667}{4000}\right]}{13.165}\right]100 = 109.49\%$$

***Correct Final Concentration for Changes in Density***

If written in terms of mass, the final concentration after mixing is equal to

$$C^{\dagger}_{4^\circ C}\left[\frac{1333}{4000}\right] + C^{\dagger}_{40^\circ C}\left[\frac{2667}{4000}\right] = C^{\dagger}_{28^\circ C}$$

or in volume units

$$\rho_{4^\circ C}C^{*}_{4^\circ C}\left[\frac{1333}{4000}\right] + \rho_{40^\circ C}C^{*}_{40^\circ C}\left[\frac{2667}{4000}\right] = \rho_{28^\circ C}C^{*}_{28^\circ C}$$

$$C^{\dagger}_{28^\circ C} = \left[\frac{\rho_{4^\circ C}C^{*}_{4^\circ C}\left[\frac{1333}{4000}\right] + \rho_{40^\circ C}C^{*}_{40^\circ C}\left[\frac{2667}{4000}\right]}{\rho_{28^\circ C}}\right]$$

$$C^{\dagger}_{28^\circ C} = \left[\frac{0.999975 \times 21.006\left[\frac{1333}{4000}\right] + 0.992220 \times 11.119\left[\frac{2667}{4000}\right]}{0.996237}\right]$$

$$C^{\dagger}_{28^\circ C} = 14.4102 \text{ mg/L}$$

$$\text{Percent saturation} = \left[\frac{14.4102}{13.165}\right]100 = 109.46\%$$

The impact of mixing two waters is much smaller than the impact of heating water. Changes in density have a negligible impact on the saturation of the mixed water.

## *Mixing of Water of Different Salinities*

The mixing of waters of different salinities may result in supersaturation of dissolved gases. This is not as serious as the impact of heating discussed in the previous section. The mixing of waters of different salinities results in supersaturation because the solubility of gases are not linear with salinity.

Ignoring the change in density with salinity, the final salinity of a mixture of waters of different salinity is

$$S_f = \frac{S_1Q_1 + S_2Q_2}{Q_1 + Q_2} \tag{3.53}$$

where

$S_f$ = final salinity of mixture (°C)
$S_1$ = salinity of stream #1 (°C)
$S_2$ = salinity of stream #2 (°C)
$Q_1$ = flow of stream #1 ($m^3/s$)
$Q_2$ = flow of stream #2 ($m^3/s$).

The percent saturation for a single gas of the resulting water is

$$\text{Percent saturation} = \left[\frac{C_1\frac{Q_1}{Q_T} + C_2\frac{Q_2}{Q_T}}{C_f^*}\right]100 \tag{3.54}$$

where

$C_1$ = measured concentration (mg/L) of dissolved gas in stream #1
$C_2$ = measured concentration (mg/L) of dissolved gas in stream #2
$C_t^*$ = saturation concentration (mg/L) at final temperature
$Q_T$ = total flow ($m^3/s$).

If it is assumed that $C_1 = C_1^*$ and $C_2 = C_2^*$, Eq. (3.43) can be written as

$$\text{Percent saturation} = \left[\frac{C_1^*\frac{Q_1}{Q_T} + C_2^*\frac{Q_2}{Q_T}}{C_f^*}\right]100 \tag{3.55}$$

where

$C_1^*$ = saturation concentration (mg/L) of dissolved gas at $S_1$
$C_2^*$ = saturation concentration (mg/L) of dissolved gas at $S_2$.

Equation (3.44) can be used without measurement of dissolved gas concentrations in streams 1 and 2. The $\Delta P$s produced by mixing two equal flows of water at $S_o$ and $S_o + \Delta S$ are presented in Table 3.12{176} for marine conditions. The final concentration of each gas is equal to $(C_1^* + C_2^*)/2$ and the Bunsen coefficient is evaluated at $T_f$:

$$\Delta P = \frac{(C^*_{O_2,S_1} + C^*_{O_2,S_2})}{2\beta_{S_f}}(0.5318) + \frac{(C^*_{N_2,S_1} + C^*_{N_2,S_2})}{2\beta_{S_f}}(0.6078) + \frac{(C^*_{Ar,S_1} + C^*_{Ar,S_2})}{2\beta_{S_f}}(0.4260) + \frac{(C^*_{CO_2,S_1} + C^*_{CO_2,S_2})}{2\beta_{S_f}}(0.3845) + P_{wv}^{S_f} - BP \tag{3.56}$$

**Table 3.12** $\Delta P$ in mmHg Produced by Mixing Equal Flows of Water at Salinities 0 g/kg and $\Delta S$ (Barometric Pressure = 760 mmHg)

| Temperature (°C) | ΔS (g/kg) | | | | | | | | | |
|---|---|---|---|---|---|---|---|---|---|---|
| | 0 | 5 | 10 | 15 | 20 | 25 | 30 | 35 | 40 | 45 |
| 0 | 0.0 | 0.1 | 0.5 | 1.1 | 2.0 | 3.1 | 4.5 | 6.2 | 8.1 | 10.2 |
| 1 | 0.0 | 0.1 | 0.5 | 1.1 | 2.0 | 3.1 | 4.4 | 6.1 | 7.9 | 10.1 |
| 2 | 0.0 | 0.1 | 0.5 | 1.1 | 1.9 | 3.0 | 4.4 | 5.9 | 7.8 | 9.9 |
| 3 | 0.0 | 0.1 | 0.5 | 1.1 | 1.9 | 3.0 | 4.3 | 5.8 | 7.6 | 9.7 |
| 4 | 0.0 | 0.1 | 0.5 | 1.0 | 1.9 | 2.9 | 4.2 | 5.7 | 7.5 | 9.5 |
| 5 | 0.0 | 0.1 | 0.5 | 1.0 | 1.8 | 2.9 | 4.1 | 5.6 | 7.4 | 9.3 |
| 6 | 0.0 | 0.1 | 0.4 | 1.0 | 1.8 | 2.8 | 4.0 | 5.5 | 7.2 | 9.2 |
| 7 | 0.0 | 0.1 | 0.4 | 1.0 | 1.8 | 2.8 | 4.0 | 5.4 | 7.1 | 9.0 |
| 8 | 0.0 | 0.1 | 0.4 | 1.0 | 1.7 | 2.7 | 3.9 | 5.3 | 7.0 | 8.8 |
| 9 | 0.0 | 0.1 | 0.4 | 1.0 | 1.7 | 2.7 | 3.8 | 5.2 | 6.8 | 8.7 |
| 10 | 0.0 | 0.1 | 0.4 | 0.9 | 1.7 | 2.6 | 3.8 | 5.1 | 6.7 | 8.5 |
| 11 | 0.0 | 0.1 | 0.4 | 0.9 | 1.6 | 2.6 | 3.7 | 5.0 | 6.6 | 8.4 |
| 12 | 0.0 | 0.1 | 0.4 | 0.9 | 1.6 | 2.5 | 3.6 | 5.0 | 6.5 | 8.2 |
| 13 | 0.0 | 0.1 | 0.4 | 0.9 | 1.6 | 2.5 | 3.6 | 4.9 | 6.4 | 8.1 |
| 14 | 0.0 | 0.1 | 0.4 | 0.9 | 1.6 | 2.4 | 3.5 | 4.8 | 6.3 | 8.0 |
| 15 | 0.0 | 0.1 | 0.4 | 0.9 | 1.5 | 2.4 | 3.5 | 4.7 | 6.2 | 7.8 |
| 16 | 0.0 | 0.1 | 0.4 | 0.8 | 1.5 | 2.4 | 3.4 | 4.6 | 6.1 | 7.7 |
| 17 | 0.0 | 0.1 | 0.4 | 0.8 | 1.5 | 2.3 | 3.3 | 4.6 | 6.0 | 7.6 |
| 18 | 0.0 | 0.1 | 0.4 | 0.8 | 1.5 | 2.3 | 3.3 | 4.5 | 5.9 | 7.4 |
| 19 | 0.0 | 0.1 | 0.4 | 0.8 | 1.4 | 2.2 | 3.2 | 4.4 | 5.8 | 7.3 |
| 20 | 0.0 | 0.1 | 0.4 | 0.8 | 1.4 | 2.2 | 3.2 | 4.4 | 5.7 | 7.2 |
| 21 | 0.0 | 0.1 | 0.4 | 0.8 | 1.4 | 2.2 | 3.1 | 4.3 | 5.6 | 7.1 |
| 22 | 0.0 | 0.1 | 0.3 | 0.8 | 1.4 | 2.1 | 3.1 | 4.2 | 5.5 | 7.0 |
| 23 | 0.0 | 0.1 | 0.3 | 0.8 | 1.4 | 2.1 | 3.1 | 4.2 | 5.4 | 6.9 |
| 24 | 0.0 | 0.1 | 0.3 | 0.8 | 1.3 | 2.1 | 3.0 | 4.1 | 5.4 | 6.8 |
| 25 | 0.0 | 0.1 | 0.3 | 0.7 | 1.3 | 2.1 | 3.0 | 4.0 | 5.3 | 6.7 |
| 26 | 0.0 | 0.1 | 0.3 | 0.7 | 1.3 | 2.0 | 2.9 | 4.0 | 5.2 | 6.6 |
| 27 | 0.0 | 0.1 | 0.3 | 0.7 | 1.3 | 2.0 | 2.9 | 3.9 | 5.1 | 6.5 |
| 28 | 0.0 | 0.1 | 0.3 | 0.7 | 1.3 | 2.0 | 2.8 | 3.9 | 5.1 | 6.4 |
| 29 | 0.0 | 0.1 | 0.3 | 0.7 | 1.2 | 1.9 | 2.8 | 3.8 | 5.0 | 6.3 |
| 30 | 0.0 | 0.1 | 0.3 | 0.7 | 1.2 | 1.9 | 2.8 | 3.8 | 4.9 | 6.3 |
| 31 | 0.0 | 0.1 | 0.3 | 0.7 | 1.2 | 1.9 | 2.7 | 3.7 | 4.9 | 6.2 |
| 32 | 0.0 | 0.1 | 0.3 | 0.7 | 1.2 | 1.9 | 2.7 | 3.7 | 4.8 | 6.1 |
| 33 | 0.0 | 0.1 | 0.3 | 0.7 | 1.2 | 1.8 | 2.7 | 3.6 | 4.7 | 6.0 |
| 34 | 0.0 | 0.1 | 0.3 | 0.7 | 1.2 | 1.8 | 2.6 | 3.6 | 4.7 | 5.9 |
| 35 | 0.0 | 0.1 | 0.3 | 0.7 | 1.2 | 1.8 | 2.6 | 3.5 | 4.6 | 5.9 |
| 36 | 0.0 | 0.1 | 0.3 | 0.6 | 1.1 | 1.8 | 2.6 | 3.5 | 4.6 | 5.8 |
| 37 | 0.0 | 0.1 | 0.3 | 0.6 | 1.1 | 1.8 | 2.5 | 3.5 | 4.5 | 5.7 |
| 38 | 0.0 | 0.1 | 0.3 | 0.6 | 1.1 | 1.7 | 2.5 | 3.4 | 4.5 | 5.7 |
| 39 | 0.0 | 0.1 | 0.3 | 0.6 | 1.1 | 1.7 | 2.5 | 3.4 | 4.4 | 5.6 |
| 40 | 0.0 | 0.1 | 0.3 | 0.6 | 1.1 | 1.7 | 2.5 | 3.3 | 4.4 | 5.5 |

## Example 3-16

Consider equal flows of water with salinities equal to 5 g/kg and 35 g/kg. If both waters are saturated with nitrogen gas at 28°C, compute the nitrogen saturation of the mixed water. Correct gas concentrations for changes in volume due to mixing.

### Data Needed

| Salinity (g/kg) | $C^*_{N_2}$ (mg/L) | Source | Density (kg/L) | Source |
|---|---|---|---|---|
| 5 | 12.776 | Table 2.13 | 0.999978 | Table 7.2 |
| 35 | 10.670 | Table 2.13 | 1.026163 | Table 7.2 |

### *Compute Final Salinity and $C^*_{N_2}$*

$$S_f = \frac{S_1 Q_1 + S_2 Q_2}{Q_1 + Q_2}$$

$$S_f = \frac{5.0 \times 1_1 + 35.0 \times 1}{2} = 20 \text{ g/kg}$$

$C^*_{N_2} = 11.678$ mg/L (20 g/kg, 28°C, Table 2.13)
$\rho = 1.011161$ kg/L (20 g/kg, 28°C, Table 7.2).

### *Compute Percent Saturation*

$$\text{Percent } N_2 \text{ saturation} = \left[\frac{C_1^* \frac{Q_1}{Q_T} + C_2^* \frac{Q_2}{Q_T}}{C_f^*}\right] 100$$

$$\text{Percent } N_2 \text{ saturation} = \left[\frac{12.776\left[\frac{1}{2}\right] + 10.670\left[\frac{1}{2}\right]}{11.678}\right] 100 = 100.39\%$$

### *Correct Final Concentration for Changes in Density*

If written in terms of mass, the final concentration after mixing is equal to

$$C^\dagger_{5\text{ g/kg}}\left[\frac{1}{2}\right] + C^\dagger_{35\text{ g/kg}}\left[\frac{1}{2}\right] = C^\dagger_{20\text{ g/kg}}$$

or in volume units

$$\rho_{5\text{ g/kg}} C^*_{5\text{ g/kg}}\left[\frac{1}{2}\right] + \rho_{35\text{ g/kg}} C^*_{35\text{ g/kg}}\left[\frac{1}{2}\right] = \rho_{20\text{ g/kg}} C^*_{20\text{ g/kg}}$$

$$C^*_{20\text{ g/kg}} = \left[\frac{\rho_{5\text{ g/kg}} C^*_{5\text{ g/kg}}\left[\frac{1}{2}\right] + \rho_{35\text{ g/kg}} C^*_{35\text{ g/kg}}\left[\frac{1}{2}\right]}{\rho_{20\text{ g/kg}}}\right]$$

$$C^*_{20\ g/kg} = \left[\frac{0.999978 \times 12.776\left[\frac{1}{2}\right] + 1.026163 \times 10.670\left[\frac{1}{2}\right]}{1.011161}\right]$$

$$C^*_{28^\circ C} = 11.732\ mg/L$$

$$\text{Percent saturation} = \left[\frac{11.732}{11.678}\right]100 = 100.46\%$$

## *Bubble Entrainment*

When bubbles are carried down into the water column or when gas and water are present together at elevated pressures, gas supersaturation may be produced. In Table 1.31{54}, the equilibrium concentration of oxygen at 40 m and 25°C is 31.78 mg/L compared to 14.62 mg/L at the surface. If the ambient concentration of dissolved oxygen is less than 31.78 mg/L at 40 m, oxygen will be transferred into the water from a bubble. Notice that the concentration of dissolved oxygen is not supersaturated with respect to the equilibrium concentration at this depth but is highly supersaturated with respect to the surface concentration of 14.62 mg/L. Gas supersaturation may be produced by this mechanism at dams, falls, and rapids. The use of diffused aeration or airlift pumps can also produce lethal dissolved gas concentrations (Colt and Westers, 1982; Cornacchia and Colt, 1984). Breaking waves in the marine environment can produce large numbers of small bubbles that can be carried down into the water column and dissolved (Schudlich and Emerson, 1996).

The same mechanism will produce gas supersaturation if gas is present in a pressurized water distribution system. This condition may result from leaks on the suction side of the pumps, clogging of intake structures so that flowing water does not completely fill the pipe, or an intake pipe that is not completely submerged. Air entrainment is more serious in seawater because of smaller bubbles that result in a significant increase in the overall gas transfer rate (Kils, 1976/1977).

The dissolved gas concentration resulting from bubble entrainment depends primarily on the depth of bubble submergence, the amount of air entrained, and the degree of mixing and turbulence. No general procedure is available for the computation of dissolved gas concentration resulting from bubble entrainment. In aquacultural applications using diffused aeration, $\Delta P$s in the range of 18–44 mmHg/m of diffuser submergence have been produced.

### Example 3-17

In a modeling study for a new dam, bubbles are observed to be carried down to 8 m during spring runoff flows. If it is assumed that a $\Delta P$ of 30 mmHg/m of bubble submergence will be produced, compute the $\Delta P$ below the dam.

$$\Delta P = 30\ mmHg/m \times 8\ m = 240\ mmHg$$

# 4 Solubility of Noble Gases in the Atmosphere

The solubilities of helium, neon, krypton, and xenon in terms of standard air solubility concentrations ($C_o^\dagger$) and Bunsen coefficients ($\beta$) have been computed as functions of salinity and temperature from the equations presented in Appendix A. Values are given for freshwater and marine conditions. These tabular data may be found as follows:

| Gas | $C_o^\dagger$ in nmol/kg | | $\beta$ in L/(L atm) | |
|---|---|---|---|---|
| | Table | Page | Table | Page |
| Helium (He) | | | | |
| Freshwater | 4.1 | 182 | 4.3 | 184 |
| 0–40 g/kg | 4.2 | 183 | 4.4 | 185 |
| Neon (Ne) | | | | |
| Freshwater | 4.5 | 186 | 4.7 | 188 |
| 0–40 g/kg | 4.6 | 187 | 4.8 | 189 |
| Krypton (Kr) | | | | |
| Freshwater | 4.9 | 190 | 4.11 | 192 |
| 0–40 g/kg | 4.10 | 191 | 4.12 | 193 |
| Xenon (Xe) | | | | |
| Freshwater | 4.13 | 194 | 4.15 | 196 |
| 0–40 g/kg | 4.14 | 195 | 4.16 | 197 |

Once the $C_o^\dagger$ value has been determined for a given temperature and salinity, the solubility can be adjusted for depth, elevation, or barometric pressures by equations presented previously (Tables 1.4 to 1.6). Conversion factors for other solubility units have been presented in Table 1.1{4}. The Bunsen coefficient can be used to compute the solubility of a gas with an arbitrary mole fraction.

### Example 4-1

Compute the standard air solubility concentrations of helium, neon, krypton, and xenon in nmol/kg and mg/L at 20°C and 35 g/kg. Use Eq. (1.1) and the conversion factors listed in Table 1.1.

Dissolved Gas Concentration in Water. DOI: 10.1016/B978-0-12-415916-7.00004-8

**Input Data**

| Gas | $C_o^\dagger$ (nmol/kg) | Source | Conversion Factor | Source |
|---|---|---|---|---|
| He | 1.6630 | Table 4.2 | $4.0026 \times 10^{-6}$ | Table 1.1 |
| Ne | 6.827 | Table 4.6 | $20.108 \times 10^{-6}$ | Table 1.1 |
| Kr | 2.4402 | Table 4.10 | $83.80 \times 10^{-6}$ | Table 1.1 |
| Xn | 0.33359 | Table 4.14 | $131.29 \times 10^{-6}$ | Table 1.1 |

$\rho = 1.02476$ kg/L (20°C and 35 g/kg; Table 7.2).

From Eq. (1.1):

$$C_o^* = C_o^\dagger \rho_w \times \text{conversion factor}$$

He: $C_o^* = 1.6630 \text{ nmol/kg} \times 1.02476 \text{ kg/L} \times 4.0026 \times 10^{-6} = 6.821 \times 10^{-6} \text{ mg/L}$

Ne: $C_o^* = 6.827 \text{ nmol/kg} \times 1.02476 \text{ kg/L} \times 20.180 \times 10^{-6} = 1.412 \times 10^{-4} \text{ mg/L}$

Kr: $C_o^* = 2.4402 \text{ nmol/kg} \times 1.02476 \text{ kg/L} \times 83.80 \times 10^{-6} = 2.096 \times 10^{-4} \text{ mg/L}$

Xn: $C_o^* = 0.33359 \text{ nmol/kg} \times 1.02476 \text{ kg/L} \times 131.29 \times 10^{-6} = 4.488 \times 10^{-5} \text{ mg/L}$

**Example 4-2**

Compute the air solubility concentrations of helium, neon, krypton, and xenon in nmol/kg at 1573 m, 4.3°C, and freshwater conditions. Use Eq. (1.8) to compute local barometric pressure and Eq. (1.5) to adjust for differences in pressure.

**Input Data**

| Gas | $C_o^\dagger$ (nmol/kg) | Source |
|---|---|---|
| He | 2.1307 | Table 4.1 |
| Ne | 9.609 | Table 4.5 |
| Kr | 4.8325 | Table 4.9 |
| Xn | 0.76183 | Table 4.13 |

$P_{wv} = 6.228$ mmHg (Table 1.21)

From Eq. (1.8):

$$\log_{10} \text{BP} = 2.880814 - \frac{h}{19,421.3}$$

$$\log_{10} \text{BP} = 2.880814 - \frac{1573}{19,421.3}; \quad \text{BP} = 630.70 \text{ mmHg}$$

From Eq. (1.5):

$$C_p^\dagger = C_o^\dagger \left[\frac{(\text{BP} - P_{\text{wv}})}{(1 - P_{\text{wv}})}\right] \quad \text{(pressures in atm)}$$

or

$$C_p^\dagger = C_o^\dagger \left[\frac{(\text{BP} - P_{\text{wv}})}{(760 - P_{\text{wv}})}\right] \quad \text{(pressures in mmHg)}$$

$$C_p^\dagger = C_o^\dagger \left[\frac{(630.70 - 6.228)}{(760 - 6.228)}\right] = 0.82846$$

He: $C_p^\dagger = 2.1307 \times 0.82846 = 1.7652$ nmol/kg

Ne: $C_p^\dagger = 9.609 \times 0.82846 = 7.961$ nmol/kg

Kr: $C_p^\dagger = 4.8325 \times 0.82846 = 4.004$ nmol/kg

Xn: $C_p^\dagger = 0.76183 \times 0.82846 = 0.6311$ nmol/kg

## Example 4-3

Find the Bunsen coefficients for the noble gases at 13.0°C and 35 g/kg.

**Input Data**

| Gas | $\beta$ (L/(L atm)) | Source |
|---|---|---|
| He | 7.5275 | Table 4.4 |
| Ne | 9.207 | Table 4.8 |
| Kr | 5.921 | Table 4.12 |
| Xn | 0.10678 | Table 4.16 |

Note that the Bunsen coefficients for helium, neon, and krypton are scaled to fit better in tabular form:

$$\beta_{\text{He}} = 7.5275/1000 = 0.0075275 \text{ L/(L atm)}$$

$$\beta_{\text{Ne}} = 9.207/1000 = 0.009207 \text{ L/(L atm)}$$

$$\beta_{\text{Kr}} = 5.921/100 = 0.05921 \text{ L/(L atm)}$$

$$\beta_{\text{Xn}} = 0.10678 = 0.10678 \text{ L/(L atm)}$$

**Table 4.1** Standard Air Saturation Concentration of Helium as a Function of Temperature in nmol/kg (freshwater, 1 atm moist air)

| Temperature (°C) | Δt (°C) | | | | | | | | | |
|---|---|---|---|---|---|---|---|---|---|---|
| | 0.0 | 0.1 | 0.2 | 0.3 | 0.4 | 0.5 | 0.6 | 0.7 | 0.8 | 0.9 |
| 0 | 2.1865 | 2.1851 | 2.1837 | 2.1823 | 2.1809 | 2.1795 | 2.1781 | 2.1767 | 2.1753 | 2.1739 |
| 1 | 2.1725 | 2.1712 | 2.1698 | 2.1685 | 2.1671 | 2.1658 | 2.1645 | 2.1631 | 2.1618 | 2.1605 |
| 2 | 2.1592 | 2.1579 | 2.1566 | 2.1553 | 2.1540 | 2.1527 | 2.1515 | 2.1502 | 2.1489 | 2.1477 |
| 3 | 2.1464 | 2.1452 | 2.1439 | 2.1427 | 2.1415 | 2.1402 | 2.1390 | 2.1378 | 2.1366 | 2.1354 |
| 4 | 2.1342 | 2.1330 | 2.1318 | 2.1307 | 2.1295 | 2.1283 | 2.1272 | 2.1260 | 2.1248 | 2.1237 |
| 5 | 2.1226 | 2.1214 | 2.1203 | 2.1192 | 2.1180 | 2.1169 | 2.1158 | 2.1147 | 2.1136 | 2.1125 |
| 6 | 2.1114 | 2.1103 | 2.1092 | 2.1081 | 2.1071 | 2.1060 | 2.1049 | 2.1039 | 2.1028 | 2.1018 |
| 7 | 2.1007 | 2.0997 | 2.0986 | 2.0976 | 2.0966 | 2.0956 | 2.0945 | 2.0935 | 2.0925 | 2.0915 |
| 8 | 2.0905 | 2.0895 | 2.0885 | 2.0875 | 2.0866 | 2.0856 | 2.0846 | 2.0836 | 2.0827 | 2.0817 |
| 9 | 2.0807 | 2.0798 | 2.0788 | 2.0779 | 2.0770 | 2.0760 | 2.0751 | 2.0741 | 2.0732 | 2.0723 |
| 10 | 2.0714 | 2.0705 | 2.0696 | 2.0687 | 2.0678 | 2.0669 | 2.0660 | 2.0651 | 2.0642 | 2.0633 |
| 11 | 2.0624 | 2.0615 | 2.0607 | 2.0598 | 2.0589 | 2.0581 | 2.0572 | 2.0564 | 2.0555 | 2.0547 |
| 12 | 2.0538 | 2.0530 | 2.0521 | 2.0513 | 2.0505 | 2.0497 | 2.0488 | 2.0480 | 2.0472 | 2.0464 |
| 13 | 2.0456 | 2.0448 | 2.0440 | 2.0432 | 2.0424 | 2.0416 | 2.0408 | 2.0400 | 2.0392 | 2.0384 |
| 14 | 2.0377 | 2.0369 | 2.0361 | 2.0354 | 2.0346 | 2.0338 | 2.0331 | 2.0323 | 2.0316 | 2.0308 |
| 15 | 2.0301 | 2.0293 | 2.0286 | 2.0278 | 2.0271 | 2.0264 | 2.0256 | 2.0249 | 2.0242 | 2.0235 |
| 16 | 2.0228 | 2.0220 | 2.0213 | 2.0206 | 2.0199 | 2.0192 | 2.0185 | 2.0178 | 2.0171 | 2.0164 |
| 17 | 2.0157 | 2.0150 | 2.0144 | 2.0137 | 2.0130 | 2.0123 | 2.0116 | 2.0110 | 2.0103 | 2.0096 |
| 18 | 2.0090 | 2.0083 | 2.0076 | 2.0070 | 2.0063 | 2.0057 | 2.0050 | 2.0044 | 2.0037 | 2.0031 |
| 19 | 2.0024 | 2.0018 | 2.0012 | 2.0005 | 1.9999 | 1.9993 | 1.9986 | 1.9980 | 1.9974 | 1.9968 |
| 20 | 1.9962 | 1.9955 | 1.9949 | 1.9943 | 1.9937 | 1.9931 | 1.9925 | 1.9919 | 1.9913 | 1.9907 |
| 21 | 1.9901 | 1.9895 | 1.9889 | 1.9883 | 1.9877 | 1.9871 | 1.9865 | 1.9859 | 1.9854 | 1.9848 |
| 22 | 1.9842 | 1.9836 | 1.9831 | 1.9825 | 1.9819 | 1.9813 | 1.9808 | 1.9802 | 1.9796 | 1.9791 |
| 23 | 1.9785 | 1.9780 | 1.9774 | 1.9768 | 1.9763 | 1.9757 | 1.9752 | 1.9746 | 1.9741 | 1.9735 |
| 24 | 1.9730 | 1.9724 | 1.9719 | 1.9714 | 1.9708 | 1.9703 | 1.9698 | 1.9692 | 1.9687 | 1.9682 |
| 25 | 1.9676 | 1.9671 | 1.9666 | 1.9661 | 1.9655 | 1.9650 | 1.9645 | 1.9640 | 1.9634 | 1.9629 |
| 26 | 1.9624 | 1.9619 | 1.9614 | 1.9609 | 1.9604 | 1.9599 | 1.9593 | 1.9588 | 1.9583 | 1.9578 |
| 27 | 1.9573 | 1.9568 | 1.9563 | 1.9558 | 1.9553 | 1.9548 | 1.9543 | 1.9538 | 1.9534 | 1.9529 |
| 28 | 1.9524 | 1.9519 | 1.9514 | 1.9509 | 1.9504 | 1.9499 | 1.9494 | 1.9490 | 1.9485 | 1.9480 |
| 29 | 1.9475 | 1.9470 | 1.9466 | 1.9461 | 1.9456 | 1.9451 | 1.9446 | 1.9442 | 1.9437 | 1.9432 |
| 30 | 1.9428 | 1.9423 | 1.9418 | 1.9413 | 1.9409 | 1.9404 | 1.9399 | 1.9395 | 1.9390 | 1.9385 |
| 31 | 1.9381 | 1.9376 | 1.9372 | 1.9367 | 1.9362 | 1.9358 | 1.9353 | 1.9349 | 1.9344 | 1.9339 |
| 32 | 1.9335 | 1.9330 | 1.9326 | 1.9321 | 1.9317 | 1.9312 | 1.9308 | 1.9303 | 1.9299 | 1.9294 |
| 33 | 1.9290 | 1.9285 | 1.9281 | 1.9276 | 1.9272 | 1.9267 | 1.9263 | 1.9258 | 1.9254 | 1.9249 |
| 34 | 1.9245 | 1.9240 | 1.9236 | 1.9231 | 1.9227 | 1.9223 | 1.9218 | 1.9214 | 1.9209 | 1.9205 |
| 35 | 1.9200 | 1.9196 | 1.9192 | 1.9187 | 1.9183 | 1.9178 | 1.9174 | 1.9170 | 1.9165 | 1.9161 |
| 36 | 1.9156 | 1.9152 | 1.9148 | 1.9143 | 1.9139 | 1.9135 | 1.9130 | 1.9126 | 1.9121 | 1.9117 |
| 37 | 1.9113 | 1.9108 | 1.9104 | 1.9100 | 1.9095 | 1.9091 | 1.9087 | 1.9082 | 1.9078 | 1.9074 |
| 38 | 1.9069 | 1.9065 | 1.9060 | 1.9056 | 1.9052 | 1.9047 | 1.9043 | 1.9039 | 1.9034 | 1.9030 |
| 39 | 1.9026 | 1.9021 | 1.9017 | 1.9013 | 1.9008 | 1.9004 | 1.9000 | 1.8995 | 1.8991 | 1.8987 |
| 40 | 1.8982 | 1.8978 | 1.8973 | 1.8969 | 1.8965 | 1.8960 | 1.8956 | 1.8952 | 1.8947 | 1.8943 |

*Source*: Based on Weiss (1971).

**Table 4.2** Standard Air Saturation Concentration of Helium as a Function of Temperature and Salinity in nmol/kg, 0–40 g/kg (seawater, 1 atm moist air)

| **Temperature (°C)** | **Salinity (g/kg)** | | | | | | | | |
|---|---|---|---|---|---|---|---|---|---|
| | **0.0** | **5.0** | **10.0** | **15.0** | **20.0** | **25.0** | **30.0** | **35.0** | **40.0** |
| **0** | 2.1865 | 2.1214 | 2.0583 | 1.9970 | 1.9375 | 1.8798 | 1.8239 | 1.7696 | 1.7169 |
| **1** | 2.1725 | 2.1084 | 2.0461 | 1.9856 | 1.9270 | 1.8701 | 1.8148 | 1.7612 | 1.7092 |
| **2** | 2.1592 | 2.0959 | 2.0345 | 1.9748 | 1.9169 | 1.8608 | 1.8062 | 1.7533 | 1.7019 |
| **3** | 2.1464 | 2.0840 | 2.0234 | 1.9645 | 1.9074 | 1.8519 | 1.7980 | 1.7457 | 1.6950 |
| **4** | 2.1342 | 2.0726 | 2.0128 | 1.9547 | 1.8983 | 1.8435 | 1.7903 | 1.7386 | 1.6884 |
| **5** | 2.1226 | 2.0618 | 2.0027 | 1.9453 | 1.8896 | 1.8355 | 1.7829 | 1.7318 | 1.6822 |
| **6** | 2.1114 | 2.0514 | 1.9931 | 1.9364 | 1.8813 | 1.8279 | 1.7759 | 1.7254 | 1.6764 |
| **7** | 2.1007 | 2.0415 | 1.9839 | 1.9279 | 1.8735 | 1.8206 | 1.7693 | 1.7193 | 1.6708 |
| **8** | 2.0905 | 2.0320 | 1.9751 | 1.9197 | 1.8660 | 1.8137 | 1.7629 | 1.7136 | 1.6656 |
| **9** | 2.0807 | 2.0229 | 1.9667 | 1.9120 | 1.8588 | 1.8072 | 1.7569 | 1.7081 | 1.6606 |
| **10** | 2.0714 | 2.0142 | 1.9586 | 1.9046 | 1.8520 | 1.8009 | 1.7512 | 1.7029 | 1.6559 |
| **11** | 2.0624 | 2.0059 | 1.9510 | 1.8975 | 1.8455 | 1.7950 | 1.7458 | 1.6979 | 1.6514 |
| **12** | 2.0538 | 1.9980 | 1.9436 | 1.8907 | 1.8393 | 1.7893 | 1.7406 | 1.6933 | 1.6472 |
| **13** | 2.0456 | 1.9903 | 1.9366 | 1.8843 | 1.8334 | 1.7839 | 1.7357 | 1.6888 | 1.6432 |
| **14** | 2.0377 | 1.9830 | 1.9298 | 1.8781 | 1.8277 | 1.7787 | 1.7310 | 1.6846 | 1.6394 |
| **15** | 2.0301 | 1.9760 | 1.9234 | 1.8721 | 1.8223 | 1.7738 | 1.7265 | 1.6805 | 1.6358 |
| **16** | 2.0228 | 1.9693 | 1.9172 | 1.8665 | 1.8171 | 1.7690 | 1.7222 | 1.6767 | 1.6323 |
| **17** | 2.0157 | 1.9628 | 1.9112 | 1.8610 | 1.8121 | 1.7645 | 1.7182 | 1.6730 | 1.6291 |
| **18** | 2.0090 | 1.9565 | 1.9055 | 1.8558 | 1.8073 | 1.7602 | 1.7143 | 1.6695 | 1.6260 |
| **19** | 2.0024 | 1.9505 | 1.9000 | 1.8507 | 1.8028 | 1.7560 | 1.7105 | 1.6662 | 1.6230 |
| **20** | 1.9962 | 1.9448 | 1.8947 | 1.8459 | 1.7984 | 1.7520 | 1.7069 | 1.6630 | 1.6201 |
| **21** | 1.9901 | 1.9392 | 1.8896 | 1.8412 | 1.7941 | 1.7482 | 1.7035 | 1.6599 | 1.6174 |
| **22** | 1.9842 | 1.9338 | 1.8846 | 1.8367 | 1.7900 | 1.7445 | 1.7001 | 1.6569 | 1.6148 |
| **23** | 1.9785 | 1.9285 | 1.8798 | 1.8323 | 1.7860 | 1.7409 | 1.6969 | 1.6541 | 1.6123 |
| **24** | 1.9730 | 1.9235 | 1.8752 | 1.8281 | 1.7822 | 1.7375 | 1.6938 | 1.6513 | 1.6099 |
| **25** | 1.9676 | 1.9185 | 1.8707 | 1.8240 | 1.7785 | 1.7341 | 1.6908 | 1.6486 | 1.6075 |
| **26** | 1.9624 | 1.9137 | 1.8663 | 1.8200 | 1.7748 | 1.7308 | 1.6879 | 1.6460 | 1.6052 |
| **27** | 1.9573 | 1.9091 | 1.8620 | 1.8161 | 1.7713 | 1.7276 | 1.6850 | 1.6435 | 1.6030 |
| **28** | 1.9524 | 1.9045 | 1.8578 | 1.8123 | 1.7679 | 1.7245 | 1.6823 | 1.6410 | 1.6008 |
| **29** | 1.9475 | 1.9000 | 1.8537 | 1.8086 | 1.7645 | 1.7215 | 1.6795 | 1.6386 | 1.5986 |
| **30** | 1.9428 | 1.8957 | 1.8497 | 1.8049 | 1.7612 | 1.7185 | 1.6768 | 1.6362 | 1.5965 |
| **31** | 1.9381 | 1.8914 | 1.8458 | 1.8013 | 1.7579 | 1.7155 | 1.6742 | 1.6338 | 1.5944 |
| **32** | 1.9335 | 1.8871 | 1.8419 | 1.7977 | 1.7546 | 1.7126 | 1.6715 | 1.6315 | 1.5923 |
| **33** | 1.9290 | 1.8830 | 1.8381 | 1.7942 | 1.7514 | 1.7097 | 1.6689 | 1.6291 | 1.5903 |
| **34** | 1.9245 | 1.8788 | 1.8343 | 1.7907 | 1.7483 | 1.7068 | 1.6663 | 1.6268 | 1.5882 |
| **35** | 1.9200 | 1.8747 | 1.8305 | 1.7873 | 1.7451 | 1.7039 | 1.6637 | 1.6244 | 1.5861 |
| **36** | 1.9156 | 1.8707 | 1.8267 | 1.7838 | 1.7419 | 1.7010 | 1.6611 | 1.6221 | 1.5840 |
| **37** | 1.9113 | 1.8666 | 1.8230 | 1.7804 | 1.7388 | 1.6981 | 1.6584 | 1.6197 | 1.5818 |
| **38** | 1.9069 | 1.8626 | 1.8192 | 1.7769 | 1.7356 | 1.6952 | 1.6558 | 1.6172 | 1.5796 |
| **39** | 1.9026 | 1.8585 | 1.8155 | 1.7734 | 1.7324 | 1.6923 | 1.6531 | 1.6148 | 1.5774 |
| **40** | 1.8982 | 1.8545 | 1.8117 | 1.7699 | 1.7291 | 1.6893 | 1.6503 | 1.6123 | 1.5751 |

*Source*: Based on Weiss (1971).

**Table 4.3** Bunsen Coefficient ($\beta$) for Helium as a Function of Temperature (L real gas/(L atm)) × 1000 (freshwater) (At 10°C, the table value is 8.9718; the Bunsen coefficient is 8.9718/1000 or 0.0089718 L/(L atm))

| Temperature (°C) | Δt (°C) | | | | | | | | | |
|---|---|---|---|---|---|---|---|---|---|---|
| | 0.0 | 0.1 | 0.2 | 0.3 | 0.4 | 0.5 | 0.6 | 0.7 | 0.8 | 0.9 |
| **0** | 9.4126 | 9.4070 | 9.4015 | 9.3960 | 9.3905 | 9.3851 | 9.3796 | 9.3742 | 9.3689 | 9.3635 |
| **1** | 9.3582 | 9.3529 | 9.3476 | 9.3424 | 9.3371 | 9.3319 | 9.3268 | 9.3216 | 9.3165 | 9.3114 |
| **2** | 9.3063 | 9.3012 | 9.2962 | 9.2912 | 9.2862 | 9.2812 | 9.2763 | 9.2714 | 9.2665 | 9.2616 |
| **3** | 9.2568 | 9.2519 | 9.2471 | 9.2424 | 9.2376 | 9.2329 | 9.2282 | 9.2235 | 9.2188 | 9.2142 |
| **4** | 9.2096 | 9.2050 | 9.2004 | 9.1958 | 9.1913 | 9.1868 | 9.1823 | 9.1779 | 9.1734 | 9.1690 |
| **5** | 9.1646 | 9.1602 | 9.1559 | 9.1516 | 9.1473 | 9.1430 | 9.1387 | 9.1345 | 9.1302 | 9.1261 |
| **6** | 9.1219 | 9.1177 | 9.1136 | 9.1095 | 9.1054 | 9.1013 | 9.0973 | 9.0932 | 9.0892 | 9.0852 |
| **7** | 9.0813 | 9.0773 | 9.0734 | 9.0695 | 9.0656 | 9.0618 | 9.0579 | 9.0541 | 9.0503 | 9.0465 |
| **8** | 9.0428 | 9.0390 | 9.0353 | 9.0316 | 9.0279 | 9.0243 | 9.0206 | 9.0170 | 9.0134 | 9.0098 |
| **9** | 9.0063 | 9.0028 | 8.9992 | 8.9957 | 8.9923 | 8.9888 | 8.9854 | 8.9819 | 8.9786 | 8.9752 |
| **10** | 8.9718 | 8.9685 | 8.9652 | 8.9618 | 8.9586 | 8.9553 | 8.9521 | 8.9488 | 8.9456 | 8.9424 |
| **11** | 8.9393 | 8.9361 | 8.9330 | 8.9299 | 8.9268 | 8.9237 | 8.9207 | 8.9176 | 8.9146 | 8.9116 |
| **12** | 8.9086 | 8.9057 | 8.9027 | 8.8998 | 8.8969 | 8.8940 | 8.8912 | 8.8883 | 8.8855 | 8.8827 |
| **13** | 8.8799 | 8.8771 | 8.8743 | 8.8716 | 8.8689 | 8.8662 | 8.8635 | 8.8608 | 8.8581 | 8.8555 |
| **14** | 8.8529 | 8.8503 | 8.8477 | 8.8452 | 8.8426 | 8.8401 | 8.8376 | 8.8351 | 8.8326 | 8.8301 |
| **15** | 8.8277 | 8.8253 | 8.8229 | 8.8205 | 8.8181 | 8.8158 | 8.8134 | 8.8111 | 8.8088 | 8.8065 |
| **16** | 8.8043 | 8.8020 | 8.7998 | 8.7976 | 8.7954 | 8.7932 | 8.7910 | 8.7889 | 8.7867 | 8.7846 |
| **17** | 8.7825 | 8.7804 | 8.7784 | 8.7763 | 8.7743 | 8.7723 | 8.7703 | 8.7683 | 8.7663 | 8.7644 |
| **18** | 8.7624 | 8.7605 | 8.7586 | 8.7567 | 8.7549 | 8.7530 | 8.7512 | 8.7494 | 8.7476 | 8.7458 |
| **19** | 8.7440 | 8.7422 | 8.7405 | 8.7388 | 8.7371 | 8.7354 | 8.7337 | 8.7320 | 8.7304 | 8.7288 |
| **20** | 8.7272 | 8.7256 | 8.7240 | 8.7224 | 8.7209 | 8.7193 | 8.7178 | 8.7163 | 8.7148 | 8.7134 |
| **21** | 8.7119 | 8.7105 | 8.7090 | 8.7076 | 8.7062 | 8.7049 | 8.7035 | 8.7021 | 8.7008 | 8.6995 |
| **22** | 8.6982 | 8.6969 | 8.6956 | 8.6944 | 8.6931 | 8.6919 | 8.6907 | 8.6895 | 8.6883 | 8.6871 |
| **23** | 8.6860 | 8.6849 | 8.6837 | 8.6826 | 8.6815 | 8.6805 | 8.6794 | 8.6783 | 8.6773 | 8.6763 |
| **24** | 8.6753 | 8.6743 | 8.6733 | 8.6724 | 8.6714 | 8.6705 | 8.6696 | 8.6687 | 8.6678 | 8.6669 |
| **25** | 8.6661 | 8.6652 | 8.6644 | 8.6636 | 8.6628 | 8.6620 | 8.6612 | 8.6605 | 8.6597 | 8.6590 |
| **26** | 8.6583 | 8.6576 | 8.6569 | 8.6562 | 8.6556 | 8.6549 | 8.6543 | 8.6537 | 8.6531 | 8.6525 |
| **27** | 8.6519 | 8.6513 | 8.6508 | 8.6503 | 8.6497 | 8.6492 | 8.6488 | 8.6483 | 8.6478 | 8.6474 |
| **28** | 8.6469 | 8.6465 | 8.6461 | 8.6457 | 8.6453 | 8.6450 | 8.6446 | 8.6443 | 8.6439 | 8.6436 |
| **29** | 8.6433 | 8.6431 | 8.6428 | 8.6425 | 8.6423 | 8.6421 | 8.6418 | 8.6416 | 8.6414 | 8.6413 |
| **30** | 8.6411 | 8.6410 | 8.6408 | 8.6407 | 8.6406 | 8.6405 | 8.6404 | 8.6403 | 8.6403 | 8.6402 |
| **31** | 8.6402 | 8.6402 | 8.6402 | 8.6402 | 8.6402 | 8.6403 | 8.6403 | 8.6404 | 8.6405 | 8.6405 |
| **32** | 8.6406 | 8.6408 | 8.6409 | 8.6410 | 8.6412 | 8.6413 | 8.6415 | 8.6417 | 8.6419 | 8.6421 |
| **33** | 8.6424 | 8.6426 | 8.6429 | 8.6432 | 8.6434 | 8.6437 | 8.6440 | 8.6444 | 8.6447 | 8.6450 |
| **34** | 8.6454 | 8.6458 | 8.6462 | 8.6466 | 8.6470 | 8.6474 | 8.6478 | 8.6483 | 8.6487 | 8.6492 |
| **35** | 8.6497 | 8.6502 | 8.6507 | 8.6512 | 8.6518 | 8.6523 | 8.6529 | 8.6535 | 8.6540 | 8.6546 |
| **36** | 8.6553 | 8.6559 | 8.6565 | 8.6572 | 8.6578 | 8.6585 | 8.6592 | 8.6599 | 8.6606 | 8.6613 |
| **37** | 8.6621 | 8.6628 | 8.6636 | 8.6643 | 8.6651 | 8.6659 | 8.6667 | 8.6676 | 8.6684 | 8.6692 |
| **38** | 8.6701 | 8.6710 | 8.6718 | 8.6727 | 8.6736 | 8.6746 | 8.6755 | 8.6764 | 8.6774 | 8.6784 |
| **39** | 8.6793 | 8.6803 | 8.6813 | 8.6824 | 8.6834 | 8.6844 | 8.6855 | 8.6865 | 8.6876 | 8.6887 |
| **40** | 8.6898 | 8.6909 | 8.6920 | 8.6932 | 8.6943 | 8.6955 | 8.6967 | 8.6978 | 8.6990 | 8.7002 |

*Source*: Based on Weiss (1971).

**Table 4.4** Bunsen Coefficient ($\beta$) for Helium as a Function of Temperature and Salinity, 0–40 g/kg (L real gas/(L atm)) × 1000 (At 10°C and 35 g/kg, the table value is 7.5754; the Bunsen coefficient is 7.5754/1000 or 0.0075754 L/(L atm))

| Temperature (°C) | Salinity (g/kg) | | | | | | | | |
|---|---|---|---|---|---|---|---|---|---|
| | 0.0 | 5.0 | 10.0 | 15.0 | 20.0 | 25.0 | 30.0 | 35.0 | 40.0 |
| 0 | 9.4126 | 9.1685 | 8.9308 | 8.6993 | 8.4737 | 8.2540 | 8.0400 | 7.8316 | 7.6285 |
| 1 | 9.3582 | 9.1176 | 8.8832 | 8.6549 | 8.4324 | 8.2156 | 8.0044 | 7.7986 | 7.5981 |
| 2 | 9.3063 | 9.0691 | 8.8379 | 8.6126 | 8.3930 | 8.1791 | 7.9706 | 7.7674 | 7.5694 |
| 3 | 9.2568 | 9.0228 | 8.7947 | 8.5724 | 8.3557 | 8.1444 | 7.9386 | 7.7379 | 7.5423 |
| 4 | 9.2096 | 8.9787 | 8.7536 | 8.5342 | 8.3202 | 8.1116 | 7.9083 | 7.7100 | 7.5167 |
| 5 | 9.1646 | 8.9368 | 8.7146 | 8.4979 | 8.2866 | 8.0806 | 7.8797 | 7.6838 | 7.4927 |
| 6 | 9.1219 | 8.8969 | 8.6775 | 8.4636 | 8.2548 | 8.0513 | 7.8527 | 7.6591 | 7.4702 |
| 7 | 9.0813 | 8.8591 | 8.6425 | 8.4311 | 8.2248 | 8.0237 | 7.8274 | 7.6360 | 7.4492 |
| 8 | 9.0428 | 8.8234 | 8.6093 | 8.4004 | 8.1966 | 7.9977 | 7.8037 | 7.6143 | 7.4296 |
| 9 | 9.0063 | 8.7895 | 8.5780 | 8.3715 | 8.1700 | 7.9733 | 7.7814 | 7.5941 | 7.4114 |
| 10 | 8.9718 | 8.7576 | 8.5485 | 8.3443 | 8.1451 | 7.9506 | 7.7607 | 7.5754 | 7.3945 |
| 11 | 8.9393 | 8.7275 | 8.5207 | 8.3188 | 8.1218 | 7.9293 | 7.7415 | 7.5581 | 7.3790 |
| 12 | 8.9086 | 8.6992 | 8.4947 | 8.2950 | 8.1000 | 7.9096 | 7.7237 | 7.5421 | 7.3648 |
| 13 | 8.8799 | 8.6727 | 8.4704 | 8.2728 | 8.0798 | 7.8914 | 7.7073 | 7.5275 | 7.3519 |
| 14 | 8.8529 | 8.6480 | 8.4478 | 8.2522 | 8.0612 | 7.8746 | 7.6923 | 7.5142 | 7.3402 |
| 15 | 8.8277 | 8.6249 | 8.4267 | 8.2331 | 8.0440 | 7.8592 | 7.6786 | 7.5022 | 7.3298 |
| 16 | 8.8043 | 8.6035 | 8.4073 | 8.2156 | 8.0282 | 7.8452 | 7.6663 | 7.4914 | 7.3206 |
| 17 | 8.7825 | 8.5837 | 8.3894 | 8.1995 | 8.0139 | 7.8325 | 7.6552 | 7.4819 | 7.3126 |
| 18 | 8.7624 | 8.5655 | 8.3731 | 8.1849 | 8.0010 | 7.8212 | 7.6454 | 7.4736 | 7.3057 |
| 19 | 8.7440 | 8.5489 | 8.3582 | 8.1717 | 7.9894 | 7.8112 | 7.6369 | 7.4665 | 7.3000 |
| 20 | 8.7272 | 8.5338 | 8.3448 | 8.1600 | 7.9792 | 7.8025 | 7.6296 | 7.4606 | 7.2954 |
| 21 | 8.7119 | 8.5203 | 8.3329 | 8.1496 | 7.9703 | 7.7950 | 7.6235 | 7.4559 | 7.2919 |
| 22 | 8.6982 | 8.5082 | 8.3223 | 8.1405 | 7.9627 | 7.7888 | 7.6186 | 7.4522 | 7.2894 |
| 23 | 8.6860 | 8.4975 | 8.3132 | 8.1328 | 7.9564 | 7.7838 | 7.6149 | 7.4497 | 7.2881 |
| 24 | 8.6753 | 8.4883 | 8.3054 | 8.1264 | 7.9513 | 7.7800 | 7.6123 | 7.4483 | 7.2877 |
| 25 | 8.6661 | 8.4805 | 8.2990 | 8.1213 | 7.9475 | 7.7773 | 7.6108 | 7.4479 | 7.2885 |
| 26 | 8.6583 | 8.4741 | 8.2939 | 8.1175 | 7.9448 | 7.7759 | 7.6105 | 7.4486 | 7.2902 |
| 27 | 8.6519 | 8.4691 | 8.2901 | 8.1149 | 7.9434 | 7.7755 | 7.6112 | 7.4504 | 7.2929 |
| 28 | 8.6469 | 8.4653 | 8.2876 | 8.1135 | 7.9431 | 7.7763 | 7.6130 | 7.4532 | 7.2966 |
| 29 | 8.6433 | 8.4629 | 8.2863 | 8.1134 | 7.9440 | 7.7782 | 7.6159 | 7.4569 | 7.3013 |
| 30 | 8.6411 | 8.4618 | 8.2863 | 8.1144 | 7.9461 | 7.7812 | 7.6198 | 7.4617 | 7.3070 |
| 31 | 8.6402 | 8.4620 | 8.2875 | 8.1166 | 7.9492 | 7.7853 | 7.6248 | 7.4675 | 7.3135 |
| 32 | 8.6406 | 8.4635 | 8.2900 | 8.1200 | 7.9535 | 7.7905 | 7.6307 | 7.4743 | 7.3211 |
| 33 | 8.6424 | 8.4662 | 8.2936 | 8.1245 | 7.9589 | 7.7967 | 7.6377 | 7.4820 | 7.3295 |
| 34 | 8.6454 | 8.4701 | 8.2984 | 8.1302 | 7.9654 | 7.8039 | 7.6457 | 7.4907 | 7.3388 |
| 35 | 8.6497 | 8.4753 | 8.3044 | 8.1370 | 7.9729 | 7.8122 | 7.6547 | 7.5003 | 7.3491 |
| 36 | 8.6553 | 8.4817 | 8.3116 | 8.1449 | 7.9815 | 7.8214 | 7.6646 | 7.5109 | 7.3602 |
| 37 | 8.6621 | 8.4892 | 8.3199 | 8.1539 | 7.9912 | 7.8317 | 7.6755 | 7.5223 | 7.3722 |
| 38 | 8.6701 | 8.4980 | 8.3293 | 8.1639 | 8.0019 | 7.8430 | 7.6873 | 7.5347 | 7.3851 |
| 39 | 8.6793 | 8.5079 | 8.3398 | 8.1751 | 8.0136 | 7.8553 | 7.7001 | 7.5480 | 7.3989 |
| 40 | 8.6898 | 8.5190 | 8.3515 | 8.1873 | 8.0263 | 7.8685 | 7.7138 | 7.5621 | 7.4134 |

*Source*: Based on Weiss (1971).

**Table 4.5** Standard Air Saturation Concentration of Neon as a Function of Temperature in nmol/kg (freshwater, 1 atm moist air)

| Temperature (°C) | Δt (°C) | | | | | | | | | |
|---|---|---|---|---|---|---|---|---|---|---|
| | **0.0** | **0.1** | **0.2** | **0.3** | **0.4** | **0.5** | **0.6** | **0.7** | **0.8** | **0.9** |
| **0** | 10.084 | 10.072 | 10.060 | 10.049 | 10.037 | 10.025 | 10.014 | 10.002 | 9.991 | 9.979 |
| **1** | 9.968 | 9.956 | 9.945 | 9.934 | 9.922 | 9.911 | 9.900 | 9.889 | 9.878 | 9.866 |
| **2** | 9.855 | 9.844 | 9.833 | 9.822 | 9.811 | 9.800 | 9.789 | 9.779 | 9.768 | 9.757 |
| **3** | 9.746 | 9.735 | 9.725 | 9.714 | 9.703 | 9.693 | 9.682 | 9.672 | 9.661 | 9.651 |
| **4** | 9.640 | 9.630 | 9.619 | 9.609 | 9.599 | 9.589 | 9.578 | 9.568 | 9.558 | 9.548 |
| **5** | 9.538 | 9.527 | 9.517 | 9.507 | 9.497 | 9.487 | 9.477 | 9.467 | 9.458 | 9.448 |
| **6** | 9.438 | 9.428 | 9.418 | 9.409 | 9.399 | 9.389 | 9.380 | 9.370 | 9.360 | 9.351 |
| **7** | 9.341 | 9.332 | 9.322 | 9.313 | 9.303 | 9.294 | 9.285 | 9.275 | 9.266 | 9.257 |
| **8** | 9.247 | 9.238 | 9.229 | 9.220 | 9.211 | 9.202 | 9.193 | 9.184 | 9.175 | 9.166 |
| **9** | 9.157 | 9.148 | 9.139 | 9.130 | 9.121 | 9.112 | 9.103 | 9.095 | 9.086 | 9.077 |
| **10** | 9.069 | 9.060 | 9.051 | 9.043 | 9.034 | 9.026 | 9.017 | 9.009 | 9.000 | 8.992 |
| **11** | 8.983 | 8.975 | 8.966 | 8.958 | 8.950 | 8.942 | 8.933 | 8.925 | 8.917 | 8.909 |
| **12** | 8.900 | 8.892 | 8.884 | 8.876 | 8.868 | 8.860 | 8.852 | 8.844 | 8.836 | 8.828 |
| **13** | 8.820 | 8.813 | 8.805 | 8.797 | 8.789 | 8.781 | 8.774 | 8.766 | 8.758 | 8.750 |
| **14** | 8.743 | 8.735 | 8.728 | 8.720 | 8.713 | 8.705 | 8.698 | 8.690 | 8.683 | 8.675 |
| **15** | 8.668 | 8.660 | 8.653 | 8.646 | 8.638 | 8.631 | 8.624 | 8.617 | 8.609 | 8.602 |
| **16** | 8.595 | 8.588 | 8.581 | 8.574 | 8.567 | 8.560 | 8.553 | 8.546 | 8.539 | 8.532 |
| **17** | 8.525 | 8.518 | 8.511 | 8.504 | 8.497 | 8.491 | 8.484 | 8.477 | 8.470 | 8.464 |
| **18** | 8.457 | 8.450 | 8.444 | 8.437 | 8.430 | 8.424 | 8.417 | 8.411 | 8.404 | 8.398 |
| **19** | 8.391 | 8.385 | 8.378 | 8.372 | 8.366 | 8.359 | 8.353 | 8.347 | 8.340 | 8.334 |
| **20** | 8.328 | 8.322 | 8.315 | 8.309 | 8.303 | 8.297 | 8.291 | 8.285 | 8.279 | 8.273 |
| **21** | 8.267 | 8.261 | 8.255 | 8.249 | 8.243 | 8.237 | 8.231 | 8.225 | 8.219 | 8.213 |
| **22** | 8.208 | 8.202 | 8.196 | 8.190 | 8.185 | 8.179 | 8.173 | 8.168 | 8.162 | 8.156 |
| **23** | 8.151 | 8.145 | 8.140 | 8.134 | 8.128 | 8.123 | 8.117 | 8.112 | 8.107 | 8.101 |
| **24** | 8.096 | 8.090 | 8.085 | 8.080 | 8.074 | 8.069 | 8.064 | 8.059 | 8.053 | 8.048 |
| **25** | 8.043 | 8.038 | 8.033 | 8.027 | 8.022 | 8.017 | 8.012 | 8.007 | 8.002 | 7.997 |
| **26** | 7.992 | 7.987 | 7.982 | 7.977 | 7.972 | 7.967 | 7.962 | 7.958 | 7.953 | 7.948 |
| **27** | 7.943 | 7.938 | 7.934 | 7.929 | 7.924 | 7.919 | 7.915 | 7.910 | 7.905 | 7.901 |
| **28** | 7.896 | 7.892 | 7.887 | 7.883 | 7.878 | 7.873 | 7.869 | 7.865 | 7.860 | 7.856 |
| **29** | 7.851 | 7.847 | 7.842 | 7.838 | 7.834 | 7.829 | 7.825 | 7.821 | 7.817 | 7.812 |
| **30** | 7.808 | 7.804 | 7.800 | 7.795 | 7.791 | 7.787 | 7.783 | 7.779 | 7.775 | 7.771 |
| **31** | 7.767 | 7.763 | 7.759 | 7.755 | 7.751 | 7.747 | 7.743 | 7.739 | 7.735 | 7.731 |
| **32** | 7.727 | 7.723 | 7.720 | 7.716 | 7.712 | 7.708 | 7.704 | 7.701 | 7.697 | 7.693 |
| **33** | 7.690 | 7.686 | 7.682 | 7.679 | 7.675 | 7.671 | 7.668 | 7.664 | 7.661 | 7.657 |
| **34** | 7.654 | 7.650 | 7.647 | 7.643 | 7.640 | 7.636 | 7.633 | 7.630 | 7.626 | 7.623 |
| **35** | 7.619 | 7.616 | 7.613 | 7.610 | 7.606 | 7.603 | 7.600 | 7.597 | 7.593 | 7.590 |
| **36** | 7.587 | 7.584 | 7.581 | 7.578 | 7.575 | 7.571 | 7.568 | 7.565 | 7.562 | 7.559 |
| **37** | 7.556 | 7.553 | 7.550 | 7.547 | 7.545 | 7.542 | 7.539 | 7.536 | 7.533 | 7.530 |
| **38** | 7.527 | 7.524 | 7.522 | 7.519 | 7.516 | 7.513 | 7.511 | 7.508 | 7.505 | 7.503 |
| **39** | 7.500 | 7.497 | 7.495 | 7.492 | 7.490 | 7.487 | 7.484 | 7.482 | 7.479 | 7.477 |
| **40** | 7.474 | 7.472 | 7.469 | 7.467 | 7.464 | 7.462 | 7.460 | 7.457 | 7.455 | 7.453 |

*Source*: Based on Hamme and Emerson (2004).

**Table 4.6** Standard Air Saturation Concentration of Neon as a Function of Temperature and Salinity in nmol/kg, 0–40 g/kg (seawater, 1 atm moist air)

| Temperature (°C) | Salinity (g/kg) | | | | | | | | |
|---|---|---|---|---|---|---|---|---|---|
| | 0.0 | 5.0 | 10.0 | 15.0 | 20.0 | 25.0 | 30.0 | 35.0 | 40.0 |
| **0** | 10.084 | 9.766 | 9.459 | 9.161 | 8.873 | 8.593 | 8.323 | 8.061 | 7.807 |
| **1** | 9.968 | 9.656 | 9.353 | 9.061 | 8.777 | 8.502 | 8.236 | 7.978 | 7.728 |
| **2** | 9.855 | 9.549 | 9.251 | 8.963 | 8.684 | 8.414 | 8.152 | 7.898 | 7.652 |
| **3** | 9.746 | 9.444 | 9.152 | 8.869 | 8.594 | 8.328 | 8.070 | 7.821 | 7.578 |
| **4** | 9.640 | 9.344 | 9.056 | 8.777 | 8.507 | 8.245 | 7.991 | 7.745 | 7.507 |
| **5** | 9.538 | 9.246 | 8.963 | 8.688 | 8.422 | 8.165 | 7.915 | 7.672 | 7.438 |
| **6** | 9.438 | 9.151 | 8.872 | 8.602 | 8.340 | 8.087 | 7.840 | 7.602 | 7.370 |
| **7** | 9.341 | 9.059 | 8.784 | 8.519 | 8.261 | 8.011 | 7.769 | 7.533 | 7.305 |
| **8** | 9.247 | 8.969 | 8.699 | 8.438 | 8.184 | 7.938 | 7.699 | 7.467 | 7.243 |
| **9** | 9.157 | 8.883 | 8.617 | 8.359 | 8.109 | 7.867 | 7.631 | 7.403 | 7.182 |
| **10** | 9.069 | 8.799 | 8.537 | 8.283 | 8.037 | 7.798 | 7.566 | 7.341 | 7.123 |
| **11** | 8.983 | 8.718 | 8.460 | 8.210 | 7.967 | 7.732 | 7.503 | 7.281 | 7.066 |
| **12** | 8.900 | 8.639 | 8.385 | 8.139 | 7.899 | 7.667 | 7.442 | 7.223 | 7.011 |
| **13** | 8.820 | 8.563 | 8.313 | 8.070 | 7.834 | 7.605 | 7.383 | 7.167 | 6.958 |
| **14** | 8.743 | 8.489 | 8.242 | 8.003 | 7.771 | 7.545 | 7.326 | 7.113 | 6.907 |
| **15** | 8.668 | 8.418 | 8.175 | 7.939 | 7.710 | 7.487 | 7.271 | 7.061 | 6.857 |
| **16** | 8.595 | 8.349 | 8.109 | 7.876 | 7.650 | 7.431 | 7.218 | 7.011 | 6.810 |
| **17** | 8.525 | 8.282 | 8.046 | 7.816 | 7.593 | 7.377 | 7.167 | 6.962 | 6.764 |
| **18** | 8.457 | 8.217 | 7.984 | 7.758 | 7.538 | 7.325 | 7.117 | 6.915 | 6.719 |
| **19** | 8.391 | 8.155 | 7.925 | 7.702 | 7.485 | 7.274 | 7.070 | 6.870 | 6.677 |
| **20** | 8.328 | 8.095 | 7.868 | 7.648 | 7.434 | 7.226 | 7.024 | 6.827 | 6.636 |
| **21** | 8.267 | 8.037 | 7.813 | 7.596 | 7.385 | 7.179 | 6.980 | 6.785 | 6.597 |
| **22** | 8.208 | 7.981 | 7.760 | 7.546 | 7.337 | 7.134 | 6.937 | 6.745 | 6.559 |
| **23** | 8.151 | 7.927 | 7.709 | 7.497 | 7.291 | 7.091 | 6.897 | 6.707 | 6.523 |
| **24** | 8.096 | 7.875 | 7.660 | 7.451 | 7.248 | 7.050 | 6.857 | 6.670 | 6.488 |
| **25** | 8.043 | 7.825 | 7.613 | 7.406 | 7.205 | 7.010 | 6.820 | 6.635 | 6.455 |
| **26** | 7.992 | 7.777 | 7.567 | 7.363 | 7.165 | 6.972 | 6.784 | 6.602 | 6.424 |
| **27** | 7.943 | 7.731 | 7.524 | 7.322 | 7.126 | 6.936 | 6.750 | 6.569 | 6.394 |
| **28** | 7.896 | 7.686 | 7.482 | 7.283 | 7.089 | 6.901 | 6.717 | 6.539 | 6.365 |
| **29** | 7.851 | 7.644 | 7.442 | 7.245 | 7.054 | 6.868 | 6.686 | 6.510 | 6.338 |
| **30** | 7.808 | 7.603 | 7.404 | 7.209 | 7.020 | 6.836 | 6.657 | 6.482 | 6.312 |
| **31** | 7.767 | 7.564 | 7.367 | 7.175 | 6.988 | 6.806 | 6.629 | 6.456 | 6.288 |
| **32** | 7.727 | 7.527 | 7.332 | 7.143 | 6.958 | 6.778 | 6.602 | 6.431 | 6.265 |
| **33** | 7.690 | 7.492 | 7.299 | 7.112 | 6.929 | 6.751 | 6.577 | 6.408 | 6.243 |
| **34** | 7.654 | 7.458 | 7.268 | 7.082 | 6.901 | 6.725 | 6.554 | 6.386 | 6.223 |
| **35** | 7.619 | 7.426 | 7.238 | 7.054 | 6.876 | 6.701 | 6.531 | 6.366 | 6.204 |
| **36** | 7.587 | 7.396 | 7.210 | 7.028 | 6.851 | 6.679 | 6.511 | 6.347 | 6.187 |
| **37** | 7.556 | 7.367 | 7.183 | 7.004 | 6.829 | 6.658 | 6.491 | 6.329 | 6.171 |
| **38** | 7.527 | 7.340 | 7.158 | 6.981 | 6.807 | 6.638 | 6.473 | 6.313 | 6.156 |
| **39** | 7.500 | 7.315 | 7.135 | 6.959 | 6.787 | 6.620 | 6.457 | 6.298 | 6.143 |
| **40** | 7.474 | 7.291 | 7.113 | 6.939 | 6.769 | 6.604 | 6.442 | 6.284 | 6.131 |

*Source*: Based on Hamme and Emerson (2004).

**Table 4.7** Bunsen Coefficient ($\beta$) for Neon as a Function of Temperature (L real gas/(L atm)) × 1000 (freshwatr) (At 10°C, the table value is 11.319; the Bunsen coefficient is 11.319/1000 or 0.0011319 L/(L atm))

| Temperature (°C) | Δt (°C) | | | | | | | | | |
|---|---|---|---|---|---|---|---|---|---|---|
| | **0.0** | **0.1** | **0.2** | **0.3** | **0.4** | **0.5** | **0.6** | **0.7** | **0.8** | **0.9** |
| **0** | 12.511 | 12.497 | 12.483 | 12.469 | 12.456 | 12.442 | 12.428 | 12.415 | 12.401 | 12.387 |
| **1** | 12.374 | 12.360 | 12.347 | 12.333 | 12.320 | 12.307 | 12.293 | 12.280 | 12.267 | 12.254 |
| **2** | 12.241 | 12.227 | 12.214 | 12.201 | 12.188 | 12.176 | 12.163 | 12.150 | 12.137 | 12.124 |
| **3** | 12.112 | 12.099 | 12.086 | 12.074 | 12.061 | 12.049 | 12.036 | 12.024 | 12.011 | 11.999 |
| **4** | 11.987 | 11.974 | 11.962 | 11.950 | 11.938 | 11.926 | 11.914 | 11.902 | 11.890 | 11.878 |
| **5** | 11.866 | 11.854 | 11.842 | 11.830 | 11.818 | 11.807 | 11.795 | 11.783 | 11.772 | 11.760 |
| **6** | 11.749 | 11.737 | 11.726 | 11.714 | 11.703 | 11.692 | 11.681 | 11.669 | 11.658 | 11.647 |
| **7** | 11.636 | 11.625 | 11.614 | 11.603 | 11.592 | 11.581 | 11.570 | 11.559 | 11.548 | 11.537 |
| **8** | 11.527 | 11.516 | 11.505 | 11.495 | 11.484 | 11.473 | 11.463 | 11.452 | 11.442 | 11.431 |
| **9** | 11.421 | 11.411 | 11.400 | 11.390 | 11.380 | 11.370 | 11.360 | 11.349 | 11.339 | 11.329 |
| **10** | 11.319 | 11.309 | 11.299 | 11.290 | 11.280 | 11.270 | 11.260 | 11.250 | 11.241 | 11.231 |
| **11** | 11.221 | 11.212 | 11.202 | 11.193 | 11.183 | 11.174 | 11.164 | 11.155 | 11.145 | 11.136 |
| **12** | 11.127 | 11.117 | 11.108 | 11.099 | 11.090 | 11.081 | 11.072 | 11.063 | 11.054 | 11.045 |
| **13** | 11.036 | 11.027 | 11.018 | 11.009 | 11.000 | 10.992 | 10.983 | 10.974 | 10.966 | 10.957 |
| **14** | 10.948 | 10.940 | 10.931 | 10.923 | 10.914 | 10.906 | 10.898 | 10.889 | 10.881 | 10.873 |
| **15** | 10.864 | 10.856 | 10.848 | 10.840 | 10.832 | 10.824 | 10.816 | 10.808 | 10.800 | 10.792 |
| **16** | 10.784 | 10.776 | 10.768 | 10.760 | 10.753 | 10.745 | 10.737 | 10.730 | 10.722 | 10.714 |
| **17** | 10.707 | 10.699 | 10.692 | 10.684 | 10.677 | 10.670 | 10.662 | 10.655 | 10.648 | 10.640 |
| **18** | 10.633 | 10.626 | 10.619 | 10.612 | 10.605 | 10.598 | 10.591 | 10.584 | 10.577 | 10.570 |
| **19** | 10.563 | 10.556 | 10.549 | 10.542 | 10.536 | 10.529 | 10.522 | 10.516 | 10.509 | 10.502 |
| **20** | 10.496 | 10.489 | 10.483 | 10.476 | 10.470 | 10.464 | 10.457 | 10.451 | 10.445 | 10.438 |
| **21** | 10.432 | 10.426 | 10.420 | 10.414 | 10.408 | 10.402 | 10.395 | 10.389 | 10.384 | 10.378 |
| **22** | 10.372 | 10.366 | 10.360 | 10.354 | 10.349 | 10.343 | 10.337 | 10.331 | 10.326 | 10.320 |
| **23** | 10.315 | 10.309 | 10.304 | 10.298 | 10.293 | 10.287 | 10.282 | 10.277 | 10.271 | 10.266 |
| **24** | 10.261 | 10.256 | 10.251 | 10.245 | 10.240 | 10.235 | 10.230 | 10.225 | 10.220 | 10.215 |
| **25** | 10.210 | 10.206 | 10.201 | 10.196 | 10.191 | 10.187 | 10.182 | 10.177 | 10.173 | 10.168 |
| **26** | 10.163 | 10.159 | 10.154 | 10.150 | 10.145 | 10.141 | 10.137 | 10.132 | 10.128 | 10.124 |
| **27** | 10.120 | 10.115 | 10.111 | 10.107 | 10.103 | 10.099 | 10.095 | 10.091 | 10.087 | 10.083 |
| **28** | 10.079 | 10.075 | 10.071 | 10.068 | 10.064 | 10.060 | 10.056 | 10.053 | 10.049 | 10.045 |
| **29** | 10.042 | 10.038 | 10.035 | 10.031 | 10.028 | 10.025 | 10.021 | 10.018 | 10.015 | 10.011 |
| **30** | 10.008 | 10.005 | 10.002 | 9.999 | 9.996 | 9.993 | 9.990 | 9.987 | 9.984 | 9.981 |
| **31** | 9.978 | 9.975 | 9.972 | 9.969 | 9.967 | 9.964 | 9.961 | 9.959 | 9.956 | 9.953 |
| **32** | 9.951 | 9.948 | 9.946 | 9.944 | 9.941 | 9.939 | 9.936 | 9.934 | 9.932 | 9.930 |
| **33** | 9.928 | 9.925 | 9.923 | 9.921 | 9.919 | 9.917 | 9.915 | 9.913 | 9.911 | 9.909 |
| **34** | 9.908 | 9.906 | 9.904 | 9.902 | 9.901 | 9.899 | 9.897 | 9.896 | 9.894 | 9.893 |
| **35** | 9.891 | 9.890 | 9.889 | 9.887 | 9.886 | 9.885 | 9.883 | 9.882 | 9.881 | 9.880 |
| **36** | 9.879 | 9.878 | 9.877 | 9.876 | 9.875 | 9.874 | 9.873 | 9.872 | 9.871 | 9.871 |
| **37** | 9.870 | 9.869 | 9.868 | 9.868 | 9.867 | 9.867 | 9.866 | 9.866 | 9.865 | 9.865 |
| **38** | 9.865 | 9.864 | 9.864 | 9.864 | 9.864 | 9.864 | 9.863 | 9.863 | 9.863 | 9.863 |
| **39** | 9.863 | 9.863 | 9.864 | 9.864 | 9.864 | 9.864 | 9.865 | 9.865 | 9.865 | 9.866 |
| **40** | 9.866 | 9.867 | 9.867 | 9.868 | 9.868 | 9.869 | 9.870 | 9.870 | 9.871 | 9.872 |

*Source*: Based on Hamme and Emerson (2004).

**Table 4.8** Bunsen Coefficient ($\beta$) for Neon as a Function of Temperature and Salinity, 0−40 g/kg (L real gas/(L atm)) × 1000 (At 10°C and 35 g/kg, the table value is 9.411; the Bunsen coefficient is 9.411/1000 or 0.009411 L/(L atm))

| Temperature (°C) | Salinity (g/kg) | | | | | | | | |
|---|---|---|---|---|---|---|---|---|---|
| | 0.0 | 5.0 | 10.0 | 15.0 | 20.0 | 25.0 | 30.0 | 35.0 | 40.0 |
| 0 | 12.511 | 12.166 | 11.831 | 11.504 | 11.186 | 10.877 | 10.576 | 10.283 | 9.998 |
| 1 | 12.374 | 12.035 | 11.704 | 11.383 | 11.070 | 10.766 | 10.470 | 10.181 | 9.901 |
| 2 | 12.241 | 11.907 | 11.582 | 11.266 | 10.958 | 10.658 | 10.367 | 10.083 | 9.807 |
| 3 | 12.112 | 11.783 | 11.464 | 11.153 | 10.850 | 10.555 | 10.268 | 9.988 | 9.717 |
| 4 | 11.987 | 11.664 | 11.349 | 11.043 | 10.745 | 10.454 | 10.172 | 9.897 | 9.629 |
| 5 | 11.866 | 11.548 | 11.239 | 10.937 | 10.643 | 10.357 | 10.079 | 9.808 | 9.544 |
| 6 | 11.749 | 11.436 | 11.131 | 10.835 | 10.545 | 10.264 | 9.990 | 9.723 | 9.463 |
| 7 | 11.636 | 11.328 | 11.028 | 10.736 | 10.451 | 10.173 | 9.903 | 9.640 | 9.384 |
| 8 | 11.527 | 11.223 | 10.928 | 10.640 | 10.360 | 10.086 | 9.820 | 9.561 | 9.309 |
| 9 | 11.421 | 11.123 | 10.832 | 10.548 | 10.272 | 10.002 | 9.740 | 9.484 | 9.236 |
| 10 | 11.319 | 11.025 | 10.739 | 10.459 | 10.187 | 9.921 | 9.663 | 9.411 | 9.166 |
| 11 | 11.221 | 10.932 | 10.649 | 10.374 | 10.105 | 9.844 | 9.589 | 9.340 | 9.098 |
| 12 | 11.127 | 10.841 | 10.563 | 10.291 | 10.027 | 9.769 | 9.517 | 9.272 | 9.034 |
| 13 | 11.036 | 10.755 | 10.480 | 10.212 | 9.951 | 9.697 | 9.449 | 9.207 | 8.972 |
| 14 | 10.948 | 10.671 | 10.400 | 10.137 | 9.879 | 9.628 | 9.384 | 9.145 | 8.913 |
| 15 | 10.864 | 10.591 | 10.324 | 10.064 | 9.810 | 9.562 | 9.321 | 9.086 | 8.856 |
| 16 | 10.784 | 10.514 | 10.251 | 9.994 | 9.744 | 9.500 | 9.261 | 9.029 | 8.802 |
| 17 | 10.707 | 10.441 | 10.181 | 9.928 | 9.681 | 9.440 | 9.204 | 8.975 | 8.751 |
| 18 | 10.633 | 10.371 | 10.115 | 9.865 | 9.621 | 9.382 | 9.150 | 8.924 | 8.703 |
| 19 | 10.563 | 10.304 | 10.051 | 9.804 | 9.563 | 9.328 | 9.099 | 8.875 | 8.657 |
| 20 | 10.496 | 10.240 | 9.991 | 9.747 | 9.509 | 9.277 | 9.050 | 8.829 | 8.614 |
| 21 | 10.432 | 10.180 | 9.933 | 9.693 | 9.458 | 9.228 | 9.005 | 8.786 | 8.573 |
| 22 | 10.372 | 10.123 | 9.879 | 9.641 | 9.409 | 9.183 | 8.962 | 8.746 | 8.535 |
| 23 | 10.315 | 10.069 | 9.828 | 9.593 | 9.364 | 9.140 | 8.921 | 8.708 | 8.499 |
| 24 | 10.261 | 10.018 | 9.780 | 9.548 | 9.321 | 9.100 | 8.884 | 8.673 | 8.467 |
| 25 | 10.210 | 9.970 | 9.735 | 9.506 | 9.282 | 9.063 | 8.849 | 8.640 | 8.436 |
| 26 | 10.163 | 9.926 | 9.694 | 9.467 | 9.245 | 9.029 | 8.817 | 8.611 | 8.409 |
| 27 | 10.120 | 9.885 | 9.655 | 9.431 | 9.212 | 8.997 | 8.788 | 8.584 | 8.384 |
| 28 | 10.079 | 9.847 | 9.620 | 9.398 | 9.181 | 8.969 | 8.762 | 8.559 | 8.362 |
| 29 | 10.042 | 9.812 | 9.588 | 9.368 | 9.153 | 8.944 | 8.738 | 8.538 | 8.342 |
| 30 | 10.008 | 9.781 | 9.559 | 9.341 | 9.129 | 8.921 | 8.718 | 8.519 | 8.325 |
| 31 | 9.978 | 9.753 | 9.533 | 9.318 | 9.107 | 8.902 | 8.700 | 8.504 | 8.311 |
| 32 | 9.951 | 9.728 | 9.510 | 9.297 | 9.089 | 8.885 | 8.686 | 8.491 | 8.300 |
| 33 | 9.928 | 9.707 | 9.491 | 9.280 | 9.074 | 8.872 | 8.674 | 8.481 | 8.292 |
| 34 | 9.908 | 9.689 | 9.476 | 9.266 | 9.062 | 8.861 | 8.665 | 8.474 | 8.286 |
| 35 | 9.891 | 9.675 | 9.463 | 9.256 | 9.053 | 8.854 | 8.660 | 8.470 | 8.284 |
| 36 | 9.879 | 9.664 | 9.454 | 9.249 | 9.047 | 8.850 | 8.658 | 8.469 | 8.285 |
| 37 | 9.870 | 9.657 | 9.449 | 9.245 | 9.045 | 8.850 | 8.659 | 8.471 | 8.288 |
| 38 | 9.865 | 9.654 | 9.447 | 9.245 | 9.047 | 8.853 | 8.663 | 8.477 | 8.295 |
| 39 | 9.863 | 9.654 | 9.449 | 9.248 | 9.052 | 8.859 | 8.670 | 8.486 | 8.305 |
| 40 | 9.866 | 9.658 | 9.455 | 9.256 | 9.060 | 8.869 | 8.682 | 8.498 | 8.318 |

*Source*: Based on Hamme and Emerson (2004).

**Table 4.9** Standard Air Saturation Concentration of Krypton as a Function of Temperature in nmol/kg (freshwater, 1 atm moist air)

| Temperature (°C) | Δt (°C) | | | | | | | | | |
|---|---|---|---|---|---|---|---|---|---|---|
| | 0.0 | 0.1 | 0.2 | 0.3 | 0.4 | 0.5 | 0.6 | 0.7 | 0.8 | 0.9 |
| **0** | 5.5533 | 5.5348 | 5.5164 | 5.4981 | 5.4799 | 5.4617 | 5.4436 | 5.4257 | 5.4078 | 5.3900 |
| **1** | 5.3723 | 5.3546 | 5.3371 | 5.3196 | 5.3022 | 5.2850 | 5.2677 | 5.2506 | 5.2336 | 5.2166 |
| **2** | 5.1997 | 5.1829 | 5.1662 | 5.1495 | 5.1329 | 5.1164 | 5.1000 | 5.0837 | 5.0674 | 5.0513 |
| **3** | 5.0351 | 5.0191 | 5.0032 | 4.9873 | 4.9715 | 4.9557 | 4.9401 | 4.9245 | 4.9090 | 4.8935 |
| **4** | 4.8782 | 4.8629 | 4.8476 | 4.8325 | 4.8174 | 4.8024 | 4.7874 | 4.7725 | 4.7577 | 4.7430 |
| **5** | 4.7283 | 4.7137 | 4.6991 | 4.6847 | 4.6703 | 4.6559 | 4.6416 | 4.6274 | 4.6133 | 4.5992 |
| **6** | 4.5852 | 4.5712 | 4.5573 | 4.5435 | 4.5297 | 4.5160 | 4.5024 | 4.4888 | 4.4753 | 4.4618 |
| **7** | 4.4484 | 4.4351 | 4.4218 | 4.4086 | 4.3954 | 4.3823 | 4.3692 | 4.3563 | 4.3433 | 4.3305 |
| **8** | 4.3176 | 4.3049 | 4.2922 | 4.2795 | 4.2670 | 4.2544 | 4.2419 | 4.2295 | 4.2172 | 4.2048 |
| **9** | 4.1926 | 4.1804 | 4.1682 | 4.1561 | 4.1441 | 4.1321 | 4.1202 | 4.1083 | 4.0964 | 4.0846 |
| **10** | 4.0729 | 4.0612 | 4.0496 | 4.0380 | 4.0265 | 4.0150 | 4.0036 | 3.9922 | 3.9808 | 3.9696 |
| **11** | 3.9583 | 3.9471 | 3.9360 | 3.9249 | 3.9139 | 3.9029 | 3.8919 | 3.8810 | 3.8701 | 3.8593 |
| **12** | 3.8486 | 3.8378 | 3.8272 | 3.8165 | 3.8060 | 3.7954 | 3.7849 | 3.7745 | 3.7641 | 3.7537 |
| **13** | 3.7434 | 3.7331 | 3.7229 | 3.7127 | 3.7025 | 3.6924 | 3.6824 | 3.6723 | 3.6624 | 3.6524 |
| **14** | 3.6425 | 3.6327 | 3.6229 | 3.6131 | 3.6034 | 3.5937 | 3.5840 | 3.5744 | 3.5648 | 3.5553 |
| **15** | 3.5458 | 3.5363 | 3.5269 | 3.5175 | 3.5082 | 3.4989 | 3.4896 | 3.4804 | 3.4712 | 3.4621 |
| **16** | 3.4530 | 3.4439 | 3.4349 | 3.4258 | 3.4169 | 3.4080 | 3.3991 | 3.3902 | 3.3814 | 3.3726 |
| **17** | 3.3638 | 3.3551 | 3.3464 | 3.3378 | 3.3292 | 3.3206 | 3.3121 | 3.3036 | 3.2951 | 3.2867 |
| **18** | 3.2782 | 3.2699 | 3.2615 | 3.2532 | 3.2450 | 3.2367 | 3.2285 | 3.2203 | 3.2122 | 3.2041 |
| **19** | 3.1960 | 3.1879 | 3.1799 | 3.1719 | 3.1640 | 3.1561 | 3.1482 | 3.1403 | 3.1325 | 3.1247 |
| **20** | 3.1169 | 3.1092 | 3.1015 | 3.0938 | 3.0862 | 3.0785 | 3.0710 | 3.0634 | 3.0559 | 3.0484 |
| **21** | 3.0409 | 3.0334 | 3.0260 | 3.0186 | 3.0113 | 3.0040 | 2.9967 | 2.9894 | 2.9821 | 2.9749 |
| **22** | 2.9677 | 2.9606 | 2.9534 | 2.9463 | 2.9392 | 2.9322 | 2.9252 | 2.9182 | 2.9112 | 2.9042 |
| **23** | 2.8973 | 2.8904 | 2.8836 | 2.8767 | 2.8699 | 2.8631 | 2.8563 | 2.8496 | 2.8429 | 2.8362 |
| **24** | 2.8295 | 2.8229 | 2.8163 | 2.8097 | 2.8031 | 2.7966 | 2.7900 | 2.7836 | 2.7771 | 2.7706 |
| **25** | 2.7642 | 2.7578 | 2.7514 | 2.7451 | 2.7388 | 2.7325 | 2.7262 | 2.7199 | 2.7137 | 2.7075 |
| **26** | 2.7013 | 2.6951 | 2.6890 | 2.6829 | 2.6768 | 2.6707 | 2.6646 | 2.6586 | 2.6526 | 2.6466 |
| **27** | 2.6406 | 2.6347 | 2.6287 | 2.6228 | 2.6170 | 2.6111 | 2.6053 | 2.5994 | 2.5936 | 2.5879 |
| **28** | 2.5821 | 2.5764 | 2.5707 | 2.5650 | 2.5593 | 2.5536 | 2.5480 | 2.5424 | 2.5368 | 2.5312 |
| **29** | 2.5257 | 2.5201 | 2.5146 | 2.5091 | 2.5037 | 2.4982 | 2.4928 | 2.4873 | 2.4819 | 2.4766 |
| **30** | 2.4712 | 2.4659 | 2.4605 | 2.4552 | 2.4499 | 2.4447 | 2.4394 | 2.4342 | 2.4290 | 2.4238 |
| **31** | 2.4186 | 2.4134 | 2.4083 | 2.4032 | 2.3981 | 2.3930 | 2.3879 | 2.3829 | 2.3778 | 2.3728 |
| **32** | 2.3678 | 2.3628 | 2.3579 | 2.3529 | 2.3480 | 2.3431 | 2.3382 | 2.3333 | 2.3284 | 2.3236 |
| **33** | 2.3187 | 2.3139 | 2.3091 | 2.3043 | 2.2996 | 2.2948 | 2.2901 | 2.2854 | 2.2807 | 2.2760 |
| **34** | 2.2713 | 2.2667 | 2.2620 | 2.2574 | 2.2528 | 2.2482 | 2.2436 | 2.2390 | 2.2345 | 2.2300 |
| **35** | 2.2254 | 2.2209 | 2.2165 | 2.2120 | 2.2075 | 2.2031 | 2.1987 | 2.1942 | 2.1898 | 2.1855 |
| **36** | 2.1811 | 2.1767 | 2.1724 | 2.1681 | 2.1638 | 2.1595 | 2.1552 | 2.1509 | 2.1467 | 2.1424 |
| **37** | 2.1382 | 2.1340 | 2.1298 | 2.1256 | 2.1214 | 2.1172 | 2.1131 | 2.1090 | 2.1048 | 2.1007 |
| **38** | 2.0966 | 2.0926 | 2.0885 | 2.0844 | 2.0804 | 2.0764 | 2.0724 | 2.0684 | 2.0644 | 2.0604 |
| **39** | 2.0564 | 2.0525 | 2.0485 | 2.0446 | 2.0407 | 2.0368 | 2.0329 | 2.0290 | 2.0252 | 2.0213 |
| **40** | 2.0175 | 2.0137 | 2.0099 | 2.0060 | 2.0023 | 1.9985 | 1.9947 | 1.9910 | 1.9872 | 1.9835 |

*Source*: Based on Weiss and Kyser (1978).

**Table 4.10** Standard Air Saturation Concentration of Krypton as a Function of Temperature and Salinity in nmol/kg, 0–40 g/kg (seawater, 1 atm moist air)

| Temperature (°C) | Salinity (g/kg) | | | | | | | | |
|---|---|---|---|---|---|---|---|---|---|
| | **0.0** | **5.0** | **10.0** | **15.0** | **20.0** | **25.0** | **30.0** | **35.0** | **40.0** |
| **0** | 5.5533 | 5.3384 | 5.1319 | 4.9333 | 4.7424 | 4.5589 | 4.3825 | 4.2129 | 4.0499 |
| **1** | 5.3723 | 5.1655 | 4.9667 | 4.7755 | 4.5917 | 4.4150 | 4.2451 | 4.0817 | 3.9246 |
| **2** | 5.1997 | 5.0006 | 4.8092 | 4.6251 | 4.4480 | 4.2778 | 4.1140 | 3.9565 | 3.8051 |
| **3** | 5.0351 | 4.8434 | 4.6590 | 4.4816 | 4.3110 | 4.1469 | 3.9890 | 3.8371 | 3.6910 |
| **4** | 4.8782 | 4.6934 | 4.5157 | 4.3447 | 4.1802 | 4.0220 | 3.8697 | 3.7231 | 3.5822 |
| **5** | 4.7283 | 4.5503 | 4.3789 | 4.2140 | 4.0554 | 3.9027 | 3.7557 | 3.6143 | 3.4782 |
| **6** | 4.5852 | 4.4135 | 4.2482 | 4.0892 | 3.9361 | 3.7887 | 3.6468 | 3.5103 | 3.3789 |
| **7** | 4.4484 | 4.2828 | 4.1233 | 3.9698 | 3.8220 | 3.6798 | 3.5428 | 3.4109 | 3.2839 |
| **8** | 4.3176 | 4.1578 | 4.0039 | 3.8557 | 3.7130 | 3.5756 | 3.4432 | 3.3158 | 3.1930 |
| **9** | 4.1926 | 4.0383 | 3.8897 | 3.7465 | 3.6087 | 3.4759 | 3.3480 | 3.2248 | 3.1061 |
| **10** | 4.0729 | 3.9239 | 3.7804 | 3.6420 | 3.5088 | 3.3804 | 3.2568 | 3.1376 | 3.0228 |
| **11** | 3.9583 | 3.8144 | 3.6757 | 3.5420 | 3.4132 | 3.2890 | 3.1694 | 3.0542 | 2.9431 |
| **12** | 3.8486 | 3.7094 | 3.5754 | 3.4461 | 3.3215 | 3.2015 | 3.0857 | 2.9742 | 2.8667 |
| **13** | 3.7434 | 3.6089 | 3.4792 | 3.3542 | 3.2337 | 3.1175 | 3.0055 | 2.8975 | 2.7934 |
| **14** | 3.6425 | 3.5125 | 3.3870 | 3.2661 | 3.1495 | 3.0370 | 2.9286 | 2.8240 | 2.7231 |
| **15** | 3.5458 | 3.4200 | 3.2986 | 3.1816 | 3.0686 | 2.9598 | 2.8547 | 2.7534 | 2.6557 |
| **16** | 3.4530 | 3.3312 | 3.2137 | 3.1004 | 2.9911 | 2.8856 | 2.7838 | 2.6857 | 2.5910 |
| **17** | 3.3638 | 3.2460 | 3.1322 | 3.0225 | 2.9166 | 2.8144 | 2.7158 | 2.6206 | 2.5288 |
| **18** | 3.2782 | 3.1641 | 3.0540 | 2.9476 | 2.8450 | 2.7460 | 2.6504 | 2.5581 | 2.4690 |
| **19** | 3.1960 | 3.0855 | 2.9787 | 2.8757 | 2.7762 | 2.6802 | 2.5875 | 2.4980 | 2.4116 |
| **20** | 3.1169 | 3.0098 | 2.9064 | 2.8065 | 2.7101 | 2.6170 | 2.5271 | 2.4402 | 2.3564 |
| **21** | 3.0409 | 2.9371 | 2.8369 | 2.7400 | 2.6465 | 2.5562 | 2.4689 | 2.3847 | 2.3033 |
| **22** | 2.9677 | 2.8671 | 2.7699 | 2.6760 | 2.5853 | 2.4976 | 2.4130 | 2.3312 | 2.2521 |
| **23** | 2.8973 | 2.7998 | 2.7055 | 2.6144 | 2.5264 | 2.4413 | 2.3591 | 2.2797 | 2.2029 |
| **24** | 2.8295 | 2.7349 | 2.6435 | 2.5551 | 2.4696 | 2.3870 | 2.3072 | 2.2301 | 2.1555 |
| **25** | 2.7642 | 2.6724 | 2.5837 | 2.4979 | 2.4150 | 2.3348 | 2.2572 | 2.1823 | 2.1098 |
| **26** | 2.7013 | 2.6122 | 2.5261 | 2.4428 | 2.3623 | 2.2844 | 2.2091 | 2.1362 | 2.0658 |
| **27** | 2.6406 | 2.5542 | 2.4706 | 2.3897 | 2.3115 | 2.2358 | 2.1626 | 2.0918 | 2.0233 |
| **28** | 2.5821 | 2.4982 | 2.4170 | 2.3385 | 2.2625 | 2.1889 | 2.1178 | 2.0490 | 1.9824 |
| **29** | 2.5257 | 2.4442 | 2.3653 | 2.2890 | 2.2152 | 2.1437 | 2.0746 | 2.0076 | 1.9429 |
| **30** | 2.4712 | 2.3921 | 2.3155 | 2.2413 | 2.1695 | 2.1001 | 2.0328 | 1.9677 | 1.9047 |
| **31** | 2.4186 | 2.3417 | 2.2673 | 2.1952 | 2.1255 | 2.0579 | 1.9925 | 1.9292 | 1.8679 |
| **32** | 2.3678 | 2.2931 | 2.2208 | 2.1508 | 2.0829 | 2.0172 | 1.9536 | 1.8920 | 1.8323 |
| **33** | 2.3187 | 2.2462 | 2.1759 | 2.1078 | 2.0418 | 1.9779 | 1.9160 | 1.8560 | 1.7979 |
| **34** | 2.2713 | 2.2008 | 2.1324 | 2.0662 | 2.0020 | 1.9399 | 1.8796 | 1.8212 | 1.7647 |
| **35** | 2.2254 | 2.1569 | 2.0904 | 2.0260 | 1.9636 | 1.9031 | 1.8445 | 1.7876 | 1.7325 |
| **36** | 2.1811 | 2.1144 | 2.0498 | 1.9871 | 1.9264 | 1.8675 | 1.8104 | 1.7551 | 1.7015 |
| **37** | 2.1382 | 2.0734 | 2.0105 | 1.9495 | 1.8904 | 1.8331 | 1.7775 | 1.7236 | 1.6714 |
| **38** | 2.0966 | 2.0336 | 1.9724 | 1.9131 | 1.8556 | 1.7998 | 1.7457 | 1.6932 | 1.6423 |
| **39** | 2.0564 | 1.9951 | 1.9356 | 1.8779 | 1.8219 | 1.7676 | 1.7148 | 1.6637 | 1.6141 |
| **40** | 2.0175 | 1.9578 | 1.8999 | 1.8438 | 1.7892 | 1.7363 | 1.6850 | 1.6352 | 1.5868 |

*Source*: Based on Weiss and Kyser (1978).

**Table 4.11** Bunsen Coefficient ($\beta$) for Krypton as a Function of Temperature (L real gas/(L atm)) × 100 (freshwater) (At 10°C, the table value is 8.086; the Bunsen coefficient is 8.086/100 or 0.08086 L/(L atm))

| Temperature (°C) | Δt (°C) | | | | | | | | | |
|---|---|---|---|---|---|---|---|---|---|---|
| | **0.0** | **0.1** | **0.2** | **0.3** | **0.4** | **0.5** | **0.6** | **0.7** | **0.8** | **0.9** |
| **0** | 10.979 | 10.942 | 10.906 | 10.870 | 10.834 | 10.799 | 10.763 | 10.728 | 10.693 | 10.658 |
| **1** | 10.623 | 10.588 | 10.554 | 10.520 | 10.485 | 10.451 | 10.418 | 10.384 | 10.351 | 10.317 |
| **2** | 10.284 | 10.251 | 10.218 | 10.186 | 10.153 | 10.121 | 10.089 | 10.057 | 10.025 | 9.993 |
| **3** | 9.962 | 9.930 | 9.899 | 9.868 | 9.837 | 9.806 | 9.776 | 9.745 | 9.715 | 9.684 |
| **4** | 9.654 | 9.625 | 9.595 | 9.565 | 9.536 | 9.506 | 9.477 | 9.448 | 9.419 | 9.390 |
| **5** | 9.362 | 9.333 | 9.305 | 9.276 | 9.248 | 9.220 | 9.192 | 9.165 | 9.137 | 9.110 |
| **6** | 9.082 | 9.055 | 9.028 | 9.001 | 8.974 | 8.947 | 8.921 | 8.894 | 8.868 | 8.842 |
| **7** | 8.816 | 8.790 | 8.764 | 8.738 | 8.713 | 8.687 | 8.662 | 8.636 | 8.611 | 8.586 |
| **8** | 8.561 | 8.536 | 8.512 | 8.487 | 8.463 | 8.438 | 8.414 | 8.390 | 8.366 | 8.342 |
| **9** | 8.318 | 8.295 | 8.271 | 8.248 | 8.224 | 8.201 | 8.178 | 8.155 | 8.132 | 8.109 |
| **10** | 8.086 | 8.064 | 8.041 | 8.019 | 7.996 | 7.974 | 7.952 | 7.930 | 7.908 | 7.886 |
| **11** | 7.865 | 7.843 | 7.821 | 7.800 | 7.779 | 7.757 | 7.736 | 7.715 | 7.694 | 7.673 |
| **12** | 7.653 | 7.632 | 7.611 | 7.591 | 7.570 | 7.550 | 7.530 | 7.510 | 7.490 | 7.470 |
| **13** | 7.450 | 7.430 | 7.410 | 7.391 | 7.371 | 7.352 | 7.332 | 7.313 | 7.294 | 7.275 |
| **14** | 7.256 | 7.237 | 7.218 | 7.199 | 7.180 | 7.162 | 7.143 | 7.125 | 7.106 | 7.088 |
| **15** | 7.070 | 7.052 | 7.034 | 7.016 | 6.998 | 6.980 | 6.962 | 6.945 | 6.927 | 6.910 |
| **16** | 6.892 | 6.875 | 6.857 | 6.840 | 6.823 | 6.806 | 6.789 | 6.772 | 6.755 | 6.738 |
| **17** | 6.722 | 6.705 | 6.689 | 6.672 | 6.656 | 6.639 | 6.623 | 6.607 | 6.591 | 6.574 |
| **18** | 6.558 | 6.543 | 6.527 | 6.511 | 6.495 | 6.479 | 6.464 | 6.448 | 6.433 | 6.417 |
| **19** | 6.402 | 6.387 | 6.371 | 6.356 | 6.341 | 6.326 | 6.311 | 6.296 | 6.281 | 6.267 |
| **20** | 6.252 | 6.237 | 6.223 | 6.208 | 6.194 | 6.179 | 6.165 | 6.151 | 6.136 | 6.122 |
| **21** | 6.108 | 6.094 | 6.080 | 6.066 | 6.052 | 6.038 | 6.025 | 6.011 | 5.997 | 5.984 |
| **22** | 5.970 | 5.956 | 5.943 | 5.930 | 5.916 | 5.903 | 5.890 | 5.877 | 5.864 | 5.850 |
| **23** | 5.837 | 5.824 | 5.812 | 5.799 | 5.786 | 5.773 | 5.760 | 5.748 | 5.735 | 5.723 |
| **24** | 5.710 | 5.698 | 5.685 | 5.673 | 5.661 | 5.649 | 5.636 | 5.624 | 5.612 | 5.600 |
| **25** | 5.588 | 5.576 | 5.564 | 5.552 | 5.541 | 5.529 | 5.517 | 5.505 | 5.494 | 5.482 |
| **26** | 5.471 | 5.459 | 5.448 | 5.436 | 5.425 | 5.414 | 5.402 | 5.391 | 5.380 | 5.369 |
| **27** | 5.358 | 5.347 | 5.336 | 5.325 | 5.314 | 5.303 | 5.292 | 5.282 | 5.271 | 5.260 |
| **28** | 5.249 | 5.239 | 5.228 | 5.218 | 5.207 | 5.197 | 5.186 | 5.176 | 5.166 | 5.155 |
| **29** | 5.145 | 5.135 | 5.125 | 5.115 | 5.105 | 5.095 | 5.085 | 5.075 | 5.065 | 5.055 |
| **30** | 5.045 | 5.035 | 5.025 | 5.016 | 5.006 | 4.996 | 4.987 | 4.977 | 4.968 | 4.958 |
| **31** | 4.949 | 4.939 | 4.930 | 4.920 | 4.911 | 4.902 | 4.893 | 4.883 | 4.874 | 4.865 |
| **32** | 4.856 | 4.847 | 4.838 | 4.829 | 4.820 | 4.811 | 4.802 | 4.793 | 4.784 | 4.775 |
| **33** | 4.767 | 4.758 | 4.749 | 4.741 | 4.732 | 4.723 | 4.715 | 4.706 | 4.698 | 4.689 |
| **34** | 4.681 | 4.672 | 4.664 | 4.656 | 4.647 | 4.639 | 4.631 | 4.623 | 4.614 | 4.606 |
| **35** | 4.598 | 4.590 | 4.582 | 4.574 | 4.566 | 4.558 | 4.550 | 4.542 | 4.534 | 4.526 |
| **36** | 4.518 | 4.511 | 4.503 | 4.495 | 4.487 | 4.480 | 4.472 | 4.464 | 4.457 | 4.449 |
| **37** | 4.442 | 4.434 | 4.427 | 4.419 | 4.412 | 4.404 | 4.397 | 4.390 | 4.382 | 4.375 |
| **38** | 4.368 | 4.361 | 4.353 | 4.346 | 4.339 | 4.332 | 4.325 | 4.318 | 4.311 | 4.304 |
| **39** | 4.297 | 4.290 | 4.283 | 4.276 | 4.269 | 4.262 | 4.255 | 4.248 | 4.242 | 4.235 |
| **40** | 4.228 | 4.221 | 4.215 | 4.208 | 4.201 | 4.195 | 4.188 | 4.182 | 4.175 | 4.169 |

*Source*: Based on Weiss and Kyser (1978).

**Table 4.12** Bunsen Coefficient ($\beta$) for Krypton as a Function of Temperature and Salinity, 0–40 g/kg (L real gas/(L atm)) × 100 (At 10°C and 35 g/kg, the table value is 6.398; the Bunsen coefficient is 6.398/100 or 0.06398 L/(L atm))

| Temperature (°C) | Salinity (g/kg) | | | | | | | | |
|---|---|---|---|---|---|---|---|---|---|
| | **0.0** | **5.0** | **10.0** | **15.0** | **20.0** | **25.0** | **30.0** | **35.0** | **40.0** |
| **0** | 10.979 | 10.596 | 10.226 | 9.869 | 9.525 | 9.193 | 8.872 | 8.562 | 8.264 |
| **1** | 10.623 | 10.254 | 9.899 | 9.555 | 9.224 | 8.904 | 8.595 | 8.297 | 8.009 |
| **2** | 10.284 | 9.929 | 9.587 | 9.256 | 8.937 | 8.628 | 8.331 | 8.043 | 7.766 |
| **3** | 9.962 | 9.620 | 9.290 | 8.971 | 8.664 | 8.366 | 8.079 | 7.802 | 7.535 |
| **4** | 9.654 | 9.325 | 9.007 | 8.700 | 8.403 | 8.117 | 7.840 | 7.572 | 7.314 |
| **5** | 9.362 | 9.044 | 8.737 | 8.441 | 8.155 | 7.878 | 7.611 | 7.353 | 7.104 |
| **6** | 9.082 | 8.776 | 8.480 | 8.194 | 7.918 | 7.651 | 7.393 | 7.144 | 6.903 |
| **7** | 8.816 | 8.520 | 8.235 | 7.959 | 7.692 | 7.434 | 7.185 | 6.944 | 6.712 |
| **8** | 8.561 | 8.276 | 8.000 | 7.734 | 7.476 | 7.227 | 6.987 | 6.754 | 6.529 |
| **9** | 8.318 | 8.043 | 7.777 | 7.519 | 7.270 | 7.030 | 6.797 | 6.572 | 6.354 |
| **10** | 8.086 | 7.820 | 7.563 | 7.314 | 7.073 | 6.841 | 6.616 | 6.398 | 6.187 |
| **11** | 7.865 | 7.607 | 7.359 | 7.118 | 6.885 | 6.660 | 6.442 | 6.232 | 6.028 |
| **12** | 7.653 | 7.404 | 7.163 | 6.931 | 6.705 | 6.487 | 6.277 | 6.073 | 5.875 |
| **13** | 7.450 | 7.209 | 6.976 | 6.751 | 6.533 | 6.322 | 6.118 | 5.921 | 5.730 |
| **14** | 7.256 | 7.023 | 6.798 | 6.580 | 6.369 | 6.164 | 5.967 | 5.775 | 5.590 |
| **15** | 7.070 | 6.845 | 6.627 | 6.415 | 6.211 | 6.013 | 5.821 | 5.636 | 5.456 |
| **16** | 6.892 | 6.674 | 6.463 | 6.258 | 6.060 | 5.868 | 5.682 | 5.503 | 5.328 |
| **17** | 6.722 | 6.510 | 6.306 | 6.107 | 5.915 | 5.729 | 5.549 | 5.375 | 5.206 |
| **18** | 6.558 | 6.354 | 6.155 | 5.963 | 5.777 | 5.596 | 5.422 | 5.252 | 5.088 |
| **19** | 6.402 | 6.203 | 6.011 | 5.825 | 5.644 | 5.469 | 5.299 | 5.135 | 4.976 |
| **20** | 6.252 | 6.059 | 5.873 | 5.692 | 5.517 | 5.347 | 5.182 | 5.023 | 4.868 |
| **21** | 6.108 | 5.921 | 5.740 | 5.565 | 5.395 | 5.230 | 5.070 | 4.915 | 4.764 |
| **22** | 5.970 | 5.789 | 5.613 | 5.443 | 5.277 | 5.117 | 4.962 | 4.811 | 4.665 |
| **23** | 5.837 | 5.662 | 5.491 | 5.325 | 5.165 | 5.009 | 4.858 | 4.712 | 4.570 |
| **24** | 5.710 | 5.539 | 5.374 | 5.213 | 5.057 | 4.906 | 4.759 | 4.616 | 4.478 |
| **25** | 5.588 | 5.422 | 5.261 | 5.105 | 4.953 | 4.806 | 4.663 | 4.525 | 4.391 |
| **26** | 5.471 | 5.309 | 5.153 | 5.001 | 4.854 | 4.711 | 4.572 | 4.437 | 4.306 |
| **27** | 5.358 | 5.201 | 5.049 | 4.901 | 4.758 | 4.619 | 4.484 | 4.353 | 4.225 |
| **28** | 5.249 | 5.097 | 4.949 | 4.806 | 4.666 | 4.531 | 4.399 | 4.272 | 4.148 |
| **29** | 5.145 | 4.997 | 4.853 | 4.713 | 4.578 | 4.446 | 4.318 | 4.194 | 4.073 |
| **30** | 5.045 | 4.901 | 4.761 | 4.625 | 4.493 | 4.364 | 4.240 | 4.119 | 4.001 |
| **31** | 4.949 | 4.808 | 4.672 | 4.540 | 4.411 | 4.286 | 4.165 | 4.047 | 3.932 |
| **32** | 4.856 | 4.719 | 4.587 | 4.458 | 4.333 | 4.211 | 4.092 | 3.977 | 3.866 |
| **33** | 4.767 | 4.634 | 4.505 | 4.379 | 4.257 | 4.138 | 4.023 | 3.911 | 3.802 |
| **34** | 4.681 | 4.551 | 4.426 | 4.303 | 4.184 | 4.069 | 3.956 | 3.847 | 3.741 |
| **35** | 4.598 | 4.472 | 4.350 | 4.230 | 4.114 | 4.002 | 3.892 | 3.785 | 3.682 |
| **36** | 4.518 | 4.396 | 4.276 | 4.160 | 4.047 | 3.937 | 3.830 | 3.726 | 3.625 |
| **37** | 4.442 | 4.322 | 4.206 | 4.093 | 3.982 | 3.875 | 3.771 | 3.669 | 3.570 |
| **38** | 4.368 | 4.251 | 4.138 | 4.027 | 3.920 | 3.815 | 3.713 | 3.614 | 3.518 |
| **39** | 4.297 | 4.183 | 4.072 | 3.965 | 3.860 | 3.758 | 3.658 | 3.562 | 3.467 |
| **40** | 4.228 | 4.117 | 4.009 | 3.904 | 3.802 | 3.702 | 3.605 | 3.511 | 3.419 |

*Source*: Based on Weiss and Kyser (1978).

**Table 4.13** Standard Air Saturation Concentration of Xenon as a Function of Temperature in nmol/kg (freshwater, 1 atm moist air)

| Temperature (°C) | Δt (°C) | | | | | | | | | |
|---|---|---|---|---|---|---|---|---|---|---|
| | 0.0 | 0.1 | 0.2 | 0.3 | 0.4 | 0.5 | 0.6 | 0.7 | 0.8 | 0.9 |
| **0** | .90466 | .90097 | .89730 | .89365 | .89001 | .88640 | .88281 | .87923 | .87567 | .87213 |
| **1** | .86861 | .86510 | .86162 | .85815 | .85470 | .85126 | .84785 | .84445 | .84107 | .83770 |
| **2** | .83435 | .83102 | .82771 | .82441 | .82113 | .81787 | .81462 | .81139 | .80817 | .80497 |
| **3** | .80179 | .79862 | .79547 | .79234 | .78922 | .78611 | .78302 | .77995 | .77689 | .77385 |
| **4** | .77082 | .76781 | .76481 | .76183 | .75887 | .75591 | .75297 | .75005 | .74714 | .74425 |
| **5** | .74137 | .73850 | .73565 | .73281 | .72999 | .72718 | .72438 | .72160 | .71883 | .71608 |
| **6** | .71334 | .71061 | .70790 | .70520 | .70251 | .69983 | .69717 | .69452 | .69189 | .68927 |
| **7** | .68666 | .68406 | .68147 | .67890 | .67634 | .67380 | .67126 | .66874 | .66623 | .66373 |
| **8** | .66125 | .65878 | .65631 | .65387 | .65143 | .64900 | .64659 | .64419 | .64180 | .63942 |
| **9** | .63705 | .63469 | .63235 | .63001 | .62769 | .62538 | .62308 | .62079 | .61851 | .61625 |
| **10** | .61399 | .61175 | .60951 | .60729 | .60507 | .60287 | .60068 | .59850 | .59632 | .59416 |
| **11** | .59201 | .58987 | .58774 | .58562 | .58351 | .58141 | .57932 | .57724 | .57517 | .57311 |
| **12** | .57106 | .56902 | .56699 | .56496 | .56295 | .56095 | .55896 | .55697 | .55500 | .55303 |
| **13** | .55108 | .54913 | .54719 | .54526 | .54334 | .54143 | .53953 | .53764 | .53575 | .53388 |
| **14** | .53201 | .53015 | .52830 | .52646 | .52463 | .52281 | .52099 | .51919 | .51739 | .51560 |
| **15** | .51382 | .51204 | .51028 | .50852 | .50677 | .50503 | .50330 | .50158 | .49986 | .49815 |
| **16** | .49645 | .49476 | .49307 | .49140 | .48973 | .48806 | .48641 | .48476 | .48312 | .48149 |
| **17** | .47987 | .47825 | .47664 | .47504 | .47345 | .47186 | .47028 | .46871 | .46714 | .46558 |
| **18** | .46403 | .46249 | .46095 | .45942 | .45789 | .45638 | .45487 | .45337 | .45187 | .45038 |
| **19** | .44890 | .44742 | .44595 | .44449 | .44304 | .44159 | .44014 | .43871 | .43728 | .43585 |
| **20** | .43444 | .43303 | .43162 | .43023 | .42883 | .42745 | .42607 | .42470 | .42333 | .42197 |
| **21** | .42061 | .41927 | .41792 | .41659 | .41526 | .41393 | .41261 | .41130 | .40999 | .40869 |
| **22** | .40740 | .40611 | .40482 | .40355 | .40227 | .40101 | .39975 | .39849 | .39724 | .39600 |
| **23** | .39476 | .39352 | .39230 | .39107 | .38986 | .38864 | .38744 | .38624 | .38504 | .38385 |
| **24** | .38266 | .38148 | .38031 | .37914 | .37797 | .37681 | .37566 | .37451 | .37337 | .37223 |
| **25** | .37109 | .36996 | .36884 | .36772 | .36660 | .36549 | .36439 | .36329 | .36219 | .36110 |
| **26** | .36002 | .35893 | .35786 | .35679 | .35572 | .35466 | .35360 | .35255 | .35150 | .35045 |
| **27** | .34941 | .34838 | .34735 | .34632 | .34530 | .34428 | .34327 | .34226 | .34125 | .34025 |
| **28** | .33926 | .33827 | .33728 | .33630 | .33532 | .33434 | .33337 | .33241 | .33145 | .33049 |
| **29** | .32953 | .32858 | .32764 | .32670 | .32576 | .32483 | .32390 | .32297 | .32205 | .32113 |
| **30** | .32022 | .31931 | .31840 | .31750 | .31660 | .31571 | .31482 | .31393 | .31305 | .31217 |
| **31** | .31129 | .31042 | .30955 | .30868 | .30782 | .30697 | .30611 | .30526 | .30442 | .30357 |
| **32** | .30273 | .30190 | .30107 | .30024 | .29941 | .29859 | .29777 | .29696 | .29614 | .29534 |
| **33** | .29453 | .29373 | .29293 | .29214 | .29135 | .29056 | .28977 | .28899 | .28821 | .28744 |
| **34** | .28667 | .28590 | .28513 | .28437 | .28361 | .28286 | .28211 | .28136 | .28061 | .27987 |
| **35** | .27913 | .27839 | .27766 | .27693 | .27620 | .27547 | .27475 | .27403 | .27332 | .27260 |
| **36** | .27190 | .27119 | .27048 | .26978 | .26909 | .26839 | .26770 | .26701 | .26632 | .26564 |
| **37** | .26496 | .26428 | .26360 | .26293 | .26226 | .26159 | .26093 | .26027 | .25961 | .25895 |
| **38** | .25830 | .25765 | .25700 | .25636 | .25571 | .25507 | .25444 | .25380 | .25317 | .25254 |
| **39** | .25191 | .25129 | .25067 | .25005 | .24943 | .24882 | .24820 | .24760 | .24699 | .24638 |
| **40** | .24578 | .24518 | .24459 | .24399 | .24340 | .24281 | .24222 | .24164 | .24106 | .24048 |

*Source*: Based on Hamme and Severinghaus (2007), Wood and Caputi (1966).

**Table 4.14** Standard Air Saturation Concentration of Xenon as a Function of Temperature and Salinity in nmol/kg, 0–40 g/kg (seawater, 1 atm moist air)

| Temperature (°C) | Salinity (g/kg) | | | | | | | | |
|---|---|---|---|---|---|---|---|---|---|
| | 0.0 | 5.0 | 1.0 | 15.0 | 2.0 | 25.0 | 30.0 | 35.0 | 40.0 |
| 0 | .90466 | .86385 | .82489 | .78769 | .75216 | .71823 | .68584 | .65491 | .62537 |
| 1 | .86861 | .82978 | .79269 | .75726 | .72341 | .69107 | .66018 | .63067 | .60248 |
| 2 | .83435 | .79739 | .76207 | .72831 | .69605 | .66522 | .63575 | .60759 | .58067 |
| 3 | .80179 | .76660 | .73295 | .70077 | .67001 | .64060 | .61248 | .58560 | .55989 |
| 4 | .77082 | .73730 | .70523 | .67456 | .64522 | .61716 | .59032 | .56464 | .54009 |
| 5 | .74137 | .70942 | .67885 | .64960 | .62161 | .59483 | .56920 | .54467 | .52120 |
| 6 | .71334 | .68289 | .65374 | .62583 | .59912 | .57354 | .54906 | .52562 | .50319 |
| 7 | .68666 | .65762 | .62981 | .60318 | .57768 | .55325 | .52986 | .50745 | .48600 |
| 8 | .66125 | .63356 | .60702 | .58160 | .55724 | .53390 | .51154 | .49012 | .46959 |
| 9 | .63705 | .61063 | .58530 | .56102 | .53775 | .51545 | .49407 | .47358 | .45393 |
| 10 | .61399 | .58877 | .56459 | .54140 | .51916 | .49784 | .47739 | .45778 | .43898 |
| 11 | .59201 | .56794 | .54484 | .52268 | .50142 | .48103 | .46146 | .44270 | .42469 |
| 12 | .57106 | .54806 | .52600 | .50482 | .48449 | .46498 | .44626 | .42829 | .41104 |
| 13 | .55108 | .52911 | .50802 | .48777 | .46832 | .44965 | .43173 | .41452 | .39799 |
| 14 | .53201 | .51102 | .49085 | .47149 | .45288 | .43501 | .41785 | .40136 | .38552 |
| 15 | .51382 | .49375 | .47447 | .45594 | .43813 | .42102 | .40458 | .38878 | .37359 |
| 16 | .49645 | .47726 | .45882 | .44108 | .42404 | .40765 | .39189 | .37674 | .36218 |
| 17 | .47987 | .46152 | .44386 | .42689 | .41056 | .39486 | .37976 | .36523 | .35127 |
| 18 | .46403 | .44647 | .42958 | .41332 | .39768 | .38263 | .36815 | .35422 | .34082 |
| 19 | .44890 | .43209 | .41592 | .40035 | .38536 | .37093 | .35705 | .34368 | .33082 |
| 20 | .43444 | .41835 | .40286 | .38794 | .37357 | .35974 | .34642 | .33359 | .32124 |
| 21 | .42061 | .40521 | .39037 | .37607 | .36230 | .34903 | .33625 | .32393 | .31207 |
| 22 | .40740 | .39264 | .37842 | .36471 | .35151 | .33877 | .32650 | .31468 | .30328 |
| 23 | .39476 | .38062 | .36699 | .35385 | .34117 | .32895 | .31717 | .30581 | .29486 |
| 24 | .38266 | .36911 | .35605 | .34344 | .33128 | .31955 | .30824 | .29732 | .28679 |
| 25 | .37109 | .35810 | .34557 | .33347 | .32180 | .31054 | .29967 | .28918 | .27906 |
| 26 | .36002 | .34756 | .33554 | .32393 | .31273 | .30191 | .29146 | .28138 | .27165 |
| 27 | .34941 | .33747 | .32593 | .31479 | .30403 | .29363 | .28360 | .27390 | .26454 |
| 28 | .33926 | .32780 | .31673 | .30603 | .29569 | .28570 | .27605 | .26673 | .25772 |
| 29 | .32953 | .31854 | .30791 | .29763 | .28770 | .27810 | .26882 | .25985 | .25118 |
| 30 | .32022 | .30966 | .29945 | .28958 | .28004 | .27081 | .26188 | .25325 | .24490 |
| 31 | .31129 | .30116 | .29135 | .28187 | .27269 | .26381 | .25523 | .24692 | .23888 |
| 32 | .30273 | .29300 | .28358 | .27447 | .26565 | .25711 | .24884 | .24084 | .23310 |
| 33 | .29453 | .28518 | .27613 | .26737 | .25889 | .25067 | .24271 | .23501 | .22755 |
| 34 | .28667 | .27769 | .26899 | .26056 | .25240 | .24449 | .23683 | .22942 | .22223 |
| 35 | .27913 | .27050 | .26214 | .25403 | .24618 | .23857 | .23119 | .22404 | .21712 |
| 36 | .27190 | .26360 | .25556 | .24776 | .24021 | .23288 | .22577 | .21889 | .21221 |
| 37 | .26496 | .25698 | .24925 | .24175 | .23447 | .22742 | .22057 | .21393 | .20750 |
| 38 | .25830 | .25063 | .24319 | .23597 | .22897 | .22217 | .21558 | .20918 | .20297 |
| 39 | .25191 | .24454 | .23738 | .23043 | .22369 | .21714 | .21078 | .20461 | .19862 |
| 40 | .24578 | .23869 | .23180 | .22511 | .21861 | .21231 | .20618 | .20023 | .19445 |

*Source*: Based on Hamme and Severinghaus (2007), Wood and Caputi (1966).

**Table 4.15** Bunsen Coefficient ($\beta$) for Xenon as a Function of Temperature (L real gas/(L atm)) (freshwater)

| Temperature (°C) | $\Delta t$ (°C) | | | | | | | | | |
|---|---|---|---|---|---|---|---|---|---|---|
| | **0.0** | **0.1** | **0.2** | **0.3** | **0.4** | **0.5** | **0.6** | **0.7** | **0.8** | **0.9** |
| **0** | .22508 | .22417 | .22327 | .22237 | .22148 | .22059 | .21971 | .21883 | .21796 | .21709 |
| **1** | .21622 | .21536 | .21450 | .21365 | .21280 | .21196 | .21112 | .21028 | .20945 | .20862 |
| **2** | .20780 | .20698 | .20617 | .20536 | .20455 | .20375 | .20295 | .20216 | .20137 | .20058 |
| **3** | .19980 | .19902 | .19825 | .19748 | .19671 | .19595 | .19519 | .19443 | .19368 | .19294 |
| **4** | .19219 | .19145 | .19071 | .18998 | .18925 | .18853 | .18781 | .18709 | .18637 | .18566 |
| **5** | .18495 | .18425 | .18355 | .18285 | .18216 | .18147 | .18078 | .18010 | .17942 | .17874 |
| **6** | .17807 | .17740 | .17673 | .17607 | .17541 | .17475 | .17410 | .17345 | .17280 | .17216 |
| **7** | .17152 | .17088 | .17024 | .16961 | .16898 | .16836 | .16774 | .16712 | .16650 | .16589 |
| **8** | .16528 | .16467 | .16407 | .16346 | .16287 | .16227 | .16168 | .16109 | .16050 | .15992 |
| **9** | .15934 | .15876 | .15818 | .15761 | .15704 | .15647 | .15591 | .15535 | .15479 | .15423 |
| **10** | .15368 | .15313 | .15258 | .15204 | .15149 | .15095 | .15041 | .14988 | .14935 | .14882 |
| **11** | .14829 | .14777 | .14724 | .14672 | .14621 | .14569 | .14518 | .14467 | .14416 | .14366 |
| **12** | .14315 | .14265 | .14216 | .14166 | .14117 | .14068 | .14019 | .13970 | .13922 | .13874 |
| **13** | .13826 | .13778 | .13731 | .13684 | .13637 | .13590 | .13543 | .13497 | .13451 | .13405 |
| **14** | .13359 | .13314 | .13269 | .13224 | .13179 | .13134 | .13090 | .13046 | .13002 | .12958 |
| **15** | .12915 | .12871 | .12828 | .12785 | .12742 | .12700 | .12658 | .12615 | .12574 | .12532 |
| **16** | .12490 | .12449 | .12408 | .12367 | .12326 | .12286 | .12245 | .12205 | .12165 | .12125 |
| **17** | .12086 | .12046 | .12007 | .11968 | .11929 | .11890 | .11852 | .11813 | .11775 | .11737 |
| **18** | .11699 | .11662 | .11624 | .11587 | .11550 | .11513 | .11476 | .11440 | .11403 | .11367 |
| **19** | .11331 | .11295 | .11259 | .11224 | .11188 | .11153 | .11118 | .11083 | .11048 | .11014 |
| **20** | .10979 | .10945 | .10911 | .10877 | .10843 | .10810 | .10776 | .10743 | .10710 | .10677 |
| **21** | .10644 | .10611 | .10579 | .10546 | .10514 | .10482 | .10450 | .10418 | .10386 | .10355 |
| **22** | .10323 | .10292 | .10261 | .10230 | .10199 | .10169 | .10138 | .10108 | .10078 | .10048 |
| **23** | .10018 | .09988 | .09958 | .09929 | .09899 | .09870 | .09841 | .09812 | .09783 | .09754 |
| **24** | .09726 | .09697 | .09669 | .09641 | .09613 | .09585 | .09557 | .09529 | .09502 | .09474 |
| **25** | .09447 | .09420 | .09393 | .09366 | .09339 | .09312 | .09286 | .09259 | .09233 | .09207 |
| **26** | .09181 | .09155 | .09129 | .09103 | .09078 | .09052 | .09027 | .09001 | .08976 | .08951 |
| **27** | .08926 | .08902 | .08877 | .08852 | .08828 | .08804 | .08779 | .08755 | .08731 | .08707 |
| **28** | .08684 | .08660 | .08636 | .08613 | .08590 | .08566 | .08543 | .08520 | .08497 | .08475 |
| **29** | .08452 | .08429 | .08407 | .08384 | .08362 | .08340 | .08318 | .08296 | .08274 | .08252 |
| **30** | .08231 | .08209 | .08188 | .08166 | .08145 | .08124 | .08103 | .08082 | .08061 | .08040 |
| **31** | .08019 | .07999 | .07978 | .07958 | .07937 | .07917 | .07897 | .07877 | .07857 | .07837 |
| **32** | .07818 | .07798 | .07778 | .07759 | .07740 | .07720 | .07701 | .07682 | .07663 | .07644 |
| **33** | .07625 | .07606 | .07588 | .07569 | .07551 | .07532 | .07514 | .07496 | .07477 | .07459 |
| **34** | .07441 | .07424 | .07406 | .07388 | .07370 | .07353 | .07335 | .07318 | .07301 | .07283 |
| **35** | .07266 | .07249 | .07232 | .07215 | .07198 | .07182 | .07165 | .07148 | .07132 | .07115 |
| **36** | .07099 | .07083 | .07067 | .07050 | .07034 | .07018 | .07003 | .06987 | .06971 | .06955 |
| **37** | .06940 | .06924 | .06909 | .06893 | .06878 | .06863 | .06848 | .06833 | .06818 | .06803 |
| **38** | .06788 | .06773 | .06758 | .06744 | .06729 | .06715 | .06700 | .06686 | .06672 | .06658 |
| **39** | .06643 | .06629 | .06615 | .06601 | .06588 | .06574 | .06560 | .06546 | .06533 | .06519 |
| **40** | .06506 | .06492 | .06479 | .06466 | .06453 | .06440 | .06426 | .06413 | .06401 | .06388 |

*Source*: Based on Hamme and Severinghaus (2007), Wood and Caputi (1966).

**Table 4.16** Bunsen Coefficient ($\beta$) for Xenon as a Function of Temperature and Salinity, 0–40 g/kg (L real gas/(L atm))

| Temperature (°C) | Salinity (g/kg) | | | | | | | | |
|---|---|---|---|---|---|---|---|---|---|
| | **0.0** | **5.0** | **10.0** | **15.0** | **20.0** | **25.0** | **30.0** | **35.0** | **40.0** |
| **0** | 0.22508 | 0.21580 | 0.20689 | 0.19835 | 0.19015 | 0.18229 | 0.17475 | 0.16753 | 0.16060 |
| **1** | 0.21622 | 0.20739 | 0.19891 | 0.19077 | 0.18296 | 0.17547 | 0.16828 | 0.16139 | 0.15478 |
| **2** | 0.20780 | 0.19939 | 0.19132 | 0.18357 | 0.17612 | 0.16898 | 0.16212 | 0.15555 | 0.14923 |
| **3** | 0.19980 | 0.19179 | 0.18410 | 0.17671 | 0.16961 | 0.16280 | 0.15626 | 0.14998 | 0.14395 |
| **4** | 0.19219 | 0.18456 | 0.17723 | 0.17019 | 0.16342 | 0.15692 | 0.15067 | 0.14468 | 0.13892 |
| **5** | 0.18495 | 0.17769 | 0.17070 | 0.16398 | 0.15752 | 0.15132 | 0.14535 | 0.13962 | 0.13412 |
| **6** | 0.17807 | 0.17114 | 0.16447 | 0.15806 | 0.15190 | 0.14598 | 0.14028 | 0.13481 | 0.12955 |
| **7** | 0.17152 | 0.16491 | 0.15855 | 0.15243 | 0.14655 | 0.14089 | 0.13545 | 0.13022 | 0.12519 |
| **8** | 0.16528 | 0.15897 | 0.15291 | 0.14707 | 0.14145 | 0.13604 | 0.13084 | 0.12584 | 0.12103 |
| **9** | 0.15934 | 0.15332 | 0.14753 | 0.14195 | 0.13659 | 0.13142 | 0.12645 | 0.12166 | 0.11706 |
| **10** | 0.15368 | 0.14794 | 0.14241 | 0.13708 | 0.13195 | 0.12701 | 0.12226 | 0.11768 | 0.11327 |
| **11** | 0.14829 | 0.14281 | 0.13753 | 0.13243 | 0.12753 | 0.12281 | 0.11826 | 0.11387 | 0.10965 |
| **12** | 0.14315 | 0.13792 | 0.13287 | 0.12800 | 0.12331 | 0.11880 | 0.11444 | 0.11025 | 0.10620 |
| **13** | 0.13826 | 0.13326 | 0.12843 | 0.12378 | 0.11929 | 0.11497 | 0.11080 | 0.10678 | 0.10291 |
| **14** | 0.13359 | 0.12881 | 0.12420 | 0.11975 | 0.11546 | 0.11132 | 0.10732 | 0.10347 | 0.09976 |
| **15** | 0.12915 | 0.12457 | 0.12016 | 0.11590 | 0.11179 | 0.10783 | 0.10400 | 0.10031 | 0.09675 |
| **16** | 0.12490 | 0.12053 | 0.11631 | 0.11223 | 0.10830 | 0.10450 | 0.10083 | 0.09730 | 0.09388 |
| **17** | 0.12086 | 0.11667 | 0.11263 | 0.10873 | 0.10496 | 0.10132 | 0.09781 | 0.09441 | 0.09114 |
| **18** | 0.11699 | 0.11299 | 0.10912 | 0.10538 | 0.10177 | 0.09828 | 0.09491 | 0.09166 | 0.08851 |
| **19** | 0.11331 | 0.10948 | 0.10577 | 0.10219 | 0.09873 | 0.09538 | 0.09215 | 0.08903 | 0.08601 |
| **20** | 0.10979 | 0.10612 | 0.10257 | 0.09914 | 0.09582 | 0.09261 | 0.08951 | 0.08651 | 0.08361 |
| **21** | 0.10644 | 0.10292 | 0.09952 | 0.09623 | 0.09305 | 0.08997 | 0.08699 | 0.08411 | 0.08132 |
| **22** | 0.10323 | 0.09987 | 0.09660 | 0.09345 | 0.09039 | 0.08744 | 0.08458 | 0.08181 | 0.07914 |
| **23** | 0.10018 | 0.09695 | 0.09382 | 0.09079 | 0.08786 | 0.08502 | 0.08227 | 0.07962 | 0.07704 |
| **24** | 0.09726 | 0.09416 | 0.09116 | 0.08825 | 0.08544 | 0.08271 | 0.08007 | 0.07752 | 0.07504 |
| **25** | 0.09447 | 0.09150 | 0.08862 | 0.08583 | 0.08313 | 0.08051 | 0.07797 | 0.07551 | 0.07313 |
| **26** | 0.09181 | 0.08896 | 0.08619 | 0.08351 | 0.08092 | 0.07840 | 0.07596 | 0.07360 | 0.07131 |
| **27** | 0.08926 | 0.08653 | 0.08387 | 0.08130 | 0.07880 | 0.07638 | 0.07404 | 0.07176 | 0.06956 |
| **28** | 0.08684 | 0.08421 | 0.08166 | 0.07919 | 0.07679 | 0.07446 | 0.07220 | 0.07001 | 0.06789 |
| **29** | 0.08452 | 0.08200 | 0.07955 | 0.07717 | 0.07486 | 0.07262 | 0.07045 | 0.06834 | 0.06630 |
| **30** | 0.08231 | 0.07988 | 0.07753 | 0.07524 | 0.07302 | 0.07087 | 0.06877 | 0.06674 | 0.06477 |
| **31** | .08019 | 0.07786 | 0.07560 | 0.07340 | 0.07126 | 0.06919 | 0.06717 | 0.06522 | 0.06332 |
| **32** | .07818 | 0.07594 | 0.07376 | 0.07164 | 0.06959 | 0.06759 | 0.06565 | 0.06376 | 0.06193 |
| **33** | .07625 | 0.07410 | 0.07200 | 0.06996 | 0.06798 | 0.06606 | 0.06419 | 0.06237 | 0.06060 |
| **34** | .07441 | 0.07234 | 0.07032 | 0.06836 | 0.06645 | 0.06460 | 0.06280 | 0.06104 | 0.05934 |
| **35** | .07266 | 0.07067 | 0.06872 | 0.06684 | 0.06500 | 0.06321 | 0.06147 | 0.05978 | 0.05813 |
| **36** | .07099 | 0.06907 | 0.06720 | 0.06538 | 0.06361 | 0.06188 | 0.06020 | 0.05857 | 0.05698 |
| **37** | .06940 | 0.06755 | 0.06575 | 0.06399 | 0.06228 | 0.06062 | 0.05900 | 0.05742 | 0.05588 |
| **38** | .06788 | 0.06610 | 0.06436 | 0.06267 | 0.06102 | 0.05941 | 0.05785 | 0.05633 | 0.05484 |
| **39** | .06643 | 0.06472 | 0.06304 | 0.06141 | 0.05982 | 0.05827 | 0.05676 | 0.05528 | 0.05385 |
| **40** | .06506 | 0.06340 | 0.06179 | 0.06021 | 0.05868 | 0.05718 | 0.05572 | 0.05430 | 0.05291 |

*Source*: Based on Hamme and Severinghaus (2007), Wood and Caputi (1966).

# 5 Solubility of Trace Gases in the Atmosphere

The solubilities of hydrogen, methane, and nitrous oxide in terms of standard air solubility concentrations ($C_o^\dagger$) and Bunsen coefficients ($\beta$) have been computed as functions of salinity and temperature from the equations presented in Appendix A. Values are given for freshwater and marine conditions. These tabular data may be found as follows:

| Gas | $C_o^\dagger$ in nmol/kg | | $\beta$ in L/(L atm) | |
|---|---|---|---|---|
| | **Table** | **Page** | **Table** | **Page** |
| Hydrogen ($H_2$) | | | | |
| Freshwater | 5.1 | 201 | 5.3 | 203 |
| 0–40 g/kg | 5.2 | 202 | 5.4 | 204 |
| Methane ($CH_4$) | | | | |
| Freshwater | 5.5 | 205 | 5.7 | 207 |
| 0–40 g/kg | 5.6 | 206 | 5.8 | 208 |
| Nitrous oxide ($N_2O$) | | | | |
| Freshwater | 5.9 | 209 | 5.11 | 211 |
| 0–40 g/kg | 5.10 | 210 | 5.12 | 212 |

Once the $C_o^\dagger$ value has been determined for a given temperature and salinity, the solubility can be adjusted for depth, elevation, or barometric pressures by equations presented previously. Conversion factors for other solubility units have been presented in Table 1.1{4}. The Bunsen coefficient can be used to compute the solubility of a gas with an arbitrary mole fraction.

**Dissolved Gas Concentration in Water. DOI: 10.1016/B978-0-12-415916-7.00005-X**

### Example 5-1

Compute the gas tension of the trace gases if each gas has a concentration of 10.5 mg/L, temperature = 32°C, and freshwater conditions. Use Eq. (1.19) and the Bunsen coefficients.

**Input Data**

| Parameter | Value | Source |
|---|---|---|
| $\beta_{H_2}$ | 0.01682 | Table 5.3 |
| $\beta_{CH_2}$ | 0.02810 | Table 5.7 |
| $\beta_{N_2O}$ | 0.4566 | Table 5.11 |
| $A_{H_2}$ | 8.4558 | Table D-1 |
| $A_{CH_4}$ | 1.0593 | Table D-1 |
| $A_{N_2O}$ | 0.3841 | Table D-1 |

$P_{wv} = 6.228$ mmHg (Table 1.21).
From Eq. (1.19):

$$\text{Gas tension (mmHg)} = C\left[\frac{A_i}{\beta_i}\right]$$

$$P^{l}_{H_2} = 10.5\left[\frac{8.4558}{0.01682}\right] = 5278.59 \text{ mmHg}$$

$$P^{l}_{CH_4} = 10.5\left[\frac{1.0593}{02810}\right] = 395.82 \text{ mmHg}$$

$$P^{l}_{N_2O} = 10.5\left[\frac{0.3841}{0.4566}\right] = 8.83 \text{ mmHg}$$

Small concentrations of hydrogen or methane gases could make a significant contribution to the total gas pressure.

**Table 5.1** Standard Air Saturation Concentration of Hydrogen as a Function of Temperature in nmol/kg (freshwater, 1 atm moist air)

| Temperature (°C) | $\Delta t$ (°C) | | | | | | | | | |
|---|---|---|---|---|---|---|---|---|---|---|
| | 0.0 | 0.1 | 0.2 | 0.3 | 0.4 | 0.5 | 0.6 | 0.7 | 0.8 | 0.9 |
| 0 | 0.53363 | 0.53293 | 0.53224 | 0.53155 | 0.53086 | 0.53017 | 0.52949 | 0.52881 | 0.52813 | 0.52745 |
| 1 | 0.52678 | 0.52610 | 0.52543 | 0.52477 | 0.52410 | 0.52344 | 0.52278 | 0.52212 | 0.52147 | 0.52081 |
| 2 | 0.52016 | 0.51951 | 0.51887 | 0.51822 | 0.51758 | 0.51694 | 0.51631 | 0.51567 | 0.51504 | 0.51441 |
| 3 | 0.51378 | 0.51315 | 0.51253 | 0.51191 | 0.51129 | 0.51067 | 0.51006 | 0.50944 | 0.50883 | 0.50822 |
| 4 | 0.50762 | 0.50701 | 0.50641 | 0.50581 | 0.50521 | 0.50462 | 0.50402 | 0.50343 | 0.50284 | 0.50225 |
| 5 | 0.50167 | 0.50108 | 0.50050 | 0.49992 | 0.49934 | 0.49877 | 0.49819 | 0.49762 | 0.49705 | 0.49648 |
| 6 | 0.49592 | 0.49535 | 0.49479 | 0.49423 | 0.49367 | 0.49312 | 0.49256 | 0.49201 | 0.49146 | 0.49091 |
| 7 | 0.49036 | 0.48982 | 0.48928 | 0.48874 | 0.48820 | 0.48766 | 0.48712 | 0.48659 | 0.48606 | 0.48553 |
| 8 | 0.48500 | 0.48447 | 0.48395 | 0.48342 | 0.48290 | 0.48238 | 0.48186 | 0.48135 | 0.48083 | 0.48032 |
| 9 | 0.47981 | 0.47930 | 0.47879 | 0.47828 | 0.47778 | 0.47728 | 0.47678 | 0.47628 | 0.47578 | 0.47528 |
| 10 | 0.47479 | 0.47430 | 0.47381 | 0.47332 | 0.47283 | 0.47234 | 0.47186 | 0.47137 | 0.47089 | 0.47041 |
| 11 | 0.46993 | 0.46946 | 0.46898 | 0.46851 | 0.46804 | 0.46757 | 0.46710 | 0.46663 | 0.46616 | 0.46570 |
| 12 | 0.46523 | 0.46477 | 0.46431 | 0.46385 | 0.46340 | 0.46294 | 0.46249 | 0.46203 | 0.46158 | 0.46113 |
| 13 | 0.46069 | 0.46024 | 0.45979 | 0.45935 | 0.45891 | 0.45846 | 0.45802 | 0.45759 | 0.45715 | 0.45671 |
| 14 | 0.45628 | 0.45585 | 0.45541 | 0.45498 | 0.45456 | 0.45413 | 0.45370 | 0.45328 | 0.45285 | 0.45243 |
| 15 | 0.45201 | 0.45159 | 0.45117 | 0.45075 | 0.45034 | 0.44992 | 0.44951 | 0.44910 | 0.44869 | 0.44828 |
| 16 | 0.44787 | 0.44746 | 0.44706 | 0.44665 | 0.44625 | 0.44585 | 0.44545 | 0.44505 | 0.44465 | 0.44425 |
| 17 | 0.44386 | 0.44346 | 0.44307 | 0.44268 | 0.44228 | 0.44189 | 0.44151 | 0.44112 | 0.44073 | 0.44035 |
| 18 | 0.43996 | 0.43958 | 0.43920 | 0.43882 | 0.43844 | 0.43806 | 0.43768 | 0.43730 | 0.43693 | 0.43655 |
| 19 | 0.43618 | 0.43581 | 0.43544 | 0.43507 | 0.43470 | 0.43433 | 0.43396 | 0.43360 | 0.43323 | 0.43287 |
| 20 | 0.43251 | 0.43215 | 0.43178 | 0.43142 | 0.43107 | 0.43071 | 0.43035 | 0.43000 | 0.42964 | 0.42929 |
| 21 | 0.42894 | 0.42858 | 0.42823 | 0.42788 | 0.42754 | 0.42719 | 0.42684 | 0.42649 | 0.42615 | 0.42581 |
| 22 | 0.42546 | 0.42512 | 0.42478 | 0.42444 | 0.42410 | 0.42376 | 0.42342 | 0.42309 | 0.42275 | 0.42242 |
| 23 | 0.42208 | 0.42175 | 0.42142 | 0.42109 | 0.42076 | 0.42043 | 0.42010 | 0.41977 | 0.41944 | 0.41912 |
| 24 | 0.41879 | 0.41846 | 0.41814 | 0.41782 | 0.41750 | 0.41717 | 0.41685 | 0.41653 | 0.41622 | 0.41590 |
| 25 | 0.41558 | 0.41526 | 0.41495 | 0.41463 | 0.41432 | 0.41400 | 0.41369 | 0.41338 | 0.41307 | 0.41276 |
| 26 | 0.41245 | 0.41214 | 0.41183 | 0.41152 | 0.41122 | 0.41091 | 0.41060 | 0.41030 | 0.40999 | 0.40969 |
| 27 | 0.40939 | 0.40909 | 0.40878 | 0.40848 | 0.40818 | 0.40788 | 0.40759 | 0.40729 | 0.40699 | 0.40669 |
| 28 | 0.40640 | 0.40610 | 0.40581 | 0.40551 | 0.40522 | 0.40493 | 0.40463 | 0.40434 | 0.40405 | 0.40376 |
| 29 | 0.40347 | 0.40318 | 0.40289 | 0.40261 | 0.40232 | 0.40203 | 0.40174 | 0.40146 | 0.40117 | 0.40089 |
| 30 | 0.40060 | 0.40032 | 0.40004 | 0.39976 | 0.39947 | 0.39919 | 0.39891 | 0.39863 | 0.39835 | 0.39807 |
| 31 | 0.39779 | 0.39751 | 0.39724 | 0.39696 | 0.39668 | 0.39641 | 0.39613 | 0.39586 | 0.39558 | 0.39531 |
| 32 | 0.39503 | 0.39476 | 0.39449 | 0.39421 | 0.39394 | 0.39367 | 0.39340 | 0.39313 | 0.39286 | 0.39259 |
| 33 | 0.39232 | 0.39205 | 0.39178 | 0.39151 | 0.39124 | 0.39097 | 0.39071 | 0.39044 | 0.39017 | 0.38991 |
| 34 | 0.38964 | 0.38938 | 0.38911 | 0.38885 | 0.38858 | 0.38832 | 0.38806 | 0.38779 | 0.38753 | 0.38727 |
| 35 | 0.38701 | 0.38674 | 0.38648 | 0.38622 | 0.38596 | 0.38570 | 0.38544 | 0.38518 | 0.38492 | 0.38466 |
| 36 | 0.38440 | 0.38414 | 0.38389 | 0.38363 | 0.38337 | 0.38311 | 0.38285 | 0.38260 | 0.38234 | 0.38208 |
| 37 | 0.38183 | 0.38157 | 0.38131 | 0.38106 | 0.38080 | 0.38055 | 0.38029 | 0.38004 | 0.37978 | 0.37953 |
| 38 | 0.37928 | 0.37902 | 0.37877 | 0.37851 | 0.37826 | 0.37801 | 0.37775 | 0.37750 | 0.37725 | 0.37700 |
| 39 | 0.37674 | 0.37649 | 0.37624 | 0.37599 | 0.37573 | 0.37548 | 0.37523 | 0.37498 | 0.37473 | 0.37448 |
| 40 | 0.37423 | 0.37397 | 0.37372 | 0.37347 | 0.37322 | 0.37297 | 0.37272 | 0.37247 | 0.37222 | 0.37197 |

*Source*: Based on Weiss (1971).

**Table 5.2** Standard Air Saturation Concentration of Hydrogen as a Function of Temperature and Salinity in nmol/kg, 0–40 g/kg (seawater, 1 atm moist air)

| Temperature (°C) | Salinity (g/kg) | | | | | | | | |
|---|---|---|---|---|---|---|---|---|---|
| | 0.0 | 5.0 | 10.0 | 15.0 | 20.0 | 25.0 | 30.0 | 35.0 | 40.0 |
| 0 | 0.53363 | 0.51722 | 0.50133 | 0.48595 | 0.47105 | 0.45661 | 0.44262 | 0.42906 | 0.41593 |
| 1 | 0.52678 | 0.51071 | 0.49516 | 0.48009 | 0.46550 | 0.45135 | 0.43763 | 0.42434 | 0.41146 |
| 2 | 0.52016 | 0.50443 | 0.48920 | 0.47444 | 0.46014 | 0.44627 | 0.43283 | 0.41979 | 0.40715 |
| 3 | 0.51378 | 0.49838 | 0.48345 | 0.46899 | 0.45497 | 0.44137 | 0.42818 | 0.41540 | 0.40299 |
| 4 | 0.50762 | 0.49253 | 0.47790 | 0.46372 | 0.44998 | 0.43664 | 0.42370 | 0.41116 | 0.39898 |
| 5 | 0.50167 | 0.48688 | 0.47254 | 0.45864 | 0.44516 | 0.43207 | 0.41938 | 0.40706 | 0.39511 |
| 6 | 0.49592 | 0.48142 | 0.46737 | 0.45373 | 0.44050 | 0.42766 | 0.41520 | 0.40311 | 0.39137 |
| 7 | 0.49036 | 0.47615 | 0.46236 | 0.44899 | 0.43600 | 0.42340 | 0.41117 | 0.39929 | 0.38776 |
| 8 | 0.48500 | 0.47105 | 0.45753 | 0.44440 | 0.43166 | 0.41929 | 0.40727 | 0.39561 | 0.38428 |
| 9 | 0.47981 | 0.46613 | 0.45286 | 0.43997 | 0.42746 | 0.41531 | 0.40351 | 0.39204 | 0.38091 |
| 10 | 0.47479 | 0.46137 | 0.44834 | 0.43569 | 0.42340 | 0.41146 | 0.39986 | 0.38860 | 0.37765 |
| 11 | 0.46993 | 0.45676 | 0.44397 | 0.43154 | 0.41947 | 0.40774 | 0.39634 | 0.38527 | 0.37450 |
| 12 | 0.46523 | 0.45230 | 0.43973 | 0.42753 | 0.41567 | 0.40414 | 0.39294 | 0.38205 | 0.37146 |
| 13 | 0.46069 | 0.44798 | 0.43564 | 0.42365 | 0.41199 | 0.40066 | 0.38964 | 0.37893 | 0.36852 |
| 14 | 0.45628 | 0.44380 | 0.43167 | 0.41989 | 0.40843 | 0.39729 | 0.38645 | 0.37591 | 0.36567 |
| 15 | 0.45201 | 0.43975 | 0.42783 | 0.41624 | 0.40498 | 0.39402 | 0.38336 | 0.37299 | 0.36291 |
| 16 | 0.44787 | 0.43582 | 0.42410 | 0.41271 | 0.40163 | 0.39085 | 0.38036 | 0.37016 | 0.36023 |
| 17 | 0.44386 | 0.43201 | 0.42049 | 0.40929 | 0.39839 | 0.38778 | 0.37746 | 0.36742 | 0.35764 |
| 18 | 0.43996 | 0.42831 | 0.41699 | 0.40597 | 0.39524 | 0.38481 | 0.37465 | 0.36476 | 0.35513 |
| 19 | 0.43618 | 0.42473 | 0.41358 | 0.40274 | 0.39219 | 0.38192 | 0.37191 | 0.36218 | 0.35270 |
| 20 | 0.43251 | 0.42124 | 0.41028 | 0.39961 | 0.38922 | 0.37911 | 0.36926 | 0.35967 | 0.35033 |
| 21 | 0.42894 | 0.41785 | 0.40707 | 0.39657 | 0.38634 | 0.37638 | 0.36668 | 0.35724 | 0.34803 |
| 22 | 0.42546 | 0.41456 | 0.40394 | 0.39360 | 0.38354 | 0.37373 | 0.36417 | 0.35487 | 0.34580 |
| 23 | 0.42208 | 0.41135 | 0.40090 | 0.39072 | 0.38081 | 0.37115 | 0.36173 | 0.35256 | 0.34363 |
| 24 | 0.41879 | 0.40822 | 0.39794 | 0.38791 | 0.37815 | 0.36863 | 0.35936 | 0.35032 | 0.34151 |
| 25 | 0.41558 | 0.40518 | 0.39505 | 0.38518 | 0.37556 | 0.36618 | 0.35704 | 0.34813 | 0.33945 |
| 26 | 0.41245 | 0.40221 | 0.39223 | 0.38251 | 0.37303 | 0.36379 | 0.35478 | 0.34600 | 0.33743 |
| 27 | 0.40939 | 0.39930 | 0.38948 | 0.37990 | 0.37056 | 0.36145 | 0.35257 | 0.34391 | 0.33547 |
| 28 | 0.40640 | 0.39646 | 0.38678 | 0.37735 | 0.36814 | 0.35917 | 0.35041 | 0.34187 | 0.33354 |
| 29 | 0.40347 | 0.39369 | 0.38415 | 0.37485 | 0.36578 | 0.35693 | 0.34830 | 0.33987 | 0.33166 |
| 30 | 0.40060 | 0.39096 | 0.38157 | 0.37240 | 0.36346 | 0.35473 | 0.34622 | 0.33792 | 0.32981 |
| 31 | 0.39779 | 0.38829 | 0.37903 | 0.37000 | 0.36118 | 0.35258 | 0.34419 | 0.33599 | 0.32800 |
| 32 | 0.39503 | 0.38567 | 0.37654 | 0.36764 | 0.35895 | 0.35046 | 0.34218 | 0.33410 | 0.32621 |
| 33 | 0.39232 | 0.38309 | 0.37410 | 0.36532 | 0.35675 | 0.34838 | 0.34021 | 0.33224 | 0.32446 |
| 34 | 0.38964 | 0.38055 | 0.37168 | 0.36303 | 0.35458 | 0.34633 | 0.33827 | 0.33040 | 0.32272 |
| 35 | 0.38701 | 0.37805 | 0.36930 | 0.36077 | 0.35243 | 0.34430 | 0.33635 | 0.32859 | 0.32101 |
| 36 | 0.38440 | 0.37557 | 0.36695 | 0.35853 | 0.35032 | 0.34229 | 0.33445 | 0.32679 | 0.31931 |
| 37 | 0.38183 | 0.37312 | 0.36462 | 0.35632 | 0.34822 | 0.34030 | 0.33256 | 0.32501 | 0.31762 |
| 38 | 0.37928 | 0.37069 | 0.36231 | 0.35413 | 0.34613 | 0.33832 | 0.33069 | 0.32323 | 0.31595 |
| 39 | 0.37674 | 0.36828 | 0.36002 | 0.35195 | 0.34406 | 0.33636 | 0.32883 | 0.32147 | 0.31428 |
| 40 | 0.37423 | 0.36588 | 0.35774 | 0.34978 | 0.34200 | 0.33440 | 0.32697 | 0.31971 | 0.31261 |

*Source*: Based on Weiss (1971).

**Table 5.3** Bunsen Coefficient ($\beta$) for Hydrogen as a Function of Temperature (L real gas/(L atm)) (freshwater)

| Temperature (°C) | Δt (°C) | | | | | | | | | |
|---|---|---|---|---|---|---|---|---|---|---|
| | **0.0** | **0.1** | **0.2** | **0.3** | **0.4** | **0.5** | **0.6** | **0.7** | **0.8** | **0.9** |
| **0** | 0.02189 | 0.02186 | 0.02183 | 0.02181 | 0.02178 | 0.02175 | 0.02173 | 0.02170 | 0.02167 | 0.02165 |
| **1** | 0.02162 | 0.02159 | 0.02157 | 0.02154 | 0.02151 | 0.02149 | 0.02146 | 0.02144 | 0.02141 | 0.02138 |
| **2** | 0.02136 | 0.02133 | 0.02131 | 0.02128 | 0.02126 | 0.02123 | 0.02121 | 0.02118 | 0.02116 | 0.02113 |
| **3** | 0.02111 | 0.02108 | 0.02106 | 0.02103 | 0.02101 | 0.02099 | 0.02096 | 0.02094 | 0.02091 | 0.02089 |
| **4** | 0.02087 | 0.02084 | 0.02082 | 0.02080 | 0.02077 | 0.02075 | 0.02073 | 0.02070 | 0.02068 | 0.02066 |
| **5** | 0.02063 | 0.02061 | 0.02059 | 0.02057 | 0.02054 | 0.02052 | 0.02050 | 0.02048 | 0.02045 | 0.02043 |
| **6** | 0.02041 | 0.02039 | 0.02037 | 0.02034 | 0.02032 | 0.02030 | 0.02028 | 0.02026 | 0.02024 | 0.02022 |
| **7** | 0.02019 | 0.02017 | 0.02015 | 0.02013 | 0.02011 | 0.02009 | 0.02007 | 0.02005 | 0.02003 | 0.02001 |
| **8** | 0.01999 | 0.01997 | 0.01995 | 0.01992 | 0.01990 | 0.01988 | 0.01986 | 0.01984 | 0.01983 | 0.01981 |
| **9** | 0.01979 | 0.01977 | 0.01975 | 0.01973 | 0.01971 | 0.01969 | 0.01967 | 0.01965 | 0.01963 | 0.01961 |
| **10** | 0.01959 | 0.01957 | 0.01955 | 0.01954 | 0.01952 | 0.01950 | 0.01948 | 0.01946 | 0.01944 | 0.01943 |
| **11** | 0.01941 | 0.01939 | 0.01937 | 0.01935 | 0.01933 | 0.01932 | 0.01930 | 0.01928 | 0.01926 | 0.01925 |
| **12** | 0.01923 | 0.01921 | 0.01919 | 0.01918 | 0.01916 | 0.01914 | 0.01912 | 0.01911 | 0.01909 | 0.01907 |
| **13** | 0.01906 | 0.01904 | 0.01902 | 0.01901 | 0.01899 | 0.01897 | 0.01896 | 0.01894 | 0.01892 | 0.01891 |
| **14** | 0.01889 | 0.01887 | 0.01886 | 0.01884 | 0.01883 | 0.01881 | 0.01879 | 0.01878 | 0.01876 | 0.01875 |
| **15** | 0.01873 | 0.01872 | 0.01870 | 0.01868 | 0.01867 | 0.01865 | 0.01864 | 0.01862 | 0.01861 | 0.01859 |
| **16** | 0.01858 | 0.01856 | 0.01855 | 0.01853 | 0.01852 | 0.01850 | 0.01849 | 0.01847 | 0.01846 | 0.01844 |
| **17** | 0.01843 | 0.01842 | 0.01840 | 0.01839 | 0.01837 | 0.01836 | 0.01834 | 0.01833 | 0.01832 | 0.01830 |
| **18** | 0.01829 | 0.01827 | 0.01826 | 0.01825 | 0.01823 | 0.01822 | 0.01821 | 0.01819 | 0.01818 | 0.01817 |
| **19** | 0.01815 | 0.01814 | 0.01813 | 0.01811 | 0.01810 | 0.01809 | 0.01807 | 0.01806 | 0.01805 | 0.01803 |
| **20** | 0.01802 | 0.01801 | 0.01800 | 0.01798 | 0.01797 | 0.01796 | 0.01795 | 0.01793 | 0.01792 | 0.01791 |
| **21** | 0.01790 | 0.01788 | 0.01787 | 0.01786 | 0.01785 | 0.01783 | 0.01782 | 0.01781 | 0.01780 | 0.01779 |
| **22** | 0.01777 | 0.01776 | 0.01775 | 0.01774 | 0.01773 | 0.01772 | 0.01770 | 0.01769 | 0.01768 | 0.01767 |
| **23** | 0.01766 | 0.01765 | 0.01764 | 0.01763 | 0.01761 | 0.01760 | 0.01759 | 0.01758 | 0.01757 | 0.01756 |
| **24** | 0.01755 | 0.01754 | 0.01753 | 0.01752 | 0.01751 | 0.01749 | 0.01748 | 0.01747 | 0.01746 | 0.01745 |
| **25** | 0.01744 | 0.01743 | 0.01742 | 0.01741 | 0.01740 | 0.01739 | 0.01738 | 0.01737 | 0.01736 | 0.01735 |
| **26** | 0.01734 | 0.01733 | 0.01732 | 0.01731 | 0.01730 | 0.01729 | 0.01728 | 0.01727 | 0.01726 | 0.01725 |
| **27** | 0.01724 | 0.01723 | 0.01722 | 0.01721 | 0.01721 | 0.01720 | 0.01719 | 0.01718 | 0.01717 | 0.01716 |
| **28** | 0.01715 | 0.01714 | 0.01713 | 0.01712 | 0.01711 | 0.01710 | 0.01710 | 0.01709 | 0.01708 | 0.01707 |
| **29** | 0.01706 | 0.01705 | 0.01704 | 0.01704 | 0.01703 | 0.01702 | 0.01701 | 0.01700 | 0.01699 | 0.01698 |
| **30** | 0.01698 | 0.01697 | 0.01696 | 0.01695 | 0.01694 | 0.01694 | 0.01693 | 0.01692 | 0.01691 | 0.01690 |
| **31** | 0.01690 | 0.01689 | 0.01688 | 0.01687 | 0.01686 | 0.01686 | 0.01685 | 0.01684 | 0.01683 | 0.01683 |
| **32** | 0.01682 | 0.01681 | 0.01680 | 0.01680 | 0.01679 | 0.01678 | 0.01677 | 0.01677 | 0.01676 | 0.01675 |
| **33** | 0.01675 | 0.01674 | 0.01673 | 0.01672 | 0.01672 | 0.01671 | 0.01670 | 0.01670 | 0.01669 | 0.01668 |
| **34** | 0.01668 | 0.01667 | 0.01666 | 0.01666 | 0.01665 | 0.01664 | 0.01664 | 0.01663 | 0.01662 | 0.01662 |
| **35** | 0.01661 | 0.01660 | 0.01660 | 0.01659 | 0.01658 | 0.01658 | 0.01657 | 0.01657 | 0.01656 | 0.01655 |
| **36** | 0.01655 | 0.01654 | 0.01654 | 0.01653 | 0.01652 | 0.01652 | 0.01651 | 0.01651 | 0.01650 | 0.01649 |
| **37** | 0.01649 | 0.01648 | 0.01648 | 0.01647 | 0.01647 | 0.01646 | 0.01645 | 0.01645 | 0.01644 | 0.01644 |
| **38** | 0.01643 | 0.01643 | 0.01642 | 0.01642 | 0.01641 | 0.01641 | 0.01640 | 0.01640 | 0.01639 | 0.01639 |
| **39** | 0.01638 | 0.01638 | 0.01637 | 0.01637 | 0.01636 | 0.01636 | 0.01635 | 0.01635 | 0.01634 | 0.01634 |
| **40** | 0.01633 | 0.01633 | 0.01632 | 0.01632 | 0.01631 | 0.01631 | 0.01630 | 0.01630 | 0.01629 | 0.01629 |

*Source*: Based on Weiss (1971).

**Table 5.4** Bunsen Coefficient ($\beta$) for Hydrogen as a Function of Temperature and Salinity, 0–40 g/kg (L real gas/(L atm))

| Temperature (°C) | Salinity (g/kg) | | | | | | | | |
|---|---|---|---|---|---|---|---|---|---|
| | **0.0** | **5.0** | **10.0** | **15.0** | **20.0** | **25.0** | **30.0** | **35.0** | **40.0** |
| **0** | 0.02189 | 0.02130 | 0.02073 | 0.02017 | 0.01963 | 0.01911 | 0.01859 | 0.01810 | 0.01761 |
| **1** | 0.02162 | 0.02104 | 0.02048 | 0.01994 | 0.01941 | 0.01889 | 0.01839 | 0.01790 | 0.01743 |
| **2** | 0.02136 | 0.02080 | 0.02025 | 0.01971 | 0.01920 | 0.01869 | 0.01820 | 0.01772 | 0.01725 |
| **3** | 0.02111 | 0.02056 | 0.02002 | 0.01950 | 0.01899 | 0.01849 | 0.01801 | 0.01754 | 0.01708 |
| **4** | 0.02087 | 0.02033 | 0.01980 | 0.01929 | 0.01879 | 0.01830 | 0.01783 | 0.01737 | 0.01692 |
| **5** | 0.02063 | 0.02010 | 0.01959 | 0.01909 | 0.01860 | 0.01812 | 0.01766 | 0.01720 | 0.01676 |
| **6** | 0.02041 | 0.01989 | 0.01939 | 0.01889 | 0.01841 | 0.01795 | 0.01749 | 0.01705 | 0.01661 |
| **7** | 0.02019 | 0.01969 | 0.01919 | 0.01871 | 0.01824 | 0.01778 | 0.01733 | 0.01689 | 0.01647 |
| **8** | 0.01999 | 0.01949 | 0.01900 | 0.01853 | 0.01806 | 0.01761 | 0.01717 | 0.01675 | 0.01633 |
| **9** | 0.01979 | 0.01930 | 0.01882 | 0.01835 | 0.01790 | 0.01746 | 0.01703 | 0.01660 | 0.01619 |
| **10** | 0.01959 | 0.01911 | 0.01864 | 0.01819 | 0.01774 | 0.01731 | 0.01688 | 0.01647 | 0.01607 |
| **11** | 0.01941 | 0.01894 | 0.01848 | 0.01803 | 0.01759 | 0.01716 | 0.01675 | 0.01634 | 0.01594 |
| **12** | 0.01923 | 0.01877 | 0.01831 | 0.01787 | 0.01744 | 0.01702 | 0.01661 | 0.01621 | 0.01582 |
| **13** | 0.01906 | 0.01860 | 0.01816 | 0.01772 | 0.01730 | 0.01689 | 0.01649 | 0.01609 | 0.01571 |
| **14** | 0.01889 | 0.01844 | 0.01801 | 0.01758 | 0.01717 | 0.01676 | 0.01636 | 0.01598 | 0.01560 |
| **15** | 0.01873 | 0.01829 | 0.01786 | 0.01745 | 0.01704 | 0.01664 | 0.01625 | 0.01587 | 0.01550 |
| **16** | 0.01858 | 0.01815 | 0.01772 | 0.01731 | 0.01691 | 0.01652 | 0.01614 | 0.01576 | 0.01539 |
| **17** | 0.01843 | 0.01801 | 0.01759 | 0.01719 | 0.01679 | 0.01641 | 0.01603 | 0.01566 | 0.01530 |
| **18** | 0.01829 | 0.01787 | 0.01746 | 0.01707 | 0.01668 | 0.01630 | 0.01592 | 0.01556 | 0.01521 |
| **19** | 0.01815 | 0.01774 | 0.01734 | 0.01695 | 0.01657 | 0.01619 | 0.01583 | 0.01547 | 0.01512 |
| **20** | 0.01802 | 0.01762 | 0.01722 | 0.01684 | 0.01646 | 0.01609 | 0.01573 | 0.01538 | 0.01503 |
| **21** | 0.01790 | 0.01750 | 0.01711 | 0.01673 | 0.01636 | 0.01600 | 0.01564 | 0.01529 | 0.01495 |
| **22** | 0.01777 | 0.01738 | 0.01700 | 0.01663 | 0.01626 | 0.01590 | 0.01555 | 0.01521 | 0.01488 |
| **23** | 0.01766 | 0.01727 | 0.01690 | 0.01653 | 0.01617 | 0.01582 | 0.01547 | 0.01513 | 0.01480 |
| **24** | 0.01755 | 0.01717 | 0.01680 | 0.01643 | 0.01608 | 0.01573 | 0.01539 | 0.01506 | 0.01473 |
| **25** | 0.01744 | 0.01707 | 0.01670 | 0.01634 | 0.01599 | 0.01565 | 0.01532 | 0.01499 | 0.01467 |
| **26** | 0.01734 | 0.01697 | 0.01661 | 0.01626 | 0.01591 | 0.01557 | 0.01524 | 0.01492 | 0.01460 |
| **27** | 0.01724 | 0.01688 | 0.01652 | 0.01618 | 0.01584 | 0.01550 | 0.01518 | 0.01486 | 0.01454 |
| **28** | 0.01715 | 0.01679 | 0.01644 | 0.01610 | 0.01576 | 0.01543 | 0.01511 | 0.01479 | 0.01449 |
| **29** | 0.01706 | 0.01671 | 0.01636 | 0.01602 | 0.01569 | 0.01537 | 0.01505 | 0.01474 | 0.01443 |
| **30** | 0.01698 | 0.01663 | 0.01629 | 0.01595 | 0.01562 | 0.01530 | 0.01499 | 0.01468 | 0.01438 |
| **31** | 0.01690 | 0.01655 | 0.01621 | 0.01588 | 0.01556 | 0.01525 | 0.01493 | 0.01463 | 0.01433 |
| **32** | 0.01682 | 0.01648 | 0.01615 | 0.01582 | 0.01550 | 0.01519 | 0.01488 | 0.01458 | 0.01429 |
| **33** | 0.01675 | 0.01641 | 0.01608 | 0.01576 | 0.01545 | 0.01514 | 0.01483 | 0.01454 | 0.01425 |
| **34** | 0.01668 | 0.01634 | 0.01602 | 0.01570 | 0.01539 | 0.01509 | 0.01479 | 0.01449 | 0.01421 |
| **35** | 0.01661 | 0.01628 | 0.01596 | 0.01565 | 0.01534 | 0.01504 | 0.01474 | 0.01445 | 0.01417 |
| **36** | 0.01655 | 0.01622 | 0.01591 | 0.01560 | 0.01529 | 0.01500 | 0.01470 | 0.01442 | 0.01414 |
| **37** | 0.01649 | 0.01617 | 0.01586 | 0.01555 | 0.01525 | 0.01495 | 0.01467 | 0.01438 | 0.01410 |
| **38** | 0.01643 | 0.01612 | 0.01581 | 0.01551 | 0.01521 | 0.01492 | 0.01463 | 0.01435 | 0.01407 |
| **39** | 0.01638 | 0.01607 | 0.01576 | 0.01546 | 0.01517 | 0.01488 | 0.01460 | 0.01432 | 0.01405 |
| **40** | 0.01633 | 0.01602 | 0.01572 | 0.01542 | 0.01513 | 0.01485 | 0.01457 | 0.01429 | 0.01402 |

*Source*: Based on Weiss (1971).

**Table 5.5** Standard Air Saturation Concentration of Methane as a Function of Temperature in nmol/kg (freshwater, 1 atm moist air)

| Temperature (°C) | Δt (°C) | | | | | | | | | |
|---|---|---|---|---|---|---|---|---|---|---|
| | 0.0 | 0.1 | 0.2 | 0.3 | 0.4 | 0.5 | 0.6 | 0.7 | 0.8 | 0.9 |
| 0 | 4.5347 | 4.5206 | 4.5065 | 4.4925 | 4.4785 | 4.4647 | 4.4509 | 4.4372 | 4.4235 | 4.4099 |
| 1 | 4.3964 | 4.3829 | 4.3695 | 4.3562 | 4.3430 | 4.3298 | 4.3166 | 4.3036 | 4.2906 | 4.2776 |
| 2 | 4.2648 | 4.2520 | 4.2392 | 4.2265 | 4.2139 | 4.2013 | 4.1888 | 4.1764 | 4.1640 | 4.1517 |
| 3 | 4.1394 | 4.1272 | 4.1151 | 4.1030 | 4.0910 | 4.0790 | 4.0671 | 4.0552 | 4.0435 | 4.0317 |
| 4 | 4.0200 | 4.0084 | 3.9968 | 3.9853 | 3.9739 | 3.9624 | 3.9511 | 3.9398 | 3.9285 | 3.9174 |
| 5 | 3.9062 | 3.8951 | 3.8841 | 3.8731 | 3.8622 | 3.8513 | 3.8405 | 3.8297 | 3.8190 | 3.8083 |
| 6 | 3.7977 | 3.7871 | 3.7766 | 3.7661 | 3.7557 | 3.7453 | 3.7349 | 3.7247 | 3.7144 | 3.7042 |
| 7 | 3.6941 | 3.6840 | 3.6739 | 3.6639 | 3.6540 | 3.6441 | 3.6342 | 3.6244 | 3.6146 | 3.6049 |
| 8 | 3.5952 | 3.5856 | 3.5760 | 3.5664 | 3.5569 | 3.5474 | 3.5380 | 3.5286 | 3.5193 | 3.5100 |
| 9 | 3.5007 | 3.4915 | 3.4824 | 3.4732 | 3.4641 | 3.4551 | 3.4461 | 3.4371 | 3.4282 | 3.4193 |
| 10 | 3.4105 | 3.4017 | 3.3929 | 3.3842 | 3.3755 | 3.3668 | 3.3582 | 3.3496 | 3.3411 | 3.3326 |
| 11 | 3.3242 | 3.3157 | 3.3073 | 3.2990 | 3.2907 | 3.2824 | 3.2742 | 3.2660 | 3.2578 | 3.2497 |
| 12 | 3.2416 | 3.2335 | 3.2255 | 3.2175 | 3.2096 | 3.2016 | 3.1937 | 3.1859 | 3.1781 | 3.1703 |
| 13 | 3.1625 | 3.1548 | 3.1471 | 3.1395 | 3.1319 | 3.1243 | 3.1167 | 3.1092 | 3.1017 | 3.0943 |
| 14 | 3.0869 | 3.0795 | 3.0721 | 3.0648 | 3.0575 | 3.0502 | 3.0430 | 3.0358 | 3.0286 | 3.0214 |
| 15 | 3.0143 | 3.0073 | 3.0002 | 2.9932 | 2.9862 | 2.9792 | 2.9723 | 2.9654 | 2.9585 | 2.9516 |
| 16 | 2.9448 | 2.9380 | 2.9313 | 2.9245 | 2.9178 | 2.9111 | 2.9045 | 2.8979 | 2.8913 | 2.8847 |
| 17 | 2.8781 | 2.8716 | 2.8651 | 2.8587 | 2.8522 | 2.8458 | 2.8394 | 2.8331 | 2.8268 | 2.8205 |
| 18 | 2.8142 | 2.8079 | 2.8017 | 2.7955 | 2.7893 | 2.7831 | 2.7770 | 2.7709 | 2.7648 | 2.7588 |
| 19 | 2.7527 | 2.7467 | 2.7408 | 2.7348 | 2.7289 | 2.7230 | 2.7171 | 2.7112 | 2.7054 | 2.6995 |
| 20 | 2.6938 | 2.6880 | 2.6822 | 2.6765 | 2.6708 | 2.6651 | 2.6595 | 2.6538 | 2.6482 | 2.6426 |
| 21 | 2.6370 | 2.6315 | 2.6260 | 2.6205 | 2.6150 | 2.6095 | 2.6041 | 2.5987 | 2.5933 | 2.5879 |
| 22 | 2.5825 | 2.5772 | 2.5719 | 2.5666 | 2.5613 | 2.5560 | 2.5508 | 2.5456 | 2.5404 | 2.5352 |
| 23 | 2.5301 | 2.5249 | 2.5198 | 2.5147 | 2.5096 | 2.5046 | 2.4995 | 2.4945 | 2.4895 | 2.4845 |
| 24 | 2.4796 | 2.4746 | 2.4697 | 2.4648 | 2.4599 | 2.4550 | 2.4502 | 2.4453 | 2.4405 | 2.4357 |
| 25 | 2.4309 | 2.4262 | 2.4214 | 2.4167 | 2.4120 | 2.4073 | 2.4026 | 2.3980 | 2.3933 | 2.3887 |
| 26 | 2.3841 | 2.3795 | 2.3749 | 2.3703 | 2.3658 | 2.3613 | 2.3568 | 2.3523 | 2.3478 | 2.3433 |
| 27 | 2.3389 | 2.3344 | 2.3300 | 2.3256 | 2.3212 | 2.3169 | 2.3125 | 2.3082 | 2.3039 | 2.2996 |
| 28 | 2.2953 | 2.2910 | 2.2867 | 2.2825 | 2.2783 | 2.2740 | 2.2698 | 2.2656 | 2.2615 | 2.2573 |
| 29 | 2.2532 | 2.2490 | 2.2449 | 2.2408 | 2.2367 | 2.2327 | 2.2286 | 2.2246 | 2.2205 | 2.2165 |
| 30 | 2.2125 | 2.2085 | 2.2045 | 2.2006 | 2.1966 | 2.1927 | 2.1887 | 2.1848 | 2.1809 | 2.1771 |
| 31 | 2.1732 | 2.1693 | 2.1655 | 2.1616 | 2.1578 | 2.1540 | 2.1502 | 2.1464 | 2.1427 | 2.1389 |
| 32 | 2.1351 | 2.1314 | 2.1277 | 2.1240 | 2.1203 | 2.1166 | 2.1129 | 2.1093 | 2.1056 | 2.1020 |
| 33 | 2.0983 | 2.0947 | 2.0911 | 2.0875 | 2.0839 | 2.0804 | 2.0768 | 2.0732 | 2.0697 | 2.0662 |
| 34 | 2.0627 | 2.0592 | 2.0557 | 2.0522 | 2.0487 | 2.0452 | 2.0418 | 2.0384 | 2.0349 | 2.0315 |
| 35 | 2.0281 | 2.0247 | 2.0213 | 2.0179 | 2.0146 | 2.0112 | 2.0078 | 2.0045 | 2.0012 | 1.9979 |
| 36 | 1.9945 | 1.9912 | 1.9880 | 1.9847 | 1.9814 | 1.9781 | 1.9749 | 1.9716 | 1.9684 | 1.9652 |
| 37 | 1.9620 | 1.9588 | 1.9556 | 1.9524 | 1.9492 | 1.9460 | 1.9429 | 1.9397 | 1.9366 | 1.9334 |
| 38 | 1.9303 | 1.9272 | 1.9241 | 1.9210 | 1.9179 | 1.9148 | 1.9117 | 1.9087 | 1.9056 | 1.9026 |
| 39 | 1.8995 | 1.8965 | 1.8935 | 1.8905 | 1.8874 | 1.8844 | 1.8814 | 1.8785 | 1.8755 | 1.8725 |
| 40 | 1.8696 | 1.8666 | 1.8636 | 1.8607 | 1.8578 | 1.8549 | 1.8519 | 1.8490 | 1.8461 | 1.8432 |

*Source*: Based on Weiss (1971).

**Table 5.6** Standard Air Saturation Concentration of Methane as a Function of Temperature and Salinity in nmol/kg, 0–40 g/kg (seawater, 1 atm moist air)

| Temperature (°C) | Salinity (g/kg) | | | | | | | | |
|---|---|---|---|---|---|---|---|---|---|
| | **0.0** | **5.0** | **10.0** | **15.0** | **20.0** | **25.0** | **30.0** | **35.0** | **40.0** |
| **0** | 4.5347 | 4.3550 | 4.1826 | 4.0172 | 3.8583 | 3.7058 | 3.5594 | 3.4188 | 3.2838 |
| **1** | 4.3964 | 4.2236 | 4.0578 | 3.8985 | 3.7456 | 3.5988 | 3.4578 | 3.3223 | 3.1921 |
| **2** | 4.2648 | 4.0985 | 3.9389 | 3.7855 | 3.6383 | 3.4968 | 3.3608 | 3.2302 | 3.1047 |
| **3** | 4.1394 | 3.9793 | 3.8256 | 3.6778 | 3.5359 | 3.3995 | 3.2683 | 3.1423 | 3.0211 |
| **4** | 4.0200 | 3.8658 | 3.7176 | 3.5751 | 3.4382 | 3.3066 | 3.1800 | 3.0583 | 2.9413 |
| **5** | 3.9062 | 3.7575 | 3.6145 | 3.4771 | 3.3450 | 3.2179 | 3.0957 | 2.9781 | 2.8651 |
| **6** | 3.7977 | 3.6541 | 3.5162 | 3.3835 | 3.2559 | 3.1331 | 3.0150 | 2.9014 | 2.7921 |
| **7** | 3.6941 | 3.5555 | 3.4223 | 3.2941 | 3.1708 | 3.0522 | 2.9380 | 2.8281 | 2.7223 |
| **8** | 3.5952 | 3.4613 | 3.3326 | 3.2087 | 3.0895 | 2.9747 | 2.8642 | 2.7579 | 2.6555 |
| **9** | 3.5007 | 3.3713 | 3.2468 | 3.1270 | 3.0116 | 2.9006 | 2.7936 | 2.6906 | 2.5915 |
| **10** | 3.4105 | 3.2853 | 3.1648 | 3.0489 | 2.9372 | 2.8296 | 2.7260 | 2.6262 | 2.5301 |
| **11** | 3.3242 | 3.2030 | 3.0864 | 2.9741 | 2.8659 | 2.7617 | 2.6613 | 2.5645 | 2.4713 |
| **12** | 3.2416 | 3.1243 | 3.0113 | 2.9024 | 2.7976 | 2.6965 | 2.5992 | 2.5053 | 2.4149 |
| **13** | 3.1625 | 3.0488 | 2.9393 | 2.8338 | 2.7321 | 2.6341 | 2.5396 | 2.4485 | 2.3607 |
| **14** | 3.0869 | 2.9766 | 2.8704 | 2.7680 | 2.6693 | 2.5742 | 2.4825 | 2.3940 | 2.3087 |
| **15** | 3.0143 | 2.9074 | 2.8043 | 2.7049 | 2.6091 | 2.5167 | 2.4276 | 2.3417 | 2.2588 |
| **16** | 2.9448 | 2.8410 | 2.7409 | 2.6444 | 2.5513 | 2.4615 | 2.3749 | 2.2914 | 2.2108 |
| **17** | 2.8781 | 2.7773 | 2.6800 | 2.5863 | 2.4958 | 2.4085 | 2.3243 | 2.2430 | 2.1646 |
| **18** | 2.8142 | 2.7161 | 2.6216 | 2.5304 | 2.4424 | 2.3575 | 2.2756 | 2.1965 | 2.1202 |
| **19** | 2.7527 | 2.6574 | 2.5655 | 2.4768 | 2.3911 | 2.3085 | 2.2287 | 2.1517 | 2.0774 |
| **20** | 2.6938 | 2.6010 | 2.5115 | 2.4252 | 2.3418 | 2.2613 | 2.1836 | 2.1086 | 2.0362 |
| **21** | 2.6370 | 2.5468 | 2.4596 | 2.3755 | 2.2943 | 2.2159 | 2.1402 | 2.0671 | 1.9965 |
| **22** | 2.5825 | 2.4946 | 2.4097 | 2.3278 | 2.2486 | 2.1722 | 2.0984 | 2.0271 | 1.9582 |
| **23** | 2.5301 | 2.4444 | 2.3616 | 2.2817 | 2.2046 | 2.1300 | 2.0580 | 1.9885 | 1.9213 |
| **24** | 2.4796 | 2.3960 | 2.3153 | 2.2374 | 2.1621 | 2.0894 | 2.0191 | 1.9512 | 1.8856 |
| **25** | 2.4309 | 2.3494 | 2.2707 | 2.1946 | 2.1212 | 2.0502 | 1.9816 | 1.9152 | 1.8512 |
| **26** | 2.3841 | 2.3045 | 2.2277 | 2.1534 | 2.0816 | 2.0123 | 1.9453 | 1.8805 | 1.8179 |
| **27** | 2.3389 | 2.2612 | 2.1861 | 2.1136 | 2.0435 | 1.9757 | 1.9102 | 1.8469 | 1.7857 |
| **28** | 2.2953 | 2.2194 | 2.1460 | 2.0751 | 2.0066 | 1.9404 | 1.8763 | 1.8144 | 1.7545 |
| **29** | 2.2532 | 2.1790 | 2.1073 | 2.0380 | 1.9710 | 1.9062 | 1.8435 | 1.7829 | 1.7244 |
| **30** | 2.2125 | 2.1399 | 2.0698 | 2.0020 | 1.9365 | 1.8731 | 1.8118 | 1.7525 | 1.6951 |
| **31** | 2.1732 | 2.1022 | 2.0336 | 1.9672 | 1.9031 | 1.8410 | 1.7810 | 1.7229 | 1.6668 |
| **32** | 2.1351 | 2.0657 | 1.9985 | 1.9335 | 1.8707 | 1.8099 | 1.7512 | 1.6943 | 1.6393 |
| **33** | 2.0983 | 2.0303 | 1.9645 | 1.9009 | 1.8393 | 1.7798 | 1.7222 | 1.6665 | 1.6126 |
| **34** | 2.0627 | 1.9960 | 1.9316 | 1.8692 | 1.8089 | 1.7506 | 1.6941 | 1.6395 | 1.5866 |
| **35** | 2.0281 | 1.9628 | 1.8996 | 1.8385 | 1.7794 | 1.7221 | 1.6668 | 1.6132 | 1.5614 |
| **36** | 1.9945 | 1.9305 | 1.8686 | 1.8086 | 1.7506 | 1.6945 | 1.6402 | 1.5877 | 1.5368 |
| **37** | 1.9620 | 1.8991 | 1.8384 | 1.7796 | 1.7227 | 1.6677 | 1.6144 | 1.5628 | 1.5129 |
| **38** | 1.9303 | 1.8687 | 1.8091 | 1.7514 | 1.6955 | 1.6415 | 1.5892 | 1.5386 | 1.4896 |
| **39** | 1.8995 | 1.8390 | 1.7805 | 1.7239 | 1.6690 | 1.6160 | 1.5646 | 1.5149 | 1.4668 |
| **40** | 1.8696 | 1.8101 | 1.7527 | 1.6970 | 1.6432 | 1.5911 | 1.5407 | 1.4918 | 1.4445 |

*Source*: Based on Weiss (1971).

**Table 5.7** Bunsen Coefficient ($\beta$) for Methane as a Function of Temperature (L real gas/(L atm)) (freshwater)

| Temperature (°C) | $\Delta t$ (°C) | | | | | | | | | |
|---|---|---|---|---|---|---|---|---|---|---|
| | **0.0** | **0.1** | **0.2** | **0.3** | **0.4** | **0.5** | **0.6** | **0.7** | **0.8** | **0.9** |
| **0** | 0.05749 | 0.05732 | 0.05714 | 0.05697 | 0.05679 | 0.05662 | 0.05645 | 0.05628 | 0.05611 | 0.05594 |
| **1** | 0.05577 | 0.05560 | 0.05543 | 0.05527 | 0.05510 | 0.05494 | 0.05477 | 0.05461 | 0.05445 | 0.05429 |
| **2** | 0.05413 | 0.05397 | 0.05381 | 0.05365 | 0.05349 | 0.05334 | 0.05318 | 0.05303 | 0.05287 | 0.05272 |
| **3** | 0.05257 | 0.05241 | 0.05226 | 0.05211 | 0.05196 | 0.05181 | 0.05166 | 0.05152 | 0.05137 | 0.05122 |
| **4** | 0.05108 | 0.05093 | 0.05079 | 0.05065 | 0.05050 | 0.05036 | 0.05022 | 0.05008 | 0.04994 | 0.04980 |
| **5** | 0.04966 | 0.04952 | 0.04939 | 0.04925 | 0.04911 | 0.04898 | 0.04884 | 0.04871 | 0.04857 | 0.04844 |
| **6** | 0.04831 | 0.04818 | 0.04805 | 0.04792 | 0.04779 | 0.04766 | 0.04753 | 0.04740 | 0.04727 | 0.04715 |
| **7** | 0.04702 | 0.04690 | 0.04677 | 0.04665 | 0.04652 | 0.04640 | 0.04628 | 0.04616 | 0.04603 | 0.04591 |
| **8** | 0.04579 | 0.04567 | 0.04555 | 0.04544 | 0.04532 | 0.04520 | 0.04508 | 0.04497 | 0.04485 | 0.04474 |
| **9** | 0.04462 | 0.04451 | 0.04439 | 0.04428 | 0.04417 | 0.04405 | 0.04394 | 0.04383 | 0.04372 | 0.04361 |
| **10** | 0.04350 | 0.04339 | 0.04328 | 0.04317 | 0.04307 | 0.04296 | 0.04285 | 0.04275 | 0.04264 | 0.04254 |
| **11** | 0.04243 | 0.04233 | 0.04222 | 0.04212 | 0.04202 | 0.04192 | 0.04181 | 0.04171 | 0.04161 | 0.04151 |
| **12** | 0.04141 | 0.04131 | 0.04121 | 0.04111 | 0.04101 | 0.04092 | 0.04082 | 0.04072 | 0.04063 | 0.04053 |
| **13** | 0.04043 | 0.04034 | 0.04024 | 0.04015 | 0.04006 | 0.03996 | 0.03987 | 0.03978 | 0.03968 | 0.03959 |
| **14** | 0.03950 | 0.03941 | 0.03932 | 0.03923 | 0.03914 | 0.03905 | 0.03896 | 0.03887 | 0.03878 | 0.03870 |
| **15** | 0.03861 | 0.03852 | 0.03844 | 0.03835 | 0.03826 | 0.03818 | 0.03809 | 0.03801 | 0.03792 | 0.03784 |
| **16** | 0.03776 | 0.03767 | 0.03759 | 0.03751 | 0.03742 | 0.03734 | 0.03726 | 0.03718 | 0.03710 | 0.03702 |
| **17** | 0.03694 | 0.03686 | 0.03678 | 0.03670 | 0.03662 | 0.03654 | 0.03647 | 0.03639 | 0.03631 | 0.03623 |
| **18** | 0.03616 | 0.03608 | 0.03601 | 0.03593 | 0.03585 | 0.03578 | 0.03570 | 0.03563 | 0.03556 | 0.03548 |
| **19** | 0.03541 | 0.03534 | 0.03526 | 0.03519 | 0.03512 | 0.03505 | 0.03498 | 0.03490 | 0.03483 | 0.03476 |
| **20** | 0.03469 | 0.03462 | 0.03455 | 0.03448 | 0.03441 | 0.03435 | 0.03428 | 0.03421 | 0.03414 | 0.03407 |
| **21** | 0.03401 | 0.03394 | 0.03387 | 0.03381 | 0.03374 | 0.03367 | 0.03361 | 0.03354 | 0.03348 | 0.03341 |
| **22** | 0.03335 | 0.03328 | 0.03322 | 0.03316 | 0.03309 | 0.03303 | 0.03297 | 0.03290 | 0.03284 | 0.03278 |
| **23** | 0.03272 | 0.03266 | 0.03260 | 0.03253 | 0.03247 | 0.03241 | 0.03235 | 0.03229 | 0.03223 | 0.03217 |
| **24** | 0.03211 | 0.03206 | 0.03200 | 0.03194 | 0.03188 | 0.03182 | 0.03176 | 0.03171 | 0.03165 | 0.03159 |
| **25** | 0.03154 | 0.03148 | 0.03142 | 0.03137 | 0.03131 | 0.03126 | 0.03120 | 0.03115 | 0.03109 | 0.03104 |
| **26** | 0.03098 | 0.03093 | 0.03087 | 0.03082 | 0.03077 | 0.03071 | 0.03066 | 0.03061 | 0.03055 | 0.03050 |
| **27** | 0.03045 | 0.03040 | 0.03035 | 0.03029 | 0.03024 | 0.03019 | 0.03014 | 0.03009 | 0.03004 | 0.02999 |
| **28** | 0.02994 | 0.02989 | 0.02984 | 0.02979 | 0.02974 | 0.02969 | 0.02964 | 0.02959 | 0.02955 | 0.02950 |
| **29** | 0.02945 | 0.02940 | 0.02935 | 0.02931 | 0.02926 | 0.02921 | 0.02917 | 0.02912 | 0.02907 | 0.02903 |
| **30** | 0.02898 | 0.02893 | 0.02889 | 0.02884 | 0.02880 | 0.02875 | 0.02871 | 0.02866 | 0.02862 | 0.02857 |
| **31** | 0.02853 | 0.02849 | 0.02844 | 0.02840 | 0.02835 | 0.02831 | 0.02827 | 0.02823 | 0.02818 | 0.02814 |
| **32** | 0.02810 | 0.02806 | 0.02801 | 0.02797 | 0.02793 | 0.02789 | 0.02785 | 0.02781 | 0.02776 | 0.02772 |
| **33** | 0.02768 | 0.02764 | 0.02760 | 0.02756 | 0.02752 | 0.02748 | 0.02744 | 0.02740 | 0.02736 | 0.02732 |
| **34** | 0.02729 | 0.02725 | 0.02721 | 0.02717 | 0.02713 | 0.02709 | 0.02705 | 0.02702 | 0.02698 | 0.02694 |
| **35** | 0.02690 | 0.02687 | 0.02683 | 0.02679 | 0.02676 | 0.02672 | 0.02668 | 0.02665 | 0.02661 | 0.02657 |
| **36** | 0.02654 | 0.02650 | 0.02647 | 0.02643 | 0.02640 | 0.02636 | 0.02633 | 0.02629 | 0.02626 | 0.02622 |
| **37** | 0.02619 | 0.02615 | 0.02612 | 0.02608 | 0.02605 | 0.02602 | 0.02598 | 0.02595 | 0.02592 | 0.02588 |
| **38** | 0.02585 | 0.02582 | 0.02578 | 0.02575 | 0.02572 | 0.02569 | 0.02566 | 0.02562 | 0.02559 | 0.02556 |
| **39** | 0.02553 | 0.02550 | 0.02546 | 0.02543 | 0.02540 | 0.02537 | 0.02534 | 0.02531 | 0.02528 | 0.02525 |
| **40** | 0.02522 | 0.02519 | 0.02516 | 0.02513 | 0.02510 | 0.02507 | 0.02504 | 0.02501 | 0.02498 | 0.02495 |

*Source*: Based on Weiss (1971).

**Table 5.8** Bunsen Coefficient ($\beta$) for Methane as a Function of Temperature and Salinity, 0–40 g/kg (L real gas/(L atm))

| Temperature (°C) | Salinity (g/kg) | | | | | | | | |
|---|---|---|---|---|---|---|---|---|---|
| | **0.0** | **5.0** | **10.0** | **15.0** | **20.0** | **25.0** | **30.0** | **35.0** | **40.0** |
| **0** | 0.05749 | 0.05544 | 0.05346 | 0.05155 | 0.04971 | 0.04793 | 0.04622 | 0.04457 | 0.04297 |
| **1** | 0.05577 | 0.05379 | 0.05189 | 0.05005 | 0.04828 | 0.04657 | 0.04492 | 0.04332 | 0.04179 |
| **2** | 0.05413 | 0.05223 | 0.05039 | 0.04862 | 0.04691 | 0.04527 | 0.04368 | 0.04214 | 0.04066 |
| **3** | 0.05257 | 0.05073 | 0.04897 | 0.04726 | 0.04561 | 0.04403 | 0.04249 | 0.04101 | 0.03958 |
| **4** | 0.05108 | 0.04931 | 0.04761 | 0.04596 | 0.04438 | 0.04284 | 0.04136 | 0.03993 | 0.03855 |
| **5** | 0.04966 | 0.04796 | 0.04632 | 0.04473 | 0.04320 | 0.04171 | 0.04029 | 0.03890 | 0.03757 |
| **6** | 0.04831 | 0.04667 | 0.04508 | 0.04355 | 0.04207 | 0.04064 | 0.03926 | 0.03792 | 0.03663 |
| **7** | 0.04702 | 0.04544 | 0.04390 | 0.04242 | 0.04099 | 0.03961 | 0.03827 | 0.03698 | 0.03573 |
| **8** | 0.04579 | 0.04426 | 0.04278 | 0.04135 | 0.03996 | 0.03863 | 0.03733 | 0.03608 | 0.03488 |
| **9** | 0.04462 | 0.04314 | 0.04171 | 0.04032 | 0.03898 | 0.03769 | 0.03643 | 0.03522 | 0.03405 |
| **10** | 0.04350 | 0.04207 | 0.04068 | 0.03934 | 0.03804 | 0.03679 | 0.03558 | 0.03440 | 0.03327 |
| **11** | 0.04243 | 0.04104 | 0.03970 | 0.03840 | 0.03714 | 0.03593 | 0.03475 | 0.03362 | 0.03252 |
| **12** | 0.04141 | 0.04007 | 0.03876 | 0.03750 | 0.03629 | 0.03511 | 0.03397 | 0.03286 | 0.03180 |
| **13** | 0.04043 | 0.03913 | 0.03787 | 0.03665 | 0.03546 | 0.03432 | 0.03321 | 0.03214 | 0.03111 |
| **14** | 0.03950 | 0.03824 | 0.03701 | 0.03583 | 0.03468 | 0.03357 | 0.03249 | 0.03145 | 0.03044 |
| **15** | 0.03861 | 0.03738 | 0.03619 | 0.03504 | 0.03393 | 0.03285 | 0.03180 | 0.03079 | 0.02981 |
| **16** | 0.03776 | 0.03656 | 0.03541 | 0.03429 | 0.03320 | 0.03216 | 0.03114 | 0.03016 | 0.02920 |
| **17** | 0.03694 | 0.03578 | 0.03466 | 0.03357 | 0.03251 | 0.03149 | 0.03050 | 0.02955 | 0.02862 |
| **18** | 0.03616 | 0.03503 | 0.03394 | 0.03288 | 0.03185 | 0.03086 | 0.02990 | 0.02896 | 0.02806 |
| **19** | 0.03541 | 0.03431 | 0.03325 | 0.03222 | 0.03122 | 0.03025 | 0.02931 | 0.02840 | 0.02752 |
| **20** | 0.03469 | 0.03362 | 0.03259 | 0.03158 | 0.03061 | 0.02967 | 0.02875 | 0.02787 | 0.02701 |
| **21** | 0.03401 | 0.03296 | 0.03195 | 0.03098 | 0.03003 | 0.02911 | 0.02822 | 0.02735 | 0.02651 |
| **22** | 0.03335 | 0.03233 | 0.03135 | 0.03039 | 0.02947 | 0.02857 | 0.02770 | 0.02686 | 0.02604 |
| **23** | 0.03272 | 0.03173 | 0.03077 | 0.02983 | 0.02893 | 0.02805 | 0.02721 | 0.02638 | 0.02558 |
| **24** | 0.03211 | 0.03115 | 0.03021 | 0.02930 | 0.02842 | 0.02756 | 0.02673 | 0.02592 | 0.02514 |
| **25** | 0.03154 | 0.03059 | 0.02967 | 0.02878 | 0.02792 | 0.02709 | 0.02627 | 0.02549 | 0.02472 |
| **26** | 0.03098 | 0.03006 | 0.02916 | 0.02829 | 0.02745 | 0.02663 | 0.02583 | 0.02506 | 0.02432 |
| **27** | 0.03045 | 0.02955 | 0.02867 | 0.02782 | 0.02699 | 0.02619 | 0.02541 | 0.02466 | 0.02393 |
| **28** | 0.02994 | 0.02905 | 0.02820 | 0.02736 | 0.02655 | 0.02577 | 0.02501 | 0.02427 | 0.02355 |
| **29** | 0.02945 | 0.02858 | 0.02774 | 0.02693 | 0.02613 | 0.02537 | 0.02462 | 0.02390 | 0.02319 |
| **30** | 0.02898 | 0.02813 | 0.02731 | 0.02651 | 0.02573 | 0.02498 | 0.02425 | 0.02354 | 0.02285 |
| **31** | 0.02853 | 0.02770 | 0.02689 | 0.02611 | 0.02534 | 0.02460 | 0.02389 | 0.02319 | 0.02251 |
| **32** | 0.02810 | 0.02728 | 0.02649 | 0.02572 | 0.02497 | 0.02425 | 0.02354 | 0.02286 | 0.02219 |
| **33** | 0.02768 | 0.02688 | 0.02610 | 0.02535 | 0.02461 | 0.02390 | 0.02321 | 0.02254 | 0.02189 |
| **34** | 0.02729 | 0.02650 | 0.02573 | 0.02499 | 0.02427 | 0.02357 | 0.02289 | 0.02223 | 0.02159 |
| **35** | 0.02690 | 0.02613 | 0.02538 | 0.02465 | 0.02394 | 0.02325 | 0.02258 | 0.02193 | 0.02130 |
| **36** | 0.02654 | 0.02578 | 0.02504 | 0.02432 | 0.02362 | 0.02295 | 0.02229 | 0.02165 | 0.02103 |
| **37** | 0.02619 | 0.02544 | 0.02471 | 0.02401 | 0.02332 | 0.02265 | 0.02200 | 0.02138 | 0.02076 |
| **38** | 0.02585 | 0.02511 | 0.02440 | 0.02370 | 0.02303 | 0.02237 | 0.02173 | 0.02111 | 0.02051 |
| **39** | 0.02553 | 0.02480 | 0.02410 | 0.02341 | 0.02274 | 0.02210 | 0.02147 | 0.02086 | 0.02027 |
| **40** | 0.02522 | 0.02450 | 0.02381 | 0.02313 | 0.02247 | 0.02184 | 0.02122 | 0.02061 | 0.02003 |

*Source*: Based on Weiss (1971).

**Table 5.9** Standard Air Saturation Concentration of Nitrous Oxide as a Function of Temperature in nmol/kg (freshwater, 1 atm moist air)

| Temperature (°C) | Δt (°C) | | | | | | | | | |
|---|---|---|---|---|---|---|---|---|---|---|
| | 0.0 | 0.1 | 0.2 | 0.3 | 0.4 | 0.5 | 0.6 | 0.7 | 0.8 | 0.9 |
| 0 | 18.728 | 18.649 | 18.570 | 18.492 | 18.414 | 18.336 | 18.259 | 18.182 | 18.106 | 18.030 |
| 1 | 17.955 | 17.880 | 17.805 | 17.731 | 17.658 | 17.584 | 17.511 | 17.439 | 17.367 | 17.295 |
| 2 | 17.224 | 17.153 | 17.083 | 17.012 | 16.943 | 16.873 | 16.805 | 16.736 | 16.668 | 16.600 |
| 3 | 16.533 | 16.466 | 16.399 | 16.333 | 16.267 | 16.201 | 16.136 | 16.071 | 16.006 | 15.942 |
| 4 | 15.878 | 15.815 | 15.752 | 15.689 | 15.626 | 15.564 | 15.503 | 15.441 | 15.380 | 15.319 |
| 5 | 15.259 | 15.199 | 15.139 | 15.079 | 15.020 | 14.961 | 14.903 | 14.845 | 14.787 | 14.729 |
| 6 | 14.672 | 14.615 | 14.558 | 14.502 | 14.446 | 14.390 | 14.334 | 14.279 | 14.224 | 14.170 |
| 7 | 14.115 | 14.061 | 14.007 | 13.954 | 13.901 | 13.848 | 13.795 | 13.743 | 13.691 | 13.639 |
| 8 | 13.587 | 13.536 | 13.485 | 13.434 | 13.384 | 13.333 | 13.283 | 13.234 | 13.184 | 13.135 |
| 9 | 13.086 | 13.037 | 12.989 | 12.941 | 12.893 | 12.845 | 12.798 | 12.750 | 12.703 | 12.657 |
| 10 | 12.610 | 12.564 | 12.518 | 12.472 | 12.426 | 12.381 | 12.336 | 12.291 | 12.246 | 12.202 |
| 11 | 12.158 | 12.114 | 12.070 | 12.026 | 11.983 | 11.940 | 11.897 | 11.854 | 11.812 | 11.770 |
| 12 | 11.728 | 11.686 | 11.644 | 11.603 | 11.561 | 11.520 | 11.480 | 11.439 | 11.398 | 11.358 |
| 13 | 11.318 | 11.278 | 11.239 | 11.199 | 11.160 | 11.121 | 11.082 | 11.043 | 11.005 | 10.967 |
| 14 | 10.929 | 10.891 | 10.853 | 10.815 | 10.778 | 10.741 | 10.704 | 10.667 | 10.630 | 10.594 |
| 15 | 10.557 | 10.521 | 10.485 | 10.449 | 10.414 | 10.378 | 10.343 | 10.308 | 10.273 | 10.238 |
| 16 | 10.204 | 10.169 | 10.135 | 10.101 | 10.067 | 10.033 | 9.999 | 9.966 | 9.932 | 9.899 |
| 17 | 9.866 | 9.833 | 9.801 | 9.768 | 9.736 | 9.703 | 9.671 | 9.639 | 9.608 | 9.576 |
| 18 | 9.544 | 9.513 | 9.482 | 9.451 | 9.420 | 9.389 | 9.358 | 9.328 | 9.297 | 9.267 |
| 19 | 9.237 | 9.207 | 9.177 | 9.147 | 9.118 | 9.089 | 9.059 | 9.030 | 9.001 | 8.972 |
| 20 | 8.943 | 8.915 | 8.886 | 8.858 | 8.830 | 8.801 | 8.773 | 8.746 | 8.718 | 8.690 |
| 21 | 8.663 | 8.635 | 8.608 | 8.581 | 8.554 | 8.527 | 8.500 | 8.474 | 8.447 | 8.421 |
| 22 | 8.394 | 8.368 | 8.342 | 8.316 | 8.290 | 8.265 | 8.239 | 8.213 | 8.188 | 8.163 |
| 23 | 8.138 | 8.112 | 8.087 | 8.063 | 8.038 | 8.013 | 7.989 | 7.964 | 7.940 | 7.916 |
| 24 | 7.892 | 7.868 | 7.844 | 7.820 | 7.796 | 7.773 | 7.749 | 7.726 | 7.702 | 7.679 |
| 25 | 7.656 | 7.633 | 7.610 | 7.587 | 7.565 | 7.542 | 7.519 | 7.497 | 7.475 | 7.452 |
| 26 | 7.430 | 7.408 | 7.386 | 7.364 | 7.343 | 7.321 | 7.299 | 7.278 | 7.256 | 7.235 |
| 27 | 7.214 | 7.192 | 7.171 | 7.150 | 7.129 | 7.109 | 7.088 | 7.067 | 7.047 | 7.026 |
| 28 | 7.006 | 6.985 | 6.965 | 6.945 | 6.925 | 6.905 | 6.885 | 6.865 | 6.846 | 6.826 |
| 29 | 6.806 | 6.787 | 6.767 | 6.748 | 6.729 | 6.709 | 6.690 | 6.671 | 6.652 | 6.633 |
| 30 | 6.615 | 6.596 | 6.577 | 6.558 | 6.540 | 6.521 | 6.503 | 6.485 | 6.466 | 6.448 |
| 31 | 6.430 | 6.412 | 6.394 | 6.376 | 6.358 | 6.341 | 6.323 | 6.305 | 6.288 | 6.270 |
| 32 | 6.253 | 6.236 | 6.218 | 6.201 | 6.184 | 6.167 | 6.150 | 6.133 | 6.116 | 6.099 |
| 33 | 6.082 | 6.066 | 6.049 | 6.032 | 6.016 | 5.999 | 5.983 | 5.967 | 5.950 | 5.934 |
| 34 | 5.918 | 5.902 | 5.886 | 5.870 | 5.854 | 5.838 | 5.822 | 5.806 | 5.791 | 5.775 |
| 35 | 5.759 | 5.744 | 5.728 | 5.713 | 5.698 | 5.682 | 5.667 | 5.652 | 5.637 | 5.622 |
| 36 | 5.607 | 5.592 | 5.577 | 5.562 | 5.547 | 5.532 | 5.518 | 5.503 | 5.488 | 5.474 |
| 37 | 5.459 | 5.445 | 5.430 | 5.416 | 5.402 | 5.388 | 5.373 | 5.359 | 5.345 | 5.331 |
| 38 | 5.317 | 5.303 | 5.289 | 5.275 | 5.261 | 5.248 | 5.234 | 5.220 | 5.207 | 5.193 |
| 39 | 5.180 | 5.166 | 5.153 | 5.139 | 5.126 | 5.113 | 5.099 | 5.086 | 5.073 | 5.060 |
| 40 | 5.047 | 5.034 | 5.021 | 5.008 | 4.995 | 4.982 | 4.969 | 4.956 | 4.943 | 4.931 |

*Source*: Based on Weiss (1971).

**Table 5.10** Standard Air Saturation Concentration of Nitrous Oxide as a Function of Temperature and Salinity in nmol/kg, 0–40 g/kg (seawater, 1 atm moist air)

| Temperature (°C) | Salinity (g/kg) | | | | | | | | |
|---|---|---|---|---|---|---|---|---|---|
| | **0.0** | **5.0** | **10.0** | **15.0** | **20.0** | **25.0** | **30.0** | **35.0** | **40.0** |
| **0** | 18.728 | 18.098 | 17.490 | 16.902 | 16.335 | 15.787 | 15.257 | 14.746 | 14.251 |
| **1** | 17.955 | 17.355 | 16.777 | 16.218 | 15.677 | 15.155 | 14.651 | 14.164 | 13.693 |
| **2** | 17.224 | 16.653 | 16.102 | 15.570 | 15.055 | 14.558 | 14.077 | 13.612 | 13.163 |
| **3** | 16.533 | 15.989 | 15.464 | 14.957 | 14.466 | 13.992 | 13.533 | 13.090 | 12.661 |
| **4** | 15.878 | 15.360 | 14.860 | 14.376 | 13.908 | 13.455 | 13.018 | 12.595 | 12.185 |
| **5** | 15.259 | 14.765 | 14.287 | 13.825 | 13.379 | 12.947 | 12.529 | 12.124 | 11.733 |
| **6** | 14.672 | 14.200 | 13.744 | 13.303 | 12.877 | 12.464 | 12.065 | 11.678 | 11.304 |
| **7** | 14.115 | 13.665 | 13.229 | 12.808 | 12.400 | 12.006 | 11.624 | 11.254 | 10.896 |
| **8** | 13.587 | 13.157 | 12.740 | 12.338 | 11.948 | 11.570 | 11.205 | 10.851 | 10.508 |
| **9** | 13.086 | 12.674 | 12.276 | 11.891 | 11.517 | 11.156 | 10.806 | 10.467 | 10.139 |
| **10** | 12.610 | 12.216 | 11.835 | 11.466 | 11.108 | 10.762 | 10.427 | 10.102 | 9.788 |
| **11** | 12.158 | 11.780 | 11.415 | 11.062 | 10.719 | 10.387 | 10.066 | 9.755 | 9.453 |
| **12** | 11.728 | 11.366 | 11.016 | 10.677 | 10.349 | 10.031 | 9.722 | 9.423 | 9.134 |
| **13** | 11.318 | 10.972 | 10.636 | 10.311 | 9.996 | 9.691 | 9.395 | 9.108 | 8.830 |
| **14** | 10.929 | 10.596 | 10.274 | 9.962 | 9.660 | 9.366 | 9.082 | 8.807 | 8.539 |
| **15** | 10.557 | 10.238 | 9.929 | 9.629 | 9.339 | 9.057 | 8.784 | 8.519 | 8.262 |
| **16** | 10.204 | 9.897 | 9.600 | 9.312 | 9.033 | 8.762 | 8.499 | 8.245 | 7.998 |
| **17** | 9.866 | 9.572 | 9.286 | 9.009 | 8.741 | 8.480 | 8.227 | 7.982 | 7.745 |
| **18** | 9.544 | 9.261 | 8.986 | 8.720 | 8.461 | 8.211 | 7.968 | 7.732 | 7.503 |
| **19** | 9.237 | 8.964 | 8.700 | 8.444 | 8.195 | 7.953 | 7.719 | 7.492 | 7.271 |
| **20** | 8.943 | 8.681 | 8.426 | 8.179 | 7.940 | 7.707 | 7.481 | 7.262 | 7.050 |
| **21** | 8.663 | 8.410 | 8.165 | 7.927 | 7.696 | 7.471 | 7.254 | 7.042 | 6.837 |
| **22** | 8.394 | 8.151 | 7.914 | 7.685 | 7.462 | 7.246 | 7.036 | 6.832 | 6.634 |
| **23** | 8.138 | 7.902 | 7.674 | 7.453 | 7.238 | 7.029 | 6.827 | 6.630 | 6.439 |
| **24** | 7.892 | 7.665 | 7.445 | 7.231 | 7.023 | 6.822 | 6.626 | 6.436 | 6.252 |
| **25** | 7.656 | 7.437 | 7.224 | 7.018 | 6.818 | 6.623 | 6.434 | 6.250 | 6.072 |
| **26** | 7.430 | 7.219 | 7.013 | 6.814 | 6.620 | 6.432 | 6.250 | 6.072 | 5.900 |
| **27** | 7.214 | 7.009 | 6.811 | 6.618 | 6.431 | 6.249 | 6.072 | 5.901 | 5.734 |
| **28** | 7.006 | 6.808 | 6.616 | 6.430 | 6.249 | 6.073 | 5.902 | 5.736 | 5.574 |
| **29** | 6.806 | 6.615 | 6.429 | 6.249 | 6.074 | 5.903 | 5.738 | 5.577 | 5.421 |
| **30** | 6.615 | 6.429 | 6.250 | 6.075 | 5.905 | 5.740 | 5.580 | 5.424 | 5.273 |
| **31** | 6.430 | 6.251 | 6.077 | 5.908 | 5.743 | 5.584 | 5.428 | 5.277 | 5.131 |
| **32** | 6.253 | 6.079 | 5.911 | 5.747 | 5.587 | 5.433 | 5.282 | 5.136 | 4.994 |
| **33** | 6.082 | 5.914 | 5.751 | 5.592 | 5.437 | 5.287 | 5.141 | 4.999 | 4.861 |
| **34** | 5.918 | 5.755 | 5.596 | 5.442 | 5.292 | 5.147 | 5.005 | 4.868 | 4.734 |
| **35** | 5.759 | 5.601 | 5.447 | 5.298 | 5.153 | 5.011 | 4.874 | 4.740 | 4.610 |
| **36** | 5.607 | 5.453 | 5.304 | 5.159 | 5.018 | 4.881 | 4.747 | 4.617 | 4.491 |
| **37** | 5.459 | 5.310 | 5.165 | 5.024 | 4.888 | 4.754 | 4.625 | 4.499 | 4.376 |
| **38** | 5.317 | 5.172 | 5.032 | 4.895 | 4.762 | 4.632 | 4.506 | 4.384 | 4.265 |
| **39** | 5.180 | 5.039 | 4.902 | 4.769 | 4.640 | 4.514 | 4.392 | 4.273 | 4.157 |
| **40** | 5.047 | 4.910 | 4.777 | 4.648 | 4.522 | 4.400 | 4.281 | 4.165 | 4.053 |

*Source*: Based on Weiss (1971).

**Table 5.11** Bunsen Coefficient ($\beta$) for Nitrous Oxide as a Function of Temperature (L real gas/(L atm)) (freshwater)

| Temperature (°C) | Δt (°C) | | | | | | | | | |
|---|---|---|---|---|---|---|---|---|---|---|
| | **0.0** | **0.1** | **0.2** | **0.3** | **0.4** | **0.5** | **0.6** | **0.7** | **0.8** | **0.9** |
| **0** | 1.3197 | 1.3142 | 1.3087 | 1.3032 | 1.2978 | 1.2923 | 1.2870 | 1.2816 | 1.2763 | 1.2710 |
| **1** | 1.2658 | 1.2605 | 1.2553 | 1.2502 | 1.2450 | 1.2399 | 1.2348 | 1.2298 | 1.2248 | 1.2198 |
| **2** | 1.2148 | 1.2099 | 1.2049 | 1.2001 | 1.1952 | 1.1904 | 1.1856 | 1.1808 | 1.1760 | 1.1713 |
| **3** | 1.1666 | 1.1619 | 1.1573 | 1.1526 | 1.1480 | 1.1435 | 1.1389 | 1.1344 | 1.1299 | 1.1254 |
| **4** | 1.1210 | 1.1166 | 1.1122 | 1.1078 | 1.1034 | 1.0991 | 1.0948 | 1.0905 | 1.0863 | 1.0820 |
| **5** | 1.0778 | 1.0736 | 1.0694 | 1.0653 | 1.0612 | 1.0571 | 1.0530 | 1.0489 | 1.0449 | 1.0409 |
| **6** | 1.0369 | 1.0329 | 1.0290 | 1.0251 | 1.0212 | 1.0173 | 1.0134 | 1.0096 | 1.0057 | 1.0019 |
| **7** | 0.9981 | 0.9944 | 0.9906 | 0.9869 | 0.9832 | 0.9795 | 0.9759 | 0.9722 | 0.9686 | 0.9650 |
| **8** | 0.9614 | 0.9578 | 0.9543 | 0.9507 | 0.9472 | 0.9437 | 0.9402 | 0.9368 | 0.9333 | 0.9299 |
| **9** | 0.9265 | 0.9231 | 0.9197 | 0.9164 | 0.9130 | 0.9097 | 0.9064 | 0.9031 | 0.8999 | 0.8966 |
| **10** | 0.8934 | 0.8902 | 0.8870 | 0.8838 | 0.8806 | 0.8775 | 0.8743 | 0.8712 | 0.8681 | 0.8650 |
| **11** | 0.8619 | 0.8589 | 0.8558 | 0.8528 | 0.8498 | 0.8468 | 0.8438 | 0.8409 | 0.8379 | 0.8350 |
| **12** | 0.8321 | 0.8292 | 0.8263 | 0.8234 | 0.8205 | 0.8177 | 0.8148 | 0.8120 | 0.8092 | 0.8064 |
| **13** | 0.8036 | 0.8009 | 0.7981 | 0.7954 | 0.7927 | 0.7900 | 0.7873 | 0.7846 | 0.7819 | 0.7793 |
| **14** | 0.7766 | 0.7740 | 0.7714 | 0.7688 | 0.7662 | 0.7636 | 0.7610 | 0.7585 | 0.7559 | 0.7534 |
| **15** | 0.7509 | 0.7484 | 0.7459 | 0.7434 | 0.7410 | 0.7385 | 0.7361 | 0.7336 | 0.7312 | 0.7288 |
| **16** | 0.7264 | 0.7240 | 0.7217 | 0.7193 | 0.7169 | 0.7146 | 0.7123 | 0.7100 | 0.7077 | 0.7054 |
| **17** | 0.7031 | 0.7008 | 0.6986 | 0.6963 | 0.6941 | 0.6918 | 0.6896 | 0.6874 | 0.6852 | 0.6830 |
| **18** | 0.6809 | 0.6787 | 0.6765 | 0.6744 | 0.6723 | 0.6701 | 0.6680 | 0.6659 | 0.6638 | 0.6617 |
| **19** | 0.6597 | 0.6576 | 0.6555 | 0.6535 | 0.6515 | 0.6494 | 0.6474 | 0.6454 | 0.6434 | 0.6414 |
| **20** | 0.6394 | 0.6375 | 0.6355 | 0.6336 | 0.6316 | 0.6297 | 0.6278 | 0.6258 | 0.6239 | 0.6220 |
| **21** | 0.6201 | 0.6183 | 0.6164 | 0.6145 | 0.6127 | 0.6108 | 0.6090 | 0.6072 | 0.6053 | 0.6035 |
| **22** | 0.6017 | 0.5999 | 0.5981 | 0.5964 | 0.5946 | 0.5928 | 0.5911 | 0.5893 | 0.5876 | 0.5859 |
| **23** | 0.5841 | 0.5824 | 0.5807 | 0.5790 | 0.5773 | 0.5756 | 0.5740 | 0.5723 | 0.5706 | 0.5690 |
| **24** | 0.5673 | 0.5657 | 0.5641 | 0.5624 | 0.5608 | 0.5592 | 0.5576 | 0.5560 | 0.5544 | 0.5528 |
| **25** | 0.5513 | 0.5497 | 0.5481 | 0.5466 | 0.5450 | 0.5435 | 0.5420 | 0.5404 | 0.5389 | 0.5374 |
| **26** | 0.5359 | 0.5344 | 0.5329 | 0.5314 | 0.5299 | 0.5285 | 0.5270 | 0.5255 | 0.5241 | 0.5226 |
| **27** | 0.5212 | 0.5198 | 0.5183 | 0.5169 | 0.5155 | 0.5141 | 0.5127 | 0.5113 | 0.5099 | 0.5085 |
| **28** | 0.5071 | 0.5058 | 0.5044 | 0.5030 | 0.5017 | 0.5003 | 0.4990 | 0.4976 | 0.4963 | 0.4950 |
| **29** | 0.4937 | 0.4924 | 0.4910 | 0.4897 | 0.4884 | 0.4872 | 0.4859 | 0.4846 | 0.4833 | 0.4820 |
| **30** | 0.4808 | 0.4795 | 0.4783 | 0.4770 | 0.4758 | 0.4745 | 0.4733 | 0.4721 | 0.4708 | 0.4696 |
| **31** | 0.4684 | 0.4672 | 0.4660 | 0.4648 | 0.4636 | 0.4624 | 0.4613 | 0.4601 | 0.4589 | 0.4577 |
| **32** | 0.4566 | 0.4554 | 0.4543 | 0.4531 | 0.4520 | 0.4508 | 0.4497 | 0.4486 | 0.4475 | 0.4463 |
| **33** | 0.4452 | 0.4441 | 0.4430 | 0.4419 | 0.4408 | 0.4397 | 0.4386 | 0.4376 | 0.4365 | 0.4354 |
| **34** | 0.4343 | 0.4333 | 0.4322 | 0.4312 | 0.4301 | 0.4291 | 0.4280 | 0.4270 | 0.4259 | 0.4249 |
| **35** | 0.4239 | 0.4229 | 0.4219 | 0.4208 | 0.4198 | 0.4188 | 0.4178 | 0.4168 | 0.4158 | 0.4149 |
| **36** | 0.4139 | 0.4129 | 0.4119 | 0.4109 | 0.4100 | 0.4090 | 0.4081 | 0.4071 | 0.4061 | 0.4052 |
| **37** | 0.4042 | 0.4033 | 0.4024 | 0.4014 | 0.4005 | 0.3996 | 0.3987 | 0.3977 | 0.3968 | 0.3959 |
| **38** | 0.3950 | 0.3941 | 0.3932 | 0.3923 | 0.3914 | 0.3905 | 0.3896 | 0.3888 | 0.3879 | 0.3870 |
| **39** | 0.3861 | 0.3853 | 0.3844 | 0.3835 | 0.3827 | 0.3818 | 0.3810 | 0.3801 | 0.3793 | 0.3784 |
| **40** | 0.3776 | 0.3768 | 0.3759 | 0.3751 | 0.3743 | 0.3735 | 0.3726 | 0.3718 | 0.3710 | 0.3702 |

*Source*: Based on Weiss (1971).

**Table 5.12** Bunsen Coefficient ($\beta$) for Nitrous Oxide as a Function of Temperature and Salinity, 0–40 g/kg (L real gas/(L atm))

| Temperature (°C) | Salinity (g/kg) | | | | | | | | |
|---|---|---|---|---|---|---|---|---|---|
| | **0.0** | **5.0** | **10.0** | **15.0** | **20.0** | **25.0** | **30.0** | **35.0** | **40.0** |
| **0** | 1.3197 | 1.2804 | 1.2424 | 1.2054 | 1.1696 | 1.1348 | 1.1010 | 1.0683 | 1.0365 |
| **1** | 1.2658 | 1.2284 | 1.1922 | 1.1570 | 1.1229 | 1.0898 | 1.0577 | 1.0265 | 0.9962 |
| **2** | 1.2148 | 1.1793 | 1.1448 | 1.1113 | 1.0788 | 1.0472 | 1.0166 | 0.9869 | 0.9580 |
| **3** | 1.1666 | 1.1327 | 1.0999 | 1.0680 | 1.0370 | 1.0069 | 0.9777 | 0.9493 | 0.9218 |
| **4** | 1.1210 | 1.0887 | 1.0574 | 1.0270 | 0.9974 | 0.9687 | 0.9408 | 0.9137 | 0.8875 |
| **5** | 1.0778 | 1.0470 | 1.0171 | 0.9881 | 0.9599 | 0.9325 | 0.9059 | 0.8800 | 0.8549 |
| **6** | 1.0369 | 1.0075 | 0.9790 | 0.9513 | 0.9243 | 0.8981 | 0.8727 | 0.8480 | 0.8239 |
| **7** | 0.9981 | 0.9701 | 0.9428 | 0.9163 | 0.8906 | 0.8655 | 0.8412 | 0.8176 | 0.7946 |
| **8** | 0.9614 | 0.9346 | 0.9085 | 0.8831 | 0.8585 | 0.8346 | 0.8113 | 0.7887 | 0.7667 |
| **9** | 0.9265 | 0.9008 | 0.8759 | 0.8517 | 0.8281 | 0.8051 | 0.7829 | 0.7612 | 0.7401 |
| **10** | 0.8934 | 0.8688 | 0.8450 | 0.8217 | 0.7992 | 0.7772 | 0.7558 | 0.7351 | 0.7149 |
| **11** | 0.8619 | 0.8384 | 0.8155 | 0.7933 | 0.7716 | 0.7506 | 0.7301 | 0.7102 | 0.6908 |
| **12** | 0.8321 | 0.8095 | 0.7876 | 0.7662 | 0.7455 | 0.7253 | 0.7056 | 0.6865 | 0.6679 |
| **13** | 0.8036 | 0.7820 | 0.7610 | 0.7405 | 0.7206 | 0.7012 | 0.6823 | 0.6640 | 0.6461 |
| **14** | 0.7766 | 0.7559 | 0.7357 | 0.7160 | 0.6969 | 0.6783 | 0.6601 | 0.6425 | 0.6253 |
| **15** | 0.7509 | 0.7310 | 0.7116 | 0.6927 | 0.6743 | 0.6564 | 0.6390 | 0.6220 | 0.6055 |
| **16** | 0.7264 | 0.7073 | 0.6886 | 0.6705 | 0.6528 | 0.6356 | 0.6188 | 0.6025 | 0.5866 |
| **17** | 0.7031 | 0.6847 | 0.6667 | 0.6493 | 0.6323 | 0.6157 | 0.5996 | 0.5839 | 0.5686 |
| **18** | 0.6809 | 0.6631 | 0.6459 | 0.6291 | 0.6127 | 0.5967 | 0.5812 | 0.5661 | 0.5513 |
| **19** | 0.6597 | 0.6426 | 0.6260 | 0.6098 | 0.5940 | 0.5786 | 0.5636 | 0.5491 | 0.5349 |
| **20** | 0.6394 | 0.6230 | 0.6070 | 0.5914 | 0.5761 | 0.5613 | 0.5469 | 0.5328 | 0.5191 |
| **21** | 0.6201 | 0.6043 | 0.5888 | 0.5738 | 0.5591 | 0.5448 | 0.5309 | 0.5173 | 0.5041 |
| **22** | 0.6017 | 0.5864 | 0.5715 | 0.5570 | 0.5428 | 0.5290 | 0.5156 | 0.5025 | 0.4897 |
| **23** | 0.5841 | 0.5694 | 0.5550 | 0.5409 | 0.5273 | 0.5139 | 0.5009 | 0.4883 | 0.4759 |
| **24** | 0.5673 | 0.5531 | 0.5391 | 0.5256 | 0.5124 | 0.4995 | 0.4869 | 0.4747 | 0.4627 |
| **25** | 0.5513 | 0.5375 | 0.5240 | 0.5109 | 0.4981 | 0.4856 | 0.4735 | 0.4616 | 0.4501 |
| **26** | 0.5359 | 0.5226 | 0.5095 | 0.4969 | 0.4845 | 0.4724 | 0.4606 | 0.4492 | 0.4380 |
| **27** | 0.5212 | 0.5083 | 0.4957 | 0.4834 | 0.4714 | 0.4597 | 0.4483 | 0.4372 | 0.4264 |
| **28** | 0.5071 | 0.4946 | 0.4824 | 0.4705 | 0.4589 | 0.4476 | 0.4366 | 0.4258 | 0.4153 |
| **29** | 0.4937 | 0.4815 | 0.4697 | 0.4582 | 0.4469 | 0.4360 | 0.4252 | 0.4148 | 0.4046 |
| **30** | 0.4808 | 0.4690 | 0.4575 | 0.4464 | 0.4354 | 0.4248 | 0.4144 | 0.4043 | 0.3944 |
| **31** | 0.4684 | 0.4570 | 0.4459 | 0.4350 | 0.4244 | 0.4141 | 0.4040 | 0.3942 | 0.3846 |
| **32** | 0.4566 | 0.4455 | 0.4347 | 0.4241 | 0.4139 | 0.4038 | 0.3940 | 0.3845 | 0.3751 |
| **33** | 0.4452 | 0.4345 | 0.4240 | 0.4137 | 0.4037 | 0.3940 | 0.3844 | 0.3751 | 0.3661 |
| **34** | 0.4343 | 0.4239 | 0.4137 | 0.4037 | 0.3940 | 0.3845 | 0.3752 | 0.3662 | 0.3574 |
| **35** | 0.4239 | 0.4137 | 0.4038 | 0.3941 | 0.3846 | 0.3754 | 0.3664 | 0.3576 | 0.3490 |
| **36** | 0.4139 | 0.4040 | 0.3943 | 0.3849 | 0.3756 | 0.3667 | 0.3579 | 0.3493 | 0.3409 |
| **37** | 0.4042 | 0.3946 | 0.3852 | 0.3760 | 0.3670 | 0.3583 | 0.3497 | 0.3414 | 0.3332 |
| **38** | 0.3950 | 0.3856 | 0.3764 | 0.3675 | 0.3587 | 0.3502 | 0.3418 | 0.3337 | 0.3258 |
| **39** | 0.3861 | 0.3770 | 0.3680 | 0.3593 | 0.3507 | 0.3424 | 0.3343 | 0.3263 | 0.3186 |
| **40** | 0.3776 | 0.3687 | 0.3599 | 0.3514 | 0.3431 | 0.3349 | 0.3270 | 0.3193 | 0.3117 |

*Source*: Based on Weiss (1971).

# 6 Solubility of Gases in Brines

The major ions in estuarine and marine waters are in constant proportion: a factor that allows preparation of very accurate solubility relationships. Brines, on the other hand, have great variability in their ionic composition (Sherwood et al., 1991). Therefore, the solubility of a gas in the Dead Sea, the Great Salt Lake, or an Antarctic lake may be quite different for a given measured salinity and temperature. Most saline lakes are dominated by sodium chloride (NaCl), so solubility data for NaCl can be used when site-specific information is not available. The conversion of concentration/mass to concentration/volume units is complicated by uncertainty in the variation of density with temperature and salinity for a particular brine.

There is great interest in injecting carbon dioxide into deep saline aquifers (Portier and Rochelle, 2005) and in sequestering $CO_2$ on the ocean floor in the form of carbon dioxide hydrate (Duan et al., 2006). Therefore, the solubility relationships for carbon dioxide (Duan et al., 2006) cover a wider range of temperature and pressures than for oxygen (Sherwood et al., 1991). In addition, the relationships for carbon dioxide can be used to estimate the solubility of NaCl brines, seawater, and brines of arbitrary ionic composition. High-pressure carbon dioxide solubility information is not presented in this book; readers interested in this application should review Duan et al. (2006) and Duan and Sun (2003) or their excellent online calculators or downloadable programs (http://www.geochem-model.org/). Information on the solubility of noble gases has been developed by Smith and Kennedy (1983), but not in a manner consistent with the methods used in this book.

The solubilities of oxygen and carbon dioxide in terms of standard air solubility concentrations (μmol/kg and mg/L) and Bunsen coefficients (L real gas/(L atm)) have been computed as functions of salinity and temperature for NaCl brines based on the equations presented in Appendix A. These tabular data for oxygen and carbon dioxide may be found as follows:

Dissolved Gas Concentration in Water. DOI: 10.1016/B978-0-12-415916-7.00006-1

## Standard Air Solubility Concentrations—NaCl Brines

| Gas | $C_o^\dagger$ in μmol/kg | | $C_o^*$ in mg/L | |
|---|---|---|---|---|
| | Table | Page | Table | Page |
| Oxygen | | | | |
| 0–120 g/kg | 6.1 | 216 | 6.7 | 222 |
| 120–240 g/kg | 6.2 | 217 | 6.8 | 223 |
| Carbon dioxide—2010 | | | | |
| 0–120 g/kg | 6.3 | 218 | 6.9 | 224 |
| 120–240 g/kg | 6.4 | 219 | 6.10 | 225 |
| Carbon dioxide—2030 | | | | |
| 0–120 g/kg | 6.5 | 220 | 6.11 | 226 |
| 120–240 g/kg | 6.6 | 221 | 6.12 | 227 |

## Bunsen Coefficients—NaCl Brines

| Gas | Bunsen coefficient ($\beta$) (L real gas/(L atm)) | |
|---|---|---|
| | Table | Page |
| $O_2$ | | |
| 0–120 g/kg | 6.13 | 228 |
| 120–240 g/kg | 6.14 | 229 |
| $CO_2$ | | |
| 0–120 g/kg | 6.15 | 230 |
| 120–240 g/kg | 6.16 | 231 |

## Vapor Pressure of Water ($P_{wv}$)—NaCl Brines

The vapor pressures of NaCl brine in mmHg are presented in Table 6.17{232} for 0–120 g/kg and Table 6.18{233} for 120–240 g/kg.

## Density of Water—NaCl Brines

The densities of NaCl brine are presented in Table 6.19{234} for 0–120 g/kg and Table 6.20{235} for 120–240 g/kg.

## Example 6-1

Compare the standard air solubility concentration of oxygen in μmol/kg for marine conditions (Table 2.1) with that of a sodium chloride brine (Table 6.1) for 0°C and 35°C. Assume that the measured salinity is 30 g/kg for both waters.

### Input Data

| Temperature (°C) | Brine | Seawater |
|---|---|---|
| 0 | 363.19 | 361.74 |
| 35 | 182.84 | 182.20 |

At least for low salinities, the two solubilities are quite close.

**Table 6.1** Standard Air Saturation Concentration of Oxygen as a Function of Temperature and Salinity in μmol/kg, 0–120 g/kg (NaCl brine, 1 atm moist air)

| Temperature (°C) | Salinity (g/kg) | | | | | | | | |
|---|---|---|---|---|---|---|---|---|---|
| | **0** | **15** | **30** | **45** | **60** | **75** | **90** | **105** | **120** |
| **0** | 454.66 | 406.68 | 363.19 | 323.82 | 288.26 | 256.18 | 227.31 | 201.37 | 178.10 |
| **1** | 442.18 | 395.88 | 353.86 | 315.80 | 281.37 | 250.29 | 222.29 | 197.10 | 174.48 |
| **2** | 430.36 | 385.65 | 345.03 | 308.19 | 274.84 | 244.71 | 217.52 | 193.05 | 171.05 |
| **3** | 419.13 | 375.93 | 336.63 | 300.96 | 268.63 | 239.39 | 212.99 | 189.19 | 167.79 |
| **4** | 408.44 | 366.67 | 328.63 | 294.07 | 262.71 | 234.33 | 208.67 | 185.52 | 164.67 |
| **5** | 398.25 | 357.83 | 320.99 | 287.48 | 257.05 | 229.48 | 204.53 | 182.00 | 161.69 |
| **6** | 388.50 | 349.37 | 313.67 | 281.17 | 251.63 | 224.83 | 200.56 | 178.63 | 158.83 |
| **7** | 379.15 | 341.26 | 306.66 | 275.12 | 246.43 | 220.37 | 196.75 | 175.38 | 156.08 |
| **8** | 370.19 | 333.47 | 299.91 | 269.29 | 241.41 | 216.07 | 193.08 | 172.25 | 153.43 |
| **9** | 361.56 | 325.97 | 293.42 | 263.68 | 236.58 | 211.93 | 189.53 | 169.23 | 150.86 |
| **10** | 353.26 | 318.75 | 287.15 | 258.27 | 231.92 | 207.92 | 186.10 | 166.31 | 148.38 |
| **11** | 345.25 | 311.78 | 281.10 | 253.04 | 227.41 | 204.04 | 182.78 | 163.48 | 145.97 |
| **12** | 337.51 | 305.04 | 275.25 | 247.97 | 223.04 | 200.28 | 179.56 | 160.73 | 143.63 |
| **13** | 330.04 | 298.53 | 269.59 | 243.07 | 218.80 | 196.64 | 176.43 | 158.05 | 141.36 |
| **14** | 322.81 | 292.22 | 264.10 | 238.31 | 214.69 | 193.09 | 173.39 | 155.45 | 139.14 |
| **15** | 315.82 | 286.11 | 258.79 | 233.69 | 210.69 | 189.65 | 170.44 | 152.92 | 136.99 |
| **16** | 309.04 | 280.19 | 253.63 | 229.21 | 206.82 | 186.31 | 167.56 | 150.46 | 134.88 |
| **17** | 302.48 | 274.46 | 248.63 | 224.86 | 203.05 | 183.05 | 164.76 | 148.06 | 132.83 |
| **18** | 296.13 | 268.89 | 243.77 | 220.64 | 199.38 | 179.88 | 162.03 | 145.72 | 130.83 |
| **19** | 289.97 | 263.50 | 239.06 | 216.54 | 195.82 | 176.81 | 159.38 | 143.44 | 128.89 |
| **20** | 284.01 | 258.28 | 234.49 | 212.56 | 192.37 | 173.81 | 156.80 | 141.22 | 126.99 |
| **21** | 278.24 | 253.21 | 230.07 | 208.70 | 189.01 | 170.91 | 154.29 | 139.06 | 125.14 |
| **22** | 272.66 | 248.31 | 225.78 | 204.95 | 185.75 | 168.08 | 151.85 | 136.96 | 123.34 |
| **23** | 267.27 | 243.57 | 221.62 | 201.33 | 182.60 | 165.35 | 149.48 | 134.93 | 121.59 |
| **24** | 262.06 | 238.99 | 217.61 | 197.82 | 179.54 | 162.69 | 147.19 | 132.95 | 119.89 |
| **25** | 257.03 | 234.57 | 213.73 | 194.43 | 176.59 | 160.13 | 144.97 | 131.04 | 118.25 |
| **26** | 252.19 | 230.31 | 209.99 | 191.16 | 173.74 | 157.65 | 142.83 | 129.19 | 116.66 |
| **27** | 247.53 | 226.21 | 206.39 | 188.01 | 170.99 | 155.27 | 140.76 | 127.41 | 115.13 |
| **28** | 243.05 | 222.27 | 202.93 | 184.98 | 168.35 | 152.97 | 138.77 | 125.69 | 113.66 |
| **29** | 238.76 | 218.49 | 199.62 | 182.08 | 165.82 | 150.77 | 136.87 | 124.05 | 112.25 |
| **30** | 234.66 | 214.88 | 196.44 | 179.30 | 163.40 | 148.66 | 135.04 | 122.48 | 110.90 |
| **31** | 230.75 | 211.43 | 193.42 | 176.66 | 161.09 | 146.66 | 133.31 | 120.98 | 109.62 |
| **32** | 227.03 | 208.16 | 190.54 | 174.14 | 158.90 | 144.75 | 131.66 | 119.56 | 108.40 |
| **33** | 223.51 | 205.05 | 187.82 | 171.76 | 156.82 | 142.95 | 130.11 | 118.23 | 107.26 |
| **34** | 220.18 | 202.12 | 185.25 | 169.52 | 154.87 | 141.26 | 128.65 | 116.97 | 106.19 |
| **35** | 217.05 | 199.37 | 182.84 | 167.41 | 153.04 | 139.68 | 127.29 | 115.81 | 105.20 |
| **36** | 214.13 | 196.81 | 180.60 | 165.46 | 151.34 | 138.22 | 126.03 | 114.73 | 104.28 |
| **37** | 211.41 | 194.42 | 178.52 | 163.65 | 149.78 | 136.87 | 124.87 | 113.75 | 103.45 |
| **38** | 208.90 | 192.23 | 176.61 | 161.99 | 148.35 | 135.65 | 123.83 | 112.87 | 102.71 |
| **39** | 206.62 | 190.24 | 174.88 | 160.50 | 147.07 | 134.55 | 122.90 | 112.09 | 102.06 |
| **40** | 204.55 | 188.44 | 173.32 | 159.17 | 145.93 | 133.59 | 122.09 | 111.41 | 101.51 |

*Source*: Based on Eq. 16, Sherwood et al. (1991).

**Table 6.2** Standard Air Saturation Concentration of Oxygen as a Function of Temperature and Salinity in μmol/kg, 120–240 g/kg (NaCl brine, 1 atm moist air)

| Temperature (°C) | Salinity (g/kg) | | | | | | | | |
|---|---|---|---|---|---|---|---|---|---|
| | **120** | **135** | **150** | **165** | **180** | **195** | **210** | **225** | **240** |
| **0** | 178.10 | 157.26 | 138.63 | 122.01 | 107.21 | 94.05 | 82.37 | 72.03 | 62.88 |
| **1** | 174.48 | 154.21 | 136.07 | 119.87 | 105.42 | 92.57 | 81.15 | 71.02 | 62.05 |
| **2** | 171.05 | 151.31 | 133.63 | 117.83 | 103.73 | 91.16 | 79.99 | 70.07 | 61.28 |
| **3** | 167.79 | 148.56 | 131.32 | 115.89 | 102.11 | 89.82 | 78.88 | 69.16 | 60.54 |
| **4** | 164.67 | 145.93 | 129.11 | 114.05 | 100.57 | 88.55 | 77.83 | 68.30 | 59.84 |
| **5** | 161.69 | 143.41 | 127.00 | 112.28 | 99.10 | 87.33 | 76.83 | 67.48 | 59.18 |
| **6** | 158.83 | 141.00 | 124.97 | 110.58 | 97.69 | 86.16 | 75.87 | 66.70 | 58.54 |
| **7** | 156.08 | 138.68 | 123.02 | 108.95 | 96.33 | 85.03 | 74.94 | 65.94 | 57.92 |
| **8** | 153.43 | 136.44 | 121.13 | 107.37 | 95.01 | 83.94 | 74.04 | 65.20 | 57.33 |
| **9** | 150.86 | 134.27 | 119.31 | 105.84 | 93.74 | 82.89 | 73.18 | 64.50 | 56.75 |
| **10** | 148.38 | 132.17 | 117.54 | 104.36 | 92.51 | 81.87 | 72.33 | 63.81 | 56.19 |
| **11** | 145.97 | 130.13 | 115.82 | 102.92 | 91.31 | 80.87 | 71.51 | 63.14 | 55.65 |
| **12** | 143.63 | 128.15 | 114.15 | 101.52 | 90.14 | 79.90 | 70.71 | 62.48 | 55.12 |
| **13** | 141.36 | 126.22 | 112.53 | 100.15 | 89.00 | 78.96 | 69.93 | 61.84 | 54.59 |
| **14** | 139.14 | 124.35 | 110.94 | 98.82 | 87.88 | 78.03 | 69.17 | 61.21 | 54.08 |
| **15** | 136.99 | 122.51 | 109.39 | 97.52 | 86.80 | 77.12 | 68.42 | 60.60 | 53.58 |
| **16** | 134.88 | 120.73 | 107.88 | 96.25 | 85.73 | 76.24 | 67.69 | 60.00 | 53.09 |
| **17** | 132.83 | 118.98 | 106.41 | 95.00 | 84.69 | 75.37 | 66.97 | 59.41 | 52.61 |
| **18** | 130.83 | 117.28 | 104.96 | 93.79 | 83.67 | 74.52 | 66.26 | 58.83 | 52.14 |
| **19** | 128.89 | 115.62 | 103.56 | 92.60 | 82.67 | 73.69 | 65.57 | 58.26 | 51.68 |
| **20** | 126.99 | 114.00 | 102.18 | 91.44 | 81.70 | 72.87 | 64.90 | 57.70 | 51.22 |
| **21** | 125.14 | 112.43 | 100.84 | 90.31 | 80.75 | 72.08 | 64.24 | 57.16 | 50.77 |
| **22** | 123.34 | 110.89 | 99.54 | 89.21 | 79.82 | 71.30 | 63.59 | 56.63 | 50.34 |
| **23** | 121.59 | 109.40 | 98.27 | 88.13 | 78.91 | 70.55 | 62.96 | 56.11 | 49.91 |
| **24** | 119.89 | 107.95 | 97.04 | 87.09 | 78.04 | 69.81 | 62.35 | 55.60 | 49.50 |
| **25** | 118.25 | 106.54 | 95.84 | 86.08 | 77.18 | 69.10 | 61.76 | 55.11 | 49.10 |
| **26** | 116.66 | 105.19 | 94.69 | 85.10 | 76.36 | 68.41 | 61.19 | 54.64 | 48.72 |
| **27** | 115.13 | 103.88 | 93.57 | 84.16 | 75.57 | 67.74 | 60.64 | 54.19 | 48.34 |
| **28** | 113.66 | 102.62 | 92.50 | 83.25 | 74.80 | 67.11 | 60.11 | 53.75 | 47.99 |
| **29** | 112.25 | 101.41 | 91.48 | 82.38 | 74.07 | 66.50 | 59.60 | 53.33 | 47.65 |
| **30** | 110.90 | 100.26 | 90.50 | 81.55 | 73.38 | 65.92 | 59.12 | 52.94 | 47.33 |
| **31** | 109.62 | 99.17 | 89.57 | 80.77 | 72.72 | 65.37 | 58.67 | 52.57 | 47.03 |
| **32** | 108.40 | 98.13 | 88.69 | 80.03 | 72.10 | 64.86 | 58.25 | 52.23 | 46.76 |
| **33** | 107.26 | 97.16 | 87.87 | 79.34 | 71.53 | 64.38 | 57.86 | 51.91 | 46.50 |
| **34** | 106.19 | 96.25 | 87.10 | 78.70 | 70.99 | 63.94 | 57.50 | 51.62 | 46.28 |
| **35** | 105.20 | 95.41 | 86.39 | 78.11 | 70.50 | 63.54 | 57.18 | 51.37 | 46.08 |
| **36** | 104.28 | 94.64 | 85.75 | 77.57 | 70.07 | 63.19 | 56.89 | 51.15 | 45.91 |
| **37** | 103.45 | 93.94 | 85.17 | 77.10 | 69.68 | 62.88 | 56.65 | 50.96 | 45.77 |
| **38** | 102.71 | 93.32 | 84.66 | 76.68 | 69.35 | 62.62 | 56.45 | 50.81 | 45.67 |
| **39** | 102.06 | 92.79 | 84.23 | 76.34 | 69.08 | 62.41 | 56.30 | 50.71 | 45.60 |
| **40** | 101.51 | 92.34 | 83.87 | 76.06 | 68.86 | 62.26 | 56.19 | 50.64 | 45.57 |

*Source*: Based on Eq. 16, Sherwood et al. (1991).

**Table 6.3** Standard Air Saturation Concentration of Carbon Dioxide as a Function of Temperature and Salinity in μmol/kg, 0–120 g/kg (NaCl brine, 1 atm moist air; mole fraction = 390 μatm)

| Temperature (°C) | Salinity (g/kg) | | | | | | | | |
|---|---|---|---|---|---|---|---|---|---|
| | **0** | **15** | **30** | **45** | **60** | **75** | **90** | **105** | **120** |
| **0** | 29.652 | 28.009 | 26.475 | 25.042 | 23.702 | 22.449 | 21.276 | 20.179 | 19.150 |
| **1** | 28.513 | 26.943 | 25.476 | 24.106 | 22.824 | 21.626 | 20.504 | 19.453 | 18.469 |
| **2** | 27.432 | 25.931 | 24.528 | 23.217 | 21.991 | 20.843 | 19.769 | 18.763 | 17.820 |
| **3** | 26.413 | 24.976 | 23.633 | 22.378 | 21.203 | 20.104 | 19.075 | 18.111 | 17.207 |
| **4** | 25.445 | 24.069 | 22.783 | 21.580 | 20.455 | 19.401 | 18.415 | 17.490 | 16.623 |
| **5** | 24.526 | 23.208 | 21.975 | 20.822 | 19.743 | 18.733 | 17.787 | 16.900 | 16.068 |
| **6** | 23.655 | 22.391 | 21.209 | 20.103 | 19.068 | 18.098 | 17.190 | 16.338 | 15.540 |
| **7** | 22.827 | 21.615 | 20.480 | 19.419 | 18.425 | 17.494 | 16.622 | 15.804 | 15.037 |
| **8** | 22.041 | 20.877 | 19.788 | 18.768 | 17.814 | 16.920 | 16.081 | 15.295 | 14.558 |
| **9** | 21.293 | 20.175 | 19.129 | 18.149 | 17.232 | 16.372 | 15.566 | 14.810 | 14.101 |
| **10** | 20.580 | 19.505 | 18.499 | 17.558 | 16.676 | 15.849 | 15.074 | 14.346 | 13.664 |
| **11** | 19.902 | 18.868 | 17.901 | 16.995 | 16.147 | 15.351 | 14.605 | 13.905 | 13.248 |
| **12** | 19.256 | 18.261 | 17.331 | 16.459 | 15.642 | 14.876 | 14.157 | 13.483 | 12.850 |
| **13** | 18.639 | 17.682 | 16.786 | 15.946 | 15.159 | 14.422 | 13.729 | 13.080 | 12.469 |
| **14** | 18.050 | 17.128 | 16.265 | 15.457 | 14.698 | 13.987 | 13.320 | 12.694 | 12.105 |
| **15** | 17.489 | 16.601 | 15.769 | 14.990 | 14.259 | 13.573 | 12.930 | 12.325 | 11.758 |
| **16** | 16.952 | 16.096 | 15.293 | 14.542 | 13.837 | 13.175 | 12.554 | 11.971 | 11.424 |
| **17** | 16.019 | 15.214 | 14.460 | 13.753 | 13.090 | 12.468 | 11.884 | 11.336 | 10.820 |
| **18** | 15.568 | 14.790 | 14.061 | 13.378 | 12.737 | 12.135 | 11.570 | 11.039 | 10.540 |
| **19** | 15.135 | 14.383 | 13.678 | 13.017 | 12.397 | 11.814 | 11.268 | 10.754 | 10.271 |
| **20** | 14.718 | 13.991 | 13.309 | 12.669 | 12.068 | 11.505 | 10.976 | 10.478 | 10.011 |
| **21** | 14.316 | 13.612 | 12.952 | 12.332 | 11.751 | 11.206 | 10.693 | 10.212 | 9.759 |
| **22** | 13.930 | 13.248 | 12.609 | 12.009 | 11.446 | 10.918 | 10.422 | 9.955 | 9.516 |
| **23** | 13.558 | 12.897 | 12.278 | 11.698 | 11.153 | 10.641 | 10.160 | 9.707 | 9.282 |
| **24** | 13.198 | 12.558 | 11.959 | 11.396 | 10.868 | 10.372 | 9.906 | 9.467 | 9.055 |
| **25** | 12.850 | 12.230 | 11.649 | 11.104 | 10.592 | 10.112 | 9.660 | 9.235 | 8.835 |
| **26** | 12.516 | 11.916 | 11.353 | 10.824 | 10.328 | 9.861 | 9.423 | 9.011 | 8.623 |
| **27** | 12.192 | 11.611 | 11.065 | 10.552 | 10.071 | 9.619 | 9.194 | 8.794 | 8.418 |
| **28** | 11.881 | 11.316 | 10.787 | 10.290 | 9.823 | 9.384 | 8.972 | 8.584 | 8.219 |
| **29** | 11.579 | 11.032 | 10.518 | 10.036 | 9.583 | 9.157 | 8.757 | 8.381 | 8.026 |
| **30** | 11.286 | 10.755 | 10.257 | 9.789 | 9.350 | 8.937 | 8.548 | 8.183 | 7.839 |
| **31** | 11.005 | 10.489 | 10.006 | 9.552 | 9.125 | 8.724 | 8.347 | 7.992 | 7.658 |
| **32** | 10.731 | 10.231 | 9.761 | 9.320 | 8.906 | 8.517 | 8.150 | 7.806 | 7.481 |
| **33** | 10.466 | 9.980 | 9.524 | 9.096 | 8.694 | 8.316 | 7.960 | 7.625 | 7.310 |
| **34** | 10.210 | 9.738 | 9.296 | 8.880 | 8.489 | 8.122 | 7.776 | 7.451 | 7.145 |
| **35** | 9.961 | 9.503 | 9.073 | 8.669 | 8.289 | 7.932 | 7.597 | 7.281 | 6.983 |
| **36** | 9.720 | 9.275 | 8.857 | 8.465 | 8.096 | 7.749 | 7.423 | 7.116 | 6.826 |
| **37** | 9.485 | 9.052 | 8.647 | 8.265 | 7.907 | 7.570 | 7.253 | 6.954 | 6.673 |
| **38** | 9.257 | 8.837 | 8.443 | 8.072 | 7.724 | 7.396 | 7.088 | 6.798 | 6.524 |
| **39** | 9.036 | 8.628 | 8.244 | 7.884 | 7.545 | 7.227 | 6.927 | 6.645 | 6.379 |
| **40** | 8.820 | 8.424 | 8.051 | 7.701 | 7.371 | 7.062 | 6.770 | 6.496 | 6.238 |

*Source*: Based on Eq. 1, Duan et al. (2006).

**Table 6.4** Standard Air Saturation Concentration of Carbon Dioxide as a Function of Temperature and Salinity in μmol/kg, 120–240 g/kg (NaCl brine, 1 atm moist air; mole fraction = 390 μatm)

| Temperature (°C) | Salinity (g/kg) | | | | | | | | |
|---|---|---|---|---|---|---|---|---|---|
| | 120 | 135 | 150 | 165 | 180 | 195 | 210 | 225 | 240 |
| 0 | 19.150 | 18.187 | 17.283 | 16.435 | 15.639 | 14.892 | 14.190 | 13.530 | 12.910 |
| 1 | 18.469 | 17.546 | 16.680 | 15.868 | 15.106 | 14.390 | 13.717 | 13.084 | 12.489 |
| 2 | 17.820 | 16.936 | 16.107 | 15.329 | 14.598 | 13.911 | 13.266 | 12.659 | 12.088 |
| 3 | 17.207 | 16.359 | 15.564 | 14.817 | 14.116 | 13.457 | 12.838 | 12.255 | 11.707 |
| 4 | 16.623 | 15.810 | 15.047 | 14.331 | 13.657 | 13.025 | 12.430 | 11.870 | 11.344 |
| 5 | 16.068 | 15.287 | 14.555 | 13.867 | 13.220 | 12.613 | 12.041 | 11.503 | 10.997 |
| 6 | 15.540 | 14.790 | 14.086 | 13.425 | 12.804 | 12.220 | 11.670 | 11.153 | 10.667 |
| 7 | 15.037 | 14.316 | 13.640 | 13.005 | 12.407 | 11.846 | 11.317 | 10.820 | 10.351 |
| 8 | 14.558 | 13.865 | 13.215 | 12.603 | 12.029 | 11.488 | 10.980 | 10.501 | 10.050 |
| 9 | 14.101 | 13.435 | 12.809 | 12.220 | 11.667 | 11.147 | 10.657 | 10.196 | 9.762 |
| 10 | 13.664 | 13.023 | 12.420 | 11.854 | 11.321 | 10.820 | 10.348 | 9.904 | 9.485 |
| 11 | 13.248 | 12.630 | 12.050 | 11.504 | 10.991 | 10.508 | 10.054 | 9.625 | 9.222 |
| 12 | 12.850 | 12.255 | 11.696 | 11.170 | 10.675 | 10.210 | 9.771 | 9.358 | 8.969 |
| 13 | 12.469 | 11.896 | 11.357 | 10.850 | 10.373 | 9.924 | 9.501 | 9.102 | 8.727 |
| 14 | 12.105 | 11.552 | 11.032 | 10.543 | 10.083 | 9.650 | 9.241 | 8.857 | 8.494 |
| 15 | 11.758 | 11.224 | 10.722 | 10.250 | 9.806 | 9.388 | 8.993 | 8.622 | 8.272 |
| 16 | 11.424 | 10.908 | 10.424 | 9.968 | 9.539 | 9.135 | 8.754 | 8.396 | 8.057 |
| 17 | 10.820 | 10.336 | 9.880 | 9.451 | 9.047 | 8.666 | 8.308 | 7.970 | 7.651 |
| 18 | 10.540 | 10.071 | 9.630 | 9.215 | 8.824 | 8.455 | 8.108 | 7.780 | 7.472 |
| 19 | 10.271 | 9.817 | 9.390 | 8.987 | 8.609 | 8.252 | 7.915 | 7.598 | 7.299 |
| 20 | 10.011 | 9.571 | 9.157 | 8.767 | 8.400 | 8.054 | 7.728 | 7.421 | 7.131 |
| 21 | 9.759 | 9.333 | 8.932 | 8.554 | 8.199 | 7.863 | 7.547 | 7.249 | 6.968 |
| 22 | 9.516 | 9.103 | 8.715 | 8.349 | 8.004 | 7.679 | 7.373 | 7.084 | 6.811 |
| 23 | 9.282 | 8.882 | 8.505 | 8.150 | 7.816 | 7.501 | 7.204 | 6.924 | 6.659 |
| 24 | 9.055 | 8.667 | 8.302 | 7.958 | 7.634 | 7.328 | 7.040 | 6.768 | 6.511 |
| 25 | 8.835 | 8.459 | 8.105 | 7.771 | 7.456 | 7.160 | 6.880 | 6.617 | 6.367 |
| 26 | 8.623 | 8.258 | 7.915 | 7.591 | 7.286 | 6.998 | 6.727 | 6.471 | 6.229 |
| 27 | 8.418 | 8.064 | 7.730 | 7.416 | 7.120 | 6.841 | 6.577 | 6.329 | 6.094 |
| 28 | 8.219 | 7.875 | 7.552 | 7.247 | 6.959 | 6.688 | 6.432 | 6.191 | 5.963 |
| 29 | 8.026 | 7.693 | 7.378 | 7.082 | 6.803 | 6.540 | 6.292 | 6.057 | 5.836 |
| 30 | 7.839 | 7.515 | 7.210 | 6.922 | 6.651 | 6.395 | 6.154 | 5.927 | 5.712 |
| 31 | 7.658 | 7.343 | 7.047 | 6.768 | 6.504 | 6.256 | 6.022 | 5.800 | 5.591 |
| 32 | 7.481 | 7.176 | 6.888 | 6.617 | 6.361 | 6.120 | 5.892 | 5.677 | 5.474 |
| 33 | 7.310 | 7.013 | 6.734 | 6.470 | 6.222 | 5.987 | 5.766 | 5.557 | 5.360 |
| 34 | 7.145 | 6.856 | 6.585 | 6.328 | 6.087 | 5.859 | 5.644 | 5.441 | 5.249 |
| 35 | 6.983 | 6.703 | 6.439 | 6.190 | 5.955 | 5.733 | 5.524 | 5.327 | 5.140 |
| 36 | 6.826 | 6.554 | 6.297 | 6.055 | 5.827 | 5.611 | 5.408 | 5.216 | 5.035 |
| 37 | 6.673 | 6.408 | 6.159 | 5.923 | 5.701 | 5.492 | 5.294 | 5.107 | 4.931 |
| 38 | 6.524 | 6.267 | 6.024 | 5.795 | 5.579 | 5.376 | 5.183 | 5.002 | 4.830 |
| 39 | 6.379 | 6.129 | 5.893 | 5.670 | 5.460 | 5.262 | 5.075 | 4.899 | 4.732 |
| 40 | 6.238 | 5.994 | 5.765 | 5.548 | 5.344 | 5.151 | 4.969 | 4.798 | 4.635 |

*Source*: Based on Eq. 1, Duan et al. (2006).

**Table 6.5** Standard Air Saturation Concentration of Carbon Dioxide as a Function of Temperature and Salinity in μmol/kg, 0–120 g/kg (NaCl brine, 1 atm moist air; mole fraction = 440 μatm)

| Temperature (°C) | Salinity (g/kg) | | | | | | | | |
|---|---|---|---|---|---|---|---|---|---|
| | 0 | 15 | 30 | 45 | 60 | 75 | 90 | 105 | 120 |
| 0 | 33.454 | 31.600 | 29.869 | 28.252 | 26.741 | 25.327 | 24.004 | 22.766 | 21.605 |
| 1 | 32.168 | 30.397 | 28.742 | 27.196 | 25.751 | 24.398 | 23.132 | 21.947 | 20.836 |
| 2 | 30.949 | 29.255 | 27.673 | 26.194 | 24.810 | 23.516 | 22.304 | 21.169 | 20.105 |
| 3 | 29.799 | 28.178 | 26.663 | 25.246 | 23.922 | 22.682 | 21.521 | 20.433 | 19.413 |
| 4 | 28.707 | 27.154 | 25.703 | 24.347 | 23.077 | 21.889 | 20.775 | 19.732 | 18.754 |
| 5 | 27.671 | 26.183 | 24.792 | 23.492 | 22.274 | 21.135 | 20.067 | 19.066 | 18.128 |
| 6 | 26.687 | 25.261 | 23.928 | 22.680 | 21.512 | 20.418 | 19.394 | 18.433 | 17.532 |
| 7 | 25.754 | 24.386 | 23.106 | 21.908 | 20.787 | 19.737 | 18.753 | 17.830 | 16.964 |
| 8 | 24.867 | 23.553 | 22.325 | 21.175 | 20.098 | 19.089 | 18.143 | 17.256 | 16.424 |
| 9 | 24.023 | 22.762 | 21.581 | 20.476 | 19.441 | 18.471 | 17.562 | 16.709 | 15.909 |
| 10 | 23.218 | 22.006 | 20.871 | 19.809 | 18.814 | 17.881 | 17.006 | 16.186 | 15.416 |
| 11 | 22.453 | 21.287 | 20.196 | 19.174 | 18.217 | 17.319 | 16.477 | 15.688 | 14.946 |
| 12 | 21.724 | 20.602 | 19.552 | 18.569 | 17.647 | 16.783 | 15.972 | 15.212 | 14.497 |
| 13 | 21.028 | 19.949 | 18.938 | 17.991 | 17.103 | 16.271 | 15.490 | 14.756 | 14.068 |
| 14 | 20.364 | 19.324 | 18.351 | 17.438 | 16.583 | 15.781 | 15.028 | 14.321 | 13.657 |
| 15 | 19.732 | 18.730 | 17.791 | 16.912 | 16.087 | 15.313 | 14.587 | 13.906 | 13.265 |
| 16 | 19.125 | 18.159 | 17.254 | 16.406 | 15.611 | 14.864 | 14.164 | 13.506 | 12.888 |
| 17 | 18.072 | 17.164 | 16.314 | 15.516 | 14.769 | 14.067 | 13.408 | 12.789 | 12.207 |
| 18 | 17.564 | 16.686 | 15.864 | 15.093 | 14.369 | 13.691 | 13.053 | 12.454 | 11.892 |
| 19 | 17.076 | 16.227 | 15.432 | 14.686 | 13.986 | 13.329 | 12.712 | 12.133 | 11.588 |
| 20 | 16.605 | 15.784 | 15.015 | 14.293 | 13.616 | 12.980 | 12.383 | 11.822 | 11.294 |
| 21 | 16.152 | 15.357 | 14.612 | 13.914 | 13.258 | 12.642 | 12.064 | 11.521 | 11.010 |
| 22 | 15.715 | 14.946 | 14.225 | 13.549 | 12.914 | 12.318 | 11.758 | 11.231 | 10.736 |
| 23 | 15.296 | 14.551 | 13.853 | 13.197 | 12.582 | 12.005 | 11.462 | 10.952 | 10.472 |
| 24 | 14.890 | 14.169 | 13.492 | 12.857 | 12.261 | 11.702 | 11.176 | 10.681 | 10.216 |
| 25 | 14.497 | 13.798 | 13.143 | 12.528 | 11.950 | 11.408 | 10.898 | 10.419 | 9.968 |
| 26 | 14.121 | 13.443 | 12.808 | 12.212 | 11.652 | 11.126 | 10.631 | 10.166 | 9.729 |
| 27 | 13.756 | 13.099 | 12.483 | 11.905 | 11.362 | 10.852 | 10.372 | 9.921 | 9.497 |
| 28 | 13.404 | 12.767 | 12.170 | 11.609 | 11.082 | 10.587 | 10.122 | 9.684 | 9.273 |
| 29 | 13.064 | 12.446 | 11.867 | 11.323 | 10.812 | 10.331 | 9.880 | 9.455 | 9.055 |
| 30 | 12.733 | 12.134 | 11.572 | 11.044 | 10.548 | 10.082 | 9.644 | 9.232 | 8.844 |
| 31 | 12.416 | 11.834 | 11.289 | 10.776 | 10.295 | 9.842 | 9.417 | 9.017 | 8.640 |
| 32 | 12.107 | 11.542 | 11.013 | 10.515 | 10.048 | 9.609 | 9.195 | 8.807 | 8.441 |
| 33 | 11.807 | 11.260 | 10.745 | 10.262 | 9.809 | 9.382 | 8.981 | 8.603 | 8.247 |
| 34 | 11.519 | 10.987 | 10.487 | 10.018 | 9.577 | 9.163 | 8.773 | 8.406 | 8.061 |
| 35 | 11.238 | 10.721 | 10.236 | 9.780 | 9.352 | 8.949 | 8.570 | 8.214 | 7.878 |
| 36 | 10.966 | 10.464 | 9.993 | 9.550 | 9.134 | 8.743 | 8.374 | 8.028 | 7.702 |
| 37 | 10.700 | 10.213 | 9.755 | 9.325 | 8.921 | 8.540 | 8.182 | 7.846 | 7.528 |
| 38 | 10.444 | 9.970 | 9.525 | 9.107 | 8.714 | 8.344 | 7.996 | 7.669 | 7.361 |
| 39 | 10.194 | 9.734 | 9.301 | 8.895 | 8.513 | 8.153 | 7.815 | 7.497 | 7.197 |
| 40 | 9.951 | 9.504 | 9.083 | 8.688 | 8.317 | 7.967 | 7.638 | 7.329 | 7.037 |

*Source*: Based on Eq. 1, Duan et al. (2006).

**Table 6.6** Standard Air Saturation Concentration of Carbon Dioxide as a Function of Temperature and Salinity in μmol/kg, 120–240 g/kg (NaCl brine, 1 atm moist air; mole fraction = 440 μatm)

| Temperature (°C) | Salinity (g/kg) | | | | | | | | |
|---|---|---|---|---|---|---|---|---|---|
| | 120 | 135 | 150 | 165 | 180 | 195 | 210 | 225 | 240 |
| 0 | 21.605 | 20.518 | 19.499 | 18.542 | 17.644 | 16.802 | 16.010 | 15.265 | 14.565 |
| 1 | 20.836 | 19.795 | 18.819 | 17.903 | 17.042 | 16.235 | 15.475 | 14.762 | 14.090 |
| 2 | 20.105 | 19.108 | 18.172 | 17.294 | 16.469 | 15.694 | 14.966 | 14.282 | 13.637 |
| 3 | 19.413 | 18.457 | 17.559 | 16.717 | 15.926 | 15.183 | 14.484 | 13.826 | 13.208 |
| 4 | 18.754 | 17.837 | 16.976 | 16.168 | 15.408 | 14.695 | 14.024 | 13.392 | 12.798 |
| 5 | 18.128 | 17.247 | 16.421 | 15.645 | 14.915 | 14.230 | 13.585 | 12.978 | 12.407 |
| 6 | 17.532 | 16.686 | 15.892 | 15.146 | 14.446 | 13.787 | 13.167 | 12.583 | 12.034 |
| 7 | 16.964 | 16.152 | 15.389 | 14.672 | 13.998 | 13.364 | 12.768 | 12.207 | 11.678 |
| 8 | 16.424 | 15.643 | 14.909 | 14.219 | 13.571 | 12.961 | 12.387 | 11.847 | 11.338 |
| 9 | 15.909 | 15.157 | 14.451 | 13.787 | 13.163 | 12.576 | 12.024 | 11.503 | 11.013 |
| 10 | 15.416 | 14.692 | 14.012 | 13.374 | 12.773 | 12.207 | 11.675 | 11.174 | 10.701 |
| 11 | 14.946 | 14.249 | 13.595 | 12.979 | 12.400 | 11.855 | 11.342 | 10.859 | 10.404 |
| 12 | 14.497 | 13.826 | 13.195 | 12.602 | 12.044 | 11.519 | 11.024 | 10.558 | 10.119 |
| 13 | 14.068 | 13.421 | 12.813 | 12.241 | 11.703 | 11.196 | 10.719 | 10.269 | 9.845 |
| 14 | 13.657 | 13.033 | 12.447 | 11.895 | 11.376 | 10.887 | 10.426 | 9.992 | 9.583 |
| 15 | 13.265 | 12.663 | 12.097 | 11.564 | 11.063 | 10.591 | 10.146 | 9.727 | 9.332 |
| 16 | 12.888 | 12.307 | 11.761 | 11.246 | 10.762 | 10.306 | 9.877 | 9.472 | 9.090 |
| 17 | 12.207 | 11.661 | 11.146 | 10.662 | 10.207 | 9.777 | 9.373 | 8.992 | 8.632 |
| 18 | 11.892 | 11.362 | 10.865 | 10.396 | 9.955 | 9.539 | 9.147 | 8.778 | 8.430 |
| 19 | 11.588 | 11.075 | 10.593 | 10.140 | 9.712 | 9.310 | 8.930 | 8.572 | 8.234 |
| 20 | 11.294 | 10.798 | 10.331 | 9.891 | 9.477 | 9.087 | 8.719 | 8.372 | 8.045 |
| 21 | 11.010 | 10.529 | 10.077 | 9.651 | 9.250 | 8.872 | 8.515 | 8.179 | 7.862 |
| 22 | 10.736 | 10.270 | 9.832 | 9.419 | 9.030 | 8.664 | 8.318 | 7.992 | 7.684 |
| 23 | 10.472 | 10.021 | 9.596 | 9.195 | 8.818 | 8.463 | 8.128 | 7.811 | 7.513 |
| 24 | 10.216 | 9.778 | 9.366 | 8.978 | 8.612 | 8.268 | 7.942 | 7.636 | 7.346 |
| 25 | 9.968 | 9.543 | 9.144 | 8.767 | 8.412 | 8.078 | 7.762 | 7.465 | 7.184 |
| 26 | 9.729 | 9.317 | 8.929 | 8.564 | 8.220 | 7.895 | 7.589 | 7.300 | 7.027 |
| 27 | 9.497 | 9.097 | 8.721 | 8.367 | 8.033 | 7.718 | 7.420 | 7.140 | 6.875 |
| 28 | 9.273 | 8.885 | 8.520 | 8.176 | 7.851 | 7.545 | 7.257 | 6.985 | 6.727 |
| 29 | 9.055 | 8.679 | 8.324 | 7.990 | 7.675 | 7.378 | 7.098 | 6.834 | 6.584 |
| 30 | 8.844 | 8.478 | 8.134 | 7.810 | 7.504 | 7.215 | 6.943 | 6.686 | 6.444 |
| 31 | 8.640 | 8.285 | 7.950 | 7.635 | 7.338 | 7.058 | 6.794 | 6.544 | 6.308 |
| 32 | 8.441 | 8.096 | 7.771 | 7.465 | 7.177 | 6.904 | 6.647 | 6.405 | 6.176 |
| 33 | 8.247 | 7.913 | 7.597 | 7.300 | 7.019 | 6.755 | 6.505 | 6.269 | 6.047 |
| 34 | 8.061 | 7.735 | 7.429 | 7.140 | 6.867 | 6.610 | 6.367 | 6.138 | 5.922 |
| 35 | 7.878 | 7.562 | 7.264 | 6.983 | 6.718 | 6.468 | 6.232 | 6.010 | 5.799 |
| 36 | 7.702 | 7.394 | 7.104 | 6.831 | 6.574 | 6.331 | 6.101 | 5.885 | 5.680 |
| 37 | 7.528 | 7.230 | 6.948 | 6.683 | 6.432 | 6.196 | 5.973 | 5.762 | 5.563 |
| 38 | 7.361 | 7.070 | 6.796 | 6.538 | 6.295 | 6.065 | 5.848 | 5.643 | 5.450 |
| 39 | 7.197 | 6.915 | 6.648 | 6.397 | 6.160 | 5.937 | 5.726 | 5.527 | 5.339 |
| 40 | 7.037 | 6.763 | 6.504 | 6.259 | 6.029 | 5.812 | 5.606 | 5.413 | 5.230 |

*Source*: Based on Eq. 1, Duan et al. (2006).

**Table 6.7** Standard Air Saturation Concentration of Oxygen as a Function of Temperature and Salinity in mg/L, 0–120 g/kg (NaCl brine, 1 atm moist air)

| Temperature (°C) | Salinity (g/kg) | | | | | | | | |
|---|---|---|---|---|---|---|---|---|---|
| | **0** | **15** | **30** | **45** | **60** | **75** | **90** | **105** | **120** |
| **0** | 14.545 | 13.160 | 11.886 | 10.717 | 9.646 | 8.668 | 7.775 | 6.963 | 6.225 |
| **1** | 14.147 | 12.811 | 11.581 | 10.450 | 9.415 | 8.467 | 7.602 | 6.814 | 6.097 |
| **2** | 13.769 | 12.479 | 11.291 | 10.198 | 9.195 | 8.277 | 7.438 | 6.673 | 5.976 |
| **3** | 13.410 | 12.164 | 11.015 | 9.957 | 8.986 | 8.096 | 7.281 | 6.538 | 5.861 |
| **4** | 13.069 | 11.864 | 10.753 | 9.728 | 8.787 | 7.923 | 7.132 | 6.409 | 5.750 |
| **5** | 12.742 | 11.578 | 10.502 | 9.509 | 8.596 | 7.758 | 6.989 | 6.286 | 5.645 |
| **6** | 12.430 | 11.303 | 10.261 | 9.299 | 8.413 | 7.599 | 6.852 | 6.168 | 5.543 |
| **7** | 12.131 | 11.040 | 10.030 | 9.098 | 8.238 | 7.446 | 6.720 | 6.055 | 5.446 |
| **8** | 11.843 | 10.787 | 9.809 | 8.904 | 8.069 | 7.300 | 6.593 | 5.945 | 5.352 |
| **9** | 11.567 | 10.544 | 9.595 | 8.717 | 7.905 | 7.158 | 6.470 | 5.839 | 5.261 |
| **10** | 11.300 | 10.309 | 9.389 | 8.536 | 7.748 | 7.021 | 6.352 | 5.737 | 5.173 |
| **11** | 11.043 | 10.082 | 9.189 | 8.361 | 7.595 | 6.888 | 6.237 | 5.637 | 5.087 |
| **12** | 10.795 | 9.863 | 8.996 | 8.192 | 7.448 | 6.760 | 6.125 | 5.541 | 5.004 |
| **13** | 10.554 | 9.651 | 8.810 | 8.029 | 7.304 | 6.635 | 6.016 | 5.447 | 4.923 |
| **14** | 10.322 | 9.445 | 8.629 | 7.870 | 7.165 | 6.513 | 5.911 | 5.356 | 4.844 |
| **15** | 10.097 | 9.246 | 8.453 | 7.715 | 7.030 | 6.395 | 5.808 | 5.267 | 4.768 |
| **16** | 9.879 | 9.053 | 8.283 | 7.566 | 6.899 | 6.281 | 5.709 | 5.180 | 4.693 |
| **17** | 9.667 | 8.866 | 8.118 | 7.420 | 6.771 | 6.169 | 5.611 | 5.096 | 4.620 |
| **18** | 9.463 | 8.685 | 7.957 | 7.279 | 6.647 | 6.061 | 5.517 | 5.013 | 4.549 |
| **19** | 9.264 | 8.508 | 7.802 | 7.142 | 6.527 | 5.955 | 5.425 | 4.933 | 4.479 |
| **20** | 9.072 | 8.338 | 7.651 | 7.008 | 6.410 | 5.852 | 5.335 | 4.855 | 4.412 |
| **21** | 8.886 | 8.172 | 7.504 | 6.879 | 6.296 | 5.753 | 5.248 | 4.779 | 4.346 |
| **22** | 8.706 | 8.012 | 7.362 | 6.754 | 6.185 | 5.656 | 5.163 | 4.705 | 4.282 |
| **23** | 8.531 | 7.857 | 7.225 | 6.632 | 6.078 | 5.562 | 5.081 | 4.634 | 4.219 |
| **24** | 8.363 | 7.707 | 7.092 | 6.515 | 5.975 | 5.471 | 5.001 | 4.564 | 4.159 |
| **25** | 8.200 | 7.563 | 6.963 | 6.401 | 5.874 | 5.382 | 4.924 | 4.497 | 4.100 |
| **26** | 8.044 | 7.423 | 6.839 | 6.291 | 5.778 | 5.297 | 4.849 | 4.432 | 4.044 |
| **27** | 7.893 | 7.289 | 6.720 | 6.186 | 5.684 | 5.215 | 4.777 | 4.369 | 3.989 |
| **28** | 7.748 | 7.160 | 6.605 | 6.084 | 5.595 | 5.136 | 4.708 | 4.308 | 3.936 |
| **29** | 7.609 | 7.036 | 6.495 | 5.986 | 5.508 | 5.060 | 4.641 | 4.250 | 3.886 |
| **30** | 7.476 | 6.917 | 6.390 | 5.893 | 5.426 | 4.988 | 4.578 | 4.195 | 3.838 |
| **31** | 7.349 | 6.804 | 6.289 | 5.804 | 5.347 | 4.919 | 4.517 | 4.142 | 3.792 |
| **32** | 7.229 | 6.697 | 6.194 | 5.719 | 5.273 | 4.853 | 4.460 | 4.092 | 3.748 |
| **33** | 7.114 | 6.594 | 6.103 | 5.639 | 5.202 | 4.791 | 4.405 | 4.044 | 3.707 |
| **34** | 7.006 | 6.498 | 6.017 | 5.563 | 5.135 | 4.732 | 4.354 | 4.000 | 3.669 |
| **35** | 6.904 | 6.407 | 5.937 | 5.492 | 5.073 | 4.678 | 4.306 | 3.958 | 3.633 |
| **36** | 6.808 | 6.322 | 5.862 | 5.426 | 5.014 | 4.627 | 4.262 | 3.920 | 3.600 |
| **37** | 6.720 | 6.244 | 5.792 | 5.365 | 4.961 | 4.580 | 4.221 | 3.885 | 3.570 |
| **38** | 6.638 | 6.171 | 5.728 | 5.308 | 4.911 | 4.537 | 4.184 | 3.853 | 3.543 |
| **39** | 6.562 | 6.105 | 5.670 | 5.257 | 4.867 | 4.498 | 4.151 | 3.825 | 3.519 |
| **40** | 6.494 | 6.045 | 5.617 | 5.212 | 4.827 | 4.464 | 4.122 | 3.800 | 3.498 |

*Source*: Based on Eq. 16, Sherwood et al. (1991).

**Table 6.8** Standard Air Saturation Concentration of Oxygen as a Function of Temperature and Salinity in mg/L, 120–240 g/kg (NaCl brine, 1 atm moist air)

| Temperature (°C) | Salinity (g/kg) | | | | | | | | |
|---|---|---|---|---|---|---|---|---|---|
| | 120 | 135 | 150 | 165 | 180 | 195 | 210 | 225 | 240 |
| 0 | 6.225 | 5.556 | 4.951 | 4.405 | 3.912 | 3.469 | 3.070 | 2.714 | 2.394 |
| 1 | 6.097 | 5.447 | 4.858 | 4.326 | 3.845 | 3.413 | 3.024 | 2.675 | 2.362 |
| 2 | 5.976 | 5.344 | 4.770 | 4.251 | 3.782 | 3.360 | 2.979 | 2.638 | 2.332 |
| 3 | 5.861 | 5.245 | 4.686 | 4.180 | 3.722 | 3.309 | 2.937 | 2.603 | 2.303 |
| 4 | 5.750 | 5.151 | 4.606 | 4.112 | 3.665 | 3.261 | 2.897 | 2.570 | 2.276 |
| 5 | 5.645 | 5.060 | 4.529 | 4.047 | 3.610 | 3.215 | 2.859 | 2.538 | 2.249 |
| 6 | 5.543 | 4.974 | 4.455 | 3.984 | 3.558 | 3.171 | 2.822 | 2.507 | 2.224 |
| 7 | 5.446 | 4.890 | 4.384 | 3.924 | 3.507 | 3.129 | 2.787 | 2.478 | 2.200 |
| 8 | 5.352 | 4.810 | 4.316 | 3.866 | 3.458 | 3.087 | 2.752 | 2.449 | 2.176 |
| 9 | 5.261 | 4.732 | 4.249 | 3.810 | 3.410 | 3.047 | 2.719 | 2.422 | 2.154 |
| 10 | 5.173 | 4.657 | 4.185 | 3.755 | 3.364 | 3.009 | 2.687 | 2.395 | 2.132 |
| 11 | 5.087 | 4.583 | 4.123 | 3.702 | 3.319 | 2.971 | 2.655 | 2.369 | 2.110 |
| 12 | 5.004 | 4.512 | 4.062 | 3.650 | 3.275 | 2.934 | 2.624 | 2.343 | 2.089 |
| 13 | 4.923 | 4.443 | 4.002 | 3.600 | 3.233 | 2.898 | 2.594 | 2.318 | 2.069 |
| 14 | 4.844 | 4.375 | 3.944 | 3.551 | 3.191 | 2.863 | 2.565 | 2.294 | 2.048 |
| 15 | 4.768 | 4.309 | 3.888 | 3.503 | 3.150 | 2.829 | 2.536 | 2.270 | 2.029 |
| 16 | 4.693 | 4.245 | 3.833 | 3.456 | 3.110 | 2.795 | 2.508 | 2.246 | 2.009 |
| 17 | 4.620 | 4.182 | 3.779 | 3.410 | 3.071 | 2.762 | 2.480 | 2.223 | 1.990 |
| 18 | 4.549 | 4.120 | 3.726 | 3.365 | 3.033 | 2.730 | 2.453 | 2.201 | 1.971 |
| 19 | 4.479 | 4.060 | 3.675 | 3.321 | 2.996 | 2.698 | 2.426 | 2.178 | 1.953 |
| 20 | 4.412 | 4.002 | 3.625 | 3.278 | 2.959 | 2.667 | 2.400 | 2.157 | 1.935 |
| 21 | 4.346 | 3.945 | 3.576 | 3.236 | 2.923 | 2.637 | 2.375 | 2.135 | 1.917 |
| 22 | 4.282 | 3.890 | 3.528 | 3.195 | 2.889 | 2.607 | 2.350 | 2.115 | 1.900 |
| 23 | 4.219 | 3.836 | 3.482 | 3.155 | 2.855 | 2.579 | 2.326 | 2.094 | 1.883 |
| 24 | 4.159 | 3.784 | 3.437 | 3.116 | 2.822 | 2.551 | 2.302 | 2.075 | 1.866 |
| 25 | 4.100 | 3.733 | 3.393 | 3.079 | 2.790 | 2.524 | 2.279 | 2.055 | 1.850 |
| 26 | 4.044 | 3.684 | 3.351 | 3.043 | 2.759 | 2.497 | 2.257 | 2.037 | 1.835 |
| 27 | 3.989 | 3.636 | 3.310 | 3.008 | 2.729 | 2.472 | 2.236 | 2.019 | 1.820 |
| 28 | 3.936 | 3.591 | 3.270 | 2.974 | 2.700 | 2.448 | 2.215 | 2.002 | 1.806 |
| 29 | 3.886 | 3.547 | 3.233 | 2.942 | 2.673 | 2.424 | 2.196 | 1.985 | 1.792 |
| 30 | 3.838 | 3.505 | 3.197 | 2.911 | 2.646 | 2.402 | 2.177 | 1.970 | 1.780 |
| 31 | 3.792 | 3.466 | 3.163 | 2.882 | 2.621 | 2.381 | 2.159 | 1.955 | 1.768 |
| 32 | 3.748 | 3.428 | 3.130 | 2.854 | 2.598 | 2.361 | 2.143 | 1.941 | 1.756 |
| 33 | 3.707 | 3.393 | 3.100 | 2.828 | 2.576 | 2.343 | 2.127 | 1.929 | 1.746 |
| 34 | 3.669 | 3.359 | 3.072 | 2.804 | 2.556 | 2.326 | 2.113 | 1.917 | 1.737 |
| 35 | 3.633 | 3.329 | 3.045 | 2.782 | 2.537 | 2.310 | 2.101 | 1.907 | 1.728 |
| 36 | 3.600 | 3.300 | 3.021 | 2.762 | 2.520 | 2.296 | 2.089 | 1.898 | 1.721 |
| 37 | 3.570 | 3.275 | 3.000 | 2.744 | 2.505 | 2.284 | 2.079 | 1.890 | 1.715 |
| 38 | 3.543 | 3.252 | 2.981 | 2.728 | 2.492 | 2.274 | 2.071 | 1.884 | 1.711 |
| 39 | 3.519 | 3.232 | 2.964 | 2.714 | 2.481 | 2.265 | 2.065 | 1.879 | 1.707 |
| 40 | 3.498 | 3.215 | 2.950 | 2.703 | 2.473 | 2.259 | 2.060 | 1.876 | 1.706 |

*Source*: Based on Eq. 16, Sherwood et al. (1991).

**Table 6.9** Standard Air Saturation Concentration of Carbon Dioxide as a Function of Temperature and Salinity in mg/L, 0–120 g/kg (NaCl brine, 1 atm moist air; mole fraction = 390 μatm)

| Temperature (°C) | Salinity (g/kg) | | | | | | | | |
|---|---|---|---|---|---|---|---|---|---|
| | **0** | **15** | **30** | **45** | **60** | **75** | **90** | **105** | **120** |
| **0** | 1.3047 | 1.2466 | 1.1917 | 1.1398 | 1.0909 | 1.0446 | 1.0009 | 0.9597 | 0.9206 |
| **1** | 1.2546 | 1.1991 | 1.1467 | 1.0971 | 1.0504 | 1.0062 | 0.9644 | 0.9250 | 0.8877 |
| **2** | 1.2071 | 1.1541 | 1.1040 | 1.0566 | 1.0119 | 0.9696 | 0.9297 | 0.8920 | 0.8563 |
| **3** | 1.1623 | 1.1115 | 1.0636 | 1.0183 | 0.9755 | 0.9351 | 0.8969 | 0.8608 | 0.8266 |
| **4** | 1.1197 | 1.0711 | 1.0252 | 0.9819 | 0.9409 | 0.9022 | 0.8656 | 0.8311 | 0.7984 |
| **5** | 1.0793 | 1.0328 | 0.9888 | 0.9473 | 0.9081 | 0.8710 | 0.8359 | 0.8028 | 0.7715 |
| **6** | 1.0409 | 0.9964 | 0.9542 | 0.9144 | 0.8768 | 0.8413 | 0.8077 | 0.7760 | 0.7459 |
| **7** | 1.0045 | 0.9617 | 0.9213 | 0.8832 | 0.8471 | 0.8130 | 0.7808 | 0.7504 | 0.7216 |
| **8** | 0.9699 | 0.9288 | 0.8901 | 0.8535 | 0.8189 | 0.7862 | 0.7553 | 0.7260 | 0.6984 |
| **9** | 0.9369 | 0.8975 | 0.8603 | 0.8252 | 0.7919 | 0.7605 | 0.7309 | 0.7028 | 0.6763 |
| **10** | 0.9054 | 0.8676 | 0.8319 | 0.7981 | 0.7662 | 0.7361 | 0.7076 | 0.6806 | 0.6551 |
| **11** | 0.8755 | 0.8392 | 0.8049 | 0.7724 | 0.7417 | 0.7128 | 0.6854 | 0.6595 | 0.6350 |
| **12** | 0.8470 | 0.8121 | 0.7791 | 0.7479 | 0.7184 | 0.6905 | 0.6642 | 0.6393 | 0.6157 |
| **13** | 0.8198 | 0.7862 | 0.7544 | 0.7244 | 0.6961 | 0.6693 | 0.6439 | 0.6199 | 0.5973 |
| **14** | 0.7938 | 0.7614 | 0.7309 | 0.7020 | 0.6747 | 0.6489 | 0.6245 | 0.6015 | 0.5796 |
| **15** | 0.7690 | 0.7379 | 0.7085 | 0.6807 | 0.6544 | 0.6295 | 0.6060 | 0.5838 | 0.5628 |
| **16** | 0.7453 | 0.7153 | 0.6869 | 0.6601 | 0.6348 | 0.6109 | 0.5883 | 0.5669 | 0.5466 |
| **17** | 0.7041 | 0.6760 | 0.6493 | 0.6242 | 0.6004 | 0.5779 | 0.5567 | 0.5366 | 0.5176 |
| **18** | 0.6842 | 0.6570 | 0.6313 | 0.6070 | 0.5840 | 0.5623 | 0.5418 | 0.5224 | 0.5040 |
| **19** | 0.6651 | 0.6388 | 0.6139 | 0.5905 | 0.5683 | 0.5473 | 0.5275 | 0.5087 | 0.4909 |
| **20** | 0.6466 | 0.6212 | 0.5972 | 0.5745 | 0.5531 | 0.5328 | 0.5136 | 0.4955 | 0.4783 |
| **21** | 0.6288 | 0.6042 | 0.5810 | 0.5591 | 0.5384 | 0.5188 | 0.5002 | 0.4827 | 0.4661 |
| **22** | 0.6117 | 0.5879 | 0.5655 | 0.5443 | 0.5242 | 0.5053 | 0.4873 | 0.4704 | 0.4544 |
| **23** | 0.5952 | 0.5722 | 0.5505 | 0.5300 | 0.5106 | 0.4923 | 0.4749 | 0.4585 | 0.4430 |
| **24** | 0.5793 | 0.5570 | 0.5360 | 0.5162 | 0.4974 | 0.4797 | 0.4629 | 0.4470 | 0.4320 |
| **25** | 0.5639 | 0.5423 | 0.5220 | 0.5028 | 0.4846 | 0.4675 | 0.4512 | 0.4359 | 0.4213 |
| **26** | 0.5491 | 0.5282 | 0.5085 | 0.4899 | 0.4724 | 0.4557 | 0.4400 | 0.4251 | 0.4111 |
| **27** | 0.5347 | 0.5145 | 0.4955 | 0.4775 | 0.4605 | 0.4444 | 0.4291 | 0.4147 | 0.4011 |
| **28** | 0.5209 | 0.5014 | 0.4829 | 0.4654 | 0.4490 | 0.4334 | 0.4186 | 0.4047 | 0.3915 |
| **29** | 0.5075 | 0.4886 | 0.4707 | 0.4538 | 0.4378 | 0.4227 | 0.4084 | 0.3949 | 0.3822 |
| **30** | 0.4946 | 0.4762 | 0.4589 | 0.4425 | 0.4270 | 0.4124 | 0.3985 | 0.3855 | 0.3731 |
| **31** | 0.4821 | 0.4643 | 0.4475 | 0.4316 | 0.4166 | 0.4024 | 0.3890 | 0.3763 | 0.3643 |
| **32** | 0.4699 | 0.4527 | 0.4364 | 0.4210 | 0.4065 | 0.3927 | 0.3797 | 0.3674 | 0.3558 |
| **33** | 0.4581 | 0.4414 | 0.4256 | 0.4107 | 0.3966 | 0.3833 | 0.3707 | 0.3588 | 0.3475 |
| **34** | 0.4468 | 0.4306 | 0.4153 | 0.4008 | 0.3871 | 0.3742 | 0.3620 | 0.3504 | 0.3395 |
| **35** | 0.4357 | 0.4200 | 0.4052 | 0.3911 | 0.3779 | 0.3653 | 0.3535 | 0.3423 | 0.3317 |
| **36** | 0.4251 | 0.4098 | 0.3954 | 0.3818 | 0.3689 | 0.3568 | 0.3453 | 0.3344 | 0.3241 |
| **37** | 0.4146 | 0.3998 | 0.3859 | 0.3726 | 0.3602 | 0.3484 | 0.3372 | 0.3267 | 0.3167 |
| **38** | 0.4045 | 0.3902 | 0.3766 | 0.3638 | 0.3517 | 0.3402 | 0.3294 | 0.3192 | 0.3095 |
| **39** | 0.3947 | 0.3808 | 0.3676 | 0.3552 | 0.3434 | 0.3323 | 0.3218 | 0.3119 | 0.3025 |
| **40** | 0.3852 | 0.3716 | 0.3589 | 0.3468 | 0.3354 | 0.3246 | 0.3144 | 0.3048 | 0.2956 |

*Source*: Based on Eq. 1, Duan et al. (2006).

**Table 6.10** Standard Air Saturation Concentration of Carbon Dioxide as a Function of Temperature and Salinity in mg/L, 120–240 g/kg (NaCl brine, 1 atm moist air; mole fraction = 390 μatm)

| Temperature (°C) | Salinity (g/kg) | | | | | | | | |
|---|---|---|---|---|---|---|---|---|---|
| | **120** | **135** | **150** | **165** | **180** | **195** | **210** | **225** | **240** |
| **0** | 0.9206 | 0.8838 | 0.8489 | 0.8160 | 0.7848 | 0.7554 | 0.7275 | 0.7011 | 0.6761 |
| **1** | 0.8877 | 0.8524 | 0.8191 | 0.7876 | 0.7578 | 0.7296 | 0.7030 | 0.6777 | 0.6539 |
| **2** | 0.8563 | 0.8226 | 0.7907 | 0.7606 | 0.7321 | 0.7051 | 0.6796 | 0.6555 | 0.6326 |
| **3** | 0.8266 | 0.7944 | 0.7639 | 0.7350 | 0.7077 | 0.6819 | 0.6575 | 0.6344 | 0.6125 |
| **4** | 0.7984 | 0.7675 | 0.7383 | 0.7107 | 0.6845 | 0.6598 | 0.6364 | 0.6142 | 0.5933 |
| **5** | 0.7715 | 0.7419 | 0.7139 | 0.6874 | 0.6624 | 0.6387 | 0.6163 | 0.5950 | 0.5749 |
| **6** | 0.7459 | 0.7176 | 0.6907 | 0.6653 | 0.6413 | 0.6186 | 0.5971 | 0.5767 | 0.5574 |
| **7** | 0.7216 | 0.6944 | 0.6686 | 0.6443 | 0.6212 | 0.5994 | 0.5788 | 0.5592 | 0.5407 |
| **8** | 0.6984 | 0.6723 | 0.6476 | 0.6242 | 0.6021 | 0.5811 | 0.5613 | 0.5425 | 0.5248 |
| **9** | 0.6763 | 0.6512 | 0.6275 | 0.6050 | 0.5838 | 0.5637 | 0.5446 | 0.5266 | 0.5095 |
| **10** | 0.6551 | 0.6310 | 0.6082 | 0.5866 | 0.5662 | 0.5469 | 0.5286 | 0.5113 | 0.4949 |
| **11** | 0.6350 | 0.6118 | 0.5899 | 0.5691 | 0.5495 | 0.5309 | 0.5134 | 0.4967 | 0.4809 |
| **12** | 0.6157 | 0.5934 | 0.5723 | 0.5524 | 0.5335 | 0.5157 | 0.4987 | 0.4827 | 0.4676 |
| **13** | 0.5973 | 0.5758 | 0.5556 | 0.5364 | 0.5182 | 0.5010 | 0.4847 | 0.4693 | 0.4547 |
| **14** | 0.5796 | 0.5590 | 0.5395 | 0.5210 | 0.5035 | 0.4870 | 0.4713 | 0.4565 | 0.4424 |
| **15** | 0.5628 | 0.5429 | 0.5241 | 0.5063 | 0.4895 | 0.4736 | 0.4585 | 0.4442 | 0.4307 |
| **16** | 0.5466 | 0.5275 | 0.5094 | 0.4922 | 0.4760 | 0.4606 | 0.4461 | 0.4323 | 0.4193 |
| **17** | 0.5176 | 0.4996 | 0.4826 | 0.4665 | 0.4512 | 0.4368 | 0.4232 | 0.4102 | 0.3980 |
| **18** | 0.5040 | 0.4866 | 0.4702 | 0.4546 | 0.4399 | 0.4260 | 0.4128 | 0.4003 | 0.3885 |
| **19** | 0.4909 | 0.4742 | 0.4583 | 0.4432 | 0.4290 | 0.4156 | 0.4028 | 0.3908 | 0.3793 |
| **20** | 0.4783 | 0.4621 | 0.4467 | 0.4322 | 0.4185 | 0.4055 | 0.3931 | 0.3815 | 0.3705 |
| **21** | 0.4661 | 0.4504 | 0.4356 | 0.4215 | 0.4083 | 0.3957 | 0.3838 | 0.3725 | 0.3618 |
| **22** | 0.4544 | 0.4392 | 0.4248 | 0.4113 | 0.3984 | 0.3862 | 0.3747 | 0.3638 | 0.3535 |
| **23** | 0.4430 | 0.4283 | 0.4144 | 0.4013 | 0.3889 | 0.3771 | 0.3660 | 0.3554 | 0.3455 |
| **24** | 0.4320 | 0.4178 | 0.4044 | 0.3917 | 0.3796 | 0.3683 | 0.3575 | 0.3473 | 0.3377 |
| **25** | 0.4213 | 0.4076 | 0.3946 | 0.3823 | 0.3707 | 0.3596 | 0.3492 | 0.3394 | 0.3300 |
| **26** | 0.4111 | 0.3978 | 0.3852 | 0.3733 | 0.3620 | 0.3514 | 0.3413 | 0.3317 | 0.3227 |
| **27** | 0.4011 | 0.3882 | 0.3761 | 0.3645 | 0.3536 | 0.3433 | 0.3335 | 0.3243 | 0.3156 |
| **28** | 0.3915 | 0.3790 | 0.3672 | 0.3560 | 0.3455 | 0.3355 | 0.3260 | 0.3171 | 0.3086 |
| **29** | 0.3822 | 0.3701 | 0.3586 | 0.3478 | 0.3376 | 0.3279 | 0.3188 | 0.3101 | 0.3019 |
| **30** | 0.3731 | 0.3614 | 0.3503 | 0.3398 | 0.3299 | 0.3205 | 0.3117 | 0.3033 | 0.2954 |
| **31** | 0.3643 | 0.3530 | 0.3422 | 0.3321 | 0.3225 | 0.3134 | 0.3048 | 0.2967 | 0.2890 |
| **32** | 0.3558 | 0.3448 | 0.3344 | 0.3245 | 0.3152 | 0.3064 | 0.2981 | 0.2903 | 0.2828 |
| **33** | 0.3475 | 0.3368 | 0.3267 | 0.3172 | 0.3082 | 0.2997 | 0.2916 | 0.2840 | 0.2768 |
| **34** | 0.3395 | 0.3291 | 0.3194 | 0.3101 | 0.3014 | 0.2931 | 0.2853 | 0.2779 | 0.2709 |
| **35** | 0.3317 | 0.3216 | 0.3122 | 0.3032 | 0.2947 | 0.2867 | 0.2791 | 0.2720 | 0.2652 |
| **36** | 0.3241 | 0.3144 | 0.3052 | 0.2965 | 0.2883 | 0.2805 | 0.2731 | 0.2662 | 0.2596 |
| **37** | 0.3167 | 0.3072 | 0.2983 | 0.2899 | 0.2819 | 0.2744 | 0.2673 | 0.2605 | 0.2542 |
| **38** | 0.3095 | 0.3003 | 0.2917 | 0.2835 | 0.2758 | 0.2685 | 0.2616 | 0.2550 | 0.2489 |
| **39** | 0.3025 | 0.2936 | 0.2852 | 0.2773 | 0.2698 | 0.2627 | 0.2560 | 0.2497 | 0.2437 |
| **40** | 0.2956 | 0.2870 | 0.2789 | 0.2712 | 0.2639 | 0.2570 | 0.2505 | 0.2444 | 0.2386 |

*Source*: Based on Eq. 1, Duan et al. (2006).

**Table 6.11** Standard Air Saturation Concentration of Carbon Dioxide as a Function of Temperature and Salinity in mg/L, 0–120 g/kg (NaCl brine, 1 atm moist air; mole fraction = 440 μatm)

| Temperature (°C) | Salinity (g/kg) | | | | | | | | |
|---|---|---|---|---|---|---|---|---|---|
| | 0 | 15 | 30 | 45 | 60 | 75 | 90 | 105 | 120 |
| 0 | 1.4719 | 1.4064 | 1.3445 | 1.2860 | 1.2307 | 1.1786 | 1.1293 | 1.0827 | 1.0387 |
| 1 | 1.4155 | 1.3529 | 1.2937 | 1.2378 | 1.1850 | 1.1352 | 1.0881 | 1.0435 | 1.0015 |
| 2 | 1.3619 | 1.3020 | 1.2455 | 1.1921 | 1.1416 | 1.0939 | 1.0489 | 1.0063 | 0.9661 |
| 3 | 1.3113 | 1.2541 | 1.2000 | 1.1488 | 1.1006 | 1.0550 | 1.0119 | 0.9711 | 0.9326 |
| 4 | 1.2633 | 1.2085 | 1.1567 | 1.1078 | 1.0615 | 1.0179 | 0.9766 | 0.9376 | 0.9007 |
| 5 | 1.2177 | 1.1652 | 1.1156 | 1.0687 | 1.0245 | 0.9826 | 0.9431 | 0.9057 | 0.8704 |
| 6 | 1.1744 | 1.1241 | 1.0766 | 1.0317 | 0.9892 | 0.9491 | 0.9113 | 0.8754 | 0.8416 |
| 7 | 1.1333 | 1.0850 | 1.0395 | 0.9964 | 0.9557 | 0.9173 | 0.8809 | 0.8466 | 0.8141 |
| 8 | 1.0942 | 1.0479 | 1.0042 | 0.9629 | 0.9238 | 0.8870 | 0.8521 | 0.8191 | 0.7879 |
| 9 | 1.0570 | 1.0126 | 0.9706 | 0.9310 | 0.8935 | 0.8581 | 0.8246 | 0.7929 | 0.7630 |
| 10 | 1.0215 | 0.9788 | 0.9385 | 0.9005 | 0.8645 | 0.8304 | 0.7983 | 0.7679 | 0.7391 |
| 11 | 0.9878 | 0.9468 | 0.9080 | 0.8714 | 0.8368 | 0.8041 | 0.7732 | 0.7440 | 0.7164 |
| 12 | 0.9556 | 0.9162 | 0.8789 | 0.8437 | 0.8105 | 0.7790 | 0.7493 | 0.7212 | 0.6946 |
| 13 | 0.9249 | 0.8870 | 0.8511 | 0.8173 | 0.7853 | 0.7551 | 0.7265 | 0.6994 | 0.6739 |
| 14 | 0.8956 | 0.8591 | 0.8246 | 0.7920 | 0.7612 | 0.7321 | 0.7046 | 0.6786 | 0.6540 |
| 15 | 0.8676 | 0.8325 | 0.7993 | 0.7679 | 0.7383 | 0.7102 | 0.6837 | 0.6587 | 0.6350 |
| 16 | 0.8408 | 0.8070 | 0.7750 | 0.7448 | 0.7162 | 0.6892 | 0.6637 | 0.6395 | 0.6167 |
| 17 | 0.7944 | 0.7626 | 0.7326 | 0.7042 | 0.6774 | 0.6520 | 0.6281 | 0.6054 | 0.5839 |
| 18 | 0.7719 | 0.7412 | 0.7122 | 0.6848 | 0.6589 | 0.6344 | 0.6112 | 0.5893 | 0.5686 |
| 19 | 0.7503 | 0.7207 | 0.6926 | 0.6662 | 0.6411 | 0.6175 | 0.5951 | 0.5739 | 0.5539 |
| 20 | 0.7295 | 0.7008 | 0.6738 | 0.6482 | 0.6240 | 0.6011 | 0.5795 | 0.5590 | 0.5396 |
| 21 | 0.7094 | 0.6817 | 0.6555 | 0.6308 | 0.6074 | 0.5853 | 0.5644 | 0.5446 | 0.5259 |
| 22 | 0.6901 | 0.6633 | 0.6380 | 0.6141 | 0.5914 | 0.5700 | 0.5498 | 0.5307 | 0.5126 |
| 23 | 0.6715 | 0.6456 | 0.6211 | 0.5979 | 0.5761 | 0.5554 | 0.5358 | 0.5173 | 0.4998 |
| 24 | 0.6535 | 0.6284 | 0.6047 | 0.5824 | 0.5612 | 0.5412 | 0.5222 | 0.5043 | 0.4874 |
| 25 | 0.6362 | 0.6119 | 0.5889 | 0.5673 | 0.5468 | 0.5274 | 0.5091 | 0.4917 | 0.4754 |
| 26 | 0.6195 | 0.5959 | 0.5737 | 0.5528 | 0.5329 | 0.5142 | 0.4964 | 0.4797 | 0.4638 |
| 27 | 0.6033 | 0.5805 | 0.5590 | 0.5387 | 0.5195 | 0.5013 | 0.4841 | 0.4679 | 0.4525 |
| 28 | 0.5877 | 0.5656 | 0.5448 | 0.5251 | 0.5065 | 0.4889 | 0.4723 | 0.4566 | 0.4417 |
| 29 | 0.5726 | 0.5512 | 0.5311 | 0.5120 | 0.4940 | 0.4769 | 0.4608 | 0.4456 | 0.4312 |
| 30 | 0.5580 | 0.5373 | 0.5177 | 0.4992 | 0.4818 | 0.4653 | 0.4496 | 0.4349 | 0.4209 |
| 31 | 0.5439 | 0.5238 | 0.5049 | 0.4869 | 0.4700 | 0.4540 | 0.4389 | 0.4246 | 0.4110 |
| 32 | 0.5302 | 0.5107 | 0.4923 | 0.4750 | 0.4586 | 0.4431 | 0.4284 | 0.4145 | 0.4014 |
| 33 | 0.5169 | 0.4980 | 0.4802 | 0.4634 | 0.4475 | 0.4324 | 0.4182 | 0.4048 | 0.3920 |
| 34 | 0.5041 | 0.4858 | 0.4685 | 0.4522 | 0.4368 | 0.4222 | 0.4084 | 0.3953 | 0.3830 |
| 35 | 0.4916 | 0.4739 | 0.4571 | 0.4413 | 0.4263 | 0.4122 | 0.3988 | 0.3861 | 0.3742 |
| 36 | 0.4796 | 0.4623 | 0.4461 | 0.4307 | 0.4162 | 0.4025 | 0.3895 | 0.3772 | 0.3656 |
| 37 | 0.4678 | 0.4511 | 0.4353 | 0.4204 | 0.4063 | 0.3930 | 0.3804 | 0.3685 | 0.3573 |
| 38 | 0.4564 | 0.4402 | 0.4249 | 0.4104 | 0.3968 | 0.3839 | 0.3716 | 0.3601 | 0.3492 |
| 39 | 0.4453 | 0.4296 | 0.4148 | 0.4007 | 0.3875 | 0.3749 | 0.3631 | 0.3519 | 0.3413 |
| 40 | 0.4345 | 0.4193 | 0.4049 | 0.3912 | 0.3784 | 0.3662 | 0.3547 | 0.3438 | 0.3335 |

*Source*: Based on Eq. 1, Duan et al. (2006).

**Table 6.12** Standard Air Saturation Concentration of Carbon Dioxide as a Function of Temperature and Salinity in mg/L, 120–240 g/kg (NaCl brine, 1 atm moist air; mole fraction = 440 μatm)

| Temperature (°C) | Salinity (g/kg) | | | | | | | | |
|---|---|---|---|---|---|---|---|---|---|
| | 120 | 135 | 150 | 165 | 180 | 195 | 210 | 225 | 240 |
| 0 | 1.0387 | 0.9971 | 0.9578 | 0.9206 | 0.8854 | 0.8522 | 0.8207 | 0.7910 | 0.7628 |
| 1 | 1.0015 | 0.9617 | 0.9241 | 0.8886 | 0.8550 | 0.8232 | 0.7931 | 0.7646 | 0.7377 |
| 2 | 0.9661 | 0.9281 | 0.8921 | 0.8581 | 0.8260 | 0.7955 | 0.7668 | 0.7395 | 0.7137 |
| 3 | 0.9326 | 0.8962 | 0.8618 | 0.8293 | 0.7985 | 0.7694 | 0.7418 | 0.7157 | 0.6910 |
| 4 | 0.9007 | 0.8659 | 0.8329 | 0.8018 | 0.7723 | 0.7444 | 0.7180 | 0.6930 | 0.6693 |
| 5 | 0.8704 | 0.8370 | 0.8054 | 0.7756 | 0.7473 | 0.7206 | 0.6953 | 0.6713 | 0.6486 |
| 6 | 0.8416 | 0.8096 | 0.7793 | 0.7506 | 0.7235 | 0.6979 | 0.6736 | 0.6506 | 0.6289 |
| 7 | 0.8141 | 0.7834 | 0.7543 | 0.7269 | 0.7009 | 0.6763 | 0.6530 | 0.6309 | 0.6100 |
| 8 | 0.7879 | 0.7585 | 0.7306 | 0.7042 | 0.6792 | 0.6556 | 0.6333 | 0.6121 | 0.5921 |
| 9 | 0.7630 | 0.7347 | 0.7079 | 0.6826 | 0.6586 | 0.6359 | 0.6144 | 0.5941 | 0.5749 |
| 10 | 0.7391 | 0.7119 | 0.6862 | 0.6619 | 0.6388 | 0.6170 | 0.5964 | 0.5769 | 0.5584 |
| 11 | 0.7164 | 0.6902 | 0.6655 | 0.6421 | 0.6200 | 0.5990 | 0.5792 | 0.5604 | 0.5426 |
| 12 | 0.6946 | 0.6695 | 0.6457 | 0.6232 | 0.6019 | 0.5818 | 0.5627 | 0.5446 | 0.5275 |
| 13 | 0.6739 | 0.6497 | 0.6268 | 0.6051 | 0.5846 | 0.5652 | 0.5469 | 0.5295 | 0.5130 |
| 14 | 0.6540 | 0.6307 | 0.6086 | 0.5878 | 0.5681 | 0.5494 | 0.5317 | 0.5150 | 0.4992 |
| 15 | 0.6350 | 0.6125 | 0.5913 | 0.5713 | 0.5523 | 0.5343 | 0.5173 | 0.5011 | 0.4859 |
| 16 | 0.6167 | 0.5951 | 0.5747 | 0.5553 | 0.5370 | 0.5197 | 0.5033 | 0.4878 | 0.4731 |
| 17 | 0.5839 | 0.5636 | 0.5444 | 0.5263 | 0.5091 | 0.4928 | 0.4774 | 0.4628 | 0.4490 |
| 18 | 0.5686 | 0.5490 | 0.5305 | 0.5129 | 0.4963 | 0.4806 | 0.4657 | 0.4516 | 0.4383 |
| 19 | 0.5539 | 0.5350 | 0.5170 | 0.5001 | 0.4840 | 0.4688 | 0.4545 | 0.4409 | 0.4280 |
| 20 | 0.5396 | 0.5213 | 0.5040 | 0.4876 | 0.4721 | 0.4574 | 0.4436 | 0.4304 | 0.4180 |
| 21 | 0.5259 | 0.5082 | 0.4914 | 0.4756 | 0.4606 | 0.4464 | 0.4330 | 0.4203 | 0.4082 |
| 22 | 0.5126 | 0.4955 | 0.4793 | 0.4640 | 0.4495 | 0.4358 | 0.4228 | 0.4105 | 0.3988 |
| 23 | 0.4998 | 0.4832 | 0.4676 | 0.4528 | 0.4387 | 0.4255 | 0.4129 | 0.4010 | 0.3898 |
| 24 | 0.4874 | 0.4714 | 0.4562 | 0.4419 | 0.4283 | 0.4155 | 0.4033 | 0.3918 | 0.3809 |
| 25 | 0.4754 | 0.4599 | 0.4452 | 0.4313 | 0.4182 | 0.4058 | 0.3940 | 0.3829 | 0.3724 |
| 26 | 0.4638 | 0.4488 | 0.4346 | 0.4211 | 0.4084 | 0.3964 | 0.3850 | 0.3743 | 0.3641 |
| 27 | 0.4525 | 0.4380 | 0.4243 | 0.4113 | 0.3990 | 0.3873 | 0.3763 | 0.3659 | 0.3560 |
| 28 | 0.4417 | 0.4276 | 0.4143 | 0.4017 | 0.3898 | 0.3785 | 0.3678 | 0.3578 | 0.3482 |
| 29 | 0.4312 | 0.4175 | 0.4046 | 0.3924 | 0.3809 | 0.3700 | 0.3596 | 0.3499 | 0.3406 |
| 30 | 0.4209 | 0.4077 | 0.3952 | 0.3834 | 0.3722 | 0.3616 | 0.3516 | 0.3422 | 0.3332 |
| 31 | 0.4110 | 0.3982 | 0.3861 | 0.3747 | 0.3638 | 0.3536 | 0.3439 | 0.3347 | 0.3261 |
| 32 | 0.4014 | 0.3890 | 0.3773 | 0.3662 | 0.3557 | 0.3457 | 0.3363 | 0.3275 | 0.3191 |
| 33 | 0.3920 | 0.3800 | 0.3686 | 0.3579 | 0.3477 | 0.3381 | 0.3290 | 0.3204 | 0.3123 |
| 34 | 0.3830 | 0.3713 | 0.3603 | 0.3499 | 0.3400 | 0.3307 | 0.3219 | 0.3135 | 0.3057 |
| 35 | 0.3742 | 0.3629 | 0.3522 | 0.3421 | 0.3325 | 0.3235 | 0.3149 | 0.3068 | 0.2992 |
| 36 | 0.3656 | 0.3547 | 0.3443 | 0.3345 | 0.3252 | 0.3164 | 0.3082 | 0.3003 | 0.2929 |
| 37 | 0.3573 | 0.3466 | 0.3366 | 0.3271 | 0.3181 | 0.3096 | 0.3015 | 0.2939 | 0.2868 |
| 38 | 0.3492 | 0.3388 | 0.3291 | 0.3199 | 0.3111 | 0.3029 | 0.2951 | 0.2877 | 0.2808 |
| 39 | 0.3413 | 0.3312 | 0.3218 | 0.3128 | 0.3044 | 0.2964 | 0.2888 | 0.2817 | 0.2749 |
| 40 | 0.3335 | 0.3238 | 0.3146 | 0.3060 | 0.2977 | 0.2900 | 0.2827 | 0.2757 | 0.2692 |

*Source*: Based on Eq. 1, Duan et al. (2006).

**Table 6.13** Bunsen Coefficient ($\beta$) for Oxygen as a Function of Temperature and Salinity in (L real gas/(L atm)), 0–120 g/kg in (L real gas/(L atm)) (NaCl brine)

| Temperature (°C) | Salinity (g/kg) | | | | | | | | |
|---|---|---|---|---|---|---|---|---|---|
| | 0 | 15 | 30 | 45 | 60 | 75 | 90 | 105 | 120 |
| 0 | 0.04894 | 0.04427 | 0.03999 | 0.03605 | 0.03245 | 0.02915 | 0.02615 | 0.02342 | 0.02093 |
| 1 | 0.04762 | 0.04312 | 0.03897 | 0.03517 | 0.03168 | 0.02849 | 0.02558 | 0.02293 | 0.02051 |
| 2 | 0.04637 | 0.04202 | 0.03802 | 0.03433 | 0.03096 | 0.02786 | 0.02504 | 0.02246 | 0.02011 |
| 3 | 0.04518 | 0.04098 | 0.03711 | 0.03354 | 0.03027 | 0.02727 | 0.02452 | 0.02202 | 0.01973 |
| 4 | 0.04405 | 0.03999 | 0.03624 | 0.03279 | 0.02961 | 0.02670 | 0.02403 | 0.02159 | 0.01937 |
| 5 | 0.04298 | 0.03905 | 0.03542 | 0.03207 | 0.02898 | 0.02616 | 0.02356 | 0.02119 | 0.01903 |
| 6 | 0.04195 | 0.03815 | 0.03463 | 0.03138 | 0.02839 | 0.02564 | 0.02311 | 0.02080 | 0.01870 |
| 7 | 0.04097 | 0.03728 | 0.03387 | 0.03072 | 0.02781 | 0.02514 | 0.02268 | 0.02043 | 0.01838 |
| 8 | 0.04003 | 0.03645 | 0.03314 | 0.03008 | 0.02726 | 0.02466 | 0.02227 | 0.02008 | 0.01807 |
| 9 | 0.03912 | 0.03566 | 0.03244 | 0.02947 | 0.02673 | 0.02420 | 0.02187 | 0.01973 | 0.01778 |
| 10 | 0.03825 | 0.03489 | 0.03177 | 0.02888 | 0.02621 | 0.02375 | 0.02148 | 0.01940 | 0.01749 |
| 11 | 0.03741 | 0.03415 | 0.03112 | 0.02831 | 0.02572 | 0.02332 | 0.02111 | 0.01908 | 0.01721 |
| 12 | 0.03660 | 0.03344 | 0.03050 | 0.02777 | 0.02524 | 0.02290 | 0.02075 | 0.01877 | 0.01695 |
| 13 | 0.03582 | 0.03275 | 0.02989 | 0.02724 | 0.02478 | 0.02250 | 0.02040 | 0.01847 | 0.01669 |
| 14 | 0.03506 | 0.03208 | 0.02930 | 0.02672 | 0.02433 | 0.02211 | 0.02006 | 0.01817 | 0.01644 |
| 15 | 0.03434 | 0.03144 | 0.02874 | 0.02623 | 0.02389 | 0.02173 | 0.01973 | 0.01789 | 0.01619 |
| 16 | 0.03363 | 0.03082 | 0.02819 | 0.02575 | 0.02347 | 0.02137 | 0.01942 | 0.01761 | 0.01595 |
| 17 | 0.03295 | 0.03022 | 0.02766 | 0.02528 | 0.02306 | 0.02101 | 0.01911 | 0.01735 | 0.01572 |
| 18 | 0.03230 | 0.02963 | 0.02715 | 0.02483 | 0.02267 | 0.02066 | 0.01881 | 0.01709 | 0.01550 |
| 19 | 0.03166 | 0.02907 | 0.02665 | 0.02439 | 0.02229 | 0.02033 | 0.01852 | 0.01683 | 0.01528 |
| 20 | 0.03105 | 0.02853 | 0.02617 | 0.02397 | 0.02192 | 0.02001 | 0.01823 | 0.01659 | 0.01507 |
| 21 | 0.03046 | 0.02800 | 0.02571 | 0.02356 | 0.02156 | 0.01969 | 0.01796 | 0.01635 | 0.01487 |
| 22 | 0.02989 | 0.02750 | 0.02526 | 0.02317 | 0.02121 | 0.01939 | 0.01770 | 0.01612 | 0.01467 |
| 23 | 0.02934 | 0.02701 | 0.02483 | 0.02279 | 0.02088 | 0.01910 | 0.01744 | 0.01590 | 0.01448 |
| 24 | 0.02881 | 0.02654 | 0.02442 | 0.02242 | 0.02056 | 0.01882 | 0.01720 | 0.01569 | 0.01429 |
| 25 | 0.02830 | 0.02609 | 0.02402 | 0.02207 | 0.02025 | 0.01855 | 0.01696 | 0.01548 | 0.01411 |
| 26 | 0.02781 | 0.02566 | 0.02364 | 0.02173 | 0.01995 | 0.01829 | 0.01674 | 0.01529 | 0.01394 |
| 27 | 0.02735 | 0.02525 | 0.02327 | 0.02141 | 0.01967 | 0.01804 | 0.01652 | 0.01510 | 0.01378 |
| 28 | 0.02691 | 0.02485 | 0.02292 | 0.02111 | 0.01940 | 0.01780 | 0.01631 | 0.01492 | 0.01363 |
| 29 | 0.02649 | 0.02448 | 0.02259 | 0.02081 | 0.01914 | 0.01758 | 0.01612 | 0.01475 | 0.01348 |
| 30 | 0.02609 | 0.02413 | 0.02228 | 0.02054 | 0.01890 | 0.01737 | 0.01593 | 0.01459 | 0.01334 |
| 31 | 0.02571 | 0.02379 | 0.02198 | 0.02028 | 0.01867 | 0.01717 | 0.01576 | 0.01444 | 0.01322 |
| 32 | 0.02536 | 0.02348 | 0.02171 | 0.02003 | 0.01846 | 0.01698 | 0.01560 | 0.01430 | 0.01310 |
| 33 | 0.02503 | 0.02319 | 0.02145 | 0.01981 | 0.01826 | 0.01681 | 0.01545 | 0.01418 | 0.01299 |
| 34 | 0.02472 | 0.02292 | 0.02121 | 0.01960 | 0.01808 | 0.01665 | 0.01531 | 0.01406 | 0.01289 |
| 35 | 0.02444 | 0.02267 | 0.02099 | 0.01941 | 0.01792 | 0.01651 | 0.01519 | 0.01396 | 0.01280 |
| 36 | 0.02418 | 0.02244 | 0.02079 | 0.01924 | 0.01777 | 0.01638 | 0.01508 | 0.01386 | 0.01272 |
| 37 | 0.02395 | 0.02224 | 0.02062 | 0.01908 | 0.01764 | 0.01627 | 0.01499 | 0.01378 | 0.01265 |
| 38 | 0.02374 | 0.02206 | 0.02046 | 0.01895 | 0.01752 | 0.01618 | 0.01491 | 0.01372 | 0.01260 |
| 39 | 0.02356 | 0.02191 | 0.02033 | 0.01884 | 0.01743 | 0.01610 | 0.01484 | 0.01367 | 0.01256 |
| 40 | 0.02341 | 0.02178 | 0.02022 | 0.01875 | 0.01736 | 0.01604 | 0.01480 | 0.01363 | 0.01253 |

*Source*: Based on Eqs. 18 and 19, Sherwood et al. (1991).

**Table 6.14** Bunsen Coefficient ($\beta$) for Oxygen as a Function of Temperature and Salinity in (L real gas /(L atm)), 120−240 g/kg (NaCl brine)

| Temperature (°C) | Salinity (g/kg) | | | | | | | | |
|---|---|---|---|---|---|---|---|---|---|
| | 120 | 135 | 150 | 165 | 180 | 195 | 210 | 225 | 240 |
| 0 | 0.02093 | 0.01868 | 0.01665 | 0.01481 | 0.01315 | 0.01166 | 0.01032 | 0.00912 | 0.00805 |
| 1 | 0.02051 | 0.01832 | 0.01634 | 0.01455 | 0.01293 | 0.01148 | 0.01017 | 0.00899 | 0.00794 |
| 2 | 0.02011 | 0.01798 | 0.01605 | 0.01430 | 0.01272 | 0.01130 | 0.01002 | 0.00887 | 0.00784 |
| 3 | 0.01973 | 0.01766 | 0.01578 | 0.01407 | 0.01253 | 0.01114 | 0.00988 | 0.00876 | 0.00775 |
| 4 | 0.01937 | 0.01735 | 0.01551 | 0.01385 | 0.01234 | 0.01098 | 0.00975 | 0.00865 | 0.00766 |
| 5 | 0.01903 | 0.01706 | 0.01526 | 0.01364 | 0.01216 | 0.01083 | 0.00963 | 0.00855 | 0.00757 |
| 6 | 0.01870 | 0.01677 | 0.01502 | 0.01343 | 0.01199 | 0.01069 | 0.00951 | 0.00845 | 0.00749 |
| 7 | 0.01838 | 0.01650 | 0.01479 | 0.01324 | 0.01183 | 0.01055 | 0.00939 | 0.00835 | 0.00741 |
| 8 | 0.01807 | 0.01624 | 0.01457 | 0.01305 | 0.01167 | 0.01042 | 0.00928 | 0.00826 | 0.00734 |
| 9 | 0.01778 | 0.01599 | 0.01435 | 0.01287 | 0.01152 | 0.01029 | 0.00918 | 0.00817 | 0.00727 |
| 10 | 0.01749 | 0.01574 | 0.01415 | 0.01269 | 0.01137 | 0.01016 | 0.00907 | 0.00809 | 0.00720 |
| 11 | 0.01721 | 0.01551 | 0.01395 | 0.01252 | 0.01122 | 0.01004 | 0.00897 | 0.00800 | 0.00713 |
| 12 | 0.01695 | 0.01528 | 0.01375 | 0.01236 | 0.01108 | 0.00993 | 0.00888 | 0.00792 | 0.00706 |
| 13 | 0.01669 | 0.01506 | 0.01356 | 0.01220 | 0.01095 | 0.00981 | 0.00878 | 0.00785 | 0.00700 |
| 14 | 0.01644 | 0.01484 | 0.01338 | 0.01204 | 0.01082 | 0.00970 | 0.00869 | 0.00777 | 0.00693 |
| 15 | 0.01619 | 0.01463 | 0.01320 | 0.01189 | 0.01069 | 0.00959 | 0.00860 | 0.00769 | 0.00687 |
| 16 | 0.01595 | 0.01443 | 0.01302 | 0.01174 | 0.01056 | 0.00949 | 0.00851 | 0.00762 | 0.00681 |
| 17 | 0.01572 | 0.01423 | 0.01285 | 0.01159 | 0.01044 | 0.00939 | 0.00843 | 0.00755 | 0.00676 |
| 18 | 0.01550 | 0.01404 | 0.01269 | 0.01145 | 0.01032 | 0.00929 | 0.00834 | 0.00748 | 0.00670 |
| 19 | 0.01528 | 0.01385 | 0.01253 | 0.01132 | 0.01021 | 0.00919 | 0.00826 | 0.00741 | 0.00664 |
| 20 | 0.01507 | 0.01367 | 0.01237 | 0.01119 | 0.01009 | 0.00910 | 0.00818 | 0.00735 | 0.00659 |
| 21 | 0.01487 | 0.01349 | 0.01222 | 0.01106 | 0.00999 | 0.00900 | 0.00810 | 0.00728 | 0.00654 |
| 22 | 0.01467 | 0.01332 | 0.01208 | 0.01093 | 0.00988 | 0.00891 | 0.00803 | 0.00722 | 0.00648 |
| 23 | 0.01448 | 0.01316 | 0.01194 | 0.01081 | 0.00978 | 0.00883 | 0.00796 | 0.00716 | 0.00644 |
| 24 | 0.01429 | 0.01300 | 0.01180 | 0.01070 | 0.00968 | 0.00875 | 0.00789 | 0.00710 | 0.00639 |
| 25 | 0.01411 | 0.01284 | 0.01167 | 0.01058 | 0.00959 | 0.00867 | 0.00782 | 0.00705 | 0.00634 |
| 26 | 0.01394 | 0.01270 | 0.01154 | 0.01048 | 0.00949 | 0.00859 | 0.00776 | 0.00700 | 0.00630 |
| 27 | 0.01378 | 0.01256 | 0.01142 | 0.01038 | 0.00941 | 0.00852 | 0.00770 | 0.00695 | 0.00626 |
| 28 | 0.01363 | 0.01243 | 0.01131 | 0.01028 | 0.00933 | 0.00845 | 0.00764 | 0.00690 | 0.00622 |
| 29 | 0.01348 | 0.01230 | 0.01120 | 0.01019 | 0.00925 | 0.00839 | 0.00759 | 0.00686 | 0.00619 |
| 30 | 0.01334 | 0.01218 | 0.01110 | 0.01010 | 0.00918 | 0.00833 | 0.00754 | 0.00682 | 0.00615 |
| 31 | 0.01322 | 0.01207 | 0.01101 | 0.01002 | 0.00911 | 0.00827 | 0.00749 | 0.00678 | 0.00612 |
| 32 | 0.01310 | 0.01197 | 0.01092 | 0.00995 | 0.00905 | 0.00822 | 0.00745 | 0.00675 | 0.00610 |
| 33 | 0.01299 | 0.01188 | 0.01084 | 0.00989 | 0.00900 | 0.00818 | 0.00742 | 0.00672 | 0.00607 |
| 34 | 0.01289 | 0.01179 | 0.01077 | 0.00983 | 0.00895 | 0.00814 | 0.00739 | 0.00669 | 0.00606 |
| 35 | 0.01280 | 0.01172 | 0.01071 | 0.00978 | 0.00891 | 0.00810 | 0.00736 | 0.00667 | 0.00604 |
| 36 | 0.01272 | 0.01165 | 0.01066 | 0.00973 | 0.00887 | 0.00808 | 0.00734 | 0.00666 | 0.00603 |
| 37 | 0.01265 | 0.01160 | 0.01062 | 0.00970 | 0.00885 | 0.00806 | 0.00733 | 0.00665 | 0.00603 |
| 38 | 0.01260 | 0.01156 | 0.01058 | 0.00967 | 0.00883 | 0.00805 | 0.00732 | 0.00665 | 0.00603 |
| 39 | 0.01256 | 0.01153 | 0.01056 | 0.00966 | 0.00882 | 0.00804 | 0.00732 | 0.00665 | 0.00604 |
| 40 | 0.01253 | 0.01151 | 0.01055 | 0.00965 | 0.00882 | 0.00805 | 0.00733 | 0.00666 | 0.00605 |

*Source*: Based on Eqs. 18 and 19, Sherwood et al. (1991).

**Table 6.15** Bunsen Coefficient ($\beta$) for Carbon Dioxide as a Function of Temperature and Salinity in (L real gas/(L atm)), 0–120 g/kg in (L real gas/(L atm)) (NaCl brine, 1 atm moist air)

| Temperature (°C) | Salinity (g/kg) | | | | | | | | |
|---|---|---|---|---|---|---|---|---|---|
| | **0** | **15** | **30** | **45** | **60** | **75** | **90** | **105** | **120** |
| **0** | 1.7026 | 1.6268 | 1.5552 | 1.4875 | 1.4236 | 1.3633 | 1.3062 | 1.2524 | 1.2015 |
| **1** | 1.6380 | 1.5656 | 1.4971 | 1.4324 | 1.3713 | 1.3136 | 1.2591 | 1.2076 | 1.1589 |
| **2** | 1.5769 | 1.5076 | 1.4421 | 1.3802 | 1.3218 | 1.2666 | 1.2145 | 1.1652 | 1.1186 |
| **3** | 1.5190 | 1.4527 | 1.3900 | 1.3308 | 1.2749 | 1.2220 | 1.1721 | 1.1249 | 1.0803 |
| **4** | 1.4642 | 1.4006 | 1.3406 | 1.2839 | 1.2303 | 1.1797 | 1.1319 | 1.0867 | 1.0440 |
| **5** | 1.4121 | 1.3513 | 1.2938 | 1.2394 | 1.1881 | 1.1396 | 1.0937 | 1.0504 | 1.0094 |
| **6** | 1.3628 | 1.3044 | 1.2493 | 1.1972 | 1.1479 | 1.1014 | 1.0574 | 1.0159 | 0.9766 |
| **7** | 1.3160 | 1.2599 | 1.2070 | 1.1570 | 1.1098 | 1.0651 | 1.0229 | 0.9830 | 0.9453 |
| **8** | 1.2715 | 1.2177 | 1.1669 | 1.1189 | 1.0735 | 1.0306 | 0.9901 | 0.9518 | 0.9156 |
| **9** | 1.2291 | 1.1775 | 1.1287 | 1.0826 | 1.0390 | 0.9978 | 0.9589 | 0.9220 | 0.8872 |
| **10** | 1.1889 | 1.1392 | 1.0923 | 1.0480 | 1.0061 | 0.9665 | 0.9291 | 0.8937 | 0.8602 |
| **11** | 1.1506 | 1.1028 | 1.0577 | 1.0151 | 0.9748 | 0.9367 | 0.9007 | 0.8667 | 0.8345 |
| **12** | 1.1141 | 1.0682 | 1.0248 | 0.9837 | 0.9449 | 0.9083 | 0.8736 | 0.8409 | 0.8099 |
| **13** | 1.0794 | 1.0351 | 0.9933 | 0.9538 | 0.9165 | 0.8812 | 0.8478 | 0.8163 | 0.7864 |
| **14** | 1.0463 | 1.0036 | 0.9633 | 0.9253 | 0.8893 | 0.8553 | 0.8232 | 0.7928 | 0.7640 |
| **15** | 1.0147 | 0.9736 | 0.9347 | 0.8981 | 0.8634 | 0.8306 | 0.7996 | 0.7703 | 0.7426 |
| **16** | 0.9845 | 0.9449 | 0.9074 | 0.8720 | 0.8386 | 0.8070 | 0.7771 | 0.7488 | 0.7221 |
| **17** | 0.9312 | 0.8939 | 0.8587 | 0.8255 | 0.7940 | 0.7643 | 0.7362 | 0.7096 | 0.6845 |
| **18** | 0.9060 | 0.8700 | 0.8359 | 0.8038 | 0.7734 | 0.7446 | 0.7174 | 0.6917 | 0.6674 |
| **19** | 0.8819 | 0.8470 | 0.8140 | 0.7829 | 0.7535 | 0.7257 | 0.6994 | 0.6745 | 0.6510 |
| **20** | 0.8586 | 0.8249 | 0.7930 | 0.7629 | 0.7344 | 0.7075 | 0.6820 | 0.6579 | 0.6352 |
| **21** | 0.8363 | 0.8036 | 0.7727 | 0.7436 | 0.7160 | 0.6899 | 0.6653 | 0.6420 | 0.6199 |
| **22** | 0.8148 | 0.7832 | 0.7533 | 0.7250 | 0.6983 | 0.6731 | 0.6492 | 0.6266 | 0.6052 |
| **23** | 0.7942 | 0.7635 | 0.7345 | 0.7071 | 0.6813 | 0.6568 | 0.6337 | 0.6118 | 0.5911 |
| **24** | 0.7743 | 0.7446 | 0.7165 | 0.6899 | 0.6649 | 0.6411 | 0.6187 | 0.5975 | 0.5774 |
| **25** | 0.7552 | 0.7263 | 0.6991 | 0.6734 | 0.6491 | 0.6261 | 0.6043 | 0.5837 | 0.5643 |
| **26** | 0.7368 | 0.7088 | 0.6824 | 0.6574 | 0.6338 | 0.6115 | 0.5904 | 0.5705 | 0.5516 |
| **27** | 0.7190 | 0.6919 | 0.6663 | 0.6420 | 0.6192 | 0.5975 | 0.5770 | 0.5577 | 0.5394 |
| **28** | 0.7019 | 0.6756 | 0.6507 | 0.6272 | 0.6050 | 0.5840 | 0.5641 | 0.5453 | 0.5276 |
| **29** | 0.6855 | 0.6599 | 0.6358 | 0.6129 | 0.5914 | 0.5710 | 0.5517 | 0.5334 | 0.5162 |
| **30** | 0.6696 | 0.6448 | 0.6213 | 0.5991 | 0.5782 | 0.5584 | 0.5396 | 0.5219 | 0.5051 |
| **31** | 0.6543 | 0.6302 | 0.6074 | 0.5859 | 0.5655 | 0.5462 | 0.5280 | 0.5108 | 0.4945 |
| **32** | 0.6396 | 0.6161 | 0.5940 | 0.5730 | 0.5532 | 0.5345 | 0.5168 | 0.5001 | 0.4843 |
| **33** | 0.6254 | 0.6026 | 0.5810 | 0.5606 | 0.5414 | 0.5232 | 0.5060 | 0.4897 | 0.4743 |
| **34** | 0.6117 | 0.5895 | 0.5685 | 0.5487 | 0.5300 | 0.5123 | 0.4955 | 0.4797 | 0.4647 |
| **35** | 0.5984 | 0.5768 | 0.5564 | 0.5372 | 0.5189 | 0.5017 | 0.4854 | 0.4700 | 0.4555 |
| **36** | 0.5856 | 0.5646 | 0.5448 | 0.5260 | 0.5083 | 0.4915 | 0.4757 | 0.4607 | 0.4465 |
| **37** | 0.5733 | 0.5528 | 0.5335 | 0.5152 | 0.4980 | 0.4817 | 0.4662 | 0.4517 | 0.4379 |
| **38** | 0.5614 | 0.5414 | 0.5226 | 0.5048 | 0.4880 | 0.4721 | 0.4571 | 0.4429 | 0.4295 |
| **39** | 0.5499 | 0.5304 | 0.5121 | 0.4948 | 0.4784 | 0.4629 | 0.4483 | 0.4345 | 0.4214 |
| **40** | 0.5387 | 0.5198 | 0.5019 | 0.4851 | 0.4691 | 0.4540 | 0.4398 | 0.4263 | 0.4135 |

*Source*: Based on Eq. 1, Duan et al. (2006).

**Table 6.16** Bunsen Coefficient ($\beta$) for Carbon Dioxide as a Function of Temperature and Salinity in (L real gas/(L atm)), 120–240 g/kg (NaCl brine, 1 atm moist air)

| Temperature (°C) | Salinity (g/kg) | | | | | | | | |
|---|---|---|---|---|---|---|---|---|---|
| | 120 | 135 | 150 | 165 | 180 | 195 | 210 | 225 | 240 |
| 0 | 1.2015 | 1.1534 | 1.1079 | 1.0649 | 1.0242 | 0.9858 | 0.9494 | 0.9149 | 0.8823 |
| 1 | 1.1589 | 1.1129 | 1.0694 | 1.0283 | 0.9894 | 0.9526 | 0.9178 | 0.8849 | 0.8537 |
| 2 | 1.1186 | 1.0746 | 1.0330 | 0.9936 | 0.9564 | 0.9211 | 0.8878 | 0.8563 | 0.8264 |
| 3 | 1.0803 | 1.0382 | 0.9983 | 0.9606 | 0.9249 | 0.8912 | 0.8593 | 0.8291 | 0.8004 |
| 4 | 1.0440 | 1.0036 | 0.9654 | 0.9293 | 0.8951 | 0.8627 | 0.8321 | 0.8032 | 0.7757 |
| 5 | 1.0094 | 0.9707 | 0.9341 | 0.8994 | 0.8667 | 0.8356 | 0.8063 | 0.7785 | 0.7522 |
| 6 | 0.9766 | 0.9394 | 0.9043 | 0.8711 | 0.8396 | 0.8098 | 0.7817 | 0.7550 | 0.7298 |
| 7 | 0.9453 | 0.9097 | 0.8759 | 0.8440 | 0.8138 | 0.7853 | 0.7582 | 0.7326 | 0.7084 |
| 8 | 0.9156 | 0.8813 | 0.8489 | 0.8183 | 0.7893 | 0.7618 | 0.7358 | 0.7113 | 0.6880 |
| 9 | 0.8872 | 0.8543 | 0.8232 | 0.7937 | 0.7659 | 0.7395 | 0.7145 | 0.6909 | 0.6685 |
| 10 | 0.8602 | 0.8286 | 0.7986 | 0.7703 | 0.7435 | 0.7181 | 0.6941 | 0.6714 | 0.6499 |
| 11 | 0.8345 | 0.8040 | 0.7752 | 0.7480 | 0.7222 | 0.6978 | 0.6746 | 0.6528 | 0.6321 |
| 12 | 0.8099 | 0.7806 | 0.7529 | 0.7266 | 0.7018 | 0.6783 | 0.6560 | 0.6350 | 0.6150 |
| 13 | 0.7864 | 0.7582 | 0.7315 | 0.7062 | 0.6823 | 0.6597 | 0.6382 | 0.6180 | 0.5988 |
| 14 | 0.7640 | 0.7368 | 0.7111 | 0.6867 | 0.6637 | 0.6419 | 0.6212 | 0.6017 | 0.5832 |
| 15 | 0.7426 | 0.7164 | 0.6915 | 0.6681 | 0.6459 | 0.6248 | 0.6049 | 0.5861 | 0.5682 |
| 16 | 0.7221 | 0.6968 | 0.6729 | 0.6502 | 0.6288 | 0.6085 | 0.5893 | 0.5711 | 0.5539 |
| 17 | 0.6845 | 0.6607 | 0.6382 | 0.6169 | 0.5967 | 0.5777 | 0.5596 | 0.5425 | 0.5263 |
| 18 | 0.6674 | 0.6444 | 0.6226 | 0.6020 | 0.5825 | 0.5641 | 0.5466 | 0.5301 | 0.5145 |
| 19 | 0.6510 | 0.6287 | 0.6077 | 0.5877 | 0.5689 | 0.5510 | 0.5341 | 0.5181 | 0.5030 |
| 20 | 0.6352 | 0.6136 | 0.5932 | 0.5739 | 0.5557 | 0.5384 | 0.5221 | 0.5066 | 0.4919 |
| 21 | 0.6199 | 0.5991 | 0.5793 | 0.5606 | 0.5430 | 0.5262 | 0.5104 | 0.4954 | 0.4812 |
| 22 | 0.6052 | 0.5850 | 0.5659 | 0.5478 | 0.5307 | 0.5145 | 0.4992 | 0.4847 | 0.4709 |
| 23 | 0.5911 | 0.5715 | 0.5530 | 0.5354 | 0.5189 | 0.5032 | 0.4883 | 0.4743 | 0.4610 |
| 24 | 0.5774 | 0.5585 | 0.5405 | 0.5235 | 0.5074 | 0.4922 | 0.4778 | 0.4642 | 0.4513 |
| 25 | 0.5643 | 0.5459 | 0.5285 | 0.5120 | 0.4964 | 0.4817 | 0.4677 | 0.4545 | 0.4420 |
| 26 | 0.5516 | 0.5338 | 0.5169 | 0.5009 | 0.4858 | 0.4715 | 0.4579 | 0.4451 | 0.4330 |
| 27 | 0.5394 | 0.5220 | 0.5057 | 0.4902 | 0.4755 | 0.4616 | 0.4485 | 0.4361 | 0.4243 |
| 28 | 0.5276 | 0.5107 | 0.4948 | 0.4798 | 0.4656 | 0.4521 | 0.4394 | 0.4273 | 0.4159 |
| 29 | 0.5162 | 0.4998 | 0.4844 | 0.4698 | 0.4560 | 0.4429 | 0.4305 | 0.4188 | 0.4078 |
| 30 | 0.5051 | 0.4893 | 0.4743 | 0.4601 | 0.4467 | 0.4340 | 0.4220 | 0.4107 | 0.3999 |
| 31 | 0.4945 | 0.4791 | 0.4645 | 0.4508 | 0.4377 | 0.4254 | 0.4137 | 0.4027 | 0.3923 |
| 32 | 0.4843 | 0.4693 | 0.4551 | 0.4417 | 0.4291 | 0.4171 | 0.4058 | 0.3951 | 0.3849 |
| 33 | 0.4743 | 0.4598 | 0.4460 | 0.4330 | 0.4207 | 0.4090 | 0.3980 | 0.3876 | 0.3778 |
| 34 | 0.4647 | 0.4506 | 0.4372 | 0.4245 | 0.4126 | 0.4013 | 0.3906 | 0.3805 | 0.3709 |
| 35 | 0.4555 | 0.4417 | 0.4287 | 0.4164 | 0.4047 | 0.3937 | 0.3833 | 0.3735 | 0.3642 |
| 36 | 0.4465 | 0.4331 | 0.4205 | 0.4085 | 0.3971 | 0.3864 | 0.3763 | 0.3668 | 0.3577 |
| 37 | 0.4379 | 0.4248 | 0.4125 | 0.4008 | 0.3898 | 0.3794 | 0.3695 | 0.3602 | 0.3515 |
| 38 | 0.4295 | 0.4168 | 0.4048 | 0.3934 | 0.3827 | 0.3726 | 0.3630 | 0.3539 | 0.3454 |
| 39 | 0.4214 | 0.4090 | 0.3973 | 0.3863 | 0.3758 | 0.3659 | 0.3566 | 0.3478 | 0.3395 |
| 40 | 0.4135 | 0.4015 | 0.3901 | 0.3793 | 0.3691 | 0.3595 | 0.3504 | 0.3419 | 0.3338 |

*Source*: Based on Eq. 1, Duan et al. (2006).

**Table 6.17** Vapor Pressure of Water as a Function of Temperature and Salinity in mmHg, 0–120 g/kg (NaCl brine)

| Temperature (°C) | Salinity (g/kg) | | | | | | | | |
|---|---|---|---|---|---|---|---|---|---|
| | **0** | **15** | **30** | **45** | **60** | **75** | **90** | **105** | **120** |
| **0** | 4.573 | 4.533 | 4.493 | 4.451 | 4.408 | 4.363 | 4.315 | 4.265 | 4.210 |
| **1** | 4.917 | 4.874 | 4.830 | 4.786 | 4.740 | 4.691 | 4.640 | 4.585 | 4.527 |
| **2** | 5.284 | 5.237 | 5.190 | 5.143 | 5.093 | 5.041 | 4.986 | 4.927 | 4.865 |
| **3** | 5.674 | 5.624 | 5.574 | 5.523 | 5.469 | 5.413 | 5.354 | 5.291 | 5.224 |
| **4** | 6.090 | 6.037 | 5.983 | 5.927 | 5.870 | 5.810 | 5.747 | 5.679 | 5.607 |
| **5** | 6.533 | 6.475 | 6.417 | 6.358 | 6.297 | 6.232 | 6.164 | 6.092 | 6.014 |
| **6** | 7.004 | 6.942 | 6.880 | 6.816 | 6.751 | 6.682 | 6.609 | 6.531 | 6.448 |
| **7** | 7.504 | 7.438 | 7.372 | 7.304 | 7.233 | 7.159 | 7.081 | 6.998 | 6.909 |
| **8** | 8.036 | 7.965 | 7.894 | 7.821 | 7.746 | 7.666 | 7.583 | 7.494 | 7.398 |
| **9** | 8.601 | 8.525 | 8.449 | 8.371 | 8.290 | 8.205 | 8.116 | 8.020 | 7.918 |
| **10** | 9.200 | 9.119 | 9.038 | 8.954 | 8.868 | 8.777 | 8.681 | 8.579 | 8.470 |
| **11** | 9.836 | 9.750 | 9.662 | 9.573 | 9.481 | 9.384 | 9.281 | 9.172 | 9.056 |
| **12** | 10.511 | 10.418 | 10.325 | 10.229 | 10.131 | 10.027 | 9.918 | 9.801 | 9.676 |
| **13** | 11.225 | 11.126 | 11.027 | 10.925 | 10.819 | 10.709 | 10.592 | 10.467 | 10.334 |
| **14** | 11.982 | 11.877 | 11.770 | 11.662 | 11.549 | 11.431 | 11.306 | 11.173 | 11.031 |
| **15** | 12.784 | 12.671 | 12.558 | 12.442 | 12.322 | 12.196 | 12.063 | 11.921 | 11.769 |
| **16** | 13.632 | 13.512 | 13.391 | 13.267 | 13.139 | 13.005 | 12.863 | 12.712 | 12.550 |
| **17** | 14.529 | 14.401 | 14.272 | 14.140 | 14.004 | 13.861 | 13.709 | 13.548 | 13.376 |
| **18** | 15.477 | 15.341 | 15.204 | 15.063 | 14.918 | 14.765 | 14.604 | 14.433 | 14.249 |
| **19** | 16.479 | 16.334 | 16.188 | 16.039 | 15.884 | 15.722 | 15.550 | 15.367 | 15.172 |
| **20** | 17.538 | 17.384 | 17.228 | 17.069 | 16.904 | 16.731 | 16.549 | 16.354 | 16.146 |
| **21** | 18.656 | 18.491 | 18.326 | 18.157 | 17.981 | 17.798 | 17.603 | 17.396 | 17.175 |
| **22** | 19.835 | 19.660 | 19.484 | 19.304 | 19.118 | 18.923 | 18.716 | 18.496 | 18.261 |
| **23** | 21.079 | 20.893 | 20.706 | 20.515 | 20.317 | 20.109 | 19.890 | 19.656 | 19.406 |
| **24** | 22.390 | 22.193 | 21.994 | 21.791 | 21.581 | 21.360 | 21.127 | 20.879 | 20.613 |
| **25** | 23.772 | 23.563 | 23.352 | 23.136 | 22.913 | 22.679 | 22.431 | 22.168 | 21.885 |
| **26** | 25.228 | 25.006 | 24.782 | 24.553 | 24.316 | 24.068 | 23.805 | 23.525 | 23.226 |
| **27** | 26.761 | 26.525 | 26.288 | 26.045 | 25.794 | 25.530 | 25.251 | 24.955 | 24.637 |
| **28** | 28.374 | 28.124 | 27.873 | 27.615 | 27.349 | 27.069 | 26.774 | 26.459 | 26.122 |
| **29** | 30.071 | 29.806 | 29.540 | 29.267 | 28.984 | 28.688 | 28.375 | 28.041 | 27.685 |
| **30** | 31.856 | 31.575 | 31.293 | 31.004 | 30.705 | 30.391 | 30.059 | 29.706 | 29.328 |
| **31** | 33.732 | 33.435 | 33.136 | 32.830 | 32.513 | 32.180 | 31.829 | 31.455 | 31.055 |
| **32** | 35.703 | 35.389 | 35.072 | 34.748 | 34.412 | 34.061 | 33.689 | 33.293 | 32.869 |
| **33** | 37.773 | 37.440 | 37.105 | 36.763 | 36.408 | 36.036 | 35.642 | 35.223 | 34.775 |
| **34** | 39.946 | 39.595 | 39.240 | 38.878 | 38.503 | 38.109 | 37.693 | 37.250 | 36.776 |
| **35** | 42.227 | 41.855 | 41.481 | 41.098 | 40.701 | 40.285 | 39.845 | 39.377 | 38.876 |
| **36** | 44.620 | 44.227 | 43.831 | 43.426 | 43.007 | 42.568 | 42.103 | 41.608 | 41.078 |
| **37** | 47.129 | 46.714 | 46.296 | 45.868 | 45.425 | 44.961 | 44.470 | 43.948 | 43.388 |
| **38** | 49.759 | 49.321 | 48.879 | 48.428 | 47.960 | 47.470 | 46.952 | 46.400 | 45.809 |
| **39** | 52.514 | 52.052 | 51.586 | 51.110 | 50.616 | 50.099 | 49.552 | 48.970 | 48.346 |
| **40** | 55.401 | 54.913 | 54.421 | 53.919 | 53.398 | 52.853 | 52.276 | 51.661 | 51.004 |

*Source*: Based on Eq. 23, Sherwood et al. (1991).

**Table 6.18** Vapor Pressure of Water as a Function of Temperature and Salinity in mmHg, 120–240 g/kg (NaCl brine)

| Temperature (°C) | Salinity (g/kg) | | | | | | | | |
|---|---|---|---|---|---|---|---|---|---|
| | **120** | **135** | **150** | **165** | **180** | **195** | **210** | **225** | **240** |
| **0** | 4.210 | 4.152 | 4.090 | 4.022 | 3.950 | 3.872 | 3.789 | 3.700 | 3.606 |
| **1** | 4.527 | 4.464 | 4.397 | 4.325 | 4.247 | 4.163 | 4.074 | 3.979 | 3.877 |
| **2** | 4.865 | 4.797 | 4.725 | 4.647 | 4.563 | 4.474 | 4.378 | 4.275 | 4.166 |
| **3** | 5.224 | 5.152 | 5.074 | 4.990 | 4.901 | 4.804 | 4.701 | 4.591 | 4.474 |
| **4** | 5.607 | 5.529 | 5.446 | 5.356 | 5.260 | 5.156 | 5.046 | 4.928 | 4.802 |
| **5** | 6.014 | 5.931 | 5.842 | 5.745 | 5.642 | 5.531 | 5.412 | 5.286 | 5.151 |
| **6** | 6.448 | 6.359 | 6.263 | 6.159 | 6.049 | 5.930 | 5.802 | 5.667 | 5.523 |
| **7** | 6.909 | 6.813 | 6.710 | 6.600 | 6.481 | 6.353 | 6.217 | 6.072 | 5.917 |
| **8** | 7.398 | 7.296 | 7.186 | 7.067 | 6.940 | 6.804 | 6.658 | 6.502 | 6.337 |
| **9** | 7.918 | 7.809 | 7.691 | 7.564 | 7.428 | 7.282 | 7.126 | 6.959 | 6.782 |
| **10** | 8.470 | 8.353 | 8.227 | 8.091 | 7.945 | 7.789 | 7.622 | 7.444 | 7.255 |
| **11** | 9.056 | 8.930 | 8.795 | 8.650 | 8.495 | 8.328 | 8.149 | 7.959 | 7.756 |
| **12** | 9.676 | 9.542 | 9.398 | 9.243 | 9.077 | 8.899 | 8.708 | 8.504 | 8.288 |
| **13** | 10.334 | 10.191 | 10.037 | 9.872 | 9.694 | 9.504 | 9.300 | 9.082 | 8.851 |
| **14** | 11.031 | 10.878 | 10.714 | 10.538 | 10.348 | 10.145 | 9.927 | 9.695 | 9.448 |
| **15** | 11.769 | 11.606 | 11.431 | 11.243 | 11.040 | 10.823 | 10.591 | 10.343 | 10.080 |
| **16** | 12.550 | 12.376 | 12.189 | 11.988 | 11.773 | 11.541 | 11.294 | 11.030 | 10.749 |
| **17** | 13.376 | 13.191 | 12.991 | 12.777 | 12.547 | 12.301 | 12.037 | 11.755 | 11.456 |
| **18** | 14.249 | 14.052 | 13.840 | 13.611 | 13.366 | 13.104 | 12.823 | 12.523 | 12.204 |
| **19** | 15.172 | 14.962 | 14.736 | 14.493 | 14.232 | 13.952 | 13.653 | 13.334 | 12.994 |
| **20** | 16.146 | 15.923 | 15.682 | 15.424 | 15.146 | 14.848 | 14.530 | 14.190 | 13.829 |
| **21** | 17.175 | 16.937 | 16.682 | 16.407 | 16.111 | 15.795 | 15.456 | 15.094 | 14.710 |
| **22** | 18.261 | 18.008 | 17.736 | 17.444 | 17.130 | 16.793 | 16.433 | 16.049 | 15.640 |
| **23** | 19.406 | 19.137 | 18.848 | 18.538 | 18.204 | 17.846 | 17.463 | 17.055 | 16.621 |
| **24** | 20.613 | 20.328 | 20.021 | 19.691 | 19.337 | 18.956 | 18.550 | 18.116 | 17.655 |
| **25** | 21.885 | 21.583 | 21.257 | 20.906 | 20.530 | 20.126 | 19.695 | 19.234 | 18.745 |
| **26** | 23.226 | 22.904 | 22.558 | 22.187 | 21.787 | 21.359 | 20.901 | 20.412 | 19.893 |
| **27** | 24.637 | 24.296 | 23.929 | 23.535 | 23.111 | 22.657 | 22.171 | 21.652 | 21.102 |
| **28** | 26.122 | 25.760 | 25.372 | 24.954 | 24.504 | 24.023 | 23.507 | 22.958 | 22.374 |
| **29** | 27.685 | 27.301 | 26.889 | 26.446 | 25.970 | 25.459 | 24.913 | 24.331 | 23.712 |
| **30** | 29.328 | 28.922 | 28.485 | 28.016 | 27.511 | 26.970 | 26.392 | 25.775 | 25.119 |
| **31** | 31.055 | 30.625 | 30.162 | 29.665 | 29.131 | 28.559 | 27.946 | 27.293 | 26.598 |
| **32** | 32.869 | 32.414 | 31.925 | 31.399 | 30.834 | 30.227 | 29.579 | 28.887 | 28.153 |
| **33** | 34.775 | 34.294 | 33.776 | 33.219 | 32.621 | 31.980 | 31.294 | 30.562 | 29.785 |
| **34** | 36.776 | 36.267 | 35.719 | 35.131 | 34.498 | 33.820 | 33.095 | 32.321 | 31.499 |
| **35** | 38.876 | 38.338 | 37.759 | 37.137 | 36.468 | 35.751 | 34.984 | 34.166 | 33.297 |
| **36** | 41.078 | 40.510 | 39.898 | 39.241 | 38.534 | 37.777 | 36.966 | 36.102 | 35.184 |
| **37** | 43.388 | 42.788 | 42.142 | 41.447 | 40.701 | 39.901 | 39.045 | 38.132 | 37.162 |
| **38** | 45.809 | 45.175 | 44.493 | 43.760 | 42.972 | 42.128 | 41.224 | 40.260 | 39.236 |
| **39** | 48.346 | 47.677 | 46.957 | 46.184 | 45.352 | 44.461 | 43.507 | 42.490 | 41.409 |
| **40** | 51.004 | 50.298 | 49.538 | 48.722 | 47.845 | 46.904 | 45.898 | 44.825 | 43.685 |

*Source*: Based on Eq. 23, Sherwood et al. (1991).

**Table 6.19** Density of Water as a Function of Temperature and Salinity in kg/m$^3$, 0–120 g/kg (NaCl brine)

| Temperature (°C) | Salinity (g/kg) | | | | | | | | |
|---|---|---|---|---|---|---|---|---|---|
| | **0** | **15** | **30** | **45** | **60** | **75** | **90** | **105** | **120** |
| **0** | 999.8 | 1011.3 | 1022.8 | 1034.3 | 1045.8 | 1057.4 | 1069.0 | 1080.6 | 1092.4 |
| **1** | 999.9 | 1011.3 | 1022.7 | 1034.2 | 1045.7 | 1057.2 | 1068.8 | 1080.4 | 1092.1 |
| **2** | 999.9 | 1011.3 | 1022.7 | 1034.1 | 1045.5 | 1057.0 | 1068.6 | 1080.2 | 1091.9 |
| **3** | 999.9 | 1011.3 | 1022.6 | 1034.0 | 1045.4 | 1056.9 | 1068.4 | 1080.0 | 1091.6 |
| **4** | 999.9 | 1011.2 | 1022.5 | 1033.9 | 1045.2 | 1056.7 | 1068.2 | 1079.7 | 1091.3 |
| **5** | 999.9 | 1011.2 | 1022.5 | 1033.7 | 1045.1 | 1056.5 | 1067.9 | 1079.4 | 1091.1 |
| **6** | 999.9 | 1011.1 | 1022.3 | 1033.6 | 1044.9 | 1056.3 | 1067.7 | 1079.2 | 1090.8 |
| **7** | 999.9 | 1011.0 | 1022.2 | 1033.4 | 1044.7 | 1056.0 | 1067.4 | 1078.9 | 1090.4 |
| **8** | 999.8 | 1011.0 | 1022.1 | 1033.3 | 1044.5 | 1055.8 | 1067.2 | 1078.6 | 1090.1 |
| **9** | 999.8 | 1010.8 | 1021.9 | 1033.1 | 1044.3 | 1055.5 | 1066.9 | 1078.3 | 1089.8 |
| **10** | 999.7 | 1010.7 | 1021.8 | 1032.9 | 1044.1 | 1055.3 | 1066.6 | 1078.0 | 1089.5 |
| **11** | 999.6 | 1010.6 | 1021.6 | 1032.7 | 1043.8 | 1055.0 | 1066.3 | 1077.7 | 1089.1 |
| **12** | 999.5 | 1010.5 | 1021.4 | 1032.5 | 1043.6 | 1054.8 | 1066.0 | 1077.3 | 1088.8 |
| **13** | 999.4 | 1010.3 | 1021.3 | 1032.3 | 1043.3 | 1054.5 | 1065.7 | 1077.0 | 1088.4 |
| **14** | 999.3 | 1010.1 | 1021.1 | 1032.0 | 1043.1 | 1054.2 | 1065.4 | 1076.7 | 1088.0 |
| **15** | 999.1 | 1010.0 | 1020.8 | 1031.8 | 1042.8 | 1053.9 | 1065.1 | 1076.3 | 1087.7 |
| **16** | 999.0 | 1009.8 | 1020.6 | 1031.5 | 1042.5 | 1053.6 | 1064.7 | 1076.0 | 1087.3 |
| **17** | 998.8 | 1009.6 | 1020.4 | 1031.3 | 1042.2 | 1053.3 | 1064.4 | 1075.6 | 1086.9 |
| **18** | 998.6 | 1009.4 | 1020.1 | 1031.0 | 1041.9 | 1052.9 | 1064.0 | 1075.2 | 1086.5 |
| **19** | 998.5 | 1009.1 | 1019.9 | 1030.7 | 1041.6 | 1052.6 | 1063.7 | 1074.8 | 1086.1 |
| **20** | 998.3 | 1008.9 | 1019.6 | 1030.4 | 1041.3 | 1052.3 | 1063.3 | 1074.5 | 1085.7 |
| **21** | 998.0 | 1008.7 | 1019.4 | 1030.1 | 1041.0 | 1051.9 | 1062.9 | 1074.1 | 1085.3 |
| **22** | 997.8 | 1008.4 | 1019.1 | 1029.8 | 1040.6 | 1051.6 | 1062.6 | 1073.7 | 1084.9 |
| **23** | 997.6 | 1008.1 | 1018.8 | 1029.5 | 1040.3 | 1051.2 | 1062.2 | 1073.3 | 1084.5 |
| **24** | 997.3 | 1007.9 | 1018.5 | 1029.2 | 1040.0 | 1050.8 | 1061.8 | 1072.9 | 1084.1 |
| **25** | 997.1 | 1007.6 | 1018.2 | 1028.9 | 1039.6 | 1050.5 | 1061.4 | 1072.5 | 1083.6 |
| **26** | 996.8 | 1007.3 | 1017.9 | 1028.5 | 1039.3 | 1050.1 | 1061.0 | 1072.1 | 1083.2 |
| **27** | 996.6 | 1007.0 | 1017.6 | 1028.2 | 1038.9 | 1049.7 | 1060.6 | 1071.6 | 1082.8 |
| **28** | 996.3 | 1006.7 | 1017.2 | 1027.8 | 1038.5 | 1049.3 | 1060.2 | 1071.2 | 1082.3 |
| **29** | 996.0 | 1006.4 | 1016.9 | 1027.5 | 1038.2 | 1048.9 | 1059.8 | 1070.8 | 1081.9 |
| **30** | 995.7 | 1006.1 | 1016.5 | 1027.1 | 1037.8 | 1048.5 | 1059.4 | 1070.4 | 1081.5 |
| **31** | 995.4 | 1005.7 | 1016.2 | 1026.8 | 1037.4 | 1048.2 | 1059.0 | 1070.0 | 1081.0 |
| **32** | 995.1 | 1005.4 | 1015.8 | 1026.4 | 1037.0 | 1047.8 | 1058.6 | 1069.5 | 1080.6 |
| **33** | 994.7 | 1005.1 | 1015.5 | 1026.0 | 1036.6 | 1047.3 | 1058.2 | 1069.1 | 1080.1 |
| **34** | 994.4 | 1004.7 | 1015.1 | 1025.6 | 1036.2 | 1046.9 | 1057.7 | 1068.7 | 1079.7 |
| **35** | 994.0 | 1004.3 | 1014.8 | 1025.3 | 1035.8 | 1046.5 | 1057.3 | 1068.2 | 1079.2 |
| **36** | 993.7 | 1004.0 | 1014.4 | 1024.9 | 1035.4 | 1046.1 | 1056.9 | 1067.8 | 1078.8 |
| **37** | 993.3 | 1003.6 | 1014.0 | 1024.5 | 1035.0 | 1045.7 | 1056.5 | 1067.3 | 1078.3 |
| **38** | 993.0 | 1003.2 | 1013.6 | 1024.1 | 1034.6 | 1045.3 | 1056.0 | 1066.9 | 1077.9 |
| **39** | 992.6 | 1002.9 | 1013.2 | 1023.7 | 1034.2 | 1044.9 | 1055.6 | 1066.5 | 1077.4 |
| **40** | 992.2 | 1002.5 | 1012.8 | 1023.3 | 1033.8 | 1044.4 | 1055.2 | 1066.0 | 1077.0 |

*Source*: Based on Eq. 22, Sherwood et al. (1991).

**Table 6.20** Density of Water as a Function of Temperature and Salinity in kg/m$^3$, 120–240 g/kg (NaCl brine)

| Temperature (°C) | Salinity (g/kg) | | | | | | | | |
|---|---|---|---|---|---|---|---|---|---|
| | **120** | **135** | **150** | **165** | **180** | **195** | **210** | **225** | **240** |
| **0** | 1092.4 | 1104.2 | 1116.1 | 1128.2 | 1140.3 | 1152.5 | 1164.9 | 1177.4 | 1190.0 |
| **1** | 1092.1 | 1103.9 | 1115.8 | 1127.8 | 1139.9 | 1152.2 | 1164.5 | 1177.0 | 1189.6 |
| **2** | 1091.9 | 1103.7 | 1115.5 | 1127.5 | 1139.6 | 1151.8 | 1164.1 | 1176.6 | 1189.2 |
| **3** | 1091.6 | 1103.4 | 1115.2 | 1127.2 | 1139.2 | 1151.4 | 1163.7 | 1176.2 | 1188.8 |
| **4** | 1091.3 | 1103.1 | 1114.9 | 1126.8 | 1138.9 | 1151.0 | 1163.3 | 1175.8 | 1188.3 |
| **5** | 1091.1 | 1102.8 | 1114.6 | 1126.5 | 1138.5 | 1150.6 | 1162.9 | 1175.3 | 1187.9 |
| **6** | 1090.8 | 1102.4 | 1114.2 | 1126.1 | 1138.1 | 1150.2 | 1162.5 | 1174.9 | 1187.5 |
| **7** | 1090.4 | 1102.1 | 1113.9 | 1125.7 | 1137.7 | 1149.8 | 1162.1 | 1174.5 | 1187.0 |
| **8** | 1090.1 | 1101.8 | 1113.5 | 1125.3 | 1137.3 | 1149.4 | 1161.6 | 1174.0 | 1186.5 |
| **9** | 1089.8 | 1101.4 | 1113.1 | 1124.9 | 1136.9 | 1149.0 | 1161.2 | 1173.5 | 1186.1 |
| **10** | 1089.5 | 1101.0 | 1112.7 | 1124.5 | 1136.5 | 1148.5 | 1160.7 | 1173.1 | 1185.6 |
| **11** | 1089.1 | 1100.7 | 1112.4 | 1124.1 | 1136.1 | 1148.1 | 1160.3 | 1172.6 | 1185.1 |
| **12** | 1088.8 | 1100.3 | 1112.0 | 1123.7 | 1135.6 | 1147.6 | 1159.8 | 1172.1 | 1184.6 |
| **13** | 1088.4 | 1099.9 | 1111.6 | 1123.3 | 1135.2 | 1147.2 | 1159.3 | 1171.6 | 1184.1 |
| **14** | 1088.0 | 1099.5 | 1111.1 | 1122.9 | 1134.7 | 1146.7 | 1158.9 | 1171.1 | 1183.6 |
| **15** | 1087.7 | 1099.1 | 1110.7 | 1122.4 | 1134.3 | 1146.3 | 1158.4 | 1170.7 | 1183.1 |
| **16** | 1087.3 | 1098.7 | 1110.3 | 1122.0 | 1133.8 | 1145.8 | 1157.9 | 1170.2 | 1182.6 |
| **17** | 1086.9 | 1098.3 | 1109.9 | 1121.6 | 1133.4 | 1145.3 | 1157.4 | 1169.6 | 1182.0 |
| **18** | 1086.5 | 1097.9 | 1109.5 | 1121.1 | 1132.9 | 1144.8 | 1156.9 | 1169.1 | 1181.5 |
| **19** | 1086.1 | 1097.5 | 1109.0 | 1120.7 | 1132.4 | 1144.3 | 1156.4 | 1168.6 | 1181.0 |
| **20** | 1085.7 | 1097.1 | 1108.6 | 1120.2 | 1132.0 | 1143.9 | 1155.9 | 1168.1 | 1180.5 |
| **21** | 1085.3 | 1096.7 | 1108.1 | 1119.7 | 1131.5 | 1143.4 | 1155.4 | 1167.6 | 1179.9 |
| **22** | 1084.9 | 1096.2 | 1107.7 | 1119.3 | 1131.0 | 1142.9 | 1154.9 | 1167.1 | 1179.4 |
| **23** | 1084.5 | 1095.8 | 1107.2 | 1118.8 | 1130.5 | 1142.4 | 1154.4 | 1166.5 | 1178.9 |
| **24** | 1084.1 | 1095.4 | 1106.8 | 1118.3 | 1130.0 | 1141.9 | 1153.9 | 1166.0 | 1178.3 |
| **25** | 1083.6 | 1094.9 | 1106.3 | 1117.9 | 1129.5 | 1141.4 | 1153.3 | 1165.5 | 1177.8 |
| **26** | 1083.2 | 1094.5 | 1105.9 | 1117.4 | 1129.1 | 1140.9 | 1152.8 | 1164.9 | 1177.2 |
| **27** | 1082.8 | 1094.0 | 1105.4 | 1116.9 | 1128.6 | 1140.4 | 1152.3 | 1164.4 | 1176.7 |
| **28** | 1082.3 | 1093.6 | 1104.9 | 1116.4 | 1128.1 | 1139.8 | 1151.8 | 1163.9 | 1176.1 |
| **29** | 1081.9 | 1093.1 | 1104.5 | 1115.9 | 1127.6 | 1139.3 | 1151.3 | 1163.3 | 1175.6 |
| **30** | 1081.5 | 1092.7 | 1104.0 | 1115.5 | 1127.1 | 1138.8 | 1150.7 | 1162.8 | 1175.1 |
| **31** | 1081.0 | 1092.2 | 1103.5 | 1115.0 | 1126.6 | 1138.3 | 1150.2 | 1162.3 | 1174.5 |
| **32** | 1080.6 | 1091.8 | 1103.1 | 1114.5 | 1126.1 | 1137.8 | 1149.7 | 1161.8 | 1174.0 |
| **33** | 1080.1 | 1091.3 | 1102.6 | 1114.0 | 1125.6 | 1137.3 | 1149.2 | 1161.2 | 1173.4 |
| **34** | 1079.7 | 1090.8 | 1102.1 | 1113.5 | 1125.1 | 1136.8 | 1148.7 | 1160.7 | 1172.9 |
| **35** | 1079.2 | 1090.4 | 1101.6 | 1113.1 | 1124.6 | 1136.3 | 1148.2 | 1160.2 | 1172.3 |
| **36** | 1078.8 | 1089.9 | 1101.2 | 1112.6 | 1124.1 | 1135.8 | 1147.6 | 1159.6 | 1171.8 |
| **37** | 1078.3 | 1089.5 | 1100.7 | 1112.1 | 1123.6 | 1135.3 | 1147.1 | 1159.1 | 1171.3 |
| **38** | 1077.9 | 1089.0 | 1100.2 | 1111.6 | 1123.1 | 1134.8 | 1146.6 | 1158.6 | 1170.8 |
| **39** | 1077.4 | 1088.5 | 1099.8 | 1111.1 | 1122.6 | 1134.3 | 1146.1 | 1158.1 | 1170.2 |
| **40** | 1077.0 | 1088.1 | 1099.3 | 1110.7 | 1122.2 | 1133.8 | 1145.6 | 1157.6 | 1169.7 |

*Source*: Based on Eq. 22, Sherwood et al. (1991).

# 7 Physical Properties of Water

Important physical properties of water are presented in this section as functions of temperature (0–40°C) and salinity (0–40 g/kg):



The equations used to produce these tables are presented in Appendix B.

**Dissolved Gas Concentration in Water. DOI: 10.1016/B978-0-12-415916-7.00007-3**

**Table 7.1** Density of Water as a Function of Temperature (Freshwater) in kg/L

| Temperature (°C) | Δt (°C) | | | | | | | | | |
|---|---|---|---|---|---|---|---|---|---|---|
| | **0.0** | **0.1** | **0.2** | **0.3** | **0.4** | **0.5** | **0.6** | **0.7** | **0.8** | **0.9** |
| **0** | .999843 | .999849 | .999856 | .999862 | .999868 | .999874 | .999880 | .999886 | .999891 | .999896 |
| **1** | .999902 | .999906 | .999911 | .999916 | .999920 | .999924 | .999928 | .999932 | .999936 | .999940 |
| **2** | .999943 | .999946 | .999949 | .999952 | .999955 | .999957 | .999959 | .999962 | .999964 | .999965 |
| **3** | .999967 | .999969 | .999970 | .999971 | .999972 | .999973 | .999974 | .999974 | .999975 | .999975 |
| **4** | .999975 | .999975 | .999975 | .999974 | .999974 | .999973 | .999972 | .999971 | .999970 | .999968 |
| **5** | .999967 | .999965 | .999963 | .999961 | .999959 | .999957 | .999954 | .999952 | .999949 | .999946 |
| **6** | .999943 | .999940 | .999936 | .999933 | .999929 | .999925 | .999922 | .999918 | .999913 | .999909 |
| **7** | .999904 | .999900 | .999895 | .999890 | .999885 | .999879 | .999874 | .999868 | .999863 | .999857 |
| **8** | .999851 | .999845 | .999839 | .999832 | .999826 | .999819 | .999812 | .999805 | .999798 | .999791 |
| **9** | .999783 | .999776 | .999768 | .999760 | .999753 | .999744 | .999736 | .999728 | .999719 | .999711 |
| **10** | .999702 | .999693 | .999684 | .999675 | .999666 | .999656 | .999647 | .999637 | .999627 | .999617 |
| **11** | .999607 | .999597 | .999587 | .999576 | .999566 | .999555 | .999544 | .999533 | .999522 | .999511 |
| **12** | .999500 | .999488 | .999477 | .999465 | .999453 | .999441 | .999429 | .999417 | .999404 | .999392 |
| **13** | .999379 | .999366 | .999354 | .999341 | .999328 | .999314 | .999301 | .999288 | .999274 | .999260 |
| **14** | .999246 | .999232 | .999218 | .999204 | .999190 | .999175 | .999161 | .999146 | .999131 | .999117 |
| **15** | .999102 | .999086 | .999071 | .999056 | .999040 | .999025 | .999009 | .998993 | .998977 | .998961 |
| **16** | .998945 | .998929 | .998912 | .998896 | .998879 | .998862 | .998845 | .998828 | .998811 | .998794 |
| **17** | .998777 | .998759 | .998742 | .998724 | .998706 | .998689 | .998671 | .998653 | .998634 | .998616 |
| **18** | .998598 | .998579 | .998560 | .998542 | .998523 | .998504 | .998485 | .998466 | .998446 | .998427 |
| **19** | .998407 | .998388 | .998368 | .998348 | .998328 | .998308 | .998288 | .998268 | .998247 | .998227 |
| **20** | .998206 | .998186 | .998165 | .998144 | .998123 | .998102 | .998081 | .998059 | .998038 | .998016 |
| **21** | .997995 | .997973 | .997951 | .997929 | .997907 | .997885 | .997863 | .997841 | .997818 | .997796 |
| **22** | .997773 | .997750 | .997727 | .997705 | .997681 | .997658 | .997635 | .997612 | .997588 | .997565 |
| **23** | .997541 | .997517 | .997494 | .997470 | .997446 | .997422 | .997397 | .997373 | .997349 | .997324 |
| **24** | .997299 | .997275 | .997250 | .997225 | .997200 | .997175 | .997150 | .997124 | .997099 | .997074 |
| **25** | .997048 | .997022 | .996997 | .996971 | .996945 | .996919 | .996893 | .996866 | .996840 | .996813 |
| **26** | .996787 | .996760 | .996734 | .996707 | .996680 | .996653 | .996626 | .996599 | .996571 | .996544 |
| **27** | .996517 | .996489 | .996462 | .996434 | .996406 | .996378 | .996350 | .996322 | .996294 | .996266 |
| **28** | .996237 | .996209 | .996180 | .996152 | .996123 | .996094 | .996065 | .996036 | .996007 | .995978 |
| **29** | .995949 | .995919 | .995890 | .995860 | .995831 | .995801 | .995771 | .995741 | .995711 | .995681 |
| **30** | .995651 | .995621 | .995591 | .995560 | .995530 | .995499 | .995468 | .995438 | .995407 | .995376 |
| **31** | .995345 | .995314 | .995283 | .995251 | .995220 | .995189 | .995157 | .995126 | .995094 | .995062 |
| **32** | .995030 | .994998 | .994966 | .994934 | .994902 | .994870 | .994837 | .994805 | .994772 | .994740 |
| **33** | .994707 | .994674 | .994641 | .994608 | .994575 | .994542 | .994509 | .994476 | .994443 | .994409 |
| **34** | .994376 | .994342 | .994308 | .994275 | .994241 | .994207 | .994173 | .994139 | .994105 | .994070 |
| **35** | .994036 | .994002 | .993967 | .993932 | .993898 | .993863 | .993828 | .993793 | .993758 | .993723 |
| **36** | .993688 | .993653 | .993618 | .993582 | .993547 | .993512 | .993476 | .993440 | .993405 | .993369 |
| **37** | .993333 | .993297 | .993261 | .993225 | .993188 | .993152 | .993116 | .993079 | .993043 | .993006 |
| **38** | .992970 | .992933 | .992896 | .992859 | .992822 | .992785 | .992748 | .992711 | .992673 | .992636 |
| **39** | .992599 | .992561 | .992524 | .992486 | .992448 | .992410 | .992373 | .992335 | .992297 | .992259 |
| **40** | .992220 | .992182 | .992144 | .992105 | .992067 | .992028 | .991990 | .991951 | .991913 | .991874 |

*Source*: Millero and Poisson (1982).

**Table 7.2** Density of Water as a Function of Temperature and Salinity in kg/L, 0–40 g/kg

| Temperature (°C) | Salinity (g/kg) | | | | | | | | |
|---|---|---|---|---|---|---|---|---|---|
| | **0.0** | **5.0** | **10.0** | **15.0** | **20.0** | **25.0** | **30.0** | **35.0** | **40.0** |
| **0** | 0.99984 | 1.00391 | 1.00795 | 1.01199 | 1.01601 | 1.02004 | 1.02407 | 1.02811 | 1.03215 |
| **1** | 0.99990 | 1.00395 | 1.00798 | 1.01199 | 1.01600 | 1.02001 | 1.02403 | 1.02805 | 1.03207 |
| **2** | 0.99994 | 1.00398 | 1.00798 | 1.01198 | 1.01597 | 1.01997 | 1.02397 | 1.02797 | 1.03198 |
| **3** | 0.99997 | 1.00398 | 1.00797 | 1.01195 | 1.01593 | 1.01991 | 1.02390 | 1.02789 | 1.03188 |
| **4** | 0.99997 | 1.00397 | 1.00795 | 1.01191 | 1.01588 | 1.01984 | 1.02381 | 1.02779 | 1.03177 |
| **5** | 0.99997 | 1.00395 | 1.00791 | 1.01186 | 1.01581 | 1.01976 | 1.02371 | 1.02768 | 1.03164 |
| **6** | 0.99994 | 1.00391 | 1.00785 | 1.01179 | 1.01572 | 1.01966 | 1.02360 | 1.02755 | 1.03151 |
| **7** | 0.99990 | 1.00386 | 1.00779 | 1.01171 | 1.01563 | 1.01955 | 1.02348 | 1.02742 | 1.03136 |
| **8** | 0.99985 | 1.00379 | 1.00770 | 1.01161 | 1.01552 | 1.01943 | 1.02335 | 1.02727 | 1.03121 |
| **9** | 0.99978 | 1.00371 | 1.00761 | 1.01150 | 1.01540 | 1.01930 | 1.02321 | 1.02712 | 1.03104 |
| **10** | 0.99970 | 1.00361 | 1.00750 | 1.01139 | 1.01527 | 1.01916 | 1.02305 | 1.02695 | 1.03086 |
| **11** | 0.99961 | 1.00350 | 1.00738 | 1.01125 | 1.01513 | 1.01900 | 1.02289 | 1.02678 | 1.03068 |
| **12** | 0.99950 | 1.00338 | 1.00725 | 1.01111 | 1.01497 | 1.01884 | 1.02271 | 1.02659 | 1.03048 |
| **13** | 0.99938 | 1.00325 | 1.00711 | 1.01096 | 1.01481 | 1.01866 | 1.02252 | 1.02639 | 1.03027 |
| **14** | 0.99925 | 1.00311 | 1.00695 | 1.01079 | 1.01463 | 1.01848 | 1.02233 | 1.02619 | 1.03006 |
| **15** | 0.99910 | 1.00295 | 1.00678 | 1.01061 | 1.01444 | 1.01828 | 1.02212 | 1.02597 | 1.02983 |
| **16** | 0.99894 | 1.00278 | 1.00661 | 1.01043 | 1.01425 | 1.01807 | 1.02191 | 1.02575 | 1.02960 |
| **17** | 0.99878 | 1.00261 | 1.00642 | 1.01023 | 1.01404 | 1.01786 | 1.02168 | 1.02552 | 1.02936 |
| **18** | 0.99860 | 1.00242 | 1.00622 | 1.01002 | 1.01382 | 1.01763 | 1.02145 | 1.02527 | 1.02911 |
| **19** | 0.99841 | 1.00222 | 1.00601 | 1.00980 | 1.01360 | 1.01740 | 1.02121 | 1.02502 | 1.02885 |
| **20** | 0.99821 | 1.00201 | 1.00579 | 1.00958 | 1.01336 | 1.01715 | 1.02095 | 1.02476 | 1.02858 |
| **21** | 0.99799 | 1.00179 | 1.00556 | 1.00934 | 1.01312 | 1.01690 | 1.02069 | 1.02450 | 1.02831 |
| **22** | 0.99777 | 1.00156 | 1.00533 | 1.00909 | 1.01286 | 1.01664 | 1.02043 | 1.02422 | 1.02802 |
| **23** | 0.99754 | 1.00132 | 1.00508 | 1.00884 | 1.01260 | 1.01637 | 1.02015 | 1.02394 | 1.02773 |
| **24** | 0.99730 | 1.00107 | 1.00482 | 1.00857 | 1.01233 | 1.01609 | 1.01986 | 1.02364 | 1.02743 |
| **25** | 0.99705 | 1.00081 | 1.00456 | 1.00830 | 1.01205 | 1.01581 | 1.01957 | 1.02334 | 1.02713 |
| **26** | 0.99679 | 1.00054 | 1.00428 | 1.00802 | 1.01176 | 1.01551 | 1.01927 | 1.02304 | 1.02681 |
| **27** | 0.99652 | 1.00026 | 1.00400 | 1.00773 | 1.01147 | 1.01521 | 1.01896 | 1.02272 | 1.02649 |
| **28** | 0.99624 | 0.99998 | 1.00371 | 1.00743 | 1.01116 | 1.01490 | 1.01864 | 1.02240 | 1.02616 |
| **29** | 0.99595 | 0.99968 | 1.00340 | 1.00712 | 1.01085 | 1.01458 | 1.01832 | 1.02207 | 1.02583 |
| **30** | 0.99565 | 0.99938 | 1.00310 | 1.00681 | 1.01053 | 1.01425 | 1.01799 | 1.02173 | 1.02548 |
| **31** | 0.99534 | 0.99907 | 1.00278 | 1.00649 | 1.01020 | 1.01392 | 1.01765 | 1.02138 | 1.02513 |
| **32** | 0.99503 | 0.99875 | 1.00245 | 1.00616 | 1.00986 | 1.01358 | 1.01730 | 1.02103 | 1.02477 |
| **33** | 0.99471 | 0.99842 | 1.00212 | 1.00582 | 1.00952 | 1.01323 | 1.01695 | 1.02067 | 1.02441 |
| **34** | 0.99438 | 0.99808 | 1.00178 | 1.00547 | 1.00917 | 1.01287 | 1.01658 | 1.02031 | 1.02404 |
| **35** | 0.99404 | 0.99774 | 1.00143 | 1.00512 | 1.00881 | 1.01251 | 1.01622 | 1.01993 | 1.02366 |
| **36** | 0.99369 | 0.99739 | 1.00107 | 1.00476 | 1.00844 | 1.01214 | 1.01584 | 1.01955 | 1.02328 |
| **37** | 0.99333 | 0.99703 | 1.00071 | 1.00439 | 1.00807 | 1.01176 | 1.01546 | 1.01917 | 1.02289 |
| **38** | 0.99297 | 0.99666 | 1.00034 | 1.00401 | 1.00769 | 1.01138 | 1.01507 | 1.01878 | 1.02249 |
| **39** | 0.99260 | 0.99629 | 0.99996 | 1.00363 | 1.00731 | 1.01099 | 1.01468 | 1.01838 | 1.02209 |
| **40** | 0.99222 | 0.99591 | 0.99958 | 1.00324 | 1.00691 | 1.01059 | 1.01428 | 1.01797 | 1.02168 |

*Source*: Millero and Poisson (1982).

**Table 7.3** Density of Water as a Function of Temperature and Salinity in kg/L, 33−37 g/kg

| Temperature (°C) | Salinity (g/kg) | | | | | | | | |
|---|---|---|---|---|---|---|---|---|---|
| | **33.0** | **33.5** | **34.0** | **34.5** | **35.0** | **35.5** | **36.0** | **36.5** | **37.0** |
| **0** | 1.02649 | 1.02690 | 1.02730 | 1.02770 | 1.02811 | 1.02851 | 1.02891 | 1.02932 | 1.02972 |
| **1** | 1.02644 | 1.02684 | 1.02724 | 1.02764 | 1.02805 | 1.02845 | 1.02885 | 1.02925 | 1.02965 |
| **2** | 1.02637 | 1.02677 | 1.02717 | 1.02757 | 1.02797 | 1.02837 | 1.02877 | 1.02917 | 1.02958 |
| **3** | 1.02629 | 1.02669 | 1.02709 | 1.02749 | 1.02789 | 1.02828 | 1.02868 | 1.02908 | 1.02948 |
| **4** | 1.02620 | 1.02659 | 1.02699 | 1.02739 | 1.02779 | 1.02818 | 1.02858 | 1.02898 | 1.02938 |
| **5** | 1.02609 | 1.02649 | 1.02688 | 1.02728 | 1.02768 | 1.02807 | 1.02847 | 1.02887 | 1.02926 |
| **6** | 1.02597 | 1.02637 | 1.02676 | 1.02716 | 1.02755 | 1.02795 | 1.02834 | 1.02874 | 1.02913 |
| **7** | 1.02584 | 1.02624 | 1.02663 | 1.02703 | 1.02742 | 1.02781 | 1.02821 | 1.02860 | 1.02900 |
| **8** | 1.02570 | 1.02610 | 1.02649 | 1.02688 | 1.02727 | 1.02767 | 1.02806 | 1.02845 | 1.02885 |
| **9** | 1.02555 | 1.02594 | 1.02634 | 1.02673 | 1.02712 | 1.02751 | 1.02790 | 1.02829 | 1.02869 |
| **10** | 1.02539 | 1.02578 | 1.02617 | 1.02656 | 1.02695 | 1.02734 | 1.02773 | 1.02812 | 1.02852 |
| **11** | 1.02522 | 1.02561 | 1.02600 | 1.02639 | 1.02678 | 1.02717 | 1.02756 | 1.02794 | 1.02833 |
| **12** | 1.02504 | 1.02542 | 1.02581 | 1.02620 | 1.02659 | 1.02698 | 1.02737 | 1.02776 | 1.02814 |
| **13** | 1.02484 | 1.02523 | 1.02562 | 1.02601 | 1.02639 | 1.02678 | 1.02717 | 1.02756 | 1.02794 |
| **14** | 1.02464 | 1.02503 | 1.02541 | 1.02580 | 1.02619 | 1.02657 | 1.02696 | 1.02735 | 1.02773 |
| **15** | 1.02443 | 1.02482 | 1.02520 | 1.02559 | 1.02597 | 1.02636 | 1.02674 | 1.02713 | 1.02752 |
| **16** | 1.02421 | 1.02459 | 1.02498 | 1.02536 | 1.02575 | 1.02613 | 1.02652 | 1.02690 | 1.02729 |
| **17** | 1.02398 | 1.02436 | 1.02475 | 1.02513 | 1.02552 | 1.02590 | 1.02628 | 1.02667 | 1.02705 |
| **18** | 1.02374 | 1.02412 | 1.02451 | 1.02489 | 1.02527 | 1.02566 | 1.02604 | 1.02642 | 1.02681 |
| **19** | 1.02349 | 1.02388 | 1.02426 | 1.02464 | 1.02502 | 1.02540 | 1.02579 | 1.02617 | 1.02655 |
| **20** | 1.02324 | 1.02362 | 1.02400 | 1.02438 | 1.02476 | 1.02514 | 1.02553 | 1.02591 | 1.02629 |
| **21** | 1.02297 | 1.02335 | 1.02373 | 1.02411 | 1.02450 | 1.02488 | 1.02526 | 1.02564 | 1.02602 |
| **22** | 1.02270 | 1.02308 | 1.02346 | 1.02384 | 1.02422 | 1.02460 | 1.02498 | 1.02536 | 1.02574 |
| **23** | 1.02242 | 1.02280 | 1.02318 | 1.02356 | 1.02394 | 1.02431 | 1.02469 | 1.02507 | 1.02545 |
| **24** | 1.02213 | 1.02251 | 1.02289 | 1.02326 | 1.02364 | 1.02402 | 1.02440 | 1.02478 | 1.02516 |
| **25** | 1.02183 | 1.02221 | 1.02259 | 1.02297 | 1.02334 | 1.02372 | 1.02410 | 1.02448 | 1.02486 |
| **26** | 1.02153 | 1.02190 | 1.02228 | 1.02266 | 1.02304 | 1.02341 | 1.02379 | 1.02417 | 1.02455 |
| **27** | 1.02121 | 1.02159 | 1.02197 | 1.02234 | 1.02272 | 1.02310 | 1.02347 | 1.02385 | 1.02423 |
| **28** | 1.02089 | 1.02127 | 1.02165 | 1.02202 | 1.02240 | 1.02277 | 1.02315 | 1.02353 | 1.02390 |
| **29** | 1.02057 | 1.02094 | 1.02132 | 1.02169 | 1.02207 | 1.02244 | 1.02282 | 1.02319 | 1.02357 |
| **30** | 1.02023 | 1.02060 | 1.02098 | 1.02135 | 1.02173 | 1.02210 | 1.02248 | 1.02285 | 1.02323 |
| **31** | 1.01989 | 1.02026 | 1.02064 | 1.02101 | 1.02138 | 1.02176 | 1.02213 | 1.02251 | 1.02288 |
| **32** | 1.01954 | 1.01991 | 1.02028 | 1.02066 | 1.02103 | 1.02141 | 1.02178 | 1.02215 | 1.02253 |
| **33** | 1.01918 | 1.01955 | 1.01993 | 1.02030 | 1.02067 | 1.02105 | 1.02142 | 1.02179 | 1.02217 |
| **34** | 1.01882 | 1.01919 | 1.01956 | 1.01993 | 1.02031 | 1.02068 | 1.02105 | 1.02143 | 1.02180 |
| **35** | 1.01845 | 1.01882 | 1.01919 | 1.01956 | 1.01993 | 1.02031 | 1.02068 | 1.02105 | 1.02142 |
| **36** | 1.01807 | 1.01844 | 1.01881 | 1.01918 | 1.01955 | 1.01993 | 1.02030 | 1.02067 | 1.02104 |
| **37** | 1.01768 | 1.01805 | 1.01843 | 1.01880 | 1.01917 | 1.01954 | 1.01991 | 1.02028 | 1.02065 |
| **38** | 1.01729 | 1.01766 | 1.01803 | 1.01841 | 1.01878 | 1.01915 | 1.01952 | 1.01989 | 1.02026 |
| **39** | 1.01690 | 1.01727 | 1.01764 | 1.01801 | 1.01838 | 1.01875 | 1.01912 | 1.01949 | 1.01986 |
| **40** | 1.01649 | 1.01686 | 1.01723 | 1.01760 | 1.01797 | 1.01834 | 1.01871 | 1.01908 | 1.01945 |

*Source*: Millero and Poisson (1982).

**Table 7.4** Specific Weight of Water as a Function of Temperature and Salinity in $kN/m^3$ ($g = 9.80665\ m/s^2$)

| **Temperature (°C)** | **Salinity (g/kg)** | | | | | | | | |
|---|---|---|---|---|---|---|---|---|---|
| | **0.0** | **5.0** | **10.0** | **15.0** | **20.0** | **25.0** | **30.0** | **35.0** | **40.0** |
| **0** | 9.805 | 9.845 | 9.885 | 9.924 | 9.964 | 10.003 | 10.043 | 10.082 | 10.122 |
| **1** | 9.806 | 9.845 | 9.885 | 9.924 | 9.964 | 10.003 | 10.042 | 10.082 | 10.121 |
| **2** | 9.806 | 9.846 | 9.885 | 9.924 | 9.963 | 10.002 | 10.042 | 10.081 | 10.120 |
| **3** | 9.806 | 9.846 | 9.885 | 9.924 | 9.963 | 10.002 | 10.041 | 10.080 | 10.119 |
| **4** | 9.806 | 9.846 | 9.885 | 9.923 | 9.962 | 10.001 | 10.040 | 10.079 | 10.118 |
| **5** | 9.806 | 9.845 | 9.884 | 9.923 | 9.962 | 10.000 | 10.039 | 10.078 | 10.117 |
| **6** | 9.806 | 9.845 | 9.884 | 9.922 | 9.961 | 9.999 | 10.038 | 10.077 | 10.116 |
| **7** | 9.806 | 9.844 | 9.883 | 9.921 | 9.960 | 9.998 | 10.037 | 10.076 | 10.114 |
| **8** | 9.805 | 9.844 | 9.882 | 9.921 | 9.959 | 9.997 | 10.036 | 10.074 | 10.113 |
| **9** | 9.805 | 9.843 | 9.881 | 9.919 | 9.958 | 9.996 | 10.034 | 10.073 | 10.111 |
| **10** | 9.804 | 9.842 | 9.880 | 9.918 | 9.956 | 9.995 | 10.033 | 10.071 | 10.109 |
| **11** | 9.803 | 9.841 | 9.879 | 9.917 | 9.955 | 9.993 | 10.031 | 10.069 | 10.107 |
| **12** | 9.802 | 9.840 | 9.878 | 9.916 | 9.953 | 9.991 | 10.029 | 10.067 | 10.106 |
| **13** | 9.801 | 9.839 | 9.876 | 9.914 | 9.952 | 9.990 | 10.028 | 10.065 | 10.104 |
| **14** | 9.799 | 9.837 | 9.875 | 9.912 | 9.950 | 9.988 | 10.026 | 10.063 | 10.101 |
| **15** | 9.798 | 9.836 | 9.873 | 9.911 | 9.948 | 9.986 | 10.024 | 10.061 | 10.099 |
| **16** | 9.796 | 9.834 | 9.871 | 9.909 | 9.946 | 9.984 | 10.021 | 10.059 | 10.097 |
| **17** | 9.795 | 9.832 | 9.870 | 9.907 | 9.944 | 9.982 | 10.019 | 10.057 | 10.095 |
| **18** | 9.793 | 9.830 | 9.868 | 9.905 | 9.942 | 9.980 | 10.017 | 10.054 | 10.092 |
| **19** | 9.791 | 9.828 | 9.866 | 9.903 | 9.940 | 9.977 | 10.015 | 10.052 | 10.090 |
| **20** | 9.789 | 9.826 | 9.863 | 9.901 | 9.938 | 9.975 | 10.012 | 10.049 | 10.087 |
| **21** | 9.787 | 9.824 | 9.861 | 9.898 | 9.935 | 9.972 | 10.010 | 10.047 | 10.084 |
| **22** | 9.785 | 9.822 | 9.859 | 9.896 | 9.933 | 9.970 | 10.007 | 10.044 | 10.081 |
| **23** | 9.783 | 9.820 | 9.856 | 9.893 | 9.930 | 9.967 | 10.004 | 10.041 | 10.079 |
| **24** | 9.780 | 9.817 | 9.854 | 9.891 | 9.928 | 9.964 | 10.001 | 10.039 | 10.076 |
| **25** | 9.778 | 9.815 | 9.851 | 9.888 | 9.925 | 9.962 | 9.999 | 10.036 | 10.073 |
| **26** | 9.775 | 9.812 | 9.849 | 9.885 | 9.922 | 9.959 | 9.996 | 10.033 | 10.070 |
| **27** | 9.772 | 9.809 | 9.846 | 9.882 | 9.919 | 9.956 | 9.993 | 10.029 | 10.066 |
| **28** | 9.770 | 9.806 | 9.843 | 9.880 | 9.916 | 9.953 | 9.989 | 10.026 | 10.063 |
| **29** | 9.767 | 9.804 | 9.840 | 9.877 | 9.913 | 9.950 | 9.986 | 10.023 | 10.060 |
| **30** | 9.764 | 9.801 | 9.837 | 9.873 | 9.910 | 9.946 | 9.983 | 10.020 | 10.057 |
| **31** | 9.761 | 9.798 | 9.834 | 9.870 | 9.907 | 9.943 | 9.980 | 10.016 | 10.053 |
| **32** | 9.758 | 9.794 | 9.831 | 9.867 | 9.903 | 9.940 | 9.976 | 10.013 | 10.050 |
| **33** | 9.755 | 9.791 | 9.827 | 9.864 | 9.900 | 9.936 | 9.973 | 10.009 | 10.046 |
| **34** | 9.751 | 9.788 | 9.824 | 9.860 | 9.897 | 9.933 | 9.969 | 10.006 | 10.042 |
| **35** | 9.748 | 9.784 | 9.821 | 9.857 | 9.893 | 9.929 | 9.966 | 10.002 | 10.039 |
| **36** | 9.745 | 9.781 | 9.817 | 9.853 | 9.889 | 9.926 | 9.962 | 9.998 | 10.035 |
| **37** | 9.741 | 9.778 | 9.814 | 9.850 | 9.886 | 9.922 | 9.958 | 9.995 | 10.031 |
| **38** | 9.738 | 9.774 | 9.810 | 9.846 | 9.882 | 9.918 | 9.954 | 9.991 | 10.027 |
| **39** | 9.734 | 9.770 | 9.806 | 9.842 | 9.878 | 9.914 | 9.951 | 9.987 | 10.023 |
| **40** | 9.730 | 9.766 | 9.802 | 9.838 | 9.874 | 9.911 | 9.947 | 9.983 | 10.019 |

*Source*: Millero and Poisson (1982).

**Table 7.5** Vapor Pressure of Water as a Function of Temperature and Salinity in mmHg

| Temperature(°C) | Salinity (g/kg) | | | | | | | | |
|---|---|---|---|---|---|---|---|---|---|
| | 0.0 | 5.0 | 10.0 | 15.0 | 20.0 | 25.0 | 30.0 | 35.0 | 40.0 |
| 0 | 4.58 | 4.57 | 4.56 | 4.54 | 4.53 | 4.52 | 4.51 | 4.49 | 4.48 |
| 1 | 4.92 | 4.91 | 4.90 | 4.89 | 4.87 | 4.86 | 4.85 | 4.83 | 4.82 |
| 2 | 5.29 | 5.28 | 5.26 | 5.25 | 5.24 | 5.22 | 5.21 | 5.19 | 5.18 |
| 3 | 5.68 | 5.67 | 5.65 | 5.64 | 5.62 | 5.61 | 5.59 | 5.58 | 5.56 |
| 4 | 6.10 | 6.08 | 6.07 | 6.05 | 6.03 | 6.02 | 6.00 | 5.98 | 5.97 |
| 5 | 6.54 | 6.52 | 6.51 | 6.49 | 6.47 | 6.45 | 6.44 | 6.42 | 6.40 |
| 6 | 7.01 | 6.99 | 6.98 | 6.96 | 6.94 | 6.92 | 6.90 | 6.88 | 6.86 |
| 7 | 7.51 | 7.49 | 7.47 | 7.45 | 7.43 | 7.41 | 7.39 | 7.37 | 7.35 |
| 8 | 8.04 | 8.02 | 8.00 | 7.98 | 7.96 | 7.94 | 7.92 | 7.89 | 7.87 |
| 9 | 8.61 | 8.59 | 8.56 | 8.54 | 8.52 | 8.50 | 8.47 | 8.45 | 8.42 |
| 10 | 9.21 | 9.18 | 9.16 | 9.14 | 9.11 | 9.09 | 9.06 | 9.04 | 9.01 |
| 11 | 9.84 | 9.82 | 9.79 | 9.77 | 9.74 | 9.71 | 9.69 | 9.66 | 9.63 |
| 12 | 10.52 | 10.49 | 10.46 | 10.44 | 10.41 | 10.38 | 10.35 | 10.32 | 10.29 |
| 13 | 11.23 | 11.20 | 11.17 | 11.14 | 11.11 | 11.08 | 11.05 | 11.02 | 10.99 |
| 14 | 11.99 | 11.96 | 11.93 | 11.89 | 11.86 | 11.83 | 11.80 | 11.76 | 11.73 |
| 15 | 12.79 | 12.76 | 12.72 | 12.69 | 12.66 | 12.62 | 12.59 | 12.55 | 12.51 |
| 16 | 13.64 | 13.60 | 13.57 | 13.53 | 13.49 | 13.46 | 13.42 | 13.38 | 13.34 |
| 17 | 14.53 | 14.49 | 14.46 | 14.42 | 14.38 | 14.34 | 14.30 | 14.26 | 14.22 |
| 18 | 15.48 | 15.44 | 15.40 | 15.36 | 15.32 | 15.28 | 15.23 | 15.19 | 15.15 |
| 19 | 16.48 | 16.44 | 16.40 | 16.35 | 16.31 | 16.26 | 16.22 | 16.17 | 16.13 |
| 20 | 17.54 | 17.49 | 17.45 | 17.40 | 17.36 | 17.31 | 17.26 | 17.21 | 17.16 |
| 21 | 18.66 | 18.61 | 18.56 | 18.51 | 18.46 | 18.41 | 18.36 | 18.31 | 18.25 |
| 22 | 19.83 | 19.78 | 19.73 | 19.68 | 19.63 | 19.57 | 19.52 | 19.46 | 19.41 |
| 23 | 21.08 | 21.02 | 20.97 | 20.91 | 20.86 | 20.80 | 20.74 | 20.68 | 20.62 |
| 24 | 22.39 | 22.33 | 22.27 | 22.21 | 22.15 | 22.09 | 22.03 | 21.97 | 21.90 |
| 25 | 23.77 | 23.70 | 23.64 | 23.58 | 23.52 | 23.45 | 23.39 | 23.32 | 23.25 |
| 26 | 25.22 | 25.15 | 25.09 | 25.02 | 24.96 | 24.89 | 24.82 | 24.75 | 24.68 |
| 27 | 26.75 | 26.68 | 26.61 | 26.54 | 26.47 | 26.40 | 26.33 | 26.25 | 26.18 |
| 28 | 28.36 | 28.29 | 28.22 | 28.14 | 28.07 | 27.99 | 27.91 | 27.83 | 27.75 |
| 29 | 30.06 | 29.98 | 29.90 | 29.82 | 29.74 | 29.66 | 29.58 | 29.50 | 29.41 |
| 30 | 31.84 | 31.76 | 31.68 | 31.59 | 31.51 | 31.42 | 31.33 | 31.25 | 31.15 |
| 31 | 33.71 | 33.63 | 33.54 | 33.45 | 33.36 | 33.27 | 33.18 | 33.08 | 32.99 |
| 32 | 35.68 | 35.59 | 35.50 | 35.40 | 35.31 | 35.21 | 35.12 | 35.02 | 34.91 |
| 33 | 37.75 | 37.65 | 37.56 | 37.46 | 37.36 | 37.25 | 37.15 | 37.05 | 36.94 |
| 34 | 39.92 | 39.82 | 39.71 | 39.61 | 39.50 | 39.40 | 39.29 | 39.18 | 39.06 |
| 35 | 42.20 | 42.09 | 41.98 | 41.87 | 41.76 | 41.65 | 41.53 | 41.41 | 41.29 |
| 36 | 44.59 | 44.48 | 44.36 | 44.24 | 44.12 | 44.00 | 43.88 | 43.76 | 43.63 |
| 37 | 47.10 | 46.98 | 46.85 | 46.73 | 46.61 | 46.48 | 46.35 | 46.22 | 46.08 |
| 38 | 49.73 | 49.60 | 49.47 | 49.34 | 49.21 | 49.07 | 48.94 | 48.80 | 48.65 |
| 39 | 52.48 | 52.34 | 52.21 | 52.07 | 51.93 | 51.79 | 51.65 | 51.50 | 51.35 |
| 40 | 55.36 | 55.22 | 55.08 | 54.93 | 54.78 | 54.64 | 54.48 | 54.33 | 54.17 |

*Source*: Ambrose and Lawrenson (1972).

**Table 7.6** Heat Capacity ($C_p$) of Water as a Function of Temperature and Salinity in J/(g C)

| Temperature (°C) | Salinity (g/kg) | | | | | | | | |
|---|---|---|---|---|---|---|---|---|---|
| | **0.0** | **5.0** | **10.0** | **15.0** | **20.0** | **25.0** | **30.0** | **35.0** | **40.0** |
| **0** | 4.2174 | 4.1812 | 4.1466 | 4.1130 | 4.0804 | 4.0484 | 4.0172 | 3.9865 | 3.9564 |
| **1** | 4.2138 | 4.1781 | 4.1439 | 4.1108 | 4.0785 | 4.0470 | 4.0161 | 3.9858 | 3.9561 |
| **2** | 4.2105 | 4.1752 | 4.1415 | 4.1088 | 4.0769 | 4.0458 | 4.0152 | 3.9853 | 3.9559 |
| **3** | 4.2074 | 4.1726 | 4.1393 | 4.1070 | 4.0755 | 4.0447 | 4.0146 | 3.9850 | 3.9559 |
| **4** | 4.2046 | 4.1702 | 4.1374 | 4.1054 | 4.0743 | 4.0439 | 4.0141 | 3.9848 | 3.9560 |
| **5** | 4.2020 | 4.1681 | 4.1356 | 4.1041 | 4.0733 | 4.0432 | 4.0137 | 3.9847 | 3.9563 |
| **6** | 4.1996 | 4.1661 | 4.1340 | 4.1028 | 4.0724 | 4.0427 | 4.0135 | 3.9849 | 3.9567 |
| **7** | 4.1974 | 4.1643 | 4.1326 | 4.1018 | 4.0717 | 4.0423 | 4.0134 | 3.9851 | 3.9572 |
| **8** | 4.1954 | 4.1627 | 4.1313 | 4.1009 | 4.0711 | 4.0420 | 4.0135 | 3.9854 | 3.9578 |
| **9** | 4.1936 | 4.1612 | 4.1302 | 4.1001 | 4.0707 | 4.0419 | 4.0136 | 3.9858 | 3.9585 |
| **10** | 4.1919 | 4.1599 | 4.1293 | 4.0995 | 4.0704 | 4.0418 | 4.0139 | 3.9864 | 3.9593 |
| **11** | 4.1903 | 4.1587 | 4.1284 | 4.0989 | 4.0701 | 4.0419 | 4.0142 | 3.9870 | 3.9602 |
| **12** | 4.1889 | 4.1577 | 4.1277 | 4.0985 | 4.0700 | 4.0420 | 4.0146 | 3.9876 | 3.9611 |
| **13** | 4.1877 | 4.1567 | 4.1271 | 4.0982 | 4.0699 | 4.0422 | 4.0150 | 3.9883 | 3.9620 |
| **14** | 4.1865 | 4.1559 | 4.1265 | 4.0979 | 4.0699 | 4.0425 | 4.0156 | 3.9891 | 3.9630 |
| **15** | 4.1855 | 4.1552 | 4.1261 | 4.0977 | 4.0700 | 4.0428 | 4.0161 | 3.9899 | 3.9640 |
| **16** | 4.1845 | 4.1545 | 4.1257 | 4.0976 | 4.0701 | 4.0432 | 4.0167 | 3.9907 | 3.9650 |
| **17** | 4.1837 | 4.1540 | 4.1254 | 4.0975 | 4.0703 | 4.0436 | 4.0173 | 3.9915 | 3.9660 |
| **18** | 4.1829 | 4.1535 | 4.1251 | 4.0975 | 4.0705 | 4.0440 | 4.0180 | 3.9923 | 3.9671 |
| **19** | 4.1822 | 4.1530 | 4.1249 | 4.0976 | 4.0708 | 4.0445 | 4.0186 | 3.9932 | 3.9681 |
| **20** | 4.1816 | 4.1527 | 4.1248 | 4.0976 | 4.0710 | 4.0449 | 4.0193 | 3.9940 | 3.9691 |
| **21** | 4.1811 | 4.1523 | 4.1247 | 4.0977 | 4.0713 | 4.0454 | 4.0199 | 3.9948 | 3.9701 |
| **22** | 4.1806 | 4.1521 | 4.1246 | 4.0978 | 4.0716 | 4.0459 | 4.0206 | 3.9956 | 3.9711 |
| **23** | 4.1801 | 4.1518 | 4.1246 | 4.0980 | 4.0719 | 4.0464 | 4.0212 | 3.9964 | 3.9720 |
| **24** | 4.1797 | 4.1516 | 4.1246 | 4.0981 | 4.0723 | 4.0468 | 4.0218 | 3.9972 | 3.9729 |
| **25** | 4.1794 | 4.1515 | 4.1246 | 4.0983 | 4.0726 | 4.0473 | 4.0224 | 3.9979 | 3.9738 |
| **26** | 4.1791 | 4.1513 | 4.1246 | 4.0985 | 4.0729 | 4.0477 | 4.0230 | 3.9986 | 3.9746 |
| **27** | 4.1788 | 4.1512 | 4.1246 | 4.0986 | 4.0732 | 4.0482 | 4.0235 | 3.9993 | 3.9754 |
| **28** | 4.1786 | 4.1511 | 4.1247 | 4.0988 | 4.0735 | 4.0486 | 4.0241 | 3.9999 | 3.9761 |
| **29** | 4.1784 | 4.1511 | 4.1247 | 4.0990 | 4.0737 | 4.0490 | 4.0246 | 4.0005 | 3.9768 |
| **30** | 4.1782 | 4.1510 | 4.1248 | 4.0991 | 4.0740 | 4.0493 | 4.0250 | 4.0010 | 3.9774 |
| **31** | 4.1781 | 4.1510 | 4.1248 | 4.0993 | 4.0743 | 4.0497 | 4.0254 | 4.0015 | 3.9780 |
| **32** | 4.1780 | 4.1510 | 4.1249 | 4.0995 | 4.0745 | 4.0500 | 4.0258 | 4.0020 | 3.9785 |
| **33** | 4.1779 | 4.1510 | 4.1250 | 4.0996 | 4.0747 | 4.0502 | 4.0262 | 4.0024 | 3.9790 |
| **34** | 4.1779 | 4.1510 | 4.1251 | 4.0998 | 4.0749 | 4.0505 | 4.0265 | 4.0028 | 3.9794 |
| **35** | 4.1779 | 4.1511 | 4.1252 | 4.0999 | 4.0751 | 4.0507 | 4.0267 | 4.0031 | 3.9798 |
| **36** | 4.1779 | 4.1511 | 4.1253 | 4.1000 | 4.0753 | 4.0510 | 4.0270 | 4.0034 | 3.9801 |
| **37** | 4.1779 | 4.1512 | 4.1254 | 4.1002 | 4.0755 | 4.0511 | 4.0272 | 4.0036 | 3.9803 |
| **38** | 4.1780 | 4.1513 | 4.1255 | 4.1003 | 4.0756 | 4.0513 | 4.0274 | 4.0038 | 3.9806 |
| **39** | 4.1782 | 4.1515 | 4.1257 | 4.1005 | 4.0758 | 4.0515 | 4.0276 | 4.0040 | 3.9807 |
| **40** | 4.1784 | 4.1516 | 4.1258 | 4.1006 | 4.0759 | 4.0516 | 4.0277 | 4.0041 | 3.9809 |

*Source*: Millero et al. (1973).

**Table 7.7** Viscosity of Water as a Function of Temperature and Salinity in N s/m$^2$ × 10$^3$ (At 10°C and 35 g/kg, the table value is 1.3864; the viscosity is 1.3864 × 10$^{-3}$ or 0.0013864 N s/m$^2$)

| Temperature (°C) | Salinity (g/kg) | | | | | | | | |
|---|---|---|---|---|---|---|---|---|---|
| | **0.0** | **5.0** | **10.0** | **15.0** | **20.0** | **25.0** | **30.0** | **35.0** | **40.0** |
| **0** | 1.7912 | 1.8043 | 1.8175 | 1.8307 | 1.8440 | 1.8574 | 1.8709 | 1.8845 | 1.8982 |
| **1** | 1.7309 | 1.7439 | 1.7569 | 1.7698 | 1.7829 | 1.7960 | 1.8093 | 1.8226 | 1.8360 |
| **2** | 1.6738 | 1.6866 | 1.6993 | 1.7121 | 1.7249 | 1.7378 | 1.7508 | 1.7639 | 1.7771 |
| **3** | 1.6195 | 1.6323 | 1.6448 | 1.6573 | 1.6699 | 1.6826 | 1.6953 | 1.7082 | 1.7211 |
| **4** | 1.5680 | 1.5806 | 1.5929 | 1.6052 | 1.6176 | 1.6301 | 1.6426 | 1.6552 | 1.6679 |
| **5** | 1.5190 | 1.5315 | 1.5436 | 1.5557 | 1.5679 | 1.5801 | 1.5924 | 1.6048 | 1.6173 |
| **6** | 1.4724 | 1.4847 | 1.4967 | 1.5086 | 1.5206 | 1.5326 | 1.5447 | 1.5569 | 1.5691 |
| **7** | 1.4280 | 1.4402 | 1.4520 | 1.4637 | 1.4755 | 1.4873 | 1.4992 | 1.5112 | 1.5232 |
| **8** | 1.3857 | 1.3978 | 1.4094 | 1.4210 | 1.4325 | 1.4442 | 1.4559 | 1.4676 | 1.4795 |
| **9** | 1.3453 | 1.3573 | 1.3688 | 1.3801 | 1.3916 | 1.4030 | 1.4145 | 1.4261 | 1.4377 |
| **10** | 1.3068 | 1.3187 | 1.3300 | 1.3412 | 1.3524 | 1.3637 | 1.3750 | 1.3864 | 1.3978 |
| **11** | 1.2700 | 1.2818 | 1.2929 | 1.3040 | 1.3150 | 1.3261 | 1.3373 | 1.3485 | 1.3598 |
| **12** | 1.2349 | 1.2466 | 1.2575 | 1.2684 | 1.2793 | 1.2902 | 1.3012 | 1.3122 | 1.3233 |
| **13** | 1.2012 | 1.2128 | 1.2236 | 1.2344 | 1.2451 | 1.2559 | 1.2667 | 1.2776 | 1.2885 |
| **14** | 1.1690 | 1.1805 | 1.1912 | 1.2018 | 1.2124 | 1.2230 | 1.2336 | 1.2443 | 1.2551 |
| **15** | 1.1382 | 1.1496 | 1.1601 | 1.1706 | 1.1810 | 1.1915 | 1.2020 | 1.2125 | 1.2231 |
| **16** | 1.1087 | 1.1200 | 1.1304 | 1.1407 | 1.1510 | 1.1613 | 1.1716 | 1.1820 | 1.1925 |
| **17** | 1.0803 | 1.0915 | 1.1018 | 1.1120 | 1.1221 | 1.1323 | 1.1425 | 1.1528 | 1.1631 |
| **18** | 1.0532 | 1.0642 | 1.0744 | 1.0844 | 1.0945 | 1.1045 | 1.1146 | 1.1247 | 1.1348 |
| **19** | 1.0271 | 1.0381 | 1.0481 | 1.0580 | 1.0679 | 1.0778 | 1.0877 | 1.0977 | 1.1077 |
| **20** | 1.0020 | 1.0129 | 1.0228 | 1.0326 | 1.0424 | 1.0521 | 1.0619 | 1.0718 | 1.0817 |
| **21** | 0.9779 | 0.9887 | 0.9985 | 1.0082 | 1.0178 | 1.0275 | 1.0371 | 1.0469 | 1.0566 |
| **22** | 0.9547 | 0.9654 | 0.9751 | 0.9847 | 0.9942 | 1.0037 | 1.0133 | 1.0229 | 1.0325 |
| **23** | 0.9324 | 0.9431 | 0.9526 | 0.9621 | 0.9715 | 0.9809 | 0.9903 | 0.9998 | 1.0093 |
| **24** | 0.9110 | 0.9215 | 0.9310 | 0.9403 | 0.9496 | 0.9589 | 0.9682 | 0.9776 | 0.9870 |
| **25** | 0.8903 | 0.9007 | 0.9101 | 0.9193 | 0.9285 | 0.9377 | 0.9469 | 0.9561 | 0.9654 |
| **26** | 0.8704 | 0.8807 | 0.8900 | 0.8991 | 0.9082 | 0.9173 | 0.9264 | 0.9355 | 0.9447 |
| **27** | 0.8512 | 0.8614 | 0.8706 | 0.8796 | 0.8886 | 0.8976 | 0.9066 | 0.9156 | 0.9246 |
| **28** | 0.8326 | 0.8428 | 0.8519 | 0.8608 | 0.8697 | 0.8786 | 0.8875 | 0.8964 | 0.9053 |
| **29** | 0.8147 | 0.8249 | 0.8338 | 0.8427 | 0.8515 | 0.8602 | 0.8690 | 0.8778 | 0.8867 |
| **30** | 0.7975 | 0.8075 | 0.8164 | 0.8251 | 0.8338 | 0.8425 | 0.8512 | 0.8599 | 0.8686 |
| **31** | 0.7808 | 0.7908 | 0.7996 | 0.8082 | 0.8168 | 0.8254 | 0.8340 | 0.8426 | 0.8512 |
| **32** | 0.7647 | 0.7746 | 0.7833 | 0.7918 | 0.8003 | 0.8088 | 0.8173 | 0.8258 | 0.8344 |
| **33** | 0.7491 | 0.7589 | 0.7675 | 0.7760 | 0.7844 | 0.7928 | 0.8012 | 0.8097 | 0.8181 |
| **34** | 0.7340 | 0.7438 | 0.7523 | 0.7607 | 0.7690 | 0.7773 | 0.7857 | 0.7940 | 0.8024 |
| **35** | 0.7194 | 0.7291 | 0.7376 | 0.7459 | 0.7541 | 0.7624 | 0.7706 | 0.7788 | 0.7871 |
| **36** | 0.7053 | 0.7149 | 0.7233 | 0.7315 | 0.7397 | 0.7478 | 0.7560 | 0.7642 | 0.7723 |
| **37** | 0.6917 | 0.7012 | 0.7095 | 0.7177 | 0.7257 | 0.7338 | 0.7419 | 0.7499 | 0.7580 |
| **38** | 0.6784 | 0.6879 | 0.6961 | 0.7042 | 0.7122 | 0.7202 | 0.7282 | 0.7362 | 0.7442 |
| **39** | 0.6656 | 0.6750 | 0.6832 | 0.6912 | 0.6991 | 0.7070 | 0.7149 | 0.7228 | 0.7307 |
| **40** | 0.6531 | 0.6625 | 0.6706 | 0.6785 | 0.6864 | 0.6942 | 0.7020 | 0.7098 | 0.7177 |

*Source*: Riley and Skirrow (1975).

**Table 7.8** Kinematic Viscosity of Water as a Function of Temperature and Salinity in $m^2/s \times 10^6$ (At 10°C and 35 g/kg, the table value is 1.3500; the kinematic viscosity is $1.3500 \times 10^{-6}$ or 0.0000013500 $m^2/s$)

| Temperature (°C) | Salinity (g/kg) | | | | | | | | |
|---|---|---|---|---|---|---|---|---|---|
| | 0.0 | 5.0 | 10.0 | 15.0 | 20.0 | 25.0 | 30.0 | 35.0 | 40.0 |
| 0 | 1.7915 | 1.7973 | 1.8031 | 1.8090 | 1.8149 | 1.8209 | 1.8269 | 1.8330 | 1.8391 |
| 1 | 1.7311 | 1.7371 | 1.7429 | 1.7489 | 1.7548 | 1.7608 | 1.7668 | 1.7729 | 1.7790 |
| 2 | 1.6739 | 1.6800 | 1.6859 | 1.6918 | 1.6978 | 1.7038 | 1.7098 | 1.7159 | 1.7220 |
| 3 | 1.6196 | 1.6258 | 1.6318 | 1.6377 | 1.6437 | 1.6497 | 1.6558 | 1.6618 | 1.6679 |
| 4 | 1.5680 | 1.5743 | 1.5803 | 1.5863 | 1.5923 | 1.5983 | 1.6044 | 1.6104 | 1.6165 |
| 5 | 1.5190 | 1.5255 | 1.5315 | 1.5375 | 1.5435 | 1.5495 | 1.5555 | 1.5616 | 1.5677 |
| 6 | 1.4724 | 1.4790 | 1.4850 | 1.4910 | 1.4970 | 1.5030 | 1.5091 | 1.5151 | 1.5212 |
| 7 | 1.4281 | 1.4347 | 1.4408 | 1.4468 | 1.4528 | 1.4588 | 1.4648 | 1.4709 | 1.4769 |
| 8 | 1.3859 | 1.3925 | 1.3986 | 1.4046 | 1.4106 | 1.4166 | 1.4226 | 1.4287 | 1.4347 |
| 9 | 1.3456 | 1.3523 | 1.3584 | 1.3644 | 1.3704 | 1.3764 | 1.3824 | 1.3884 | 1.3944 |
| 10 | 1.3072 | 1.3140 | 1.3201 | 1.3261 | 1.3321 | 1.3381 | 1.3440 | 1.3500 | 1.3560 |
| 11 | 1.2705 | 1.2773 | 1.2835 | 1.2895 | 1.2954 | 1.3014 | 1.3074 | 1.3133 | 1.3193 |
| 12 | 1.2355 | 1.2423 | 1.2485 | 1.2545 | 1.2604 | 1.2664 | 1.2723 | 1.2783 | 1.2842 |
| 13 | 1.2020 | 1.2089 | 1.2150 | 1.2210 | 1.2269 | 1.2329 | 1.2388 | 1.2447 | 1.2506 |
| 14 | 1.1699 | 1.1769 | 1.1830 | 1.1890 | 1.1949 | 1.2008 | 1.2067 | 1.2126 | 1.2185 |
| 15 | 1.1392 | 1.1462 | 1.1523 | 1.1583 | 1.1642 | 1.1701 | 1.1760 | 1.1818 | 1.1877 |
| 16 | 1.1099 | 1.1168 | 1.1229 | 1.1289 | 1.1348 | 1.1407 | 1.1465 | 1.1524 | 1.1582 |
| 17 | 1.0817 | 1.0887 | 1.0948 | 1.1007 | 1.1066 | 1.1124 | 1.1183 | 1.1241 | 1.1299 |
| 18 | 1.0546 | 1.0617 | 1.0678 | 1.0737 | 1.0795 | 1.0854 | 1.0912 | 1.0970 | 1.1027 |
| 19 | 1.0287 | 1.0358 | 1.0418 | 1.0477 | 1.0536 | 1.0594 | 1.0651 | 1.0709 | 1.0767 |
| 20 | 1.0038 | 1.0109 | 1.0169 | 1.0228 | 1.0286 | 1.0344 | 1.0401 | 1.0459 | 1.0516 |
| 21 | 0.9799 | 0.9869 | 0.9930 | 0.9988 | 1.0046 | 1.0104 | 1.0161 | 1.0218 | 1.0275 |
| 22 | 0.9569 | 0.9639 | 0.9700 | 0.9758 | 0.9816 | 0.9873 | 0.9930 | 0.9987 | 1.0044 |
| 23 | 0.9347 | 0.9418 | 0.9478 | 0.9536 | 0.9594 | 0.9651 | 0.9708 | 0.9764 | 0.9821 |
| 24 | 0.9134 | 0.9205 | 0.9265 | 0.9323 | 0.9380 | 0.9437 | 0.9494 | 0.9550 | 0.9606 |
| 25 | 0.8929 | 0.9000 | 0.9060 | 0.9118 | 0.9175 | 0.9231 | 0.9287 | 0.9343 | 0.9399 |
| 26 | 0.8732 | 0.8803 | 0.8862 | 0.8920 | 0.8976 | 0.9033 | 0.9089 | 0.9144 | 0.9200 |
| 27 | 0.8541 | 0.8612 | 0.8671 | 0.8729 | 0.8785 | 0.8841 | 0.8897 | 0.8952 | 0.9008 |
| 28 | 0.8358 | 0.8428 | 0.8488 | 0.8545 | 0.8601 | 0.8657 | 0.8712 | 0.8767 | 0.8822 |
| 29 | 0.8180 | 0.8251 | 0.8310 | 0.8367 | 0.8423 | 0.8479 | 0.8534 | 0.8589 | 0.8643 |
| 30 | 0.8009 | 0.8080 | 0.8139 | 0.8196 | 0.8251 | 0.8307 | 0.8361 | 0.8416 | 0.8470 |
| 31 | 0.7844 | 0.7915 | 0.7974 | 0.8030 | 0.8086 | 0.8140 | 0.8195 | 0.8249 | 0.8304 |
| 32 | 0.7685 | 0.7755 | 0.7814 | 0.7870 | 0.7925 | 0.7980 | 0.8034 | 0.8088 | 0.8142 |
| 33 | 0.7531 | 0.7601 | 0.7659 | 0.7715 | 0.7770 | 0.7825 | 0.7879 | 0.7933 | 0.7986 |
| 34 | 0.7382 | 0.7452 | 0.7510 | 0.7566 | 0.7620 | 0.7675 | 0.7728 | 0.7782 | 0.7835 |
| 35 | 0.7237 | 0.7308 | 0.7365 | 0.7421 | 0.7475 | 0.7529 | 0.7583 | 0.7636 | 0.7689 |
| 36 | 0.7098 | 0.7168 | 0.7226 | 0.7281 | 0.7335 | 0.7389 | 0.7442 | 0.7495 | 0.7548 |
| 37 | 0.6963 | 0.7033 | 0.7090 | 0.7145 | 0.7199 | 0.7253 | 0.7306 | 0.7358 | 0.7411 |
| 38 | 0.6832 | 0.6902 | 0.6959 | 0.7014 | 0.7068 | 0.7121 | 0.7173 | 0.7226 | 0.7278 |
| 39 | 0.6705 | 0.6775 | 0.6832 | 0.6887 | 0.6940 | 0.6993 | 0.7045 | 0.7098 | 0.7149 |
| 40 | 0.6583 | 0.6652 | 0.6709 | 0.6763 | 0.6816 | 0.6869 | 0.6921 | 0.6973 | 0.7025 |

*Source*: Riley and Skirrow (1975), Millero and Poisson (1981).

**Table 7.9** Surface Tension of Water as a Function of Temperature and Salinity in N/m × $10^3$ (At 10°C and 35 g/kg, the table value is 74.97; the surface tension is 74.97 × $10^{-3}$ or 0.07497 N/m)

| Temperature (°C) | Salinity (g/kg) | | | | | | | | |
|---|---|---|---|---|---|---|---|---|---|
| | **0.0** | **5.0** | **10.0** | **15.0** | **20.0** | **25.0** | **30.0** | **35.0** | **40.0** |
| **0** | 75.64 | 75.75 | 75.86 | 75.97 | 76.08 | 76.19 | 76.30 | 76.41 | 76.52 |
| **1** | 75.50 | 75.61 | 75.72 | 75.83 | 75.94 | 76.05 | 76.16 | 76.27 | 76.38 |
| **2** | 75.35 | 75.46 | 75.57 | 75.68 | 75.79 | 75.90 | 76.01 | 76.13 | 76.24 |
| **3** | 75.21 | 75.32 | 75.43 | 75.54 | 75.65 | 75.76 | 75.87 | 75.98 | 76.09 |
| **4** | 75.06 | 75.17 | 75.28 | 75.40 | 75.51 | 75.62 | 75.73 | 75.84 | 75.95 |
| **5** | 74.92 | 75.03 | 75.14 | 75.25 | 75.36 | 75.47 | 75.58 | 75.69 | 75.80 |
| **6** | 74.78 | 74.89 | 75.00 | 75.11 | 75.22 | 75.33 | 75.44 | 75.55 | 75.66 |
| **7** | 74.63 | 74.74 | 74.85 | 74.96 | 75.07 | 75.18 | 75.29 | 75.41 | 75.52 |
| **8** | 74.49 | 74.60 | 74.71 | 74.82 | 74.93 | 75.04 | 75.15 | 75.26 | 75.37 |
| **9** | 74.34 | 74.45 | 74.57 | 74.68 | 74.79 | 74.90 | 75.01 | 75.12 | 75.23 |
| **10** | 74.20 | 74.31 | 74.42 | 74.53 | 74.64 | 74.75 | 74.86 | 74.97 | 75.08 |
| **11** | 74.06 | 74.17 | 74.28 | 74.39 | 74.50 | 74.61 | 74.72 | 74.83 | 74.94 |
| **12** | 73.91 | 74.02 | 74.13 | 74.24 | 74.35 | 74.46 | 74.57 | 74.69 | 74.80 |
| **13** | 73.77 | 73.88 | 73.99 | 74.10 | 74.21 | 74.32 | 74.43 | 74.54 | 74.65 |
| **14** | 73.62 | 73.73 | 73.85 | 73.96 | 74.07 | 74.18 | 74.29 | 74.40 | 74.51 |
| **15** | 73.48 | 73.59 | 73.70 | 73.81 | 73.92 | 74.03 | 74.14 | 74.25 | 74.36 |
| **16** | 73.34 | 73.45 | 73.56 | 73.67 | 73.78 | 73.89 | 74.00 | 74.11 | 74.22 |
| **17** | 73.19 | 73.30 | 73.41 | 73.52 | 73.63 | 73.74 | 73.85 | 73.97 | 74.08 |
| **18** | 73.05 | 73.16 | 73.27 | 73.38 | 73.49 | 73.60 | 73.71 | 73.82 | 73.93 |
| **19** | 72.90 | 73.01 | 73.13 | 73.24 | 73.35 | 73.46 | 73.57 | 73.68 | 73.79 |
| **20** | 72.76 | 72.87 | 72.98 | 73.09 | 73.20 | 73.31 | 73.42 | 73.53 | 73.64 |
| **21** | 72.62 | 72.73 | 72.84 | 72.95 | 73.06 | 73.17 | 73.28 | 73.39 | 73.50 |
| **22** | 72.47 | 72.58 | 72.69 | 72.80 | 72.91 | 73.02 | 73.14 | 73.25 | 73.36 |
| **23** | 72.33 | 72.44 | 72.55 | 72.66 | 72.77 | 72.88 | 72.99 | 73.10 | 73.21 |
| **24** | 72.18 | 72.29 | 72.40 | 72.52 | 72.63 | 72.74 | 72.85 | 72.96 | 73.07 |
| **25** | 72.04 | 72.15 | 72.26 | 72.37 | 72.48 | 72.59 | 72.70 | 72.81 | 72.92 |
| **26** | 71.90 | 72.01 | 72.12 | 72.23 | 72.34 | 72.45 | 72.56 | 72.67 | 72.78 |
| **27** | 71.75 | 71.86 | 71.97 | 72.08 | 72.19 | 72.30 | 72.42 | 72.53 | 72.64 |
| **28** | 71.61 | 71.72 | 71.83 | 71.94 | 72.05 | 72.16 | 72.27 | 72.38 | 72.49 |
| **29** | 71.46 | 71.57 | 71.68 | 71.80 | 71.91 | 72.02 | 72.13 | 72.24 | 72.35 |
| **30** | 71.32 | 71.43 | 71.54 | 71.65 | 71.76 | 71.87 | 71.98 | 72.09 | 72.20 |
| **31** | 71.18 | 71.29 | 71.40 | 71.51 | 71.62 | 71.73 | 71.84 | 71.95 | 72.06 |
| **32** | 71.03 | 71.14 | 71.25 | 71.36 | 71.47 | 71.58 | 71.69 | 71.81 | 71.92 |
| **33** | 70.89 | 71.00 | 71.11 | 71.22 | 71.33 | 71.44 | 71.55 | 71.66 | 71.77 |
| **34** | 70.74 | 70.85 | 70.96 | 71.08 | 71.19 | 71.30 | 71.41 | 71.52 | 71.63 |
| **35** | 70.60 | 70.71 | 70.82 | 70.93 | 71.04 | 71.15 | 71.26 | 71.37 | 71.48 |
| **36** | 70.46 | 70.57 | 70.68 | 70.79 | 70.90 | 71.01 | 71.12 | 71.23 | 71.34 |
| **37** | 70.31 | 70.42 | 70.53 | 70.64 | 70.75 | 70.86 | 70.97 | 71.09 | 71.20 |
| **38** | 70.17 | 70.28 | 70.39 | 70.50 | 70.61 | 70.72 | 70.83 | 70.94 | 71.05 |
| **39** | 70.02 | 70.13 | 70.25 | 70.36 | 70.47 | 70.58 | 70.69 | 70.80 | 70.91 |
| **40** | 69.88 | 69.99 | 70.10 | 70.21 | 70.32 | 70.43 | 70.54 | 70.65 | 70.76 |

*Source*: Riley and Skirrow (1975).

**Table 7.10** Heat of Vaporization in MJ/kg

| Temperature (°C) | Δt (°C) | | | | | | | | | |
|---|---|---|---|---|---|---|---|---|---|---|
| | **0.0** | **0.1** | **0.2** | **0.3** | **0.4** | **0.5** | **0.6** | **0.7** | **0.8** | **0.9** |
| **0** | 2.5025 | 2.5023 | 2.5021 | 2.5018 | 2.5016 | 2.5013 | 2.5011 | 2.5009 | 2.5006 | 2.5004 |
| **1** | 2.5001 | 2.4999 | 2.4997 | 2.4994 | 2.4992 | 2.4990 | 2.4987 | 2.4985 | 2.4982 | 2.4980 |
| **2** | 2.4978 | 2.4975 | 2.4973 | 2.4970 | 2.4968 | 2.4966 | 2.4963 | 2.4961 | 2.4959 | 2.4956 |
| **3** | 2.4954 | 2.4951 | 2.4949 | 2.4947 | 2.4944 | 2.4942 | 2.4939 | 2.4937 | 2.4935 | 2.4932 |
| **4** | 2.4930 | 2.4928 | 2.4925 | 2.4923 | 2.4920 | 2.4918 | 2.4916 | 2.4913 | 2.4911 | 2.4908 |
| **5** | 2.4906 | 2.4904 | 2.4901 | 2.4899 | 2.4897 | 2.4894 | 2.4892 | 2.4889 | 2.4887 | 2.4885 |
| **6** | 2.4882 | 2.4880 | 2.4877 | 2.4875 | 2.4873 | 2.4870 | 2.4868 | 2.4866 | 2.4863 | 2.4861 |
| **7** | 2.4858 | 2.4856 | 2.4854 | 2.4851 | 2.4849 | 2.4846 | 2.4844 | 2.4842 | 2.4839 | 2.4837 |
| **8** | 2.4834 | 2.4832 | 2.4830 | 2.4827 | 2.4825 | 2.4823 | 2.4820 | 2.4818 | 2.4815 | 2.4813 |
| **9** | 2.4811 | 2.4808 | 2.4806 | 2.4803 | 2.4801 | 2.4799 | 2.4796 | 2.4794 | 2.4792 | 2.4789 |
| **10** | 2.4787 | 2.4784 | 2.4782 | 2.4780 | 2.4777 | 2.4775 | 2.4772 | 2.4770 | 2.4768 | 2.4765 |
| **11** | 2.4763 | 2.4761 | 2.4758 | 2.4756 | 2.4753 | 2.4751 | 2.4749 | 2.4746 | 2.4744 | 2.4741 |
| **12** | 2.4739 | 2.4737 | 2.4734 | 2.4732 | 2.4730 | 2.4727 | 2.4725 | 2.4722 | 2.4720 | 2.4718 |
| **13** | 2.4715 | 2.4713 | 2.4710 | 2.4708 | 2.4706 | 2.4703 | 2.4701 | 2.4699 | 2.4696 | 2.4694 |
| **14** | 2.4691 | 2.4689 | 2.4687 | 2.4684 | 2.4682 | 2.4679 | 2.4677 | 2.4675 | 2.4672 | 2.4670 |
| **15** | 2.4667 | 2.4665 | 2.4663 | 2.4660 | 2.4658 | 2.4656 | 2.4653 | 2.4651 | 2.4648 | 2.4646 |
| **16** | 2.4644 | 2.4641 | 2.4639 | 2.4636 | 2.4634 | 2.4632 | 2.4629 | 2.4627 | 2.4625 | 2.4622 |
| **17** | 2.4620 | 2.4617 | 2.4615 | 2.4613 | 2.4610 | 2.4608 | 2.4605 | 2.4603 | 2.4601 | 2.4598 |
| **18** | 2.4596 | 2.4594 | 2.4591 | 2.4589 | 2.4586 | 2.4584 | 2.4582 | 2.4579 | 2.4577 | 2.4574 |
| **19** | 2.4572 | 2.4570 | 2.4567 | 2.4565 | 2.4563 | 2.4560 | 2.4558 | 2.4555 | 2.4553 | 2.4551 |
| **20** | 2.4548 | 2.4546 | 2.4543 | 2.4541 | 2.4539 | 2.4536 | 2.4534 | 2.4532 | 2.4529 | 2.4527 |
| **21** | 2.4524 | 2.4522 | 2.4520 | 2.4517 | 2.4515 | 2.4512 | 2.4510 | 2.4508 | 2.4505 | 2.4503 |
| **22** | 2.4500 | 2.4498 | 2.4496 | 2.4493 | 2.4491 | 2.4489 | 2.4486 | 2.4484 | 2.4481 | 2.4479 |
| **23** | 2.4477 | 2.4474 | 2.4472 | 2.4469 | 2.4467 | 2.4465 | 2.4462 | 2.4460 | 2.4458 | 2.4455 |
| **24** | 2.4453 | 2.4450 | 2.4448 | 2.4446 | 2.4443 | 2.4441 | 2.4438 | 2.4436 | 2.4434 | 2.4431 |
| **25** | 2.4429 | 2.4427 | 2.4424 | 2.4422 | 2.4419 | 2.4417 | 2.4415 | 2.4412 | 2.4410 | 2.4407 |
| **26** | 2.4405 | 2.4403 | 2.4400 | 2.4398 | 2.4396 | 2.4393 | 2.4391 | 2.4388 | 2.4386 | 2.4384 |
| **27** | 2.4381 | 2.4379 | 2.4376 | 2.4374 | 2.4372 | 2.4369 | 2.4367 | 2.4364 | 2.4362 | 2.4360 |
| **28** | 2.4357 | 2.4355 | 2.4353 | 2.4350 | 2.4348 | 2.4345 | 2.4343 | 2.4341 | 2.4338 | 2.4336 |
| **29** | 2.4333 | 2.4331 | 2.4329 | 2.4326 | 2.4324 | 2.4322 | 2.4319 | 2.4317 | 2.4314 | 2.4312 |
| **30** | 2.4310 | 2.4307 | 2.4305 | 2.4302 | 2.4300 | 2.4298 | 2.4295 | 2.4293 | 2.4291 | 2.4288 |
| **31** | 2.4286 | 2.4283 | 2.4281 | 2.4279 | 2.4276 | 2.4274 | 2.4271 | 2.4269 | 2.4267 | 2.4264 |
| **32** | 2.4262 | 2.4260 | 2.4257 | 2.4255 | 2.4252 | 2.4250 | 2.4248 | 2.4245 | 2.4243 | 2.4240 |
| **33** | 2.4238 | 2.4236 | 2.4233 | 2.4231 | 2.4229 | 2.4226 | 2.4224 | 2.4221 | 2.4219 | 2.4217 |
| **34** | 2.4214 | 2.4212 | 2.4209 | 2.4207 | 2.4205 | 2.4202 | 2.4200 | 2.4197 | 2.4195 | 2.4193 |
| **35** | 2.4190 | 2.4188 | 2.4186 | 2.4183 | 2.4181 | 2.4178 | 2.4176 | 2.4174 | 2.4171 | 2.4169 |
| **36** | 2.4166 | 2.4164 | 2.4162 | 2.4159 | 2.4157 | 2.4155 | 2.4152 | 2.4150 | 2.4147 | 2.4145 |
| **37** | 2.4143 | 2.4140 | 2.4138 | 2.4135 | 2.4133 | 2.4131 | 2.4128 | 2.4126 | 2.4124 | 2.4121 |
| **38** | 2.4119 | 2.4116 | 2.4114 | 2.4112 | 2.4109 | 2.4107 | 2.4104 | 2.4102 | 2.4100 | 2.4097 |
| **39** | 2.4095 | 2.4093 | 2.4090 | 2.4088 | 2.4085 | 2.4083 | 2.4081 | 2.4078 | 2.4076 | 2.4073 |
| **40** | 2.4071 | 2.4069 | 2.4066 | 2.4064 | 2.4062 | 2.4059 | 2.4057 | 2.4054 | 2.4052 | 2.4050 |

*Source*: Brooker (1967).

# References


Ambrose, D., Lawrenson, I.J. 1972. The vapor pressure of water. J. Chem. Thermody, 4, 755–761.

Beiningen, K.T. 1973. A manual for measuring dissolved oxygen and nitrogen gas concentrations in water with the Van Slyke-Neill apparatus. Fish Commission of Oregon, Portland, OR.

Benson, B.B., Krause, D. 1980. The concentration and isotopic fractionation of gases in freshwater in equilibrium with the atmosphere. Oxygen. Limn. Oceanogr., 25, 662–671.

Benson, B.B., Krause, D. 1984. The concentration and isotopic fractionation of oxygen dissolved in freshwater and seawater in equilibrium with the atmosphere. Limn. Oceanogr., 29, 620–632.

Bouck, G.R. 1982. Gasometer: an inexpensive device for continuous monitoring of dissolved gases and supersaturation. Trans. Am. Fish. Soc., 111, 505–516.

Brooker, D.B. 1967. Mathematical model of the psychrometric chart. Trans. Am. Soc. Ag. Eng., 10, 558–560, 563.

Colt, J. 1983. The computation and reporting of dissolved gas levels. Water Res., 17, 841–849.

Colt, J. 1986. Gas supersaturation—impact on the design and operation of aquatic systems. Aquacult. Eng., 5, 49–85.

Colt, J., Westers, H. 1982. Production of gas supersaturation by aeration. Trans. Am. Fish. Soc., 111, 342–360.

Cornacchia, J., Colt, J.E. 1984. The effects of dissolved gas supersaturation on larval striped bass *Morone saxatilis* (Walbaum). J. Fish Dis., 7, 15–27.

Crozier, T.E., Yamamoto, S. 1974. Solubility of hydrogen in water, seawater, and NaCl solutions. J. Chem. Eng. Data, 19, 242–244.

D'Aoust, B.G., Clark, M.J.R. 1980. Analysis of supersaturated air in natural waters and reservoirs. Trans. Am. Fish. Soc., 109, 708–724.

DOE, 1994. Handbook of Methods for the Analysis of the Various Parameters of the Carbon Dioxide System in Seawater, Version 2.1, Dickson, A.G. Goyet, C. (eds.), CDIAC-74, Oak Ridge National Laboratory, Oak Ridge, TN.

Duan, Z., Sun, R. 2003. An improved model calculating $CO_2$ solubility in pure water and aqueous NaCl solutions from 273 to 533 K and from 0 to 2000 bar. Chem. Geol., 193, 257–271.

Duan, Z., Sun, R., Zhu, C., Chou, I.-M. 2006. An improved model for the calculation of $CO_2$ solubility in aqueous solutions containing $Na^+$, $K^+$, $Ca^{2+}$, $Mg^{2+}$, and $SO_4^{2-}$. Mar. Chem., 98, 131–139.

Fickeisen, D.H., Schneider, M.J., Montgomery, J.C. 1975. A comparative evaluation of the Weiss saturometer. Trans. Am. Fish. Soc., 104, 816–820.

Goff, J.A., Gratch, S. 1946. Low-pressure properties of water from −160 to 212F. Trans. Am. Soc. Heat. Vent. Eng., 52, 95–122.

Green, E.J., Carritt, D.E. 1967. New tables for oxygen saturation of seawater. J. Mar. Res., 25, 140–147.

Hamme, R.C., Emerson, S.R. 2004. The solubility of neon, nitrogen and argon in distilled and seawater. Deep-Sea Res., 51, 1517–1528.

Hamme, R.C., Severinghaus, J.P. 2007. Trace gas disequilibra during deep-water formation. Deep-Sea Res., 54, 939–950.

Hutchinson, G.E. 1957. A Treatise on Limnology, Vol. 1, John Wiley and Sons, New York, NY.

IPCC, 2007. Climate Change 2007: The Physical Science Basis. Contribution of Working Group I to the Fourth Assessment Report of the Intergovernmental Panel on Climate Change, Solomon, S., Qin, D., Manning, M., Chen, Z., Marquis, M., Averyt, K.B., Tignor, M., Miller, H.L. (eds.), p. 996. Cambridge University Press, Cambridge.

Kils, U. 1976. The salinity effect on aeration in mariculture. Meeresforsch., 25, 210–216.

Korson, L., Drost-Hansen, W., Millero, F.J. 1969. Viscosity of water at various temperatures. J. Phys. Chem., 73, 34–39.

Lewis, E., Wallace, D. 1998. Program Developed for $CO_2$ System Calculations, Carbon Dioxide Information Analysis Center, Oak Ridge National Laboratory, Oak Ridge, TN 10/4/09 <http://cdiac.ornl.gov/oceans/co2rprt.html>.

Lutgens, F.K., Tarbuck, E.J. 1995. The Atmosphere, 6th ed., Prentice Hall, Upper Saddle River, NJ.

Millero, F.J. 1974. Seawater as a Multi-Component Electrolyte Solution. The Sea Goldberg, E.D., (ed.), Vol. 5, pp. 1–80. Wiley-Interscience, New York, NY.

Millero, F.J. 1996. Chemical Oceanography, 2nd ed., CRC Press, Boca Raton, FL.

Millero, F.J., Poisson, A. 1981. International one-atmosphere equation of state of seawater. Deep-Sea Res., 28A, 625–629.

Millero, F.J., Perron, G., Desnoyers, J.E. 1973. Heat capacity of seawater solutions from 5°C to 35°C and 0.5 to 22‰ chlorinity. J. Geophy. Res., 78, 4499–4507.

Mortimer, C.H. 1981. The oxygen content of air-saturated fresh waters over ranges of temperature and atmospheric pressure of limnological interest. Mitt. Int. Ver. Liminol., 22, 1–23.

Portier, S., Rochelle, C. 2005. Modelling $CO_2$ solubility in pure water and NaCl-type waters from 0°C to 300°C and from 1 to 300 bar. Application to the Utsira Formation at Sleipner. Chem. Geol., 217, 187–199.

Riley, J.P., Skirrow, G. (eds.), 1975. Chemical Oceanography, Vol. 2, 2nd ed. Academic Press, New York, NY.

Robinson, R.A. 1954. The vapor pressure and osmotic equivalent of sea water. J. Mar. Biol. Assoc. UK, 33, 449–455.

Schudlich, R., Emerson, S. 1996. Gas supersaturation in the surface ocean: the role of heat flux, gas exchange, and bubbles. Deep-Sea Res., 43, 569–589.

Sengers, J.M.H.L., Klein, M., Gallagher, J.S. 1972. Pressure–volume–temperature relationships of gases; virial coefficients. American Institute of Physics Handbook, Zemansky, M.W. (ed.), 3rd ed. Pages 4–204 to 4–227. McGraw-Hill, New York, NY.

Sherwood, J.E., Stagnitti, F., Kokkinn, M.J. 1991. Dissolved oxygen concentrations in hypersaline waters. Limn. Oceanogr., 36, 235–250.

Smith, S.P., Kennedy, B.M. 1983. The solubility of noble gases in water in NaCl brine. Geochim. Cosochim. Acta, 47, 503–515.

Standard Methods, 2005. Standard Methods for the Examination of Water and Wastewater, 21st ed., American Public Health Association, Washington, DC.

Stringer, E.T. 1972. Foundations of Climatology, W.H. Freedman, San Francisco, CA.

Weiss, R.F. 1971. Solubility of helium and neon in water and seawater. J. Chem. Eng. Data, 16, 235–241.

Weiss, R.F. 1974. Carbon dioxide in water and seawater: the solubility of a non-ideal gas. Mar. Chem., 2, 203–215.

Weiss, R.F., Kyser, T.K. 1978. Solubility of krypton in water and seawater. J. Chem. Eng. Data, 23, 69–72.

Weiss, R.F., Price, B.A. 1980. Nitrous oxide solubility in water and seawater. Mar. Chem., 8, 347–359.

Weitkamp, D.E., Katz, M. 1980. A review of dissolved gas supersaturation literature. Trans. Am. Fish. Soc., 109, 659–702.

Wood, D., Caputi, R. 1966. Solubilities of Kr and Xe in fresh and sea water. Technical Report, U.S. Naval Radiological Defense Laboratory, San Francisco, California, USA (quoted in Hamme and Severinghaus, 2007).

Yamamoto, S., Alcaukas, J.B., Crozier, T.E. 1976. Solubility of methane in distilled water and seawater. J. Chem. Eng. Data, 19, 242–244.

# Appendix A: Computation of Gas Solubility

The basis for the solubility data used in the text will be presented in this appendix as a function of temperature and salinity. The following abbreviations will be used in this appendix:

| Parameter | Abbreviation | Units or Value |
|---|---|---|
| Temperature | t | Celsius (°C) |
| Temperature | T | Kelvin (°C +273.15) |
| Salinity | S | g/kg |
| Density of water | ρ | kg/m$^3$ or kg/L |
| Gas constant | R | 0.08205601 atm L/(mol K) |

## Atmospheric Gases—Freshwater, Estuarine, and Marine

### *Computation of Standard Air Solubility Concentrations (μmol/kg)*

**Oxygen**: The solubility equation for oxygen is based on Eq. (22) in Benson and Krause (1984):

$$C_o^{\dagger} = 0.20946\left[\frac{F(1 - P_{wv})(1 - \theta_o)}{k_{o,s}M_w}\right] \tag{A-1}$$

where

$C_o^{\dagger}$ = standard air solubility concentrations (μmol/kg)
$F$ = a salinity factor (defined below)
$P_{wv}$ = vapor pressure of water (atm; based on Green and Carritt, 1967)
$\theta_o$ = a constant that depends on the second virial coefficient of oxygen (defined below)
$k_{o,s}$ = Henry coefficient for oxygen (atm; defined below)
$M_w$ = mean molecular mass of sea salt (67.7933 g/mol).

The detailed equations for the computation of $F$, $\theta_o$, and $k_{o,s}$ are presented below in terms of salinity ($S$), temperature ($t$°C), and temperature ($T$ °K):

$$F = 1000 - 0.716582 \times S \tag{A-2}$$

$$(1-\theta_o) = 0.999025 + 1.426 \times 10^{-5}t - 6.436 \times 10^{-3}t^2 \quad \text{(A-3)}$$

$$\ln k_{o,s} = 3.71814 + 5596.17/T - 1{,}049{,}668/T^2 + S(0.0225034 - 13.6083/T + 2565.68/T^2) \quad \text{(A-4)}$$

Values of the Henry constant are presented in Table A-1{255} as a function of temperature and salinity.

**Nitrogen and Argon**: The solubility equations for nitrogen and argon are based on Eq. (1) in Hamme and Emerson (2004):

$$\ln C = A_0 + A_1 T_s + A_2 T_s^2 + A_3 T_s^3 + S(B_0 + B_1 T_s^2 + B_2 T_s^2) \quad \text{(A-5)}$$

where

$$T_s = \ln\left(\frac{298.15 - t}{273.15 + t}\right) \quad \text{(A-6)}$$

Values of the constants for nitrogen and argon are presented below:

| Constant | Nitrogen | Argon |
|---|---|---|
| $A_0$ | 6.42931 | 2.79150 |
| $A_1$ | 2.92704 | 3.17609 |
| $A_2$ | 4.32531 | 4.13116 |
| $A_3$ | 4.69149 | 4.90379 |
| $B_0$ | $-7.44129 \times 10^{-3}$ | $-6.96233 \times 10^{-3}$ |
| $B_1$ | $-8.02566 \times 10^{-3}$ | $-7.666670 \times 10^{-3}$ |
| $B_2$ | $-1.46775 \times 10^{-2}$ | $-1.16888 \times 10^{-2}$ |

**Carbon Dioxide**: The solubility equation for carbon dioxide is developed in terms of $k_o^\dagger$ (Weiss, 1974). This parameter is equal to:

$$k_o^\dagger = \frac{\beta}{\rho \times MV} \quad \text{(A-7)}$$

where

$k_o^\dagger$ = solubility parameter (mol/(kg atm)
$\beta$ = Bunsen coefficient (L/(L atm))
$\rho$ = density of water (kg/L)
$MV$ = Molecular volume of carbon dioxide at STP (L).

**Table A-1** Henry Law Constant for Oxygen as a Function of Temperature and Salinity in atm

| Temperature (°C) | Salinity (g/kg) | | | | | | | | |
|---|---|---|---|---|---|---|---|---|---|
| | **0** | **5** | **10** | **15** | **20** | **25** | **30** | **35** | **40** |
| **0** | 25264. | 26173. | 27115. | 28090. | 29101. | 30149. | 31233. | 32358. | 33522. |
| **1** | 25974. | 26899. | 27857. | 28850. | 29878. | 30943. | 32045. | 33187. | 34369. |
| **2** | 26688. | 27630. | 28605. | 29614. | 30659. | 31741. | 32861. | 34020. | 35220. |
| **3** | 27407. | 28365. | 29356. | 30382. | 31443. | 32542. | 33679. | 34856. | 36074. |
| **4** | 28130. | 29103. | 30111. | 31153. | 32231. | 33346. | 34500. | 35694. | 36930. |
| **5** | 28856. | 29845. | 30869. | 31927. | 33021. | 34153. | 35324. | 36535. | 37787. |
| **6** | 29585. | 30590. | 31629. | 32703. | 33814. | 34962. | 36150. | 37377. | 38647. |
| **7** | 30317. | 31337. | 32392. | 33482. | 34608. | 35773. | 36976. | 38221. | 39507. |
| **8** | 31051. | 32086. | 33156. | 34262. | 35404. | 36584. | 37804. | 39065. | 40367. |
| **9** | 31787. | 32837. | 33922. | 35043. | 36201. | 37397. | 38632. | 39909. | 41228. |
| **10** | 32524. | 33589. | 34689. | 35825. | 36998. | 38210. | 39461. | 40753. | 42088. |
| **11** | 33262. | 34341. | 35456. | 36607. | 37795. | 39022. | 40289. | 41596. | 42947. |
| **12** | 34000. | 35094. | 36223. | 37389. | 38592. | 39834. | 41116. | 42439. | 43804. |
| **13** | 34739. | 35847. | 36990. | 38171. | 39388. | 40645. | 41941. | 43280. | 44660. |
| **14** | 35477. | 36599. | 37757. | 38951. | 40183. | 41454. | 42766. | 44118. | 45514. |
| **15** | 36214. | 37350. | 38522. | 39730. | 40976. | 42262. | 43587. | 44955. | 46365. |
| **16** | 36950. | 38100. | 39285. | 40507. | 41768. | 43067. | 44407. | 45788. | 47213. |
| **17** | 37685. | 38848. | 40047. | 41282. | 42556. | 43870. | 45223. | 46619. | 48057. |
| **18** | 38418. | 39594. | 40806. | 42055. | 43342. | 44669. | 46036. | 47446. | 48898. |
| **19** | 39148. | 40337. | 41562. | 42824. | 44125. | 45465. | 46846. | 48268. | 49734. |
| **20** | 39876. | 41077. | 42315. | 43590. | 44904. | 46257. | 47651. | 49087. | 50566. |
| **21** | 40600. | 41814. | 43065. | 44353. | 45679. | 47045. | 48452. | 49901. | 51393. |
| **22** | 41322. | 42548. | 43811. | 45111. | 46450. | 47828. | 49248. | 50709. | 52214. |
| **23** | 42039. | 43277. | 44552. | 45865. | 47216. | 48606. | 50038. | 51512. | 53030. |
| **24** | 42752. | 44002. | 45289. | 46613. | 47976. | 49379. | 50823. | 52309. | 53839. |
| **25** | 43461. | 44723. | 46021. | 47357. | 48732. | 50147. | 51602. | 53100. | 54642. |
| **26** | 44165. | 45438. | 46748. | 48095. | 49482. | 50908. | 52375. | 53885. | 55438. |
| **27** | 44864. | 46148. | 47469. | 48828. | 50225. | 51663. | 53142. | 54663. | 56227. |
| **28** | 45558. | 46853. | 48184. | 49554. | 50963. | 52411. | 53901. | 55433. | 57009. |
| **29** | 46245. | 47551. | 48894. | 50274. | 51693. | 53153. | 54654. | 56197. | 57783. |
| **30** | 46927. | 48243. | 49596. | 50987. | 52417. | 53887. | 55398. | 56952. | 58549. |
| **31** | 47602. | 48929. | 50292. | 51693. | 53134. | 54614. | 56136. | 57700. | 59307. |
| **32** | 48271. | 49607. | 50981. | 52392. | 53842. | 55333. | 56865. | 58439. | 60057. |
| **33** | 48933. | 50279. | 51662. | 53083. | 54544. | 56044. | 57586. | 59170. | 60797. |
| **34** | 49588. | 50943. | 52336. | 53767. | 55237. | 56747. | 58298. | 59892. | 61529. |
| **35** | 50235. | 51600. | 53002. | 54442. | 55921. | 57441. | 59002. | 60605. | 62251. |
| **36** | 50875. | 52249. | 53660. | 55109. | 56598. | 58126. | 59696. | 61309. | 62964. |
| **37** | 51507. | 52890. | 54310. | 55768. | 57265. | 58803. | 60382. | 62003. | 63668. |
| **38** | 52131. | 53522. | 54951. | 56418. | 57924. | 59470. | 61058. | 62688. | 64361. |
| **39** | 52746. | 54146. | 55583. | 57059. | 58573. | 60128. | 61724. | 63363. | 65044. |
| **40** | 53353. | 54761. | 56207. | 57691. | 59213. | 60776. | 62381. | 64027. | 65718. |

*Source*: Based on Eq. 30, Benson and Krause (1984).

The equation for the computation of $k_o^\dagger$ is

$$\begin{aligned} \ln k_o^\dagger = {} & A_1 + A_2(100/T) + A_3 \ln(T/100) + A_4 T_s^3 \\ & + S(B_1 + B_2(T/100) + B_3(T/100)^2) \end{aligned} \tag{A-8}$$

Values of the constants for Eq. (A-8) are presented below:

| **Constant** | **mol/(kg atm)** |
|---|---|
| $A_1$ | −60.2409 |
| $A_2$ | 93.4517 |
| $A_3$ | 23.3585 |
| $B_1$ | 0.023517 |
| $B_2$ | −0.023656 |
| $B_3$ | 0.0047036 |

Values of $k_o^\dagger$ and $k_o^*$ are presented in Tables A-2{257} and A-3{258}, respectively, as a function of temperature and salinity.

The standard air solubility of carbon dioxide is equal to:

$$C_o^+ = \chi k_o^\dagger (P - P_{wv}) \exp\left[P\left(\frac{B + 2\delta}{RT}\right) + \overline{v}\left(\frac{1-P}{RT}\right)\right] \tag{A-9}$$

where

$C_o^+$ = standard air solubility concentration (mol/kg)
$\chi$ = mole fraction of carbon dioxide in dry air (dimensionless)
$k_o^\dagger$ = as defined in Eq. (A-8) (mol/(kg atm))
$P$ = barometric pressure (atm)
$P_{wv}$ = vapor pressure (atm; based on Eq. (1.10) in Weiss and Price, 1980)
$B$ = second virial coefficient of carbon dioxide (cm$^3$/mol; see below)
$\delta$ = cross-virial coefficient for carbon dioxide and air (cm$^3$/mol; see below)
$\overline{v}$ = partial molar volume of carbon dioxide (32.3 cm$^3$/mol)
$R$ = gas constant (0.08205601 atm L/(mol K))
$T$ = absolute temperature (273.15 + °C).

$$B = -1636.75 + 12.0408T - 3.27957 \times 10^{-2}T^2 + 3.16528 \times 10^{-5}T^3 \tag{A-10}$$

$$\delta = 57.7 - 0.118T \tag{A-11}$$

Note that while $B$, $\delta$, and $\overline{v}$ have units of cm$^3$/mol in the above equations, they must be converted to L/mol when substituted in Eq. (A-9). The units of $C_o^+$ in Eq. (A-9) are mol/kg rather than μmol/kg.

**Table A-2** Solubility Coefficient ($k_o^\dagger$) for Carbon Dioxide as a Function of Temperature and Salinity in mol/(kg atm) × 100 (At 10°C and 35 g/kg, the table value is 4.388; the solubility coefficient is 4.388 × $10^{-2}$ or 0.04388 mol/(kg atm))

| **Temperature (°C)** | **Salinity (g/kg)** | | | | | | | | |
|---|---|---|---|---|---|---|---|---|---|
| | **0** | **5** | **10** | **15** | **20** | **25** | **30** | **35** | **40** |
| **0** | 7.758 | 7.528 | 7.305 | 7.089 | 6.880 | 6.676 | 6.479 | 6.287 | 6.101 |
| **1** | 7.458 | 7.238 | 7.024 | 6.817 | 6.616 | 6.421 | 6.232 | 6.048 | 5.870 |
| **2** | 7.174 | 6.963 | 6.758 | 6.560 | 6.367 | 6.180 | 5.999 | 5.822 | 5.651 |
| **3** | 6.904 | 6.702 | 6.506 | 6.316 | 6.131 | 5.952 | 5.777 | 5.608 | 5.444 |
| **4** | 6.649 | 6.455 | 6.267 | 6.084 | 5.907 | 5.735 | 5.568 | 5.405 | 5.248 |
| **5** | 6.407 | 6.221 | 6.040 | 5.865 | 5.695 | 5.529 | 5.369 | 5.213 | 5.062 |
| **6** | 6.177 | 5.999 | 5.825 | 5.657 | 5.493 | 5.335 | 5.180 | 5.031 | 4.885 |
| **7** | 5.959 | 5.787 | 5.621 | 5.459 | 5.302 | 5.149 | 5.001 | 4.857 | 4.718 |
| **8** | 5.752 | 5.587 | 5.427 | 5.271 | 5.120 | 4.974 | 4.831 | 4.693 | 4.558 |
| **9** | 5.554 | 5.396 | 5.242 | 5.093 | 4.948 | 4.807 | 4.670 | 4.537 | 4.407 |
| **10** | 5.367 | 5.215 | 5.067 | 4.923 | 4.784 | 4.648 | 4.516 | 4.388 | 4.263 |
| **11** | 5.189 | 5.042 | 4.900 | 4.762 | 4.627 | 4.497 | 4.370 | 4.247 | 4.127 |
| **12** | 5.019 | 4.878 | 4.741 | 4.608 | 4.479 | 4.353 | 4.231 | 4.112 | 3.997 |
| **13** | 4.857 | 4.721 | 4.590 | 4.462 | 4.337 | 4.216 | 4.098 | 3.984 | 3.873 |
| **14** | 4.703 | 4.572 | 4.446 | 4.322 | 4.202 | 4.086 | 3.972 | 3.862 | 3.755 |
| **15** | 4.556 | 4.430 | 4.308 | 4.189 | 4.074 | 3.961 | 3.852 | 3.746 | 3.643 |
| **16** | 4.416 | 4.295 | 4.177 | 4.063 | 3.951 | 3.843 | 3.738 | 3.635 | 3.536 |
| **17** | 4.282 | 4.166 | 4.052 | 3.942 | 3.834 | 3.730 | 3.628 | 3.530 | 3.433 |
| **18** | 4.155 | 4.042 | 3.933 | 3.826 | 3.723 | 3.622 | 3.524 | 3.429 | 3.336 |
| **19** | 4.033 | 3.924 | 3.819 | 3.716 | 3.616 | 3.519 | 3.425 | 3.333 | 3.243 |
| **20** | 3.916 | 3.812 | 3.710 | 3.611 | 3.515 | 3.421 | 3.330 | 3.241 | 3.154 |
| **21** | 3.805 | 3.704 | 3.606 | 3.510 | 3.417 | 3.327 | 3.239 | 3.153 | 3.070 |
| **22** | 3.699 | 3.601 | 3.507 | 3.414 | 3.325 | 3.237 | 3.152 | 3.069 | 2.989 |
| **23** | 3.597 | 3.503 | 3.412 | 3.322 | 3.236 | 3.151 | 3.069 | 2.989 | 2.911 |
| **24** | 3.499 | 3.409 | 3.321 | 3.235 | 3.151 | 3.069 | 2.990 | 2.912 | 2.837 |
| **25** | 3.406 | 3.319 | 3.233 | 3.150 | 3.070 | 2.991 | 2.914 | 2.839 | 2.766 |
| **26** | 3.317 | 3.232 | 3.150 | 3.070 | 2.992 | 2.916 | 2.841 | 2.769 | 2.699 |
| **27** | 3.231 | 3.150 | 3.070 | 2.993 | 2.917 | 2.844 | 2.772 | 2.702 | 2.634 |
| **28** | 3.149 | 3.071 | 2.994 | 2.919 | 2.846 | 2.775 | 2.705 | 2.638 | 2.572 |
| **29** | 3.071 | 2.995 | 2.920 | 2.848 | 2.778 | 2.709 | 2.642 | 2.576 | 2.512 |
| **30** | 2.995 | 2.922 | 2.850 | 2.780 | 2.712 | 2.645 | 2.580 | 2.517 | 2.455 |
| **31** | 2.923 | 2.852 | 2.783 | 2.715 | 2.649 | 2.585 | 2.522 | 2.461 | 2.401 |
| **32** | 2.854 | 2.785 | 2.718 | 2.653 | 2.589 | 2.527 | 2.466 | 2.406 | 2.349 |
| **33** | 2.787 | 2.721 | 2.656 | 2.593 | 2.531 | 2.471 | 2.412 | 2.354 | 2.298 |
| **34** | 2.723 | 2.659 | 2.596 | 2.535 | 2.476 | 2.417 | 2.360 | 2.305 | 2.250 |
| **35** | 2.662 | 2.600 | 2.539 | 2.480 | 2.422 | 2.366 | 2.311 | 2.257 | 2.204 |
| **36** | 2.603 | 2.543 | 2.484 | 2.427 | 2.371 | 2.316 | 2.263 | 2.211 | 2.160 |
| **37** | 2.546 | 2.488 | 2.431 | 2.376 | 2.322 | 2.269 | 2.217 | 2.167 | 2.118 |
| **38** | 2.492 | 2.436 | 2.381 | 2.327 | 2.275 | 2.224 | 2.174 | 2.125 | 2.077 |
| **39** | 2.439 | 2.385 | 2.332 | 2.280 | 2.229 | 2.180 | 2.131 | 2.084 | 2.038 |
| **40** | 2.389 | 2.336 | 2.285 | 2.235 | 2.186 | 2.138 | 2.091 | 2.045 | 2.000 |

*Source*: Based on constants in Table 1 in Weiss (1974).

**Table A-3** Solubility Coefficient ($k_o^*$) for Carbon Dioxide as a Function of Temperature and Salinity in mol/(L atm) × 100 (At 10°C and 35 g/kg, the table value is 4.507; the solubility coefficient is 4.507 × $10^{-2}$ or 0.04507 mol/(kg atm))

| Temperature (°C) | Salinity (g/kg) | | | | | | | | |
|---|---|---|---|---|---|---|---|---|---|
| | **0** | **5** | **10** | **15** | **20** | **25** | **30** | **35** | **40** |
| **0** | 7.758 | 7.558 | 7.364 | 7.175 | 6.990 | 6.810 | 6.635 | 6.465 | 6.298 |
| **1** | 7.458 | 7.267 | 7.081 | 6.899 | 6.723 | 6.550 | 6.382 | 6.219 | 6.060 |
| **2** | 7.174 | 6.991 | 6.813 | 6.639 | 6.469 | 6.304 | 6.143 | 5.986 | 5.833 |
| **3** | 6.905 | 6.730 | 6.558 | 6.392 | 6.229 | 6.070 | 5.916 | 5.766 | 5.619 |
| **4** | 6.650 | 6.481 | 6.317 | 6.157 | 6.001 | 5.849 | 5.701 | 5.556 | 5.416 |
| **5** | 6.408 | 6.246 | 6.088 | 5.935 | 5.785 | 5.639 | 5.497 | 5.358 | 5.223 |
| **6** | 6.178 | 6.023 | 5.871 | 5.724 | 5.580 | 5.440 | 5.303 | 5.170 | 5.040 |
| **7** | 5.959 | 5.810 | 5.665 | 5.523 | 5.385 | 5.251 | 5.119 | 4.991 | 4.867 |
| **8** | 5.751 | 5.608 | 5.469 | 5.333 | 5.200 | 5.071 | 4.945 | 4.822 | 4.702 |
| **9** | 5.554 | 5.417 | 5.282 | 5.152 | 5.024 | 4.900 | 4.779 | 4.660 | 4.545 |
| **10** | 5.366 | 5.234 | 5.105 | 4.979 | 4.857 | 4.737 | 4.621 | 4.507 | 4.396 |
| **11** | 5.187 | 5.060 | 4.936 | 4.816 | 4.698 | 4.583 | 4.470 | 4.361 | 4.254 |
| **12** | 5.017 | 4.895 | 4.776 | 4.659 | 4.546 | 4.435 | 4.327 | 4.222 | 4.119 |
| **13** | 4.855 | 4.737 | 4.623 | 4.511 | 4.402 | 4.295 | 4.191 | 4.090 | 3.991 |
| **14** | 4.700 | 4.587 | 4.477 | 4.369 | 4.264 | 4.162 | 4.062 | 3.964 | 3.869 |
| **15** | 4.553 | 4.444 | 4.338 | 4.234 | 4.133 | 4.034 | 3.938 | 3.844 | 3.752 |
| **16** | 4.412 | 4.307 | 4.205 | 4.105 | 4.008 | 3.913 | 3.820 | 3.729 | 3.641 |
| **17** | 4.278 | 4.177 | 4.078 | 3.982 | 3.889 | 3.797 | 3.708 | 3.620 | 3.535 |
| **18** | 4.149 | 4.052 | 3.958 | 3.865 | 3.775 | 3.686 | 3.600 | 3.516 | 3.434 |
| **19** | 4.027 | 3.933 | 3.842 | 3.753 | 3.666 | 3.581 | 3.498 | 3.417 | 3.337 |
| **20** | 3.910 | 3.820 | 3.732 | 3.646 | 3.562 | 3.480 | 3.400 | 3.322 | 3.245 |
| **21** | 3.798 | 3.711 | 3.626 | 3.544 | 3.463 | 3.384 | 3.306 | 3.231 | 3.157 |
| **22** | 3.691 | 3.607 | 3.526 | 3.446 | 3.368 | 3.291 | 3.217 | 3.144 | 3.073 |
| **23** | 3.589 | 3.508 | 3.429 | 3.352 | 3.277 | 3.203 | 3.131 | 3.061 | 2.992 |
| **24** | 3.491 | 3.413 | 3.337 | 3.263 | 3.190 | 3.119 | 3.050 | 2.982 | 2.915 |
| **25** | 3.397 | 3.322 | 3.249 | 3.177 | 3.107 | 3.038 | 2.971 | 2.906 | 2.842 |
| **26** | 3.307 | 3.235 | 3.164 | 3.095 | 3.027 | 2.961 | 2.897 | 2.833 | 2.771 |
| **27** | 3.221 | 3.151 | 3.083 | 3.016 | 2.951 | 2.887 | 2.825 | 2.764 | 2.704 |
| **28** | 3.138 | 3.071 | 3.005 | 2.941 | 2.878 | 2.816 | 2.756 | 2.697 | 2.639 |
| **29** | 3.059 | 2.994 | 2.931 | 2.869 | 2.808 | 2.748 | 2.690 | 2.633 | 2.578 |
| **30** | 2.983 | 2.920 | 2.859 | 2.799 | 2.741 | 2.683 | 2.627 | 2.572 | 2.518 |
| **31** | 2.910 | 2.850 | 2.791 | 2.733 | 2.676 | 2.621 | 2.567 | 2.514 | 2.462 |
| **32** | 2.840 | 2.782 | 2.725 | 2.669 | 2.615 | 2.561 | 2.509 | 2.457 | 2.407 |
| **33** | 2.773 | 2.717 | 2.662 | 2.608 | 2.555 | 2.504 | 2.453 | 2.403 | 2.355 |
| **34** | 2.708 | 2.654 | 2.601 | 2.549 | 2.498 | 2.449 | 2.400 | 2.352 | 2.305 |
| **35** | 2.646 | 2.594 | 2.543 | 2.493 | 2.444 | 2.396 | 2.348 | 2.302 | 2.257 |
| **36** | 2.587 | 2.536 | 2.487 | 2.439 | 2.391 | 2.345 | 2.299 | 2.254 | 2.211 |
| **37** | 2.529 | 2.481 | 2.433 | 2.387 | 2.341 | 2.296 | 2.252 | 2.209 | 2.166 |
| **38** | 2.474 | 2.428 | 2.382 | 2.337 | 2.292 | 2.249 | 2.207 | 2.165 | 2.124 |
| **39** | 2.421 | 2.376 | 2.332 | 2.288 | 2.246 | 2.204 | 2.163 | 2.123 | 2.083 |
| **40** | 2.370 | 2.327 | 2.284 | 2.242 | 2.201 | 2.161 | 2.121 | 2.082 | 2.044 |

*Source*: Based on constants in Table 1 in Weiss (1974).

Equation (A-9) can be written as

$$C_o^+ = \chi F^\dagger \tag{A-12}$$

where

$$F^\dagger = k_o(P - P_{wv})\exp\left[P\left(\frac{B+2\delta}{RT}\right) + \overline{v}\left(\frac{1-P}{RT}\right)\right] \tag{A-13}$$

Equations (A-12) and (A-13) can be used to compute the standard air solubility concentration or air solubility of carbon dioxide as a function of mole fraction. For the computation of standard air solubility concentration, $P$ in Eq. (A-16) should be set equal to 1 atm, and for air solubility concentration it should be set equal to the local barometric pressure (in atm).

## Computation of Bunsen Coefficients

**Oxygen**: The Bunsen coefficient for oxygen is based on Eq. (16) in Benson and Krause (1980).

$$\beta^1 = \frac{\rho}{k_o M}[1 - \theta_o(1 + P_{wv})] \tag{A-14}$$

where

$\beta^1$ = Bunsen coefficient (mol/(L atm))
$\rho$ = density of water (kg/L)
$k_o$ = Henry Law Constant (atm; see Eq. (A-4))
$M$ = molecular weight of water (18.0153 g/mol)
$\theta_o$ = a constant that depends on the second virial coefficient of oxygen (see Eq. (A-3))
$P_{wv}$ = vapor pressure of water (atm; based on Green and Carritt, 1967).

Equation (A-14) was developed for freshwater conditions. Substituting the numerical value of $M$ into Eq. (A-14) gives the Bunsen coefficient in terms of μmol/(L atm):

$$\beta^1 = 5.5508 \times 10^7 \frac{\rho}{k_o}[1 - \theta_o(1 + P_{wv})] \tag{A-15}$$

Multiplying Eq. (A-15) by the molecular volume of oxygen (22.392 L) converts the equation to the conventional units of (L real gas/(L atm)):

$$\beta^1 = 1.24294 \times 10^4 \frac{\rho}{k_o}[1 - \theta_o(1 + P_{wv})] \tag{A-16}$$

For an ideal gas, the value of the first constant in Eq. (A-16) would be equal to $1.24416 \times 10^4$.

Conversion of Eq. (A-16) to marine conditions (Benson and Krause, 1984) requires the substitution of $k_{o,s}$ for $k_o$ and the addition of a modified salinity factor ($F$) into Eq. (A-16):

$$\beta^1 = 1.24294 \times 10^4 \frac{\rho F}{1000 k_{o,s}} [1 - \theta_o(1 + P_{wv})] \tag{A-17}$$

The salinity factor in Eq. (A-2) must be divided by 1000 for consistent units. Therefore, as the salinity approaches zero, the value of $k_{o,s} \rightarrow k_o$, $F \rightarrow 1.00$, and Eq. (A-17) reduces to Eq. (A-15).

**Nitrogen and Argon**: The Bunsen coefficients for nitrogen and argon are based on Eq. (3) in Hamme and Emerson (2004):

$$\beta = \frac{\rho(C_o^\dagger \times 10^{-6}) MV}{(1 - P_{wv})\chi} \tag{A-18}$$

where

$\beta$ = Bunsen coefficient (L/(L atm))
$\rho$ = density of water (kg/L)
$C_o^\dagger$ = standard air solubility concentration (μmol/kg). Note that Eq. (A-5) returns concentration in μmol/kg $\times 10^6$
$MV$ = molecular volume (L/mol)
$P_{wv}$ = vapor pressure of water (atm; based on Eq. (1.1) in Ambrose and Lawrenson, 1972)
$\chi$ = mole fraction of gas in standard atmosphere (dimensionless).

Values of $MV$ and $\chi$ for nitrogen and argon can be found in Table D-1{287}.

**Carbon Dioxide**: The Bunsen coefficient for carbon dioxide is based on Weiss (1974). Equation (A-7) can be rewritten as

$$\beta = k_o^\dagger \times \rho \times MV \tag{A-19}$$

## Noble Atmospheric Gases—Freshwater, Estuarine, and Marine

**Helium, Krypton**: Standard air solubility concentrations (nmol/kg) for helium (Weiss, 1971) and krypton (Weiss and Kyser, 1978) are based on the following equation:

$$\begin{aligned} \ln C_o^\dagger = {} & A_1 + A_2(100/T) + A_3 \ln(T/100) + A_4(T/100) \\ & + S(B_1 + B_2(T/100) + B_3(T/100)^2) \end{aligned} \tag{A-20}$$

The original units for helium and krypton are mL/kg and are converted to nmol/kg by the following two equations:

$$C_o^\dagger\ (\mathrm{nmol/kg\text{-}helium}) = C_o^\dagger\ (\mathrm{mL/kg\text{-}helium})\left[\frac{10^9\ \mathrm{nmol/mol}}{22{,}426\ \mathrm{mL/mol}}\right] \tag{A-21}$$

$$C_o^\dagger\ (\mathrm{nmol/kg\text{-}krypton}) = C_o^\dagger\ (\mathrm{mL/kg\text{-}krypton})\left[\frac{10^9\ \mathrm{nmol/mol}}{22{,}351\ \mathrm{mL/mol}}\right] \tag{A-22}$$

The regression constants for these two gases are presented below:

| **Constant** | **Helium (He)** | | **Krypton (Kr)** | |
|---|---|---|---|---|
| | $C_o^\dagger$ **(mL/kg)** | $\beta$ **(L/(L atm))** | $C_o^\dagger$ **(mL/kg)** | $\beta$ **(L/(L atm))** |
| $A_1$ | −167.2178 | −34.6261 | −112.6840 | −57.2596 |
| $A_2$ | +216.3442 | +43.0285 | +153.5817 | +87.4242 |
| $A_3$ | +139.2032 | +14.1391 | +74.4690 | +22.9332 |
| $A_4$ | −22.6202 | — | −10.0189 | — |
| $B_1$ | −0.044781 | −0.042340 | −0.011213 | −0.008723 |
| $B_2$ | +0.023541 | +0.022624 | −0.00184 | −0.002793 |
| $B_2$ | −0.0034266 | −0.0033120 | +0.0011201 | +0.0012398 |

The Bunsen coefficient (L/(L atm)) for helium (Weiss, 1971) and krypton (Weiss and Kyser, 1978) are based on the following equation:

$$\ln \beta = A_1 + A_2(100/T) + A_3 \ln(T/100) + S(B_1 + B_2(T/100) + B_3(T/100)^2) \tag{A-23}$$

The regression constants for the two gases are presented in the previous table. Generally, $1000\beta_{\mathrm{He}}$ and $100\beta_{\mathrm{Kr}}$ are presented because of the low solubility of these gases.

**Neon, Xenon**: Standard air solubility concentrations (nmol/kg) for neon (Hamme and Emerson, 2004) and xenon (Wood and Caputi, 1966) are based on the following equations:

$$\ln C = A_0 + A_1 T_s + A_2 T_s^2 + S(B_0 + B_1 T_s^2) \tag{A-24}$$

where

$$T_s = \ln\left(\frac{298.15 - t}{273.15 + t}\right) \tag{A-25}$$

Values of the constants for neon and xenon are presented below:

| Constant | Neon (nmol/kg) | Xenon (μmol/kg) |
|---|---|---|
| $A_0$ | +2.18156 | −7.48588 |
| $A_1$ | +1.29108 | +5.08763 |
| $A_2$ | +2.12504 | +4.22078 |
| $B_0$ | $-5.9477 \times 10^{-3}$ | $-8.17791 \times 10^{-3}$ |
| $B_1$ | $-5.13896 \times 10^{-3}$ | $-1.20172 \times 10^{-2}$ |

The solubility for xenon must be multiplied by 1000 to convert to nmol/kg. The above regression equation and regression constants for xenon are based on the Wood and Caputi (1966) data as fitted by Hamme and Severinghaus (2007). This data was not published by Hamme and Severinghaus (2007), but was provided as a personal communication. Based on the analysis conducted by Hamme and Severinghaus (2007), the Wood and Caputi (1966) data is thought to be 2% high. The solubility of xenon is the most poorly defined of all the gases presented in this book, and it is expected that better solubility information will be available in the future.

The Bunsen coefficients for neon and xenon were computed from the $C_o^{\dagger}$ by the following equation:

$$\beta = \frac{\rho(C_o^{\dagger} \times 10^{-9})MV}{(1 - P_{wv})\chi} \tag{A-26}$$

The parameters in this equation are discussed under the section for Bunsen coefficients for nitrogen and argon. The only difference between the parameters for these two gases and the parameters for those in the equation found in the Bunsen coefficient section is that standard air solubility concentration for these two gases is given in nmol/kg and, therefore, this concentration must be multiplied by $10^{-9}$ to convert to mol/kg.

## Trace Atmospheric Gases—Freshwater, Estuarine, and Marine

**Hydrogen, Methane**: Bunsen coefficients for hydrogen (Crozier and Yamamoto, 1974) and methane (Yamamoto et al., 1976) are based on the following equation:

$$\ln \beta = A_1 + A_2(100/T) + A_3 \ln(T/100) + S(B_1 + B_2(T/100) + B_3(T/100)^2) \tag{A-27}$$

where

| Constant | Hydrogen (L/(L atm)) | Methane (L/(L atm)) |
|---|---|---|
| $A_1$ | −39.9611 | −67.1962 |
| $A_2$ | +53.9381 | +99.1624 |
| $A_3$ | +16.3135 | +27.9015 |
| $B_1$ | −0.036249 | −0.072909 |
| $B_2$ | +0.017566 | +0.041674 |
| $B_3$ | −0.0023010 | −0.0064603 |

Because no information is available for the standard air solubility concentration of these two gases, these parameters were computed from the Bunsen coefficients and mole fractions using a modified version of Eq. (1.13):

$$C_o^{\dagger}(\mathrm{H_2}) = \rho\left[\frac{1000K_{\mathrm{H_2}}\beta_{\mathrm{H_2}}\chi_{\mathrm{H_2}}}{22,428}\right]\left[\frac{\mathrm{BP}-P_{wv}}{760}\right]10^9 \quad \text{(A-28)}$$

where

$C_o^{\dagger}$ = standard air solubility concentration (μmol/kg)
BP = standard atmospheric pressure (760 mmHg)
$\chi_{\mathrm{H_2}}$ = mole fraction of hydrogen in the atmosphere.

The original unit of Eq. (1.13) was mg/L, therefore, it was divided by the mole volume of hydrogen (22,428 mL/mol) to give mol/L, multiplied by $10^9$ to convert to μmol/L, and, finally, multiplied by the density of water to convert to μmol/kg. For methane, Eq. (A-28) would be rewritten as:

$$C_o^{\dagger}(\mathrm{CH_4}) = \rho\left[\frac{1000K_{\mathrm{CH_4}}\beta_{\mathrm{CH_4}}\chi_{\mathrm{CH_4}}}{22,360}\right]\left[\frac{\mathrm{BP}-P_{wv}}{760}\right]10^9 \quad \text{(A-29)}$$

**Nitrous Oxide**: The solubility equation for nitrous oxide is developed in terms of $k_o^{\dagger}$ (Weiss and Price, 1980) just as the solubility equation for carbon dioxide is developed (Weiss, 1974; Weiss and Price, 1980). This parameter is equal to:

$$k_o^{\dagger} = \frac{\beta}{\rho \times MV} \quad \text{(A-30)}$$

where

$k_o^{\dagger}$ = solubility parameter (mol/(kg atm))
$\beta$ = Bunsen coefficient (L/(L atm))

$\rho$ = density of water (kg/L)
$MV$ = molecular volume of carbon dioxide at STP (L).

The equation for the computation of $k_o^\dagger$ is

$$\ln k_o^\dagger = A_1 + A_2(100/T) + A_3 \ln(T/100) + A_4 T_s^3 + S(B_1 + B_2(T/100) + B_3(T/100)^2) \quad \text{(A-31)}$$

Values of the constants for Eq. (A-31) are presented below:

| Constant | mol/(kg atm) |
|---|---|
| $A_1$ | −64.8539 |
| $A_2$ | +100.2520 |
| $A_3$ | +25.2049 |
| $B_1$ | −0.062544 |
| $B_2$ | +0.035337 |
| $B_3$ | −0.0054699 |

The standard air solubility of nitrous oxide is equal to:

$$C_o^+ = \chi k_o^\dagger (P - P_{wv}) \exp\left[P\left(\frac{B + 2\delta}{RT}\right) + \overline{v}\left(\frac{1 - P}{RT}\right)\right] \quad \text{(A-32)}$$

These parameters were defined for carbon dioxide, and only specific information for nitrous oxide will be presented in this section:

$\overline{v}$ = partial molar volume of nitrous oxide (32.3 $cm^3$/mol).

$$\frac{(B + 2\delta)}{RT} = +0.04739 - 9.4563/T - 6.427 \times 10^{-5} T \quad \text{(A-33)}$$

Note that while $B$, $\delta$, and $\overline{v}$ have units of $cm^3$/mol in the above equations, they must be converted to L/mol when substituted in Eq. (A-32). The units of $C_o^+$ in Eq. (A-32) are mol/kg rather than μmol/kg. Weiss and Price (1980) also present $F^\dagger$ and $F^*$ parameters that can be used to compute the standard air solubility concentration of nitrous oxide as a function of mole fraction. This is useful as the mole fraction of nitrous oxide is increasing in the atmosphere.

The Bunsen coefficients for nitrous oxide can be computed from Eq. (A-30):

$$\beta = k_o^\dagger \times \rho \times MV \quad \text{(A-34)}$$

# Major Atmospheric Gases—NaCl Brines

## *Computation of Standard Air Solubility Concentrations ($\mu$mol/kg)*

**Oxygen**: The solubility equation for oxygen is based on Eq. (16) in Sherwood et al. (1991):

$$C_o^{\dagger} = \frac{0.20946}{k_o}(5.5509 \times 10^{-2} - 3.8399 \times 10^{-5} S)(1 - P_{wv}) \exp(B) \tag{A-35}$$

where

$C_o^{\dagger}$ = standard air solubility concentration (mol/kg)
$k_o$ = Henry Law constant (atm)
$P_{wv}$ = vapor pressure of water (atm; based on Sherwood et al., 1991)
$B$ = second virial coefficient for oxygen (1/atm; see below).

$$B = -0.00975 + 1.426 \times 10^{-5} t - 6.436 \times 10^{-8} t^2 \tag{A-36}$$

Equation (16) in Sherwood et al. (1991) was originally written in terms of $C_o^*$, so it was necessary to delete the $\rho_s$ term.

**Carbon Dioxide**: The solubility equation for carbon dioxide is based on Eq. (1) in Duan and Sun (2006):

$$\begin{aligned} \ln(m_{CO_2}) = {} & \ln(y_{CO_2}\phi_{CO_2}P) - \mu_{CO_2}^{1(0)}/RT \\ & - 2\lambda_{CO_2-Na}(m_{Na} + m_K + 2m_{Ca} + 2m_{Mg}) \\ & - \zeta_{CO_2-Na-Cl}m_{Cl}(m_{Na} + m_K + m_{Mg} + m_{Ca}) \\ & + 0.07m_{SO_4} \end{aligned} \tag{A-37}$$

where

$P$ = total pressure (bar)
$T$ = absolute temperature (K)
$R$ = universal gas constant
$m$ = molality of components dissolved in water (mol/kg)
$y_{CO_2}$ = mole fraction of $CO_2$ in the vapor phase
$\phi_{CO_2}$ = fugacity coefficient of $CO_2$
$\mu_{CO_2}^{1(0)}/RT$ = standard chemical potential of $CO_2$ in the liquid phase
$\lambda_{CO_2-Na}$ = interaction parameter between $CO_2$ and $Na^+$
$\zeta_{CO_2-Na-Cl}$ = interaction parameter between $CO_2$ and $Na^+$, and Cl.

$y_{CO_2}$ was estimated from:

$$y_{CO_2} = \frac{P - P_{wv}}{P} \tag{A-38}$$

where

$P_{wv}$ = vapor pressure of water (bar).

$\mu_{CO_2}^{1(0)}$, $\lambda_{CO_2-Na}$, and $\zeta_{CO_2-Na-Cl}$ were computed from Eq. (7) in Duan and Sun (2003) and regression coefficient in Table 1.2. For $T > 290$ K, the constant for $\mu_{CO_2}^{1(0)}/RT$ were refitted by Duan and Sun (2006), but not reported (R. Sun, personal communication, May 26, 2010). The following regression coefficients should be used for $T > 290$K:

$C_1 = 134.72067$
$C_2 = -3.6727291E-1$
$C_3 = -14132.405$
$C_4 = 4.7809063E-4$
$C_5 = -5622.8080$
$C_6 = 7.9181559E-2$
$C_7 = -1.2283602E-2$
$C_8 = -2.9597665$
$C_9 = 6.5155997E-1$
$C_{10} = 7.4901468E-4$
$C_{11} = 0.0$

$\phi_{CO_2}$ was computed from Eq. (2) in Duan et al. (2006) and the regression coefficients in Table 1. The solubility relationships developed by Duan et al. (2006) and Duan and Sun (2003) are for pure carbon dioxide gas ($\chi_{CO_2} = 1.00$) and $C_o^\dagger$ and $C_o^*$ were computed from:

$$C_o^\dagger\ (\mu\text{mol/kg}) = 10^6 m_{CO_2}\left[\frac{P}{P-P_{wv}}\right]\left[\chi_{CO_2}(P-P_{wv})\right] \quad \text{(A-39)}$$

$$C_o^*\ (\text{mg/L}) = MW_{CO_2}\rho\, m_{CO_2}\left[\frac{P}{P-P_{wv}}\right]\left[\chi_{CO_2}(P-P_{wv})\right] \quad \text{(A-40)}$$

where

$m_{CO_2}$ = solubility of pure carbon dioxide gas (Eq. (A-37))
$P$ = total pressure (atm)
$P_{wv}$ = vapor pressure of water (atm)
$MW_{CO_2}$ = molecular weight of carbon dioxide gas (mg/mol)
$\rho$ = density of water (kg/L).

## Computation of Bunsen Coefficients

**Oxygen**: The Bunsen coefficient for oxygen is based on Eq. (19) in Sherwood et al. (1991):

$$\beta = K_o \times MV \quad \text{(A-41)}$$

where

$\beta$ = Bunsen coefficient (L/(L atm))
$K_o$ = constant (mol/(L atm); see below)
$MV$ = molecular volume (22.392 L/mol).

$$K_o = \frac{\rho}{k_o}(5.5509 \times 10^{-2} - 3.8399 \times 10^{-5} S) \quad \text{(A-42)}$$

where

$\rho$ = density of water (kg/L; based on Eq. (22) in Sherwood et al. (1991)).

**Carbon Dioxide**: The Bunsen coefficient for carbon dioxide is based on Eq. (1) in Duan and Sun (2006) for $m_{CO_2}$ and the following equation:

$$\beta\ (\mathrm{L/L\ atm}) = MV_{CO_2}\, \rho\, m_{CO_2} \left[\frac{P}{P - P_{wv}}\right] \quad \text{(A-43)}$$

where

$MV_{CO_2}$ = molecular volume of carbon dioxide gas (L/mol).

and the other parameters were defined for Eqs (A-38)–(A-40).

# Appendix B: Computation of Physical Properties of Water

The basis for physical properties of water that are used in this book will be presented in this appendix. The following abbreviations will be used in this appendix:

| Parameter | Abbreviation | Units or Value |
|---|---|---|
| Temperature | t | Celsius (°C) |
| Temperature | T | Kelvin (°C + 273.15) |
| Salinity | S | g/kg |
| Chlorinity | CL | g/kg |
| Density of water | ρ | $kg/m^3$ |
| Acceleration due to gravity | g | $9.80665\ m/s^2$ |

## Density of Water ($\rho$) in $kg/m^3$

For 0–40°C and 0.5–43 g/kg, the 1 atm equation of state of seawater is based on Millero and Poisson (1981):

$$\rho = \rho_o + AS + BS^{3/2} + CS^2 \quad \text{(B-1)}$$

$$A = A_0 + A_1t + A_2t^2 + A_3t^3 + A_4t^4 \quad \text{(B-2)}$$

$$B = B_0 + B_1t + B_2t^2 + B_3t^3 \quad \text{(B-3)}$$

$$C = C_1 \quad \text{(B-4)}$$

$$\rho_o = D_0 + D_1t + D_2t^2 + D_3t^3 + D_4t^5 \quad \text{(B-5)}$$

| Equation (B-1) | | Equation (B-2) | | Equation (B-3) | | Equation (B-4) | |
|---|---|---|---|---|---|---|---|
| *As* | Value | *Bs* | Value | *Cs* | Value | *Ds* | Value |
| $A_0$ | $+8.24493 \times 10^{-1}$ | $B_0$ | $-5.72466 \times 10^{-3}$ | $C_0$ | $4.8314 \times 10^{-4}$ | $D_0$ | 999.842594 |
| $A_1$ | $-4.0899 \times 10^{-3}$ | $B_1$ | $+1.0227 \times 10^{-4}$ | | | $D_1$ | $+6.793952 \times 10^{-2}$ |
| $A_2$ | $+7.6438 \times 10^{-5}$ | $B_2$ | $-1.6546 \times 10^{-6}$ | | | $D_2$ | $-9.095290 \times 10^{-3}$ |
| $A_3$ | $-8.2467 \times 10^{-7}$ | | | | | $D_3$ | $+1.001685 \times 10^{-4}$ |
| $A_4$ | $+5.3875 \times 10^{-9}$ | | | | | $D_4$ | $1.120083 \times 10^{-6}$ |
| | | | | | | $D_5$ | $+6.536332 \times 10^{-9}$ |

For NaCl brines over the range of 0–35°C and 0–260 g/kg, Eq. (22) in Sherwood et al. (1991) is used:

$A_0 = 0.999792$
$A_1 = 6.92234 \times 10^{-5}$
$A_2 = -8.15399 \times 10^{-6}$
$A_3 = +4.25067 \times 10^{-8}$
$A_4 = 7.68124 \times 10^{-4}$
$A_5 = -1.46445 \times 10^{-7}$
$A_6 = 1.60452 \times 10^{-8}$
$A_7 = -4.10220 \times 10^{-6}$
$A_8 = 4.04168 \times 10^{-8}$
$A_9 = 7.64930 \times 10^{-11}$
$A_{10} = 1.38698 \times 10^{-7}$
$A_{11} = -1.79894 \times 10^{-9}$

$$\rho = 1000.0\,[A_0 + A_1 t + A_2 t^2 + A_3 t^3 + A_4 S + A_5 S^2 + A_6 S^{5/2} + A_7 tS + A_8 t^2 S + A_9 t^3 S + A_{10} t S^{3/2} + A_{11} t^2 S^{3/2}] \tag{B-6}$$

The initial equation has units of g/mL and is multiplied by 1000 to convert to kg/m$^3$.

## Specific Weight of Water ($\gamma$) in kN/m$^3$

The specific weight of water is based on the density of water (Millero and Poisson, 1981) and the value of the acceleration of gravity ($g$) listed in the first table in this Appendix:

$$\gamma = \frac{\rho g}{1000} \tag{B-7}$$

## Hydrostatic Head of Water in mmHg/m and kPa/m

The hydrostatic head of water is based on the density of water (Millero and Poisson, 1981) and the appropriate pressure units:

$$\text{Hydrostatic head (mmHg}/m) = \frac{760\ \text{mmHg}}{(101,325\ \text{Pa}/\rho g)} \tag{B-8}$$

$$\text{Hydrostatic head (kPa}/m) = \frac{101.325\ \text{kPa}}{(101,325\ \text{Pa}/\rho g)} \tag{B-9}$$

## Dynamic Viscosity of Water ($\mu$) in N s/m$^2$

Dynamic viscosity of water is based on Korson et al. (1969) and Millero (1974):

$$\mu 20 = 1.0020 \times 10^{-3}\ (20°\text{C in freshwater}) \tag{B-10}$$

$$\mu_t = \mu_{20}{}^{*}10^{**}\left[\frac{1.1709(20 - t) - 0.001827(t - 20)^2}{t + 89.93}\right] \tag{B-11}$$

$$CL = (S - 0.03)/1.805 \tag{B-12}$$

$$CL_V = (CL)(\rho/1000.0) \tag{B-13}$$

$$A = 0.000366 + 5.185 \times 10^{-5}(t - 5.0) \tag{B-14}$$

$$B = 0.002756 + 3.300 \times 10^{-5}(t - 5.0) \tag{B-15}$$

$$\mu = \mu_t[1.0 + A(CL_V)^{1/2} + B(CL_V)] \tag{B-16}$$

Note that the table values are given in N s/m$^2 \times 10^{+3}$. For example, the viscosity at 20°C and 20 g/kg is $1.0424 \times 10^{-3}$ N s/m$^2$.

## Kinematic Viscosity of Water ($\nu$) in m$^2$/s

The kinematic viscosity of water is based on the dynamic viscosity (Korson et al. 1969; Millero, 1974) and the density of water (Millero and Poisson, 1981):

$$\nu = \frac{\mu}{\rho} \tag{B-17}$$

Note that the table values are given in $m^2/s \times 10^{+6}$. For example, the kinematic viscosity at 20°C and 20 g/kg is $1.0286 \times 10^{-6}$ $m^2/s$.

## Heat Capacity of Water ($C_p$) in kJ/(kg K)

Heat capacity of water is based on Millero et al. (1973):

$$C_p^o = 4.2174 - 3.720283 \times 10^{-3}t + 1.412855 \times 10^{-4}t^2 - 2.654387 \times 10^{-6}t^3 + 2.093236 \times 10^{-8}t^4 \quad \text{(B-18)}$$

$$CL = S/1.80655 \quad \text{(B-19)}$$

$$A = -(13.81 - 0.1938t + .0025t^2)/(1000.0) \quad \text{(B-20)}$$

$$B = (0.43 - .0099t + 0.0001t^2)/(1000.0) \quad \text{(B-21)}$$

$$C_p = C_p^o + A(CL) + B(CL)^{3/2} \quad \text{(B-22)}$$

## Latent Heat of Vaporization (LHV) of Water in MJ/kg

Latent heat of vaporization of water is based on Brooker (1967):

$$\text{LHV (MJ/kg)} = 2.502535259 - 0.00238576424t \quad \text{(B-23)}$$

## Surface Tension of Water ($\sigma$) in N/m

Surface tension of water in N/m is based on Riley and Skirrow (1975):

$$\sigma \text{ (N/m)} = \frac{75.64 - 0.144t + 0.0221S}{1000} \quad \text{(B-24)}$$

## Vapor Pressure of Freshwater Water ($P_{wv}$)

Four different equations are used to compute the vapor pressure of water for the solubility relationships in this book:

| Solubility Relationship | | Source of Vapor Pressure |
|---|---|---|
| **Gas** | **Reference** | **Reference** |
| $O_2$ | Benson and Krause (1984) | Green and Carritt (1967) |
| $O_2$ (brine) | Sherwood et al. (1991) | Sherwood et al. (1991) |
| $N_2$ and Ar | Hamme and Emerson (2004) | Ambrose and Lawrenson (1972) |
| $CO_2$ | Weiss (1974), Weiss and Price (1980) | Equation fitted to Goff and Gratch (1946) and Robinson (1954) |

The relationship developed by Ambrose and Lawrenson (1972) is the most complex and is preferred for typical estuarine and oceanic salinities. The relationship developed by Weiss (1974) is the simplest and useful for spreadsheet applications. The Sherwood et al. (1991) equation was developed for NaCl brines and may be used for real brines if location-specific information is not available. The original vapor pressure equations were used in the individual gas solubility equations to try to reproduce the author's results. The use of other vapor pressure relationships may change the solubility parameters in the last decimal place. The water vapor equations given below are given in their original units.

## *Vapor Pressure of Water ($P_{wv}$) in atm (Green and Carritt, 1967)*

$A = 5.370 \times 10^{-4}$
$B = 18.1973$
$C = 1.0 - 373.16/T$
$D = 3.1813 \times 10^{-7}$
$E = 26.1205$
$F = 1.0 - T/373.16$
$G = 1.8726 \times 10^{-2}$
$H = 8.03945$
$X = 5.02802$
$Y = 373.16/T$

$$P_{wv} = \{(1.0 - A \times S) \times \exp(B \times C + D \times (1.0 - \exp(E \times F)) - G \times (1.0 - \exp(H \times C)) + X \times \ln(Y))\} \quad \text{(B-25)}$$

## *Vapor Pressure of Water ($P_{wv}$) in atm (Sherwood et al., 1991)*

$A_1 = +48.4171$
$A_2 = -6821.5$
$A_3 = -5.0903$
$A_4 = -5.8785 \times 10^{-4}$
$A_5 = -1.2276 \times 10^{-8}$
$A_6 = -6.93 \times 10^{-9}$

$$\ln(P_{wv}) = A_1 + A_2(1/T) + A_3\ln(T) + A_4S + A_5S^2 + A_6S^3 \quad \text{(B-26)}$$

### *Vapor Pressure of Water ($P_{wv}$) in kPa (Ambrose and Lawrenson, 1972)*

The vapor pressure of freshwater is represented by the Chebyshev polynomial:

$$T\log_{10}P^{o}_{wv} = \tfrac{1}{2}a_0 + \left[\sum\nolimits_{k=1}^{11} a_k E_k(x)\right] \tag{B-27}$$

where

$$x = \frac{2T - 921}{375} \tag{B-28}$$

$T$ = temperature in Kelvin measured on the International Practical Temperature Scale of 1968.

The coefficients of the Chebyshev polynomial are:

| | |
|---|---|
| $a_0 = 2794.0144$ | $a_6 = 0.1371$ |
| $a_1 = 1430.6181$ | $a_7 = 0.0629$ |
| $a_2 = -18.2465$ | $a_8 = 0.0261$ |
| $a_3 = 7.6875$ | $a_9 = 0.0200$ |
| $a_4 = -0.0328$ | $a_{10} = 0.0117$ |
| $a_5 = 0.2728$ | $a_{11} = 0.0067$ |

The actual Chebyshev polynomials are equal to:

$$E_0(x) = 1$$

$$E_1(x) = x$$

$$E_2(x) = 2x^2 - 1$$

$$E_3(x) = 4x^3 - 3x$$

$$E_4(x) = 8x^4 - 8x^2 + 1$$

$$E_5(x) = 16x^5 - 20x^3 + 5x$$

$$E_6(x) = 32x^6 - 48x^4 + 18x^2 - 1$$

$$E_7(x) = 64x^7 - 112x^5 + 56x^3 - 7x$$

$$E_8(x) = 128x^8 - 256x^6 + 160x^4 - 32x^2 + 1$$

$$E_9(x) = 256x^9 - 576x^7 + 432x^5 - 120x^3 + 9x$$

$$E_{10}(x) = 512x^{10} - 1280x^8 + 1120x^6 - 400x^4 + 50x^2 - 1$$

$$E_{11}(x) = 1024x^{11} - 2816x^9 + 2816x^7 - 1232x^5 + 220x^3 - 11x$$

The vapor pressure of seawater (DOE, 1994) is related to that of pure water by:

$$P^s_{wv} = P^o_{wv}\exp(-0.018\phi\sum\nolimits_B m_B/m^\circ) \tag{B-29}$$

For seawater,

$$\sum\nolimits_B m_B/m^\circ = \frac{31.998S}{1000 - 1.005S} = \psi \tag{B-30}$$

or

$$P^s_{wv} = P^o_{wv}\exp(-0.018\phi\psi) \tag{B-31}$$

The value of the osmotic coefficient (Millero, 1974) of seawater ($\phi$) is equal to

$$\phi = 0.90799 - 0.08992\left(\frac{\psi}{2}\right) + 0.18458\left(\frac{\psi}{2}\right)^2 - 0.073958\left(\frac{\psi}{2}\right)^3 - 0.00221\left(\frac{\psi}{2}\right)^4 \tag{B-32}$$

Vapor pressure of water ($P_{wv}$) in atm (Weiss and Price, 1980):

$$P_{wv} = \exp(24.4543 - 67.4509(100/T) - 4.8489\ln(T/100) - 0.000544S) \tag{B-33}$$

# Appendix C: Computer Programs

These computer programs supplement the tables in the body of the book. They are written in FORTRAN 99 and are designed to run on a 32-bit Windows computer under XP. With the appropriate FORTRAN compilers, they could be ported to other operating systems. They use the same functions and relationships that were used to generate the tables in this book. These programs can be downloaded from: http://www.elsevierdirect.com/companion.jsp?ISBN=9780124159167

## AIRSAT

This program will compute the standard air solubility concentration and air solubility concentration of atmospheric, noble, and trace gases. It assumes moist air (see mole fraction data listed in Table D-1{287}). Because the mole fractions of carbon dioxide and nitrous oxide are increasing, the mole fractions of these two gases can be specified. If no values are provided, this program will default to the values listed in Table D-1{287}. The program computes the solubility information for a single temperature, salinity, and pressure. The program also computes the physical properties of water for this combination of values.

Two files are needed:

```
AIRSAT_DATA.TXT
AIRSAT.EXE
```

The program executes when `AIRSAT.EXE` is double-clicked. The solubility data is written to `AIRSAT.OUT`. Both `AIRSAT_DATA.TXT` and `AIRSAT.OUT` are simple text files and can be opened with WORD or Notepad. Generally, the "EXE" and "TXT" extensions are not displayed.

A sample of `AIRSAT_DATA.TXT` is presented in Table C-1{278}. The first four lines specify output for atmospheric gases, noble gases, trace gases, and physical properties of water. If you do not want the output for a particular section, "no" should be entered in columns 42–44. Barometric pressure can be specified as elevation, kPa, mmHg, mbar, or atm. Only one value is needed. The value must be entered in columns 42–49 and must contain a decimal point. The next two lines are for temperature (°C) and salinity (g/kg). This is followed by the mole fraction of carbon dioxide and nitrous oxide in μatm and the maximum values of these two parameters. The development of the solubility relationships for these gases assumes that $\chi \ll 1.0$. The program will compare the entered

**Table C-1** Sample Input for AIRSAT

```
Interested in atmospheric gases?                  XXXyesXXXXXXXXX
Interested in noble gases?                        XXXyesXXXXXXXXX
Interested in trace gases? ?                      XXXyesXXXXXXXXX
Interested in physical properties of water?       XXXyesXXXXXXXXX
Elevation (m)                                     XXX         XXXXXXXXX
Pressure (kPa)                                    XXX         XXXXXXXXX
Pressure (mmHg)                                   XXX760.00   XXXXXXXXX
Pressure (mbar)                                   XXX         XXXXXXXXX
Pressure (atm)                                    XXX         XXXXXXXXX
Temperature (°C)                                  XXX20.00000XXXXXXXXX
Salinity (g/kg)                                   XXX35.00000XXXXXXXXX
Mole fraction of CO2 (uatm)                       XXX390.0000XXXXXXXXX
Mole fraction of N2O (uatm)                       XXX0.319000XXXXXXXXX
Maximum mole fraction of CO2 (µatm)               XXX1500.000XXXXXXXXX
Maximum mole fraction of N2O (µatm)               XXX10.00000XXXXXXXXX

0123456789 0123456789 0123456789 0123456789 0123456789 012345679
0          1          2          3          4          5
```

Notes:
(1) Atmospheric pressure can be specified in terms of kPa, mmHg, mbar, or atm. Enter only one!
(2) The maximum mole fraction of carbon dioxide and nitrous oxide are approximate limits for the atmospheric concentration relationships. For higher values, use the arbitrary mole fraction program.

values of each parameter with the maximum values. The entered value for both parameters must be less than the respective maximum mole fraction value. For larger values of the mole fractions, ARBSAT should be used.

The output from the program is divided into four major sections: atmospheric gases, noble gases, trace gases, and physical properties. A sample output from the program is presented in Table C-2{279}. The solubility sections are divided into four sections:

- $C^{\dagger}$(on a mass basis)
  - µmol/kg (or nmol/L for noble and trace gases)
  - mg/kg
  - mL real gas at STP/kg
  - mL ideal gas at STP/kg
  - µg-atm/kg (oxygen only)
- $C^{*}$(on a volume basis)
  - µmol/L (or nmol/L for noble and trace gases)
  - mg/L
  - mL real gas at STP/L
  - mL ideal gas at STP/L
  - mg-atm/L (oxygen only)

**Table C-2** Output from AIRSAT

STANDARD AIR SOLUBILITY OR AIR SOLUBILITY – ATMOSPHERIC GASES
Temperature (°C) = 20.000
Salinity (g/kg) = 35.000
Elevation (m) = 0.00000
Pressure (kPa) = 101.325
Pressure (mmHg) = 760.000
Pressure (mbar) = 1013.250
Pressure (atm) = 1.00000
Mole fraction of $CO_2$ (μatm) = 390.0000
Mole fraction of $N_2O$ (μatm) = 0.3190

| PARAMETER | OXYGEN | NITROGEN | ARGON | CARBON DIOXIDE |
|---|---|---|---|---|
| C+ (MASS) | | | | |
| μmol/kg | 225.5361 | 419.7732 | 11.0745 | 12.3111 |
| mg/kg | 7.2167 | 11.7595 | 0.4424 | 0.5418 |
| mL real/kg | 5.0502 | 9.4046 | 0.2480 | 0.2741 |
| mL ideal/kg | 5.0552 | 9.4088 | 0.2482 | 0.2759 |
| μg-atm/kg | 451.0721 | | | |
| C* (VOLUME) | | | | |
| μmol/L | 231.1210 | 430.1680 | 11.3488 | 12.6160 |
| mg/L | 7.3954 | 12.0507 | 0.4534 | 0.5552 |
| mL real/L | 5.1753 | 9.6375 | 0.2541 | 0.2809 |
| mL ideal/L | 5.1803 | 9.6418 | 0.2544 | 0.2828 |
| μg-atm/L | 462.2420 | | | |
| BUNSEN COEFFICIENT | | | | |
| L real/(L atm) | 0.02527903 | 0.01262844 | 0.02783961 | 0.73946960 |
| mg/(L mmHg) | 0.04753090 | 0.02077711 | 0.06534798 | 1.92338000 |
| mg/(L kPa) | 0.35651110 | 0.15584120 | 0.49015020 | 14.42654000 |
| GAS TENSION PER mg/L | | | | |
| mmHg/(mg/L) | 21.0389 | 48.1299 | 15.3027 | 0.519918 |

STANDARD AIR SOLUBILITY OR AIR SOLUBILITY – NOBLE GASES
Temperature (°C) = 20.000
Salinity (g/kg) = 35.000
Pressure (kPa) = 101.325
Elevation (m) = 0.00000
Pressure (mmHg) = 760.000
Pressure (mbar) = 1013.250
Pressure (atm) = 1.00000
Mole fraction of $CO_2$ (μatm) = 390.0000
Mole fraction of $N_2O$ (μatm) = 0.3190

*(Continued)*

**Table C-2** (Continued)

| PARAMETER | HELIUM | NEON | KRYPTON | XENON |
|---|---|---|---|---|
| C+ (MASS) | | | | |
| nmol/kg | 1.6630 | 6.8271 | 2.4402 | 0.3336 |
| mg/kg | 0.66562E −05 | 0.13777E −03 | 0.20449E −03 | 0.43797E −04 |
| mL real/kg | 0.37294E −04 | 0.15309E −03 | 0.54542E −04 | 0.74257E −05 |
| mL ideal/kg | 0.37274E −04 | 0.15302E −03 | 0.54696E −04 | 0.74771E −05 |
| C* (VOLUME) | | | | |
| nmol/L | 1.7041 | 6.8271 | 2.5007 | 0.3419 |
| mg/L | 0.68210E −05 | 0.14118E −03 | 0.20956E −03 | 0.44882E −04 |
| mL real/L | 0.38217E −04 | 0.15688E −03 | 0.55892E −04 | 0.76096E −05 |
| mL ideal/L | 0.38197E −04 | 0.15681E −03 | 0.56050E −04 | 0.76623E −05 |
| BUNSEN COEFFICIENT | | | | |
| L real/(L atm) | 0.00746062 | 0.00882931 | 0.05022508 | 0.08651219 |
| mg/(L mmHg) | 0.00175207 | 0.01045495 | 0.24777284 | 0.67138230 |
| mg/(L kPa) | 0.01314159 | 0.07841857 | 1.85844920 | 5.03578200 |
| GAS TENSION PER mg/L | | | | |
| mmHg/(mg/L) | 570.754 | 95.6485 | 4.03595 | 1.489464 |

STANDARD AIR SOLUBILITY OR AIR SOLUBILITY – TRACE GASES
Temperature (°C) = 20.000
Salinity (g/kg) = 35.000
Elevation (m) = 0.00000
Pressure (kPa) = 101.325
Pressure (mmHg) = 760.000
Pressure (mbar) = 1013.250
Pressure (atm) = 1.00000
Mole fraction of $CO_2$ (μatm) = 390.0000
Mole fraction of $N_2O$ (μatm) = 0.3190

| PARAMETER | HYDROGEN | METHANE | NITROUS OXIDE |
|---|---|---|---|
| C + (MASS) | | | |
| nmol/kg | 0.359670 | 2.108627 | 7.26230 |
| mg/kg | 0.72502E −06 | 0.33829E −04 | 0.31964E −03 |
| mL real/kg | 0.80667E −05 | 0.47149E −04 | 0.16154E −03 |
| mL ideal/kg | 0.80616E −05 | 0.47263E −04 | 0.16278E −03 |
| nmol/L | 0.368577 | 2.160842 | 7.44214 |
| mg/L | 0.74298E −06 | 0.34666E −04 | 0.32755E −03 |
| mL real/L | 0.82664E −05 | 0.48316E −04 | 0.16554E −03 |
| mL ideal/L | 0.82613E −05 | 0.48433E −04 | 0.16681E −03 |

*(Continued)*

**Table C-2** (Continued)

| | | | |
|---|---|---|---|
| BUNSEN COEFFICIENT | | | |
| L real/(L atm) | 0.01537815 | 0.02786696 | 0.53282344 |
| mg/(L mmHg) | 0.00181867 | 0.02630824 | 1.38725500 |
| mg/(L kPa) | 0.01364114 | 0.19732804 | 10.40527000 |
| GAS TENSION PER mg/L | | | |
| mmHg/(mg/L) | 549.8528 | 38.01090 | 0.056909 |

PHYSICAL PROPERTIES OF WATER
Temperature (°C) = 20.000
Salinity (g/kg) = 35.000
Elevation (m) = 0.00000

| | KPA | MM HG | MBAR | ATM |
|---|---|---|---|---|
| Pressure | 101.3250 | 760.0000 | 1013.250 | 1.000000 |
| Vapor pressure | 2.2946 | 17.2112 | 22.946 | 0.022646 |

DENSITY OF WATER = 1024.763kg/$m^3$
SPECIFIC WEIGHT OF WATER = 10.0495kN/$m^3$
SPECIFIC WEIGHT OF WATER = 75.3774mmHg/m
HEAT CAPACITY OF WATER = 3.99401J/(g C)
VISCOSITY OF WATER = $1.07178 \times 10^{-3}$ N s/$m^2$
KINEMATIC VISCOSITY OF WATER = $1.04588 \times 10^{-6}$ $m^2$/s
SURFACE TENSION OF WATER = $73.53350 \times 10^{-3}$ N/m
HEAT OF VAPORIZATION = 2.45482MJ/kg

Bunsen coefficient ($\beta$)
L real gas/(L atm)
mg/(L mmHg)
mg/(L kPa)
Gas tension
mmHg/(mg/L)

The Bunsen coefficients for the noble gases are presented in terms of standard units. In the text, the Bunsen coefficients for helium, neon, and krypton are scaled to fit better in tabular form. To rerun another temperature, salinity, pressure, or mole fraction combination, the output file must first be deleted. To change the input data, AIRSAT_DATA.txt should be opened with Notepad or WORD. If WORD is used, the file must be saved as a text file (under the "save as" option).

## ARBSAT

This program will compute the solubility concentration of atmospheric, noble, and trace gases for arbitrary mole fractions. It assumes a moist ideal gas and can be used for mole fractions much larger than atmospheric values, but with reduced accuracy. Two files are needed:

```
ARBSAT_DATA.TXT
ARBSAT.EXE
```

The program executes when `ARBSAT.EXE` is double-clicked. The solubility data is written to `ARBSAT.OUT`. Both `ARBSAT_DATA.TXT` and `ARBSAT.OUT` are simple text files and can be opened with WORD or Notepad.

Barometric pressure can be specified as elevation, kPa, mmHg, mbar, or atm. Only one is needed. Temperature and salinity are read next. The mole fraction of each gas can be specified as either a percent or in μatm. Only one value is needed for each gas. If no mole fraction information is entered for a specific gas, the program assumes that you are not interested in that gas and no output data will be written for that gas. A sample input file and output file from the program are presented in Tables C-3{282} and C-4{283}. The output parameters are similar to AIRSAT, except instead of mmHg/(mg/L), the total gas tension is presented for each gas. To rerun another temperature, salinity, pressure, or mole fraction combination, the output

**Table C-3** Sample Input for ARBSAT

| | | | | |
|---|---|---|---|---|
| Elevation (m) | XXX | | XXX | |
| Pressure (kPa) | XXX | | XXX | |
| Pressure (mmHg) | XXX | 760.00 | XXX | |
| Pressure (mbar) | XXX | | XXX | |
| Pressure (atm) | XXX | | XXX | |
| Temperature (°C) | XXX | 20.00000 | XXX | |
| Salinity (g/kg) | XXX | 35.00000 | XXX | |
| Oxygen | XXX | 50.000 | XXX | XXX |
| Nitrogen | XXX | | XXX | XXX |
| Argon | XXX | | XXX | XXX |
| Carbon Dioxide | XXX | | XXX | XXX |
| Helium | XXX | 100.0 | XXX | XXX |
| Neon | XXX | | XXX | XXX |
| Krypton | XXX | | XXX | XXX |
| Xenon | XXX | | XXX | XXX |
| Hydrogen | XXX | | XXX | XXX |
| Methane | XXX | 10.0 | XXX | XXX |
| Nitrous Oxide | XXX | | XXX | XXX |

Notes:
(1) Atmospheric pressure can be specified in terms of elevation, kPa, mmHg, mbar, or atm. Enter only one!
(2) The mole fraction for each gas can be specified as either percent or μatm. Enter only one for each gas!

**Table C-4** Output from ARBSAT

| | |
|---|---|
| SOLUBILITY–OXYGEN | |
| Temperature (°C) = 20.000 | |
| Salinity (g/kg) = 35.000 | |
| Elevation (m) = 0.00000 | |
| Pressure (kPa) = 101.325 | |
| Pressure (mmHg) = 760.000 | |
| Pressure (mbar) = 1013.250 | |
| Pressure (atm) = 1.00000 | |
| Mole fraction0.50000000 | |
| C+ (MASS) | |
| μmol/kg | 538.351 |
| mg/kg | 17.2262 |
| mL real/kg | 12.0548 |
| mL ideal/kg | 12.0666 |
| μg-atm/kg | 1076.7030 |
| C* (VOLUME) | |
| μmol/L | 551.683 |
| mg/L | 17.6527 |
| mL real/L | 12.3533 |
| mL ideal/L | 12.3654 |
| μg-atm/L | 1103.3652 |
| BUNSEN COEFFICIENT | |
| L real/(L atm) | 0.025279032000 |
| mg/(L mmHg) | 0.047530900000 |
| mg/(L kPa) | 0.356511100000 |
| GAS TENSION | |
| mmHg | 371.3950 |
| SOLUBILITY–HELIUM | |
| Temperature (°C) = 20.000 | |
| Salinity (g/kg) = 35.000 | |
| Elevation (m) = 0.00000 | |
| Pressure (kPa) = 101.325 | |
| Pressure (mmHg) = 760.000 | |
| Pressure (mbar) = 1013.250 | |
| Pressure (atm) = 1.00000 | |
| Mole fraction1.00000000 | |

(*Continued*)

**Table C-4** (Continued)

| | | |
|---|---|---|
| C+ (MASS) | | |
| nmol/kg | | 317286.50 |
| mg/kg | | 1.2700 |
| mL real/kg | | 7.1155 |
| mL ideal/kg | | 7.1117 |
| C* (VOLUME) | | |
| nmol/L | | 325143.43 |
| mg/L | | 1.3014 |
| mL real/L | | 7.2917 |
| mL ideal/L | | 7.2878 |
| BUNSEN COEFFICIENT | | |
| L real/(L atm) | | 0.007460622100 |
| mg/(L mmHg) | | 0.001752068100 |
| mg/(L kPa) | | 0.013141592000 |
| GAS TENSION | | |
| mmHg | | 742.7902 |
| SOLUBILITY – METHANE | | |
| Temperature (°C) | = | 20.000 |
| Salinity (g/kg) | = | 35.000 |
| Elevation (m) | = | 0.00000 |
| Pressure (kPa) | = | 101.325 |
| Pressure (mmHg) | = | 760.000 |
| Pressure (mbar) | = | 1013.250 |
| Pressure (atm) | = | 1.00000 |
| Mole fraction | | 0.10000000 |
| C + (MASS) | | |
| nmol/kg | | 118862.82 |
| mg/kg | | 1.9069 |
| mL real/kg | | 2.6578 |
| mL ideal/kg | | 2.6642 |
| C* (VOLUME) | | |
| nmol/L | | 121806.22 |
| mg/L | | 1.9541 |
| mL real/L | | 2.7236 |
| mL ideal/L | | 2.7302 |

*(Continued)*

**Table C-4** (Continued)

| | |
|---|---|
| BUNSEN COEFFICIENT | |
| L real/(L atm) | 0.027866960000 |
| mg/(L mmHg) | 0.026308242000 |
| mg/(L kPa) | 0.197328040000 |
| GAS TENSION | |
| mmHg | 74.2785 |

file must first be deleted. Because of the wider range of solubilities considered in this program, the output information may be presented with more significant figures that are justified for some conditions.

# Appendix D: Supplemental Information

This appendix contains the following three tables:



**Table D-1** Properties of Gases

| Gas | Molecular Weight (g) | Molar Volume (L real gas at STP) | Second Virial Volume ($cm^3$/mol at STP)[a] | Mole Fraction in Dry (atm) | $K$[b] | $A$[c] |
|---|---|---|---|---|---|---|
| Oxygen ($O_2$) | 31.998 | 22.392 | −22 | 0.20946[d] | 1.42899 | 0.5318 |
| Nitrogen ($N_2$) | 28.014 | 22.404 | −10 | 0.78084[d] | 1.25040 | 0.6078 |
| Argon (Ar) | 39.948 | 22.393 | −21 | 0.00934[d] | 1.78395 | 0.4260 |
| Carbon dioxide ($CO_2$) | 44.009 | 22.263 | −151 | 379 μatm[e] | 1.97678 | 0.3845 |
| Helium (He) | 4.0026 | 22.426 | +12.0 | 5.24 μatm[d] | 0.17848 | 4.2582 |
| Neon (Ne) | 20.180 | 22.424 | + 10.4 | 18.18 μatm[d] | 0.89993 | 0.8445 |
| Krypton (Kr) | 83.80 | 22.351 | −62.9 | 1.14 μatm[d] | 3.74927 | 0.2027 |
| Xenon (Xe) | 131.29 | 22.260 | −153.7 | 0.09 μatm[d] | 5.89802 | 0.1289 |
| Hydrogen ($H_2$) | 2.0158 | 22.428 | +13.7 | 0.55 μatm[d] | 0.08988 | 8.4558 |
| Methane ($CH_4$) | 16.043 | 22.360 | −53.6 | 1.774 μatm[e] | 0.71749 | 1.0593 |
| Nitrous oxide ($N_2O$) | 44.013 | 22.243 | −171 | 0.319 μatm[e] | 1.97873 | 0.3841 |

[a]Based on Sengers et al. (1972) for He, Ne, Kr, Xe, $H_2$, and $CH_4$; others from the primary article for each gas.

[b]$K_i = \dfrac{\text{molecular weight (mg)}}{\text{molecular volume of real gas (mL at STP)}}$

[c]$A_i = \dfrac{760\ \text{mmHg/atm}}{(1000\ \text{mL}/L)(K\ \text{mg/ml})}$; partial pressure (mmHg) $= \left[\dfrac{C_i}{\beta_i}\right] A_i$

[d]Lutgens and Tarbuck (1995).

[e]IPPC (2007); values in 2007.

## Constants

Volume of ideal gas = 22.413996 L
$R = 0.082057463$ L atm/(K mol)
$g = 9.80665$ m/s$^2$

**Table D-2** Definitions of Key Symbols in the Text

| Symbol | Definition |
|---|---|
| $A$ | $A = 760/1000$ K, for a specific gas; $PP^1 = CA/\beta$ |
| atm | Standard atmospheric pressure, 101.325 kPa or 760 mmHg |
| $C$ | Concentration of a gas (mol/kg, mol/L, mg/L, mL/L) |
| $C_o$ | Standard air saturation concentration at 1 atm and moist air (mol/kg, mol/L, mg/L, mL/L) |
| $C_p$ | Air saturation concentration at $p$ pressure and moist air (mol/kg, mol/L, mg/L, mL/L) |
| $C_{p,x}$ | Saturation concentration at $p$ pressure, arbitrary mole fraction, and moist gas (mol/kg, mol/L, mg/L, mL/L) |
| $F$ | $F = C_o/\chi$ Relationship developed by Weiss and Price (1980) to compute solubility of gases with increasing atmospheric concentrations. |
| $K$ | Molecular weight/molecular volume at STP for a gas |
| $P$ | Pressure (atm, mmHg, psi, or kPa) |
| $P_{\text{hydro}}$ | Hydrostatic pressure of water at depth $z$ (atm, mmHg, psi, or kPa) |
| $P_{wv}$ | Vapor pressure of water (atm, mmHg, psi, or kPa) |
| $P^g$ | Partial pressure of a gas in the gas phase (mmHg) |
| $P^l$ | Gas tension of a gas in the liquid phase (mmHg) |
| $P_t$ | Total pressure at depth $z$, sum of barometric pressure and hydrostatic pressure (mmHg) |
| STP | Standard temperature and pressure (1 atm pressure and 0°C temperature) |
| TGP | Total gas pressure, sum of partial pressure of gases + water vapor (mmHg) |
| TGP% | Total gas pressure expressed as percent of local barometric pressure (%) |
| $\text{TGP}_{\text{uncomp}}$ | Uncompensated total gas pressure, total gas pressure that aquatic animals experience at depth $z$ |
| $Z$ | Depth of animal or diffuser below the water surface |
| $\beta$ | Bunsen coefficient of a gas (L/(L atm)) |
| $\gamma$ | Specific weight of water ($\rho g$) (kN/m$^3$, kPa/m, mmHg/m) |
| $\Delta P$ | Difference in pressure between total gas pressure and local barometric pressure |
| $\Delta P_i$ | Differences in partial pressure and gas tension for the $i$th gas |
| $\chi$ | Mole fraction of a gas, equal to percent composition of dry gas expressed as decimal fraction |
| † | Superscript applied to $C$, $C_o$, $C_p$, $C_{p,x}$, or $F$ to indicate that solubility is expressed on a mass basis (μmol/kg, mol/kg, mg/kg, mL/kg) |
| * | Superscript applied to $C$, $C_o$, $C_p$, $C_{p,x}$, or $F$ to indicate that solubility is expressed on a volume basis (μmol/L, mol/L, mg/L, mL/L) |

**Table D-3** Units and Conversions

| | Units | | Conversions | |
|---|---|---|---|---|
| | **Units** | **Abbreviations** | **Convert to** | **Multiply by** |
| Concentration | milligrams/kilogram | mg/kg | mg/L | ρ |
| | micromole/kilogram | μmol/kg | μmol/L | ρ |
| | nanomole/kilogram | nmol/kg | nmol/L | ρ |
| | micromole/kilogram | μmol/kg | mol/L | $\rho 10^{-6}$ |
| | nanomole/kilogram | nmol/kg | mol/L | $\rho 10^{-9}$ |
| Density ($\rho$) | kilogram/liter | kg/L | | |
| Length | millimeter | mm | m | 1/1000 |
| | feet | ft | m | 0.3048 |
| Mass | kilogram | kg | g | 1000 |
| | gram | g | kg | $10^{-3}$ |
| | milligram | mg | kg | $10^{-6}$ |
| Pressure | millimeter of Hg | mmHg | atm | 1/760 |
| | kiloPascal | kPa | atm | 1/101.325 |
| | pounds per square inch | psi | atm | 1/14.6959 |
| | millibar | mbar | atm | $0.986923 \times 10^{-3}$ |
| | bar | bar | atm | 0.986923 |
| Temperature | Celsius | °C | °F | °F = °C × 9/5+32 |
| | Fahrenheit | °F | °C | °C = (°F−32) × 5/9 |
| | Kelvin | °K | °C | °C = °K−273.15 |
| Time | second | s | | |
| Volume | cubic centimeter | $cm^3$ | L | 1/100 |
| | Liter | L | $dm^3$ | 1.000 |
| | cubic decimeter | $dm^3$ | L | 1.000 |
| | cubic meter | $m^3$ | L | 1000 |
| | gallon | gal | L | 3.785 |
| | cubic foot | $ft^3$ | L | 28.317 |
| Weight or force | pound | lb | N | 4.448 |
| | Newton | N | kN | $10^{-3}$ |

Printed in the United States
By Bookmasters